HYDROGEOLOGY

Hydrogeology

Principles and Practice

Kevin M. Hiscock

School of Environmental Sciences
University of East Anglia
United Kingdom

Blackwell
Publishing

BLACKWELL PUBLISHING
350 Main Street, Malden, MA 02148-5020, USA
9600 Garsington Road, Oxford OX4 2DQ, UK
550 Swanston Street, Carlton, Victoria 3053, Australia

First published 2005 by Blackwell Science Ltd

3   2007

Library of Congress Cataloging-in-Publication Data

Hiscock, K.M. (Kevin M.)
Hydrogeology : principles and practice / Kevin M. Hiscock.
     p.   cm.
   Includes bibliographical references and index.
   ISBN: 978-0-632-05763-4
   1. Hydrogeology.   I. Title.
   GB1003.2.H57 2005
   551.49—dc22

                              2004014013

A catalogue record for this title is available from the British Library.

Set in 10/12pt Dante
by Graphicraft Ltd, Hong Kong
Printed and bound in the United Kingdom
by TJ International Ltd, Padstow, Cornwall

For further information on
Blackwell Publishing, visit our website:
www.blackwellpublishing.com

Contents

Boxes

Preface

In embarking on writing this book on the principles and practice of hydrogeology, I have purposely aimed to reflect the development of hydrogeology as a science and its relevance to the environment. As a science, hydrogeology requires an interdisciplinary approach with applications to water resources investigations, pollution studies and environmental management. The skills of hydrogeologists are required as much by scientists and engineers as by planners and decision-makers. Within the current era of integrated river basin management, the chance to combine hydrogeology with wider catchment or watershed issues, including the challenge of adapting to climate change, has never been greater. Hence, to equip students to meet these and future challenges, the purpose of this book is to demonstrate the principles of hydrogeology and illustrate the importance of groundwater as a finite and vulnerable resource. By including fundamental material in physical, chemical, environmental isotope and contaminant hydrogeology together with practical techniques of groundwater investigation, development and protection, the content of this book should appeal to students and practising professionals in hydrogeology and environmental management. Much of the material contained here is informed by my own research interests in hydrogeology and also from teaching undergraduate and postgraduate courses in hydrology and hydrogeology within the context of environmental sciences. This experience is reflected in the choice of case studies, both European and international, used to illustrate the many aspects of hydrogeology and its connection with the natural and human environments.

Kevin Hiscock
University of East Anglia, Norwich, UK

Acknowledgements

No book is produced without the assistance of others and I am no exception in recognizing the input of colleagues, family and friends. Several people have provided help with proof reading sections and in supplying references and additional material. These people are Julian Andrews, Alison Bateman, Ros Boar, Lewis Clark, Sarah Cornell, Kate Dennis, Aidan Foley, Thomas Grischek, Norm Henderson, Cathy Hiscock, Mike Leeder, Beth Moon, Lorraine Rajasooriyar, Peter Ravenscroft, Mike Rivett and John Tellam. An enormous thank you is owed to Phillip Judge and Sheila Davies for their patient and expert preparation of the figures and Rosie Cullington for typing the many tables contained throughout. The staff and facilities of the Library at the University of East Anglia are appreciated for providing the necessary literature with which to compile this book. If this were not enough, I am indebted to Tim Atkinson, Richard Hey and Alan Kendall for helping form the content of this book through the years spent together teaching and examining undergraduate and postgraduate students in hydrology and hydrogeology in the School of Environmental Sciences at the University of East Anglia. I am also grateful to the editorial team at Blackwell Publishing for their guidance and support during the publication process and for the very helpful comments on the manuscript provided by Willy Burgess and one anonymous reviewer. Last but not least, I especially thank Cathy, Laura and Rebecca for their patience during the long hours spent writing this book.

Symbols and Abbreviations

Multiples and submultiples

Symbol	Name	Equivalent
T	tera	10^{12}
G	giga	10^{9}
M	mega	10^{6}
k	kilo	10^{3}
d	deci	10^{-1}
c	centi	10^{-2}
m	milli	10^{-3}
μ	micro	10^{-6}
n	nano	10^{-9}
p	pico	10^{-12}

Symbols and abbreviations

Symbol	Description	Units
[–]	activity	mol kg^{-1}
(–)	concentration (see Box 3.1)	mol L^{-1} or mg L^{-1}
A	area	m^2
A	radionuclide activity	
AE	actual evapotranspiration	mm
ASR	artificial storage and recovery	
(aq)	aqueous species	
atm	atmosphere (pressure)	
B	barometric efficiency	
BOD	biological oxygen demand	mg L^{-1}
Bq	becquerel (unit of radioactivity; 1 Bq = 1 disintegration per second)	
b	aquifer thickness	m
2b	fracture aperture	m
C	shape factor for determining k_i	
C	specific moisture capacity of a soil	(m of water)$^{-1}$
C, c	concentration	
°C	degrees Celsius (temperature)	
CEC	cation exchange capacity	meq (100 g)$^{-1}$

Symbol	Description	Units
CFC	chlorofluorocarbon	
Ci	curie (older unit of radioactivity; $1\ Ci = 3.7 \times 10^{10}$ disintegrations per second)	
COD	chemical oxygen demand	mg L^{-1}
D	hydraulic diffusivity	m^2 s^{-1}
D	hydrodynamic dispersion coefficient	m^2 s^{-1}
$D*$	molecular diffusion coefficient	m^2 s^{-1}
DIC	dissolved inorganic carbon	mg L^{-1}
DNAPL	dense, non-aqueous phase liquid	
DOC	dissolved organic carbon	mg L^{-1}
d	mean pore diameter	m
$\dfrac{dh}{dl}$	hydraulic gradient	
E^o	standard electrode potential	V
EC	electrical conductivity	S cm^{-1}
Eh	redox potential	V
e	void ratio	
e-	electron	
eq	chemical equivalent (see Box 3.1)	eq L^{-1}
F	Faraday constant (9.65×10^4 C mol^{-1})	
F	Darcy–Weisbach friction factor	
f_c	infiltration capacity	cm h^{-1}
f_t	infiltration rate	cm h^{-1}
f_{OC}	weight fraction organic carbon content	
G	Gibbs free energy	kJ mol^{-1}
g	gravitational acceleration	m s^{-2}
g	gram (mass)	
(g)	gas	
H	depth (head) of water measured at a flow gauging structure	m
H	enthalpy	kJ mol^{-1}
h	hydraulic head	m
I	ionic strength	mol L^{-1}
i	hydraulic gradient $\left(\dfrac{dh}{dl}\right)$	
IAEA	International Atomic Energy Agency	
IAP	ion activity product	mol^n L^{-n}
J	joule (energy, quantity of heat)	
K	equilibrium constant	mol^n L^{-n}
K (hydraulics)	hydraulic conductivity	m s^{-1}
K (temperature)	kelvin	
K_d	partition or distribution coefficient	mL g^{-1}
K_f	fracture hydraulic conductivity	m s^{-1}
K_H	Henry's law constant	Pa m^3 mol^{-1}
K_{OC}	organic carbon-water partition coefficient	
K_{OW}	octanol-water partition coefficient	
K_s	selectivity coefficient	
K_{sp}	solubility product	mol^n L^{-n}
k_i	intrinsic permeability	m^2
L	litre (volume)	
LNAPL	light, non-aqueous phase liquid	
l	length	m
MNA	monitored natural attenuation	
m	mass	kg
mol	amount of substance (see Box 3.1)	

Symbol	Description	Units
n	an integer	
n	roughness coefficient (Manning's n)	
n	porosity	
n_e	effective porosity	
P	Peclet number	
P	precipitation amount	mm
P	pressure	Pa (or N m^{-2})
P	partial pressure	atm or Pa
P_A, P_o	atmospheric pressure	atm or Pa
Pa	pascal (pressure)	
P_w	porewater pressure	Pa or m of water
PAH	polycyclic aromatic hydrocarbon	
PDB	Pee Dee Belemnite	
PE	potential evapotranspiration	mm
p	$-\log_{10}$	
ppm	parts per million	
Q	discharge	m^3 s^{-1}
Q_f	fracture flow discharge	m^3 s^{-1}
q	specific discharge or darcy velocity	m s^{-1}
R	hydraulic radius	m
R (R_d)	recharge (direct recharge)	mm a^{-1}
R	universal gas constant (8.314 J mol^{-1} K^{-1})	
RC	root constant	mm
R_d	retardation factor	
R_e	Reynolds number	
rem	roentgen equivalent man (older unit of dose equivalent; 1 rem = 0.01 Sv)	
S	entropy	J mol^{-1} K^{-1}
S	sorptivity of soil	cm (min)$^{-1/2}$
S	storativity	
S	slope	
S_p	specific retention	
S_s	specific storage	s^{-1}
S_{sp}	specific surface area	m^{-1}
S_y	specific yield	
SMD	soil moisture deficit	mm
STP	standard temperature and pressure of gases (0°C, 1 atmosphere pressure)	
Sv	sievert (unit of dose equivalent that accounts for the relative biological effects of different types of radiation; 1 Sv = 100 rem)	
s	groundwater level drawdown	m
s	solubility	g (100 g)$^{-1}$ solvent
T	transmissivity	m^2 s^{-1}
T	absolute temperature	K
TDS	total dissolved solids	mg L^{-1}
TOC	total organic carbon	mg L^{-1}
t	time	s
$t_{1/2}$	radionuclide half-life	s
V	volume	m^3
V	volt (electrical potential)	
V_s	volume of solid material	m^3
VOC	volatile organic compound	
V$_{SMOW}$	Vienna Standard Mean Ocean Water	
v	groundwater or river water velocity	m s^{-1}
$\bar{v}$	average linear velocity (groundwater)	m s^{-1}
$\bar{v}_c$	average linear contaminant velocity	m s^{-1}

Symbol	Description	Units
$\bar{v}_w$	average linear water velocity	m s^{-1}
WHO	World Health Organization	
WMO	World Meteorological Organization	
WMWL	World Meteoric Water Line	
$W(u)$	well function	
w	width	m
ZFP	zero flux plane	
z (hydraulics)	elevation	m
z (chemistry)	electrical charge	

Greek symbols

Symbol	Description	Units
α	aquifer compressibility	m^2 N^{-1} (or Pa^{-1})
α	aquifer dispersivity	m
α	isotope fractionation factor	
β	water compressibility	m^2 N^{-1} (or Pa^{-1})
γ (chemistry)	activity coefficient	kg mol^{-1}
γ (hydraulics)	specific weight	N
δ	stable isotope notation	
$\delta-$	partial negative charge	
$\delta+$	partial positive charge	
θ	volumetric moisture content	
λ	radionuclide decay constant	s^{-1}
μ	viscosity	N s m^{-2}
μ/ρ	kinematic viscosity	m^2 s^{-1}
ρ	fluid density	kg m^{-3}
ρ_b	bulk mass density	kg m^{-3}
ρ_s	particle mass density	kg m^{-3}
σ_e	effective stress	Pa (or N m^{-2})
σ_T	total stress	Pa (or N m^{-2})
τ	total competing cation concentration	meq (100 g)$^{-1}$
Σ	sum of	
Φ	fluid potential ($= h$)	m
ψ	fluid pressure	m of water
ψ_a	air entry pressure	m of water
Ω	saturation index	

Introduction

1

1.1 Scope of this book

This book is about the study of hydrogeology and the significance of groundwater in the terrestrial aquatic environment. Water is a precious natural resource, without which there would be no life on Earth. We, ourselves, are composed of two-thirds water by body weight. Our everyday lives depend on the availability of inexpensive, clean water and safe ways to dispose of it after use. Water supplies are also essential in supporting food production and industrial activity. As a source of water, groundwater obtained from beneath the Earth's surface is often cheaper, more convenient and less vulnerable to pollution than surface water.

Groundwater, because it is unnoticed underground, is often unacknowledged and undervalued resulting in adverse environmental, economic and social consequences. The over-exploitation of groundwater by uncontrolled pumping can cause detrimental effects on neighbouring boreholes and wells, land subsidence, saline water intrusion and the drying out of surface waters and wetlands. Without proper consideration for groundwater resources, groundwater pollution from uncontrolled uses of chemicals and the careless disposal of wastes on land cause serious impacts requiring difficult and expensive remediation over long periods of time. Major sources of contamination include agrochemicals, industrial and municipal wastes, tailings and process wastewater from mines, oil field brine pits, leaking underground storage tanks and pipelines, and sewage sludge and septic systems.

Achieving sustainable development of groundwater resources by the future avoidance of over-exploitation and contamination is an underlying theme of this book. By studying topics such as the properties of porous material, groundwater flow theory, well hydraulics, groundwater chemistry, environmental isotopes, contaminant hydrogeology and techniques of groundwater remediation and aquifer management, it is the responsibility of us all to manage groundwater resources to balance environmental, economic and social requirements and achieve sustainable groundwater development (Fig. 1.1).

The eight main chapters of this book aim to provide an introduction to the principles and practice of hydrogeology and to explain the role of groundwater in the aquatic environment. Chapter 1 provides a definition of hydrogeology and charts the history of the development of hydrogeology as a science. The water cycle is described and the importance of groundwater as a natural resource is explained. The legislative framework for the protection of groundwater resources is introduced with reference to industrialized and developing countries. Chapters 2 and 3 discuss the principles of physical and chemical hydrogeology that are fundamental to an understanding of the occurrence, movement and chemistry of groundwater in the Earth's crust. The relationships between geology and aquifer conditions are demonstrated both in terms of flow through porous material and rock–water interactions. Chapter 4 provides an introduction to the application of environmental isotopes in hydrogeological investigations for assessing the age of groundwater recharge and includes a section on noble gases to illustrate the identification of palaeowaters and aquifer evolution.

In the second half of this book, Chapter 5 provides an introduction to the range of field investigation techniques used in the assessment of catchment water

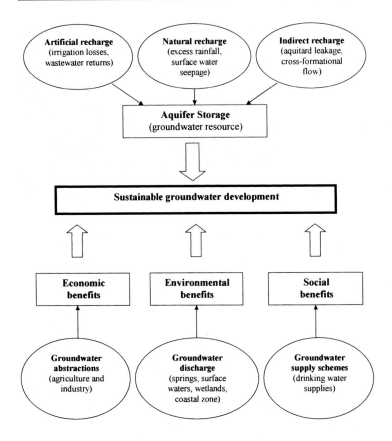

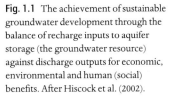

Fig. 1.1 The achievement of sustainable groundwater development through the balance of recharge inputs to aquifer storage (the groundwater resource) against discharge outputs for economic, environmental and human (social) benefits. After Hiscock et al. (2002).

resources and includes stream gauging methods, well hydraulics and tracer techniques. The protection of groundwater from surface contamination requires knowledge of solute transport processes and Chapter 6 introduces the principles of contaminant hydrogeology. Chapter 6 also covers water quality criteria and discusses the nature of contamination arising from a variety of urban, industrial and agricultural sources and in addition the causes and effects of saline intrusion in coastal regions. Following this Chapter 7 discusses methods of groundwater pollution remediation and protection, and includes sections that introduce risk assessment methods and spatial planning techniques. The final chapter, Chapter 8, returns to the topic of catchment water resources and demonstrates integrated methods for aquifer management together with consideration of groundwater interactions with rivers and wetlands, as well as the potential impacts of climate change on groundwater.

Each chapter in this book concludes with recommended further reading to help extend the reader's knowledge of hydrogeology. In addition, for students of hydrogeology, a set of discursive and numerical exercises is provided in Appendix 10 to provide practice in solving groundwater problems. The remaining appendices include data and information in support of the main chapters of this book and will be of wider application in Earth and environmental sciences.

1.2 What is hydrogeology?

Typical definitions of hydrogeology emphasize the occurrence, distribution, movement and geological interaction of water in the Earth's crust. Hydrogeology is an interdisciplinary subject and also encompasses aspects of hydrology. Hydrology has been defined as the study of the occurrence and movement of water on and over the Earth's surface independent of the seepage of groundwater and springs which sustain river flows during seasonal dry periods. However, too strict a division between the two subjects is unhelpful, particularly when trying to decipher the impact of human activities on the aquatic environment. How well we respond to the challenges of pollution of surface water and groundwater, the impacts of over-exploitation of water resources, and the possible impact of climate change will depend largely on our ability to take a holistic view of the aquatic environment.

1.3 Early examples of groundwater exploitation

The vast store of water beneath the ground surface has long been realized to be an invaluable source of water for human consumption and use. Throughout the world, springs fed by groundwater are revered for their life-giving or curative properties (Fig. 1.2), and utilization of groundwater long preceded understanding of its origin, occurrence and movement.

Groundwater development dates from ancient times, as manifest by the wells and horizontal tunnels known as qanats (ghanats) or aflaj (singular, falaj), both Arabic terms describing a small, artificial channel excavated as part of a water distribution system, which appear to have originated in Persia about 3000 years ago. Examples of such systems are found in a band across the arid regions extending from Afghanistan to Morocco. In Oman, the rural villages and aflaj-supplied oases lie at the heart of Omani culture and tradition. The system of participatory management of communal aflaj is an ancient tradition in Oman by which common-property flows are channelled and distributed to irrigation plots on a time-based system, under the management of a local community (Young 2002).

Figure 1.3 shows a cross-section along a qanat with its typical horizontal or gently sloping gallery laboriously dug through alluvial material, occasionally up to 30 km in length, and with vertical shafts dug at closely spaced intervals to provide access to the tunnel. Groundwater recharging the alluvium in the mountain foothills is fed by gravity flow from beneath the water table at the upper end of the qanat to a ground surface outlet and irrigation canal on the

Fig. 1.2 Lady's Well in Coquetdale, northern England (National Grid Reference NT 953 028). Groundwater seeping from glacial deposits at the foot of a gently sloping hillside is contained within an ornamental pool floored with loose gravel. The site has been used since Roman times as a roadside watering place and was walled round and given its present shape in either Roman or medieval times. Anglo Saxon Saint Ninian, the fifth century apostle, is associated with the site, and with other 'wells' beside Roman roads in Northumberland, and marks the spot where Saint Paulinus supposedly baptized 3000 Celtic heathens in its holy water during Easter week, 627 AD. The name of the well, Lady's Well, was adopted in the second half of the twelfth century when the nearby village of Holystone became the home of a priory of Augustinian canonesses. The well was repaired and adorned with a cross, and the statue brought from Alnwick, in the eighteenth and nineteenth centuries. Today, groundwater overflowing from the pool supplies the village of Holystone.

arid plain at its lower end (Fig. 1.4). The depth of the mother well (Fig. 1.3) is normally less than 50 m, with discharges, which vary seasonally with water-table fluctuations, seldom exceeding 3 m^3 s^{-1}.

Such early exploitation of groundwater as part of a sophisticated engineered system is also evident in the supply of water to feed the fountains of Rome (Box 1.1).

1.4 History of hydrogeology

As is evident from the above examples, exploitation of groundwater resources long preceded the founding of geology, let alone hydrogeology. Even as late as the seventeenth century it was generally assumed that water emerging from springs could not be derived from rainfall, for it was believed that the quantity was inadequate and the Earth too impervious to permit

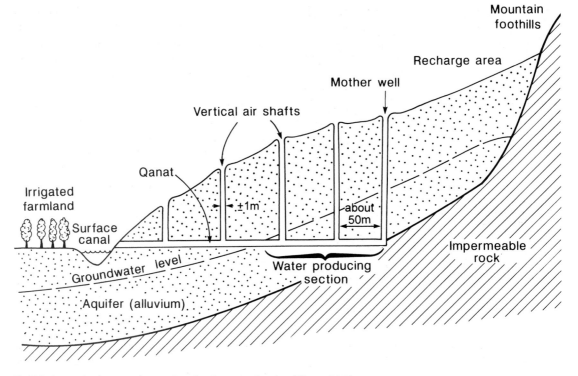

Fig. 1.3 Longitudinal section of a qanat. Based on Beaumont (1968) and Biswas (1972).

infiltration of rainwater far below the surface. A clear understanding of the hydrological cycle was achieved by the end of the seventeenth century. The French experimentalists Pierre Perrault (1611–1680) and Edme Mariotte (c. 1620–1684) made measurements of rainfall and runoff in the River Seine drainage basin, and the English astronomer Edmond Halley (1656–1742) demonstrated that evaporation of seawater was sufficient to account for all springs and stream flow (Halley 1691). Over one hundred years later, the famous chemist John Dalton (1766–1844) made further observations of the water cycle, including a consideration of the origin of springs (Dalton 1799).

One of the earliest applications of the principles of geology to the solution of hydrological problems was made by the Englishman William Smith (1769–1839), the 'father of English geology' and originator of the epoch-making Map of England (1815). During his work as a canal and colliery workings surveyor in the west of England, Smith noted the various soils and the character of the rocks from which they were

derived and used his knowledge of rock succession to locate groundwater resources to feed the summit levels of canals and supply individual houses and towns (Mather 1998).

In Britain, the industrial revolution led to a huge demand for water resources to supply new towns and cities, with Nottingham, Liverpool, Sunderland and parts of London all relying on groundwater. This explosion in demand for water gave impetus to the study of the economic aspects of geology. It was at this time that Lucas (1874) introduced the term 'hydrogeology' and produced the first real hydrogeological map (Lucas 1877). Towards the end of the nineteenth century, William Whitaker, sometimes described as the 'father of English hydrogeology', and an avid collector of well records, produced the first water supply memoir of the Geological Survey (Whitaker & Reid 1899) in which the water supply of Sussex is systematically recorded.

The drilling of many artesian wells stimulated parallel activity in France during the first half of the

Fig. 1.4 Irrigation canal supplied with water by a qanat or falaj in Oman. Photograph provided courtesy of M.R. Leeder.

nineteenth century. The French municipal hydraulic engineer Henry Darcy (1803–1858) studied the movement of water through sand and from empirical observations defined the basic equation, universally known as Darcy's law, that governs groundwater flow in most alluvial and sedimentary formations. Darcy's law is the foundation of the theoretical aspects of groundwater flow and his work was extended by another Frenchman, Arsène Dupuit (1804–1866), whose name is synonymous with the equation for axially symmetric flow towards a well in a permeable, porous material.

The pioneering work of Darcy and Dupuit was followed by the German civil engineer, Adolph Thiem (1836–1908), who made theoretical analyses of problems concerning groundwater flow towards wells and galleries, and by the Austrian Philip Forchheimer

(1852–1933) who, for the first time, applied advanced mathematics to the study of hydraulics. One of his major contributions was a determination of the relationship between equipotential surfaces and flow lines. Inspired by earlier techniques used to understand heat flow problems, and starting with Darcy's law and Dupuit's assumptions, Forchheimer derived a partial differential equation, the Laplace equation, for steady groundwater flow. Forchheimer was also the first to apply the method of mirror images to groundwater flow problems; for example, the case of a pumping well located adjacent to a river.

Much of Forchheimer's work was duplicated in the United States by Charles Slichter (1864–1946), apparently oblivious of Forchheimer's existence. However, Slichter's theoretical approach was vital to the advancement of groundwater hydrology in America at a time when the emphasis was on exploration and understanding the occurrence of groundwater. This era was consolidated by Meinzer (1923) in his book on the occurrence of groundwater in the United States. Meinzer (1928) was also the first to recognize the elastic storage behaviour of artesian aquifers. From his study of the Dakota sandstone (Meinzer & Hard 1925), it appeared that more water was pumped from the region than could be explained by the quantity of recharge at outcrop, such that the water-bearing formation must possess some elastic behaviour in releasing water contained in storage. Seven years later, Theis (1935), again using the analogy between heat flow and water flow, presented the groundbreaking mathematical solution that describes the transient behaviour of water levels in the vicinity of a pumping well.

Two additional major contributions in the advancement of physical hydrogeology were made by Hubbert and Jacob in their 1940 publications. Hubbert (1940) detailed work on the theory of natural groundwater flow in large sedimentary basins, while Jacob (1940) derived a general partial differential equation describing transient groundwater flow. Significantly, the equation described the elastic behaviour of porous rocks introduced by Meinzer over a decade earlier. Today, much of the training in groundwater flow theory and well hydraulics, and the use of computer programs to solve hydrogeological problems, is based on the work of these early hydrogeologists during the first half of the twentieth century.

The remarkable organization and engineering skills of the Roman civilization are demonstrated in the book written by Sextus Julius Frontinus and translated into English by C.E. Bennett (1969). In the year 97 AD, Frontinus was appointed to the post of water commissioner, during the tenure of which he wrote the *De Aquis*. The work is of a technical nature, written partly for his own instruction, and partly for the benefit of others. In it, Frontinus painstakingly details every aspect of the construction and maintenance of the aqueducts existing in his day.

For more than four hundred years, the city of Rome was supplied with water drawn from the River Tiber, and from wells and springs. Springs were held in high esteem, and treated with veneration. Many were believed to have healing properties, such as the springs of Juturna, part of a fountain known from the south side of the Roman Forum. As shown in Fig. 1, by the time of Frontinus, these supplies were augmented by several aqueducts, presumably giving a reliable supply of good quality water, in many cases dependent on groundwater. For example, the Vergine aqueduct brought water from the estate of Lucullus where soldiers, out hunting for water, were shown springs which, when dug out, yielded a copious supply. Frontinus records that the intake of Vergine is located in a marshy spot, surrounded by a concrete enclosure for the purpose of confining the gushing waters. The length of the water course was 14,105 paces (20.9 km). For 19.1 km of this distance the water was carried in an underground channel, and for 1.8 km above ground, of which 0.8 km was on substructures at various points, and 1.0 km on arches. The source of the Vergine spring is shown on a modern hydrogeological map (Boni et al. 1986) as issuing from permeable volcanic rocks with a mean discharge of 1.0 m^3 s^{-1} (Fig. 1). Frontinus also describes the Marcia aqueduct with its intake issuing from a tranquil pool of deep green hue. The length of the water-carrying conduit is 61,710^1/$_2$ paces (91.5 km), with 10.3 km on arches. Today, the source of the Marcia spring is known to issue from extensively fractured limestone rocks with a copious mean discharge of 5.4 m^3 s^{-1}.

After enumerating the lengths and courses of the several aqueducts, Frontinus enthuses: 'with such an array of indispensable structures carrying so many waters, compare, if you will, the idle Pyramids or the useless, though famous, works of the Greeks!' To protect the aqueducts from wilful pollution, a law was introduced such that: 'No one shall with malice pollute the waters where they issue publicly. Should any one pollute them, his fine shall be 10,000 sestertii' which, at the time, was a very large fine. Clearly, the 'polluter pays' principle was readily adopted by the Romans! Further historical and architectural details of the ancient aqueducts of Rome are given by Bono and Boni (2001).

The Vergine aqueduct is one of only two of the original aqueducts still in use. The total discharge of the ancient aqueducts was in excess of 10 m^3 s^{-1} supplying a population at the end of the first century AD of about 0.5 million. Today, Rome is supplied with 23 m^3 s^{-1} of groundwater, mainly from karst limestone aquifers, and serving a population of 3.5 million (Bono & Boni 2001). Many of the groundwater sources are springs from the karst system of the Simbruini Mountains east of Rome.

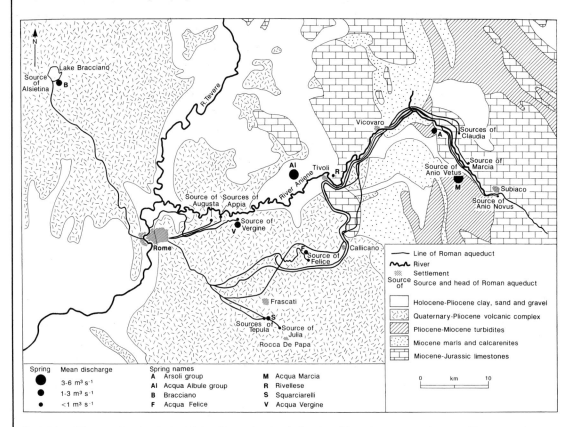

Fig. 1 Map of the general geology in the vicinity of Rome showing the location of the spring sources and routes of Roman aqueducts. Based on Bennett (1969) and Boni et al. (1986).

The development of the chemical aspects of hydrogeology stemmed from the need to provide good quality water for drinking and agricultural purposes. The objective description of the hydrochemical properties of groundwater was assisted by Piper (1944) and Stiff (1951) who presented graphical procedures for the interpretation of water analyses. Later, notable contributions were made by Chebotarev (1955), who described the natural chemical evolution of groundwater in the direction of groundwater flow, and Hem (1959), who provided extensive guidance on the study and interpretation of the chemical characteristics of natural waters. Later texts by Garrels and Christ (1965) and Stumm and Morgan (1981) provided thorough, theoretical treatments of aquatic chemistry.

By the end of the twentieth century, the previous separation of hydrogeology into physical and chemical fields of study had merged with the need to understand the fate of contaminants in the subsurface environment. Contaminants are advected and dispersed by groundwater movement and can simultaneously undergo chemical processes that act to reduce pollutant concentrations. More recently, the introduction of immiscible pollutants, such as petroleum products and organic solvents into aquifers, has led to intensive research and technical advances in the theoretical description, modelling and field investigation of multiphase systems. At the same time, environmental legislation has proliferated, and has acted as a driver in contaminant hydrogeology. Today, research efforts are directed towards understanding natural attenuation processes as part of a managed approach to restoring contaminated land and groundwater.

Hence, hydrogeology has now developed into a truly interdisciplinary subject, and students who aim to become hydrogeologists require a firm foundation in Earth sciences, physics, chemistry, biology, mathematics, statistics and computer science, together with an adequate understanding of environmental economics and law, and government policy.

1.5 The water cycle

A useful start in promoting a holistic approach to linking ground and surface waters is to adopt the hydrological cycle as a basic framework. The hydrological cycle, as depicted in Fig. 1.5, can be thought of as the continuous circulation of water near the surface of the Earth from the ocean to the atmosphere and then via precipitation, surface runoff and groundwater flow back to the ocean. Warming of the ocean by solar radiation causes water to be evaporated into the atmosphere and transported by winds to the land masses where the vapour condenses and falls as precipitation. The precipitation is either returned directly to the ocean, intercepted by vegetated surfaces and returned to the atmosphere by evapotranspiration, collected to form surface runoff, or infiltrated into the soil and underlying rocks to form groundwater. The surface runoff and groundwater flow contribute to surface streams and rivers that flow to the ocean, with pools and lakes providing temporary surface storage.

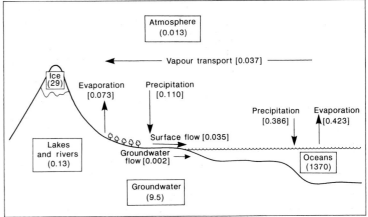

Fig. 1.5 The hydrological cycle. The global water cycle has three major pathways: precipitation, evaporation and water vapour transport. Vapour transport from sea to land is returned as runoff (surface water and groundwater flow). Numbers in () represent inventories (in 10^6 km^3) for each reservoir. Fluxes in [] are in 10^6 km^3 a^{-1}. After Berner and Berner (1987).

Table 1.1 Inventory of water at or near the Earth's surface. After Berner and Berner (1987).

Reservoir	Volume (× 10⁶ km³)	% of total
Oceans	1370	97.25
Ice caps and glaciers	29	2.05
Deep groundwater (750–4000 m)	5.3	0.38
Shallow groundwater (<750 m)	4.2	0.30
Lakes	0.125	0.01
Soil moisture	0.065	0.005
Atmosphere*	0.013	0.001
Rivers	0.0017	0.0001
Biosphere	0.0006	0.00004
Total	1408.7	100

* As liquid equivalent of water vapour.

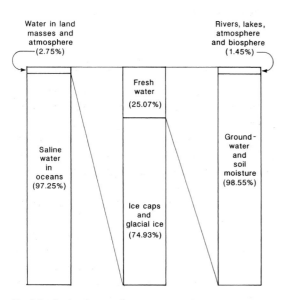

Fig. 1.6 The distribution of water at or near the Earth's surface. Only a very small amount of freshwater (<0.3% of total water) is readily available to humans and other biota. After Maurits la Riviére (1987).

Of the total water in the global cycle, Table 1.1 shows that saline water in the oceans accounts for 97.25%. Land masses and the atmosphere therefore contain 2.75%. Ice caps and glaciers hold 2.05%, groundwater to a depth of 4 km accounts for 0.68%, freshwater lakes 0.01%, soil moisture 0.005% and rivers 0.0001%. About 75% of the water in land areas

is locked in glacial ice or is saline (Fig. 1.6). The relative importance of groundwater can be realized when it is considered that, of the remaining quarter of water in land areas, around 98% is stored underground. In addition to the more accessible groundwater involved in the water cycle above a depth of 4 km, estimates of the volume of interstitial water in rock pores at even greater depths range from 53×10^6 km³ (Ambroggi 1977) to 320×10^6 km³ (Garrels et al. 1975).

Within the water cycle, and in order to conserve total water, evaporation must balance precipitation for the Earth as a whole. The average global precipitation rate, which is equal to the evaporation rate, is 496,000 km³ a⁻¹. However, as Fig. 1.5 shows, for any one portion of the Earth, evaporation and precipitation generally do not balance. The differences comprise water transported from the oceans to the continents as atmospheric water vapour and water returned to the oceans as river runoff and a small amount (~6%) of direct groundwater discharge to the oceans (Zektser & Loáiciga 1993).

The approximate breakdown of direct groundwater discharge from continents to adjacent oceans and seas is estimated as follows: Australia 24 km³ a⁻¹; Europe 153 km³ a⁻¹; Africa 236 km³ a⁻¹; Asia 328 km³ a⁻¹; the Americas 729 km³ a⁻¹; and major islands 914 km³ a⁻¹ (Zektser & Loáiciga 1993). The low contribution from the Australian continent of direct groundwater discharge, despite its relatively large territory, is attributed to the widespread occurrence of low-permeability surface rocks that cover the continent. At the other extreme, the overall proximity of recharge areas to discharge areas is the reason why major islands of the world contribute over one-third of the world's direct groundwater discharge to the oceans. The largest direct groundwater flows to oceans are found in mountainous areas of tropical and humid zones and can reach $10–15 \times 10^{-3}$ m³ s⁻¹ km⁻². The smallest direct groundwater discharge values of $0.2–0.5 \times 10^{-3}$ m³ s⁻¹ km⁻² occur in arid and arctic regions that have unfavourable recharge and permeability conditions (Zektser & Loáiciga 1993).

Taking the constant volume of water in a given reservoir and dividing by the rate of addition (or loss) of water to (from) it enables the calculation of a residence time for that reservoir. For the oceans, the volume of water present (1370×10^6 km³; see Fig. 1.5) divided by the rate of river runoff to the oceans

Table 1.2 Agents of material transport to the oceans. After Garrels et al. (1975).

Agent	% of total transport	Remarks
Rivers	89	Dissolved load 17%, suspended load 72%
Glacier ice	7	Ground rock debris plus material up to boulder size. Mainly from Antarctica and Greenland. Distributed in seas by icebergs
Groundwater	2	Dissolved materials similar to river composition. Estimate poorly constrained
Coastal erosion	1	Sediments eroded from cliffs, etc.
Volcanic	0.3(?)	Dust from explosive eruptions. Estimate poorly constrained
Wind-blown dust	0.2	Related to desert source areas and wind patterns, e.g. Sahara, major source for tropical Atlantic

$(0.037 \times 10^6 \text{ km}^3 \text{ a}^{-1})$ gives an average time that a water molecule spends in the ocean of about 37,000 years. Lakes, rivers, glaciers and shallow groundwater have residence times ranging between days and thousands of years. Because of extreme variability in volumes and precipitation and evaporation rates, no simple average residence time can be given for each of these reservoirs. As a rough calculation, and with reference to Fig. 1.5 and Table 1.1, if about 6% $(2220 \text{ km}^3 \text{ a}^{-1})$ of runoff from land is taken as active groundwater circulation, then the time taken to replenish the volume $(4.2 \times 10^6 \text{ km}^3)$ of shallow groundwater stored below the Earth's surface is of the order of 2000 years. In reality, groundwater residence times vary from about 2 weeks to 10,000 years (Nace 1971), and longer (Edmunds 2001). A similar estimation for rivers provides a value of about 20 days. These estimates, although a gross simplification of the natural variability, do serve to emphasize the potential longevity of groundwater pollution compared to more rapid flushing of contaminants from river systems.

As an agent of material transport to the oceans of products of weathering processes, groundwater probably represents only a small fraction of the total transport (Table 1.2). Rivers (89% of total transport) represent an important pathway while groundwater accounts for a poorly constrained estimate of 2% of total transport in the form of dissolved materials (Garrels et al. 1975). More recent estimates by Zektser and Loáiciga (1993) indicate that globally the transport of salts via direct groundwater discharge is approximately $1.3 \times 10^9 \text{ t a}^{-1}$, roughly equal to half of the quantity contributed by rivers to the oceans. Given a volumetric rate of direct groundwater discharge to the oceans of $2220 \text{ km}^3 \text{ a}^{-1}$, the average dissolved solids concentration is about 585 mg L^{-1}. This calculation illustrates the long residence time of groundwater in the Earth's crust where its mineral content is concentrated by dissolution.

1.6 Groundwater as a natural resource

Groundwater is an important natural resource. Worldwide, more than 2 billion people depend on groundwater for their daily supply (Kemper 2004). A large proportion of the world's agriculture and irrigation is dependent on groundwater, as are a large number of industries. Whether groundwater or surface water is exploited for water supply is largely dependent on the location of aquifers relative to the point of demand. A large urban population with a high demand for water would only be able to exploit groundwater if the aquifer, typically a sedimentary rock, has favourable storage and transmission properties, whereas in a sparsely populated rural district more limited but essential water supplies might be found in poor aquifers, such as weathered basement rock.

The relationship between population and geology can be inferred from Tables 1.3 and 1.4, which give a breakdown of water use by purpose and type (surface water and groundwater) for regions of England and Wales. Surface water abstraction for electricity generation is the largest category, but most of the freshwater abstracted for cooling purposes is returned to rivers and can be used again downstream. In terms of public water supply abstractions, groundwater is especially significant in the Southern (73% dependence

Table 1.3 Estimated abstractions from all surface water and groundwater in England and Wales by purpose and Environment Agency region for 1996. All data are given as 10^3 m^3 day^{-1}. Source: Environment Agency for England & Wales.

Region	Public water supply	Spray irrigation	Rural*	Electricity supply	Other industry†	Total
North East	2538	55	873	5142	1871	10,479
Welsh	2051	10	411	6826	565	9863
North West	1595	6	174	6148	1124	9047
Southern	1451	23	1098	4210	365	7147
South West	1284	8	2046	3573	127	7038
Thames	4130	14	406	1715	255	6520
Midlands	2602	82	71	1681	406	4842
Anglian	1803	171	97	2001	497	4569
Total	17,454	369	5176	31,296	5210	59,505

* Category includes agricultural use (excluding spray irrigation), fish farming, public water supply (private abstractions for domestic use by individual households) and other (private domestic water supply wells and boreholes, public water supply transfer licences and frost protection use).
† Category includes industrial and mineral washing uses.

Table 1.4 Estimated abstractions from groundwaters in England and Wales by purpose and Environment Agency region for 1996. All data are given as 10^3 m^3 day^{-1}. Source: Environment Agency for England & Wales.

Region	Public water supply	Spray irrigation	Rural*	Electricity supply	Other industry†	Total
Thames	1378 (33)‡	8	63	0	176	1625
Southern	1056 (73)	10	202	0	155	1423
Midlands	1024 (39)	34	14	9	138	1219
Anglian	735 (41)	68	51	0	218	1072
North East	441 (17)	40	97	0	135	713
South West	407 (32)	3	185	2	31	628
North West	262 (16)	2	9	0	121	394
Welsh	113 (6)	2	9	3	28	155
Total	5416 (31)	167	630	14	1002	7229

* † See Table 1.3.
‡ Groundwater supply as a percentage of the total surface water and groundwater supply (see also Table 1.5).

on groundwater), Anglian (41%), Midlands (39%) and Thames (33%) regions and accounts for 42% of the total public water supply in these four regions. In these densely populated regions of south-east England and the English Midlands, good quality groundwater is obtained from the high-yielding Cretaceous Chalk and Triassic sandstone aquifers.

At the European level, groundwater is again a significant economic resource. As Table 1.5 reveals, large quantities of groundwater are abstracted in France, Germany, Italy and Spain (all in excess of 5000×10^6 m^3 a^{-1}) comprising 16% of the total water abstracted in these four countries. Overall, average annual water abstraction from groundwater accounts for 20% of the total, ranging from in excess of 50% for Austria, Belgium, Denmark and Luxembourg to, respectively, only 10% and 12% for Finland and Ireland. The data given in Table 1.5 should be treated with caution given the lack of a common European procedure for estimating water resources and the fact that the data probably underestimate the contribution made by groundwater to municipal water supplies. According to a report commissioned for the European Commission (RIVM & RIZA 1991), about 75% of the inhabitants of Europe depend on groundwater for their water supply.

Table 1.5 Average annual water abstractions in European Union member states by type for the period 1980–1995. The data are ordered in terms of the percentage groundwater contributes to the total abstraction. Source: European Environment Agency Data Service.

Country	Surface water ($\times 10^6$ m^3)	Groundwater ($\times 10^6$ m^3)	Total ($\times 10^6$ m^3)	% Groundwater
Denmark	9	907	916	99
Belgium	2385	4630	7015	66
Austria	1038	1322	2360	56
Luxembourg	28	29	57	51
Portugal	4233	3065	7298	42
Greece	3470	1570	5040	31
Italy	40,000	12,000	52,000	23
United Kingdom	9344	2709	12,053	22
Sweden	2121	588	2709	22
Spain	29,901	5422	35,323	15
Germany	51,151	7711	58,862	13
France	35,195	5446	40,641	13
Netherlands	10,965	1711	12,676	13
Ireland	945	125	1070	12
Finland	3011	335	3346	10
Total	193,796	47,570	241,366	20

Note: The data given in this table should be considered with some reservation due to the lack of a common European procedure to estimate water resources.

A similar picture emerges of the importance of groundwater for the population of North America. In Canada, almost 8 million people, or 26% of the population, rely on groundwater for domestic use. Five million of these users live in rural areas where groundwater is a reliable and cheap water supply that can be conveniently abstracted close to the point of use. The remaining groundwater users are located primarily in smaller municipalities. For example, 100% of the population of Prince Edward Island and over 60% of the populations of New Brunswick and the Yukon Territory rely on groundwater for domestic supplies. In Ontario, a province where groundwater is also used predominantly for supplying municipalities, 22% of the population are reliant on groundwater.

The abstraction of fresh and saline water in the United States from 1960 to 2000 as reported by Solley et al. (1998) and Hutson et al. (2004) is shown in Fig. 1.7. The estimated total abstraction for 1995 is 1522×10^6 m^3 day^{-1} for all offstream uses (all uses except water used instream for hydroelectric power generation) and is 10% less than the 1980 peak estimate. This total has varied by less than 3% since 1985. In 2000, the estimated total water use in the United States is 1544×10^6 m^3 day^{-1}. Estimates of abstraction by source indicate that during 1995, total fresh surface water abstractions were 996×10^6 m^3 day^{-1} and total groundwater abstractions were 293×10^6 m^3 day^{-1} (or 23% of the combined freshwater abstractions). The respective figures for 2000 are 991×10^6 m^3 day^{-1} and 316×10^6 m^3 day^{-1}, with 24% of freshwater abstractions from groundwater.

Total water abstraction for public water supply in the United States in 2000 is estimated to have been 163×10^6 m^3 day^{-1}, an 8% increase since 1995. This increase compares with a 7% growth in the population for the same period. Per capita public water supply use increased from about 678 L day^{-1} in 1995 to 683 L day^{-1} in 2000, but is still less than the per capita consumption of 696 L day^{-1} recorded for 1990.

The two largest water use categories in 2000 were cooling water for thermoelectric power generation (738×10^6 m^3 day^{-1} of fresh and saline water) and irrigation (518×10^6 m^3 day^{-1} of freshwater). Of these two categories, irrigation accounts for the greater abstraction of freshwater. The area of irrigated land increased nearly 7% between 1995 and 2000 with an increase in freshwater abstraction of 2% for this water-use category. The area irrigated with sprinkler and micro-irrigation systems has continued to rise and now comprises more than half of the total.

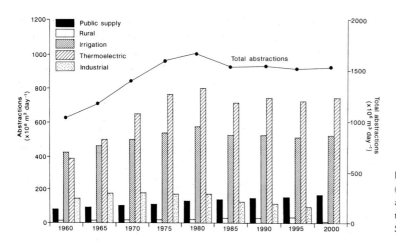

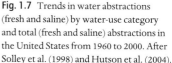

Fig. 1.7 Trends in water abstractions (fresh and saline) by water-use category and total (fresh and saline) abstractions in the United States from 1960 to 2000. After Solley et al. (1998) and Hutson et al. (2004).

In 2000, surface water was the primary source of irrigation water in the arid West and the Mountain States and groundwater was the primary source in the Central States. California, Idaho, Colorado and Nebraska combined accounted for one-half of the total irrigation water abstractions. California and Idaho accounted for 40% of surface water abstractions and California and Nebraska accounted for one-third of groundwater abstractions. In general, groundwater abstractions for irrigation have increased significantly. In 1950, groundwater accounted for 23% of total irrigation water, while in 2000 it accounted for 42%.

1.7 Management and protection of groundwater resources in the United Kingdom

Approaches to the management and protection of groundwater resources have developed in parallel with our understanding of the economic and environmental implications of groundwater exploitation. In the United Kingdom, it is interesting to follow the introduction of relevant legislation, and how this has increased hydrogeological knowledge.

Hydrogeological experience prior to 1945 rested on a general awareness of sites likely to provide favourable yields, changes in chemistry down-gradient from the point of recharge and hazards such as ground sub-sidence from groundwater over-exploitation. The Water Act 1945 provided legal control on water abstractions and this prompted an era of water resources assessment that included surveys of groundwater resources, the development of methods to assess recharge amounts (Section 5.5) and the initiation of groundwater studies. Increased abstraction from the Chalk aquifer during the 1950s and a drought in 1959 highlighted the effect of groundwater abstractions upon Chalk streams and stimulated the need for river baseflow studies (Section 5.7.1). Furthermore, the application of quantitative pumping test analysis techniques (Section 5.8.2) during this period revealed spatial variations in aquifer transmissivity and an association between transmissivity and topography.

The Water Resources Act 1963 led to the formation of 27 catchment-based authorities responsible for pollution prevention, fisheries, land drainage and water resources. The Act ushered in a decade of groundwater resources management that required the licensing of all abstractions in England and Wales. Under Section 14 of the Act, each authority was required to undertake a survey of resources and the Water Resources Board (abolished 1974) was established with the task of resource planning on a national scale. Regional groundwater schemes were developed in the context of river basin analysis for the purposes of river augmentation by groundwater, seasonal abstraction and artificial recharge. Scientific advancement in the application of numerical models to solve non-linear equations of groundwater flow permitted the prediction of future groundwater abstraction regimes.

The Water Act 1973 reflected the importance of water quality aspects and heralded the current interest in groundwater quality. The Act led to the formation of 10 catchment-based regional water authorities with responsibility for all water and sewerage services and for all parts of the water cycle. The Control of Pollution Act 1974 extended the powers of the regional water authorities in controlling effluent discharge to underground strata and limited certain activities that could lead to polluting discharges. The first aquifer protection policies were developed at this time.

The Water Act 1989 separated the water supply and regulatory functions of the regional water authorities, and the new National Rivers Authority was set up to manage water resources planning, abstraction control, pollution prevention and aquifer protection. A number of other Acts of Parliament followed including the Environmental Protection Act 1990 and the Water Resources Act 1991 that control the direct and indirect discharge of harmful substances into groundwater and are, in part, an enactment of the European Communities Directive on the Protection of Groundwater Against Certain Dangerous Substances (80/68/EEC). Further controls on discharges were implemented under the Groundwater Regulations 1998. In addition, the Water Resources Act 1991 consolidated all the provisions of the Water Resources Act 1963 in respect of the control of groundwater abstractions. In pursuing a strategy to protect both individual borehole sources and wider groundwater resources, the National Rivers Authority (1992) developed its practice and policy for the protection of groundwater with the aim of raising awareness of the vulnerability of groundwater to surface-derived pollution. Following the establishment of the Environment Agency under the Environment Act 1995 (when the National Rivers Authority, Her Majesty's Inspectorate of Pollution and the Waste Regulatory Authorities were brought together), the practice and policy document for the protection of groundwater was updated (Environment Agency 1998).

Currently, the Environment Agency is promoting a national framework for water resources protection in the context of emerging European initiatives, principally the Water Framework Directive (Section 1.8). The Water Act 2003 is one example of new legislation

to further the sustainable use of water resources and protect the environment. The Act links water abstraction licensing to local water resource availability and moves from a licensing system based on purpose of use to one based on volume consumed. The Act also introduces time-limited licences to give flexibility in making changes to abstraction rights in the face of climate change (Section 8.5) and increased demand. From 2012, licences without a time limit will be revoked, without a right to compensation, if an abstraction causes significant environmental damage.

1.8 European Union Water Framework Directive

The Water Framework Directive (WFD) establishing a framework for Community action in the field of water policy is a far-reaching piece of legislation governing water resources management and protection in the European Union (Council of European Communities 2000). The Directive (2000/60/EC) was adopted in December 2000 and requires Member States to enforce appropriate measures to achieve good ecological and chemical status of all water bodies by 2015. The purpose of the Directive is to establish a framework for the protection of inland surface waters, transitional waters (estuaries), coastal waters and groundwater to prevent further deterioration of aquatic ecosystems and, with regard to their water needs, terrestrial ecosystems and wetlands. In its implementation, the WFD requires an integrated approach to river basin management and promotes sustainable water use based on long-term protection of available water resources. A specific purpose of the WFD is to ensure the progressive reduction of pollution of groundwater and prevent its further pollution.

Article 17 of the WFD requires a proposal (2003/0210(COD)) from the Commission for a Groundwater Daughter Directive leading to the adoption of specific measures to prevent and control groundwater pollution and achieve good groundwater chemical status (Commission of the European Communities 2003). In addition, the proposal introduces measures for protecting groundwater from indirect pollution (discharges of pollutants into groundwater after percolation through the ground or subsoil). In the proposed Directive, compliance with good chemical status is based on a comparison of monitoring data with quality standards existing in EU legislation on nitrates and plant protection and biocidal products which set threshold values (maximum permissible concentrations) in groundwater for a number of pollutants. With regard to pollutants that are not covered by EU legislation, the proposed Directive requires Member States to establish threshold values defined at the national, river basin or groundwater body levels, thus taking into account the great diversity of groundwater characteristics across the EU.

The proposed Groundwater Daughter Directive sets out specific criteria for the identification of significant and sustained upward trends in pollutant concentrations, and for the definition of starting points for when action must be taken to reverse these trends. In this respect, significance is defined both on the basis of time series and environmental significance. Time series are periods of time during which a trend is detected through regular monitoring. Environmental significance describes the point at which the concentration of a pollutant starts to threaten to worsen the quality of groundwater. This point is set at 75% of the quality standard or the threshold value defined by Member States. In 2012, a comprehensive programme of measures to prevent or limit pollution of water, including groundwater, will become operational under the WFD. Monitoring results obtained through the application of the Groundwater Daughter Directive will be used to design the measures to prevent or limit pollution of groundwater.

1.9 Management and protection of groundwater resources in the United States

Groundwater management in the United States is highly fragmented, with responsibilities shared among a large number of federal, state and local programmes. At each level of government, unique legal authorities allow for the control of one or more threats to groundwater, such as groundwater contamination arising from municipal, industrial, mining and agricultural activities.

Beginning with the 1972 amendments to the federal Water Pollution Control Act, and followed by the Safe Drinking Water Act 1974, the federal

government's role in groundwater management has increased. The introduction of the Resource Conservation and Recovery Act (RCRA) 1976 and the Comprehensive Environmental Response, Compensation and Liability Act (CERCLA) 1980, established the federal government's current focus on groundwater remediation. With these acts, the federal government has directed billions of dollars in public and private resources towards cleaning up contaminated groundwater at 'Superfund' sites, RCRA corrective action facilities and leaking underground storage tanks. In 1994, the National Academy of Sciences estimated that over a trillion dollars, or approximately $4000 per person in the United States, will be spent in the next 30 years on remediating contaminated soil and groundwater.

The approach to groundwater protection at the federal level has left the management of many contaminant threats, for example hazardous materials used by light industries (such as dry cleaners, printers or car maintenance workshops), to state and local government authorities. Other groundwater threats, such as over-abstraction, are not generally addressed under federal law, but left to states and local governments to manage.

In 1984, the US Environmental Protection Agency (USEPA) created the Office of Ground Water Protection to initiate a more comprehensive groundwater resource protection approach and to lead programmes aimed at resource protection. Such programmes include the Wellhead Protection and Sole Source Aquifer Programs, which were established by Amendments to the Safe Drinking Water Act 1986. The Wellhead Protection Program (WHPP) encourages communities to protect their groundwater resources used for drinking water. The Sole Source Aquifer Program limits federal activities that could contaminate important sources of groundwater.

State groundwater management programmes are seen as critical to the future achievement of effective and sustainable protection of groundwater resources. In 1991, the USEPA established a Ground Water Strategy to place greater emphasis on comprehensive state management of groundwater as a resource through the promotion of Comprehensive State Ground Water Protection Programs (CSGWPPs) together with better alignment of federal programmes with state groundwater resource protection priorities (United States Environmental Protection Agency 1999).

1.10 Groundwater resources in developing countries

In the developing world, groundwater is extensively used for drinking water supplies, especially in smaller towns and rural areas, where it is the cheapest source. Groundwater schemes consist typically of large numbers of boreholes, often drilled on an uncontrolled basis, providing untreated, unmonitored and often unconnected supplies. Shallower dug wells continue to be constructed in some cases. Better yielding boreholes (100 L s^{-1}) are quite widely developed in larger towns to provide piped supplies. Even in these cases, raw water monitoring and treatment are often limited and intermittent. An example of the significance of groundwater in leading the economic development in rural and expanding urban areas is the Quaternary Aquifer of the North China Plain (Box 1.2).

It remains one of the greatest challenges for the future to provide the basic amenity of a safe and reliable supply of drinking water to the entire world's population. Despite the efforts of governments, charities and aid agencies, many villagers have to walk hundreds of metres to obtain drinking water from sources that may be unprotected from contamination (Fig. 1.8). Pollution sources include unsewered pit latrines to dispose of human wastes, inorganic fertilizers and pesticides used in an effort to secure self-sufficiency in food production, and industrial wastes in urban areas.

The Third World Water Forum held in Osaka, Japan, in March 2003 emphasized issues relating to the development and management of groundwater and recommended that many developing nations need to appreciate their social and economic dependency on groundwater and to invest in strengthening institutional provisions and building institutional capacity for its improved management. International development agencies and banks are urged to give higher priority to supporting realistic initiatives to strengthen governance of groundwater resources

Fig. 1.8 Collection of water for domestic use from a hand-pumped tube well drilled in Precambrian metamorphic rock in the Uda Walawe Basin, Sri Lanka.

and local aquifer management. For the future, sustainable livelihoods, food security and key ecological systems will be dependent on such initiatives.

1.11 FURTHER READING

Appleton, J.D., Fuge, R. & McCall, G.J.H. (1996) *Environmental Geochemistry and Health with Special Reference to Developing Countries*. Geological Society, London, Special Publications, **113**.

Berner, E.K. & Berner, R.A. (1987) *The Global Water Cycle: geochemistry and environment*. Prentice-Hall, Englewood Cliffs, New Jersey.

Biswas, A.K. (1972) *History of Hydrology*. North-Holland, Amsterdam.

Deming, D. (2002) *Introduction to Hydrogeology*. McGraw-Hill Higher Education, New York.

Downing, R.A. & Wilkinson, W.B. (eds) (1991) *Applied Groundwater Hydrology: a British perspective*. Clarendon Press, Oxford.

Hiscock, K.M., Rivett, M.O. & Davison, R.M. (eds) (2002) *Sustainable Groundwater Development*. Geological Society, London, Special Publications, **193**.

Jones, J.A.A. (1997) *Global Hydrology: processes, resources and environmental management*. Addison Wesley Longman, Harlow.

Kemper, K.E. (ed.) (2004) Theme issue: groundwater – from development to management. *Hydrogeology Journal* **12**(1).

Price, M. (1996) *Introducing Groundwater*, 2nd edn. Chapman & Hall, London.

Groundwater development of the Quaternary Aquifer of the North China Plain

BOX 1.2

The Quaternary Aquifer of the North China Plain represents one of the world's largest aquifer systems and underlies extensive tracts of the Hai river basin and the catchments of the adjacent Huai and Huang (Yellow) river systems (Fig. 1) and beyond. This densely populated area comprises a number of extensive plains, known collectively as the North China Plain, and includes three distinct

hydrogeological settings within the Quaternary aquifer system (Fig. 2). The semi-arid climate of north-eastern China is characterized by cold, dry winters (December–March) and hot, humid summers (July–September).

The Quaternary Aquifer supports an enormous exploitation of groundwater which has lead to large socioeconomic benefits in

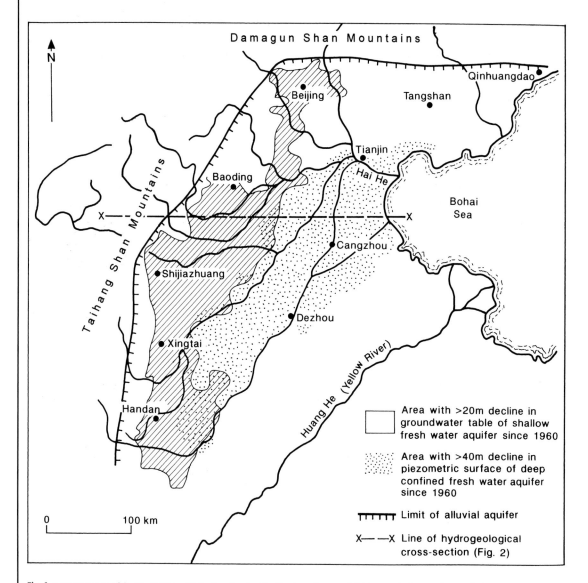

Fig. 1 Location map of the North China Plain showing the distribution of areas exhibiting marked groundwater depletion as a consequence of aquifer over-exploitation of the Quaternary aquifer system (Fig. 2). After Foster et al. (2004).

BOX
1 . 2

Continued

terms of irrigated grain production, farming employment and rural poverty alleviation, together with urban and industrial water supply provision. An estimated water supply of $27 \times 10^9 \, m^3 \, a^{-1}$ in the Hai river basin alone was derived from wells and boreholes in 1988 (MWR 1992), but such large exploitation of groundwater has led to increasing difficulties in the last few years.

Given the heavy dependence on groundwater resources in the North China Plain, a number of concerns have been identified in recent years (Fig. 1) including a falling water table in the shallow freshwater aquifer, declining water levels in the deep freshwater aquifer, aquifer salinization as a result of inadequately controlled pumping, and aquifer pollution from uncontrolled urban and indus-

trial wastewater discharges. These issues are interlinked but do not affect the three main hydrogeological settings equally (Table 1). A range of water resources management strategies are considered by Foster et al. (2004) that could contribute to reducing and eventually eliminating the current aquifer depletion and include agricultural water-saving measures, changes in land use and crop regimes, artificial aquifer recharge of excess surface runoff, re-use of treated urban wastewater, and improved institutional arrangements that deliver these water savings and technologies while at the same time limiting further exploitation of groundwater for irrigated agriculture and industrial production (Foster et al. 2004).

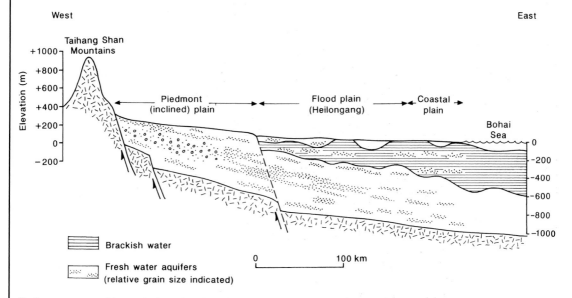

Fig. 2 Cross-section of the North China Plain showing the general hydrogeological setting of the Quaternary aquifer system which includes the gently sloping piedmont plain and associated major alluvial fans, the main alluvial plain (Heilongang) and the coastal plain around the margin of the Bohai Sea. After Foster et al. (2004).

Table 1 Key groundwater issues in the North China Plain listed according to hydrogeological setting (Fig. 2). After Foster et al. (2004).

Groundwater issue	Hydrogeological setting		
	Piedmont plain	Flood plain	Coastal plain
Falling water table of shallow freshwater aquifer	+++	+++	+
Depletion of deep freshwater aquifer	0*	+++	++
Risk of shallow aquifer and/or soil salinization	0	++	+++
Groundwater pollution from urban and industrial wastewater	+++	+	0

+++, very important; ++, important; +, minor importance; 0, not important.
* Effects of excessive abstraction may be reflected in the overlying shallow freshwater aquifer which is here in hydraulic continuity.

Physical hydrogeology

2

2.1 Introduction

The occurrence of groundwater within the Earth's crust and the emergence of springs at the ground surface are determined by the lithology of geological materials, regional geological structure, geomorphology of landforms and the availability of recharge sources. The infiltration of rainfall to the water table and the flow of groundwater in an aquifer towards a discharge area are governed by physical laws that describe changes in energy of the groundwater. In this chapter, the physical properties of aquifer storage and permeability are discussed in relation to different rock types and hydrogeological conditions. Then, starting with Darcy's law, the fundamental law of groundwater flow, the equations of steady-state and transient groundwater flow are derived for the hydraulic conditions encountered above and below the water table. Next, examples of analytical solutions to simple one-dimensional groundwater flow problems are presented and this is followed by an explanation of the influence of topography in producing various scales of groundwater movement, including patterns of local, intermediate and regional flow. The last section of this chapter deals with the occurrence of groundwater resources. The wide range of aquifer types and expected borehole yields associated with sedimentary, metamorphic and igneous rock types are described with reference to the hydrogeological units that occur in the United Kingdom.

2.2 Porosity

The porosity of a soil or rock is that fraction of a given volume of material that is occupied by void space, or interstices. Porosity, indicated by the symbol n, is usually expressed as the ratio of the volume of voids, V_v, to the total unit volume, V_t, of a soil or rock, such that $n = V_v/V_t$. Porosity can be determined in the laboratory from knowledge of the bulk mass density, ρ_b, and particle mass density, ρ_s, of the porous material (see Section 5.4.1) using the relationship:

$$n = 1 - \frac{\rho_b}{\rho_s} \qquad \text{eq. 2.1}$$

In fractured rocks, secondary or fracture porosity can be estimated by the method of scan lines using the relation $n_f = Fa$ where F is the number of joints per unit distance intersecting a straight scan line across a rock outcrop, and a is the mean aperture of the fractures.

Porosity is closely associated with the void ratio, e, the ratio of the volume of voids to the volume of the solid material, V_s, such that $e = V_v/V_s$. The relation between porosity and void ratio can be expressed as:

$$n = \frac{e}{(1 + e)} \qquad \text{eq. 2.2}$$

or

$$e = \frac{n}{(1 - n)} \qquad \text{eq. 2.3}$$

Void ratio displays a wide range of values. In soils and rocks with a total porosity ranging from 0.001 to 0.7, the corresponding void ratio range is from 0.001 to 2.3.

Fig. 2.1 Types of porosity with relation to rock texture: (a) well-sorted sedimentary deposit having high porosity; (b) poorly sorted sedimentary deposit having low porosity; (c) well-sorted sedimentary deposit consisting of pebbles that are themselves porous, so that the whole deposit has a very high porosity; (d) well-sorted sedimentary deposit whose porosity has been reduced by the deposition of mineral matter (cementation) in the interstices; (e) soluble rock made porous by solution; (f) crystalline rock made porous by fracturing. After Meinzer (1923).

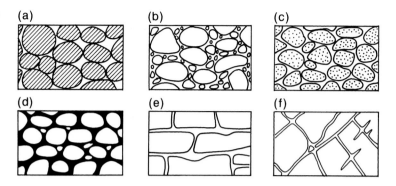

In general, unconsolidated sediments such as gravels, sands, silts and clays, which are composed of angular and rounded particles, have larger porosities than indurated, consolidated sediments such as sandstone and limestone. Crystalline igneous and metamorphic rocks have especially low porosities because the pores are merely within the intercrystal surfaces. Conversely, formations rich in platy clay minerals with very fine grain size can achieve high porosity values.

As illustrated in Fig. 2.1, porosity is controlled by the shape and arrangement of constituent grains, the degree of sorting, compaction, cementation, fracturing and solutional weathering. Porosity values range from negligibly small (0%) for unfractured to 0.1 (10%) for weathered crystalline rocks to 0.4–0.7 (40–70%) for unconsolidated clay deposits (Table 2.1).

There is a distinction between primary porosity, which is the inherent character of a soil or rock matrix that developed during its formation, and secondary porosity. Secondary porosity may develop as a result of secondary physical and chemical weathering along the bedding planes and joints of indurated sediments such as limestones and sandstones, or as a result of structurally controlled regional fracturing and near-surface weathering in hard rocks such as igneous and metamorphic rocks. Where both primary and secondary porosities are present, a dual-porosity system is recognized, for example as a result of fracturing and fissuring in porous sandstone or limestone.

Not all the water contained in the pore space of a soil or rock can be viewed as being available to groundwater flow, particularly in fine-grained or fractured aquifers. In an aquifer with a water table, the volume

Table 2.1 Range of values of hydraulic conductivity and porosity for different geological materials. Based on data contained in Freeze and Cherry (1979) and Back et al. (1988).

Geological material	Hydraulic conductivity, K (m s^{-1})	Porosity, n
Fluvial deposits (alluvium)	10^{-5}–10^{-2}	0.05–0.35
Glacial deposits		
Basal till	10^{-11}–10^{-6}	0.30–0.35
Lacustrine silt and clay	10^{-13}–10^{-9}	0.35–0.70
Outwash sand and gravel	10^{-7}–10^{-3}	0.25–0.50
Loess	10^{-11}–10^{-5}	0.35–0.50
Sandstone	10^{-10}–10^{-5}	0.05–0.35
Shales		
Unfractured	10^{-13}–10^{-9}	0–0.10
Fractured	10^{-9}–10^{-5}	0.05–0.50
Mudstone	10^{-12}–10^{-10}	0.35–0.45
Dolomite	10^{-9}–10^{-5}	0.001–0.20
Oolitic limestone	10^{-7}–10^{-6}	0.01–0.25
Chalk		
Primary	10^{-8}–10^{-5}	0.15–0.45
Secondary	10^{-5}–10^{-3}	0.005–0.02
Coral limestones	10^{-3}–10^{-1}	0.30–0.50
Karstified limestones	10^{-6}–10^{0}	0.05–0.50
Marble, fractured	10^{-8}–10^{-5}	0.001–0.02
Volcanic tuff	10^{-7}–10^{-5}	0.15–0.40
Basaltic lava	10^{-13}–10^{-2}	0–0.25
Igneous and metamorphic rocks: unfractured and fractured	10^{-13}–10^{-5}	0–0.10

of water released from groundwater storage per unit surface area of aquifer per unit decline in the water table is known as the specific yield, S_y (see Section 2.11.3). The fraction of water that is retained in the soil or rock

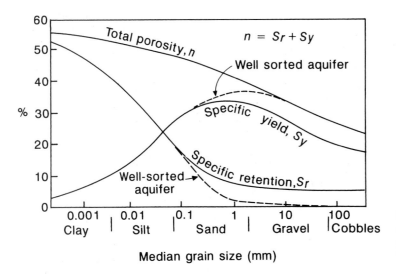

Fig. 2.2 Relation between median grain size and water storage properties of typical alluvial sediments. After Davis and De Wiest (1966).

against the force of gravity is termed the specific retention, S_r. As shown in Fig. 2.2, the sum of the specific yield and specific retention $(S_y + S_r)$ is equal to the total porosity, n. It is useful to distinguish the total porosity from the effective porosity, n_e, of a porous material. The total porosity relates to the storage capability of the material whereas the effective porosity relates to the transmissive capability of the material.

In coarse-grained rocks with large pores, the capillary films that surround the solid particles occupy only a small proportion of the pore space such that S_y and n_e will almost equal n. In fine-grained rocks and clay, capillary forces dominate such that S_r will almost equal n, but n_e will be much less than n. These variations can be described by the term specific surface area, S_{sp}, defined as the ratio of total surface area of the interstitial voids to total volume of the porous material. In sands, S_{sp} will be of the order of 1.5×10^4 m^{-1} but in montmorillonite clay it is about 1.5×10^9 m^{-1} (Marsily 1986). These properties are important in the adsorption of water molecules and dissolved ions on mineral surfaces, especially on clay.

In the case of a fissured or fractured aquifer, such as weathered limestone and crystalline rocks, water contained in the solid matrix is typically immobile and the only effective porosity is associated with the mobile water contained in the fissures and fractures. With increasing depth, the frequency of fissures and fractures decreases and the increasing overburden

pressure closes any remaining openings such that the effective porosity of these formations substantially declines.

2.3 Darcy's law and hydraulic conductivity

Water contained within the interconnected voids of soils and rocks is capable of moving, and the ability of a rock to store and transmit water constitutes its hydraulic properties. At the centre of the laws that govern the behaviour of groundwater flow in saturated material is that formulated empirically by the French municipal engineer for Dijon, Henry Darcy, in 1856. Using the type of experimental apparatus shown in Fig. 2.3, Darcy studied the flow of water through porous material contained in a column and found that the total flow, Q, is proportional to both the difference in water level, $h_1 - h_2$, measured in manometer tubes at either end of the column and the cross-sectional area of flow, A, and inversely proportional to the column length, L. When combined with the constant of proportionality, K, Darcy obtained:

$$Q = KA \frac{(h_1 - h_2)}{L} \qquad \text{eq. 2.4}$$

In general terms, Darcy's law, as it is known, can be written as:

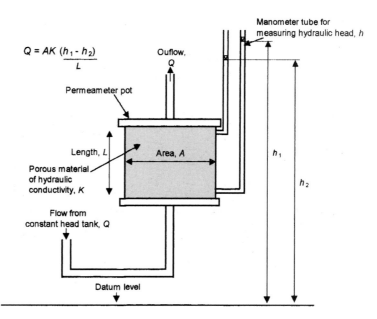

$$Q = AK \frac{(h_1 - h_2)}{L}$$

Manometer tube for measuring hydraulic head, h

Ouflow, Q

Permeameter pot

Length, L

Porous material of hydraulic conductivity, K

Area, A

h_1

h_2

Flow from constant head tank, Q

Datum level

Fig. 2.3 Permeameter apparatus for determining the hydraulic conductivity of saturated porous material using Darcy's law.

$$Q = -KA \frac{dh}{dl} \qquad \text{eq. 2.5}$$

where dh/dl represents the hydraulic gradient, with the negative sign indicating flow in the direction of decreasing hydraulic head. K is called the hydraulic conductivity of the porous material. Adopting the shorthand of dh/dl equal to i, then equation 2.5 can be written as:

$$Q = -AiK \qquad \text{eq. 2.6}$$

Hydraulic conductivity or, as it is occasionally referred to in older publications, the coefficient of permeability, has dimensions of $[L\ T^{-1}]$ and is a measure of the ease of movement of water through a porous material. Values of hydraulic conductivity display a wide range in nature, spanning 13 orders of magnitude (Table 2.1). In general, coarse-grained and fractured materials have high values of hydraulic conductivity, while fine-grained silts and clays have low values. An illustration of the relationship between hydraulic conductivity and grain size is shown for two alluvial aquifers in Fig. 2.4. In such aquifers, the sediment grain size commonly increases with depth

such that the greatest hydraulic conductivity is generally deep in the aquifer (Sharp 1988). The properties of the geological material will significantly influence the isotropy and homogeneity of the hydraulic conductivity distribution (see Section 2.4).

The hydraulic conductivity of geological materials is not only a function of the physical properties of the porous material, but also the properties of the migrating fluid, including specific weight, $\gamma (= \rho g$, where ρ is the density of the fluid and g is the gravitational acceleration), and viscosity, μ, such that:

$$K = k_i \frac{\gamma}{\mu} \qquad \text{eq. 2.7}$$

where the constant of proportionality, k_i, is termed the intrinsic permeability because it is a physical property intrinsic to the porous material alone.

The density and viscosity of water are functions of temperature and pressure but these effects are not great for the ranges of temperature and pressure encountered in most groundwater situations (see Appendix 2). A one-third increase in hydraulic conductivity is calculated using equation 2.7 for a temperature increase from $5\,^\circ C$ ($\rho = 999.965$ kg m^{-3}, $\mu = 1.5188 \times 10^{-3}$ N s m^{-2}) to $15\,^\circ C$ ($\rho = 999.099$ kg m^{-3}, $\mu = 1.1404$

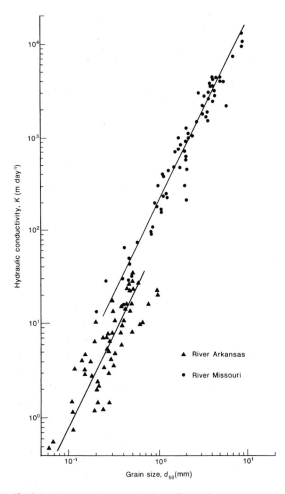

Fig. 2.4 Laboratory-determined values of hydraulic conductivity as a function of grain size for alluvial aquifers in the Rivers Missouri and Arkansas. Note the log–log scales. After Sharp (1988).

$\times 10^{-3}$ N s m^{-2}), although a groundwater flow system exhibiting such a temperature change would be considered unusual. An example is groundwater that penetrates deep in the Earth's crust, becomes heated and returns rapidly to the surface as highly mineralized hot springs. Equally, in coastal areas, saline intrusion into fresh groundwater will cause variations in fluid density such that information about both k_i and K is required in any investigation.

The intrinsic permeability is representative of the properties of the porous material alone and is related to the size of the openings through which the fluid moves. For unconsolidated sand, Krumbein and Monk (1943) derived the following empirical relationship where GM_d is the geometric mean of the grain diameter (mm) and σ is the standard deviation of the grain size in phi units ($-\log_2$(grain diameter in mm)):

$$k_i = 760(GM_d)^2 e^{-1.3\sigma} = Cd^2 \qquad \text{eq. 2.8}$$

As shown, equation 2.8 is more generally expressed as $k_i = Cd^2$ where d is equal to the mean pore diameter and C represents a dimensionless 'shape factor' assessing the contribution made by the shape of the pore openings, as influenced by the relationship between the pore and grain sizes and their effect on the tortuosity of fluid flow. Intrinsic permeability has the dimensions of $[L^2]$ and, using nomenclature common in the petroleum industry, the unit of k_i is the darcy where 1 darcy is equivalent to 9.87×10^{-13} m^2.

Now, combining equations 2.5, 2.7 and 2.8 gives a full expression of the flow through a porous material as:

$$\frac{Q}{A} = q = -K\frac{dh}{dl} = -\frac{Cd^2 \rho g}{\mu}\frac{dh}{dl} \qquad \text{eq. 2.9}$$

The quotient Q/A, or q, indicates the discharge per unit cross-sectional area of saturated porous material. The term q, referred to as the specific discharge, has the dimensions of velocity $[L\ T^{-1}]$ and is also known as the darcy velocity or darcy flux. It is important to remember that the darcy velocity is not the true, microscopic velocity of the water moving along winding flowpaths within the soil or rock. Instead, by dividing the specific discharge by the fraction of open space (in other words, effective porosity, n_e) through which groundwater flows across a given sectional area, this provides an average measure of groundwater velocity such that:

$$\frac{Q}{An_e} = \frac{q}{n_e} = \bar{v} \qquad \text{eq. 2.10}$$

where $\bar{v}$ is the average linear velocity (Fig. 2.5).

As illustrated in Box 2.1, the application of equations 2.5 and 2.10 to simple hydrogeological situations enables first estimates to be obtained for groundwater flow and velocity. More accurate calcula-

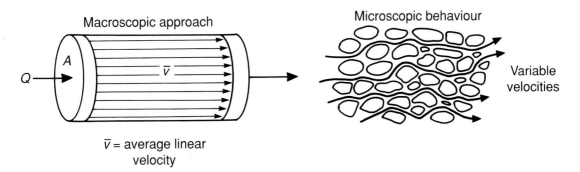

Fig. 2.5 Macroscopic (Darcian) approach to the analysis of groundwater flow contrasted with the true, microscopic behaviour of tortuous flowpaths.

tions require the use of more advanced techniques such as flow net analysis (Box 2.3) and groundwater modelling (see Further reading and Section 5.9).

2.4 Isotropy and homogeneity

Aquifer properties, such as hydraulic conductivity, are unlikely to conform to the idealized, uniform porous material whether viewed at the microscopic or regional scale. The terms isotropy, anisotropy, homogeneity and heterogeneity are used to describe the spatial variation and directional trends in aquifer property values.

If the hydraulic conductivity, K, is independent of position within a geological formation, the formation is homogeneous. If the hydraulic conductivity varies from place to place, then the formation is heterogeneous. The type of heterogeneity will depend on the geological environment that gave rise to the deposit or rock type. As shown in Fig. 2.6, layered heterogeneity is common in sedimentary rocks where each bed comprising the formation has its own hydraulic conductivity value. Strong, layered heterogeneity will be present in interbedded deposits of clay and sand. Similarly large contrasts can arise in cases of discontinuous heterogeneity caused by the presence of faults or large-scale stratigraphic features. Trending heterogeneity exists in formations such as deltas, alluvial fans and glacial outwash plains where there is sorting and grading of the material deposits. Vertical trends in hydraulic conductivity are also present in consolidated rocks where permeability is dependent on joint and fracture density.

It is widely accepted that the statistical distribution of hydraulic conductivity for a geological formation is described by a log-normal probability density function with the average hydraulic conductivity calculated as a geometric mean. Trending heterogeneity within a geological formation can be regarded as a trend in the mean hydraulic conductivity value.

An isotropic geological formation is one where the hydraulic conductivity is independent of the direction of measurement at a point in the formation. If the hydraulic conductivity varies with the direction of measurement at a point, the formation is anisotropic at that point. The principal directions of anisotropy correspond to the maximum and minimum values of hydraulic conductivity and are usually at right angles to each other. The primary cause of anisotropy on a small scale is the orientation of clay minerals in sedimentary rocks and unconsolidated sediments. In consolidated rocks, the direction of jointing or fracturing can impart strong anisotropy at various scales, from the local to regional.

Combining the above definitions, and as shown in Fig. 2.7, it is possible to recognize four possible combinations of heterogeneity and anisotropy when describing the nature of the hydraulic conductivity of a formation.

As a result of introducing anisotropy, it is also necessary to recognize that in a three-dimensional flow system the specific discharge or darcy velocity (eq. 2.9) as defined by Darcy's law, is a vector quantity with components q_x, q_y and q_z given by:

$$q_x = -K_x \frac{\partial h}{\partial x}, \quad q_y = -K_y \frac{\partial h}{\partial y}, \quad q_z = -K_z \frac{\partial h}{\partial z}$$

BOX
2.1

Application of Darcy's law to simple hydrogeological situations

The following two worked examples illustrate the application of Darcy's law to simple hydrogeological situations. In the first example, the alluvial aquifer shown in Fig. 1 is recharged by meltwater runoff from the adjacent impermeable mountains that run parallel to the axis of the valley. If the groundwater that collects in the aquifer discharges to the river, then it is possible to estimate the river flow at the exit from the valley. To solve this problem, and assuming that the river is entirely supported by groundwater discharge under steady, uniform flow conditions, the groundwater discharge (Q) can be calculated using equation 2.5 and the information given in Fig. 1, as follows:

$$Q = -KA \frac{dh}{dl}$$

$$Q = 1 \times 10^{-3} \cdot 20 \cdot 5000 \cdot 4 \times 10^{-3}$$

$$Q = 0.4 \, m^3 \, s^{-1} \qquad \text{eq. 1}$$

Accounting for both halves of the valley floodplain, the total discharge from the alluvial aquifer as river flow is 0.8 m³ s⁻¹.

In the second example (Fig. 2), a municipal waste disposal facility is situated in a former sand and gravel quarry. The waste is in contact with the water table and is directly contaminating the aquifer. The problem is to estimate the time taken for dissolved solutes to reach a spring discharge area located down-gradient of the waste tip. Assuming that the contaminant is unreactive and moves at the same rate as the steady, uniform groundwater flow, then from a consideration of equation 2.10 and the information given in Fig. 2, the average linear velocity is calculated as follows:

$$\bar{v} = \frac{q}{n_e} = \frac{K dh/dl}{n_e} \qquad \text{eq. 2}$$

$$\bar{v} = \frac{5 \times 10^{-4} \cdot 5 \times 10^{-3}}{0.25} \qquad \text{eq. 3}$$

$$\bar{v} = 1 \times 10^{-5} \, m \, s^{-1} \qquad \text{eq. 4}$$

Therefore, the time taken, t, to move a distance of 200 m in the direction of groundwater flow from the waste tip to the spring is:

$$t = \frac{200}{1 \times 10^{-5}} = 2 \times 10^7 \, s \approx 230 \, \text{days} \qquad \text{eq. 5}$$

West   East

Fig. 1 Alluvial aquifer bounded by impermeable bedrock.

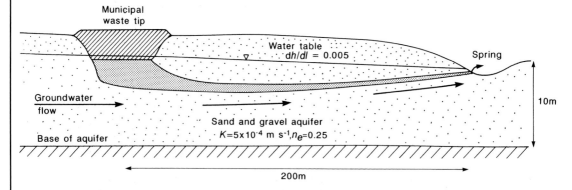

Fig. 2 Municipal waste disposal site contaminating a sand and gravel aquifer.

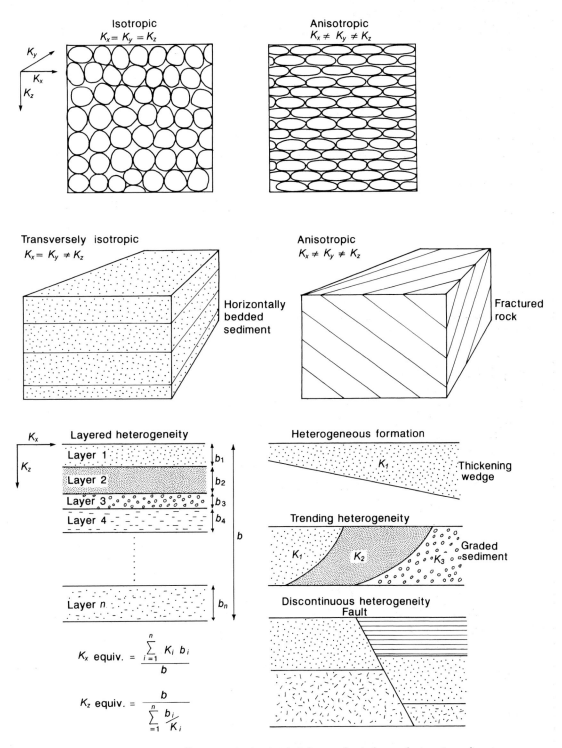

Fig. 2.6 Examples of isotropy, anisotropy and heterogeneity showing the influence of grain shape and orientation, sedimentary environment and geological structure on hydraulic conductivity. After Fetter (2001).

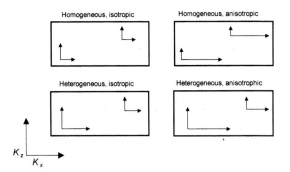

Fig. 2.7 Four possible combinations of heterogeneity and anisotropy describing the hydraulic conductivity of a porous material. After Freeze and Cherry (1979).

where K_x, K_y and K_z are the hydraulic conductivity values in the x, y and z directions.

2.5 Aquifers, aquitards and aquicludes

Natural variations in the permeability and ease of transmission of groundwater in different geological materials lead to the recognition of aquifers, aquitards and aquicludes. An aquifer is a layer or layered sequence of rock or sediment comprising one or more geological formations that contains water and is able to transmit significant quantities of water under an ordinary hydraulic gradient. Aquifers therefore have sufficient permeability to transmit groundwater that can be exploited economically from wells or springs. Good aquifers are usually developed in sands, gravels, solutionally weathered limestones and fractured sandstones.

The term aquitard is used to describe a formation of lower permeability that may transmit quantities of water that are significant in terms of regional groundwater flow, but from which negligible supplies of groundwater can be obtained. Examples of aquitards include fluvial and glacio-fluvial silts and sandy clays, sedimentary rocks with few fractures and fractured crystalline rock.

An aquiclude is a saturated geological unit of such low permeability that it is incapable of transmitting significant quantities of water under ordinary hydraulic gradients and can act as a barrier to regional groundwater flow. Aquiclude rocks include clays, shales and metamorphic rocks.

2.6 Hydraulic properties of fractured rocks

By adopting the Darcian approach to the analysis of groundwater flow described in Section 2.3, it is implicit that the physical assemblage of grains that comprise the porous material are considered as a representative continuum, and that macroscopic laws, such as Darcy's law, provide macroscopically averaged descriptions of the microscopic behaviour. In other words, Darcy's law describes groundwater flow as a flux through a porous material that is imagined to have continuous, smoothly varying properties. In reality, intergranular and fractured porous materials are highly heterogeneous when examined at a scale similar to the spacing of the dominant pore size. The consequence of this is that Darcy's law can be used successfully, but only at a scale large enough to contain a representative assemblage of pores. This is the continuum scale. At subcontinuum scales, the local pore network geometry strongly influences flow and the transport of contaminants. This is particularly relevant in fractured rocks where the dimension of the fracture spacing can impart a continuum scale that exceeds the size of many practical problems.

In fractured material such as carbonate and crystalline rocks and fissured clay sediments such as glacial tills, the conceptual model of groundwater flow can either be grossly simplified or a detailed description of the aquifer properties attempted as depicted in Fig. 2.8. With the exception of conduit flow in karst aquifers, fracture flow models generally assume that both fracture apertures and flow velocities are small such that Darcy's law applies and flow is laminar (Box 2.2). In the example of the equivalent porous material shown in Fig. 2.8b, the primary and secondary porosity and hydraulic conductivity distributions are represented as the equivalent or effective hydraulic properties of a continuous porous material. A drawback with this approach is that it is often difficult to determine the size of the representative elemental volume of material from which to define the effective hydraulic property values. Hence, the equivalent porous material approach may adequately represent the behaviour of a regional flow system but is likely to reproduce local conditions poorly.

More advanced approaches, such as the discrete fracture and dual-porosity models shown in Figs 2.8c and 2.8d, represent groundwater movement through

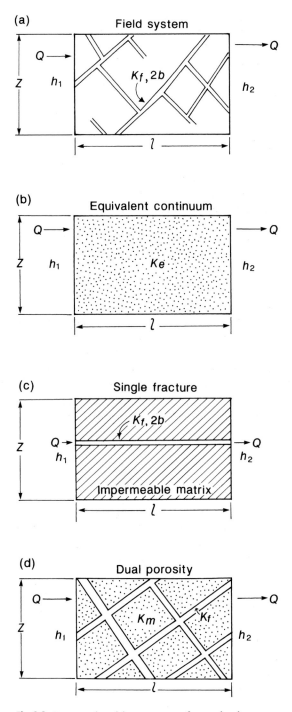

Fig. 2.8 Conceptual models to represent a fractured rock system. The fracture network of aperture 2b and with groundwater flow from left to right is shown in (a). The equivalent porous material, discrete fracture and dual porosity models representative of (a) are shown in (b), (c) and (d), respectively. After Gale (1982).

the fracture network. Flow through a single fracture may be idealized as occurring between two parallel plates with a uniform separation or fracture aperture, 2b. The relation between flow and hydraulic gradient for individual fractures under laminar flow conditions is usually considered to be governed by the 'cubic law' presented by Snow (1969) and further validated by Witherspoon et al. (1980) and Gale (1982). In this treatment, the flow rate through a fracture, Q_f, may be expressed as:

$$Q_f = -2bwK_f \frac{dh}{dl} \qquad \text{eq. 2.11}$$

where w is the width of the fracture, K_f the hydraulic conductivity of the fracture and l the length over which the hydraulic gradient is measured. The hydraulic conductivity, K_f, is calculated from:

$$K_f = \frac{\rho g (2b)^2}{12\mu} \qquad \text{eq. 2.12}$$

where ρ is fluid density, μ is fluid viscosity and g is the gravitational acceleration.

If the expression for K_f (eq. 2.12) is substituted in equation 2.11, then:

$$Q_f = -\frac{2}{3} \frac{w\rho g b^3}{\mu} \frac{dh}{dl} \qquad \text{eq. 2.13}$$

It can be seen from equation 2.13 that the flow rate increases with the cube of the fracture aperture. Use of a model based on these equations requires a description of the fracture network, including the mapping of fracture apertures and geometry, that can only be determined by careful fieldwork.

In the case of the dual-porosity model, flow through the fractures is accompanied by exchange of water and solute to and from the surrounding porous rock matrix. Exchange between the fracture network and the porous blocks may be represented by a term that describes the rate of mass transfer. In this model, both the hydraulic properties of the fracture network and porous rock matrix need to be assessed, adding to the need for field mapping and hydraulic testing.

BOX
2.2

Laminar and turbulent flows

Darcy's law applies when flow is laminar but at high flow velocities, turbulent flow occurs and Darcy's law breaks down. Under laminar conditions individual 'particles' of water move in paths parallel to the direction of flow, with no mixing or transverse component to the fluid motion. These conditions can be visualized by making an analogy between flow in a straight, cylindrical tube of constant diameter, and flow through porous granular or fissured material. At the edge of the tube, the flow velocity is zero rising to a maximum at the centre. As the flow velocity increases, so fluctuating eddies develop and transverse mixing occurs whereupon the flow becomes turbulent.

Flow rates that exceed the upper limit of Darcy's law are common in karstic limestones (Section 2.7) and dolomites and highly permeable volcanic formations. Also, the high velocities experienced close to the well screen of a pumping borehole can create turbulent conditions. The change from laminar flow at low velocities to turbulent flow at high velocities is usually related to the dimensionless Reynolds number, R_e, which expresses the ratio of inertial to viscous forces during flow. For flow through porous material, the Reynolds number is expressed as follows:

$$R_e = \frac{\rho q d}{\mu} \qquad \text{eq. 1}$$

where ρ is fluid density, μ is viscosity and q is the specific discharge (or characteristic velocity for fissured or fractured material). The characteristic length, d, can represent the mean pore diameter, mean grain diameter or, in the case of a fissure or fracture, either the hydraulic radius (cross-sectional area/wetted perimeter) or width of the fissure.

For laminar flow in granular material, Darcy's law is valid as long as values of R_e do not exceed the range 1–10. Since fully turbulent flow does not occur until velocities are high and R_e is in the range 10^2 to 10^3, the transition between the linear laminar and turbulent regimes is characterized by non-linear laminar flow. In karst aquifers, conduit flow may remain in the laminar regime in pipes up to about 0.5 m in diameter provided the flow velocity does not exceed 1×10^{-3} m s^{-1} (Fig. 1).

The following example illustrates the application of the Reynolds number in determining whether groundwater flow is laminar or turbulent. A fissure in a limestone aquifer has a width, w, of 2 m and an aperture, $2b$, of 0.1 m. A tracer dye moves along the fissure at a velocity of 0.03 m s^{-1}. From this information, the characteristic

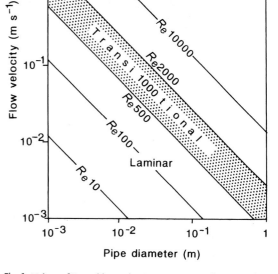

Fig. 1 Values of Reynolds number (R_e) at various velocities and conduit diameters and showing fields of different flow regimes. After Smith et al. (1976b).

length of the fissure is $(2bw)/(2(2b+w)) = (0.1 \times 2)/(2(0.1+2)) = 0.05$ m. The characteristic velocity is equal to the tracer velocity. Hence, if the kinematic viscosity, μ/ρ, at 10°C is 1.31×10^{-6} m^2 s^{-1}, then using equation 1:

$$R_e = \frac{0.03 \times 0.05}{1.31 \times 10^{-6}} = 1145 \qquad \text{eq. 2}$$

and the flow is transitional to turbulent.

2.7 Karst aquifers

The term karst is used widely to describe the distinctive landforms that develop on rock types such as limestones, gypsum and halite that are readily dissolved by water. The name karst is derived from a word meaning stony ground used to describe the Kras region, now part of Slovenia and Croatia, where distinctive karst landforms are exceptionally well developed. Karst areas are typically characterized by a lack of permanent surface streams and the presence of swallow holes (Fig. 2.9) and enclosed depressions. Rainfall runoff usually occurs underground in solutionally enlarged channels, some of which are large

Fig. 2.9 The disappearance of the upper River Fergus at An Clab, south-east Burren, County Clare, Ireland, where surface runoff from Namurian shales disappears into a swallow hole at the contact with Carboniferous limestone.

enough to form caves. Well-known karst areas include: the pinnacle karst of the Guilin area, southern China; Mammoth Caves, Kentucky, USA; the Greek Islands; the Dordogne, Vercors and Tarn areas of France; Postojna Caves in Slovenia; and The Burren, County Clare, Ireland. The karst of Ireland is described by the Karst Working Group (2000), and Ford and Williams (1989) provide an extensive treatment of karst geomorphology and hydrology in general.

Figure 2.10 shows a typical model of groundwater flow and is used here to describe groundwater conditions in the Mendip Hills karst aquifer located in the west of England. The hills extend 50 km east–west and 10 km north–south. In the west they rise above surrounding lowlands and form a broad karst plateau at about 260 m above sea level, developed in Carboniferous limestone. Structurally, the Mendips comprise four *en echelon* periclines with cores of Devonian sandstone. The dip of the limestones on the northern limbs of the folds is generally steep (60–90°) but to the south it is more gentle (20–40°). The periclines emerge from beneath younger Triassic rocks in the west, but in the east are covered by Mesozoic strata which are in the process of being removed by erosion. As a result, the karstic features are better developed and probably older in the west than in the east. The principal aquifer is the Carboniferous limestone which has been extensively exploited for water supply, primarily by spring abstraction. Spring discharges are generally flashy, with a rapid response to storms, such that abstracted water is normally stored in surface reservoirs.

As shown in Fig. 2.10, groundwater discharge is via springs located at the lowest limestone outcrop, often where the limestones dip below Triassic mudstones.

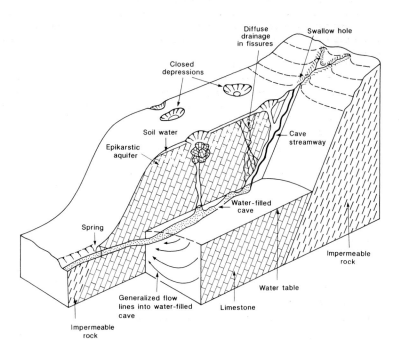

Fig. 2.10 Block diagram showing the occurrence of groundwater in karst aquifers. After an original by T.C. Atkinson.

The larger springs are fed by conduits or flooded cave systems. The conduits act as drains within the saturated zone of the aquifer, and groundwater in fissures and fractures flows towards the conduits. Within the saturated zone the conduit flow has a turbulent regime while the diffuse fissure flow obeys Darcy's law (Box 2.2).

Recharge to the aquifer can be characterized as allogenic and autogenic. Allogenic recharge comprises sinking streams which collect on sandstone and shale exposed in the core of the periclines. These streams pass directly into the conduit system through swallow holes. Autogenic recharge is either concentrated by closed depressions (dolines) or occurs as diffuse infiltration through the soil. Closed depressions are the first-order tributaries of the conduit system and focus concentrated recharge into shafts and caves. Weathering in the upper few metres of bedrock produces dense fissuring that provides storage for water in the unsaturated zone in what is sometimes referred to as the epikarstic aquifer. The epikarstic aquifer is recharged by infiltration and drains to the saturated zone via fractures and fissures, but with frequent concentration of drainage into shafts which form tributaries to cave systems.

Analysis of hydrographs (see Section 5.7), baseflow recession curves, water balances (see Section 8.2.1) and tracer tests (see Section 5.8.3) indicates that the diffuse flow component of the saturated zone in the Mendip Hills has a storativity of about 1% and a hydraulic conductivity of 10^{-4} to 10^{-3} m s^{-1} (Atkinson 1977). About 70% of the flow in the saturated zone is via conduits but these comprise less than one-thirtieth of the active storage. From direct exploration, the depth of conduit circulation beneath the water table is known to exceed 60 m, implying a total storage in the saturated zone of at least 600 mm of precipitation, roughly equivalent to one year's runoff. Significant storage also occurs in the epikarstic aquifer although the total amount is not known.

Karst aquifers can be classified according to the relative importance of diffuse flow and conduit flow, the degree of concentration of recharge and the amount of storage in the aquifer as shown in Fig. 2.11. The Mendip Hills karst aquifer has high storage, about 50% concentration of recharge into streams and closed depressions and 70% conduit flow in the saturated zone.

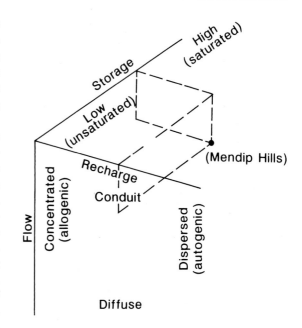

Fig. 2.11 Conceptual classification of karst aquifers from a consideration of recharge and groundwater flow mechanisms and the degree of saturated aquifer storage. After an original by T.C. Atkinson.

In karst aquifers where turbulent flow conditions can develop in solutionally developed conduits, representation of the hydraulic behaviour of the system is complicated by the difficulty in characterizing the hydraulic properties. A number of approaches are commonly used to model the behaviour of karst aquifers. The first is to assume that groundwater flow is governed by Darcy's law and then to use one of the models shown in Fig. 2.8. Further approaches to modelling flow in karst conduits is to adopt the Darcy–Weisbach pipe flow equation (eq. 5.22) or, for mature karst landscapes, to use a 'black box' model in which empirical functions are developed based on field observations of flow to reproduce input and output responses, in particular of recharge and spring flow. These functions may or may not include the usual aquifer parameters such as hydraulic conductivity, storativity and porosity. A third, hybrid approach is to use the aquifer response functions developed as for the 'black box' approach and then make use of these in an equivalent porous material model, although it must be recognized that large uncertainties remain, requiring careful field validation.

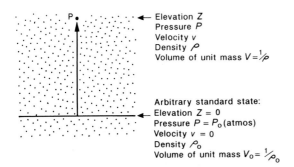

Fig. 2.12 Work done in moving a unit mass of fluid from the standard state to a point P in a groundwater flow system.

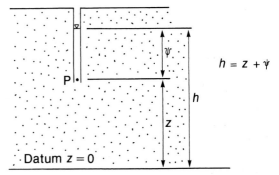

Fig. 2.13 Relation between hydraulic head, h, pressure head, ψ, and elevation head, z, at a point P in a column of porous material.

2.8 Groundwater potential and hydraulic head

As described in the previous sections of this chapter, the porosity and hydraulic conductivity of porous material characterize the distribution and ease of movement of groundwater in geological formations. When analysing the physical process of groundwater flow, analogies are drawn with the flow of heat through solids from higher to lower temperatures and the flow of electrical current from higher to lower voltages. The rates of flow of heat and electricity are proportional to the potential gradients and, in a similar way, groundwater flow is also governed by a potential gradient.

Groundwater possesses energy in mechanical, thermal and chemical forms with flow controlled by the laws of physics and thermodynamics. With reference to Fig. 2.12, the work done in moving a unit mass of fluid from the standard state to a point, P, in the flow system is composed of the following three components:

1 Potential energy (mgz) required to lift the mass to elevation, z.
2 Kinetic energy ($mv^2/2$) required to accelerate the fluid from zero velocity to velocity, v.
3 Elastic energy required to raise the fluid from pressure P_o to pressure P.

The latter quantity can be thought of as the change in potential energy per unit volume of fluid and is found from:

$$m \int_{P_o}^{P} \frac{v}{m} \, dp = m \int_{P_o}^{P} \frac{dP}{\rho} \qquad \text{eq. 2.14}$$

Given that groundwater velocities in porous material are very small, the kinetic energy term can be ignored such that at the new position, P, the fluid potential, Φ, or mechanical energy per unit mass ($m = 1$) is:

$$\Phi = gz + \int_{P_o}^{P} \frac{dP}{\rho} \qquad \text{eq. 2.15}$$

For incompressible fluids that have a constant density, and therefore are not affected by a change in pressure, then:

$$\Phi = gz + \frac{(P - P_o)}{\rho} \qquad \text{eq. 2.16}$$

To relate the fluid potential to the hydraulic head measured by Darcy in his experiment (Fig. 2.3), Fig. 2.13 demonstrates that the fluid pressure at position P in a column containing porous material is found as follows:

$$P = \rho g \psi + P_o \qquad \text{eq. 2.17}$$

where ψ is the height of the water column above P and P_o is atmospheric pressure (the pressure at the standard state).

It can be seen that $\psi = h - z$ and so, substituting in equation 2.17:

$$P = \rho g(h - z) + P_o \qquad \text{eq. 2.18}$$

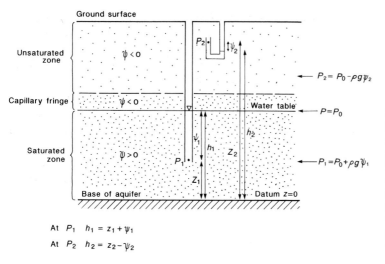

Fig. 2.14 Condition of pressure head, ψ, for the unsaturated and saturated zones of an aquifer. At the water table, fluid pressure is equal to atmospheric (P_o) and by convention is set equal to zero. Note also that the unsaturated or vadose zone is the region of a geological formation containing solid, water and air phases while in the saturated or phreatic zone, pore spaces of the solid material are all water-filled.

By substituting this expression for pressure into the equation for fluid potential, equation 2.16, then:

$$\Phi = gz + \frac{[\rho g(h - z) + P_o - P_o]}{\rho} \qquad \text{eq. 2.19}$$

and, thus:

$$\Phi = [gz + gh - gz] = gh \qquad \text{eq. 2.20}$$

The result of equation 2.20 provides a significant relationship in hydrogeology: the fluid potential, Φ, at any point in a porous material can simply be found from the product of hydraulic head and acceleration due to gravity. Since gravity is, for all practical purposes, almost constant near the Earth's surface, Φ is almost exactly correlated with h. The significance is that hydraulic head is a measurable, physical quantity and is therefore a suitable measure of fluid potential, Φ.

Returning to the analogy with heat and electricity, where rates of flow are governed by potential gradients, it is now shown that groundwater flow is driven by a fluid potential gradient equivalent to a hydraulic head gradient. In short, groundwater flows from regions of higher to lower hydraulic head.

With reference to equations 2.16 and 2.20, and, by convention, setting the atmospheric pressure, P_o, to zero, then:

$$gh = gz + \frac{P}{\rho} \qquad \text{eq. 2.21}$$

The pressure at point P in Fig. 2.13 is equal to $\rho g\psi$, and so it can be shown by substitution in equation 2.21 and by dividing through by g that:

$$h = z + \psi \qquad \text{eq. 2.22}$$

Equation 2.22 confirms that the hydraulic head at a point within a saturated porous material is the sum of the elevation head, z, and pressure head, ψ, thus providing a relationship that is basic to an understanding of groundwater flow. This expression is equally valid for the unsaturated and saturated zones of porous material but it is necessary to recognize, as shown in Fig. 2.14, that the pressure head term, ψ, is a negative quantity in the unsaturated zone as a result of adopting the convention of setting atmospheric pressure to zero and working in gauge pressures. From this, it follows that at the level of the water table the water pressure is equal to zero (i.e. atmospheric pressure). In the capillary fringe above the water table, the aquifer material is completely saturated, but because of capillary suction drawing water up from the water table, the porewater pressure is negative, that is less than atmospheric pressure (for further discussion, see Section 5.4.1). The capillary fringe varies in thickness depending on the diameter of the pore space and ranges from a few centimetres for coarse-grained material to several metres for fine-grained deposits.

2.9 Interpretation of hydraulic head and groundwater conditions

2.9.1 Groundwater flow direction

Measurements of hydraulic head, normally achieved by the installation of a piezometer or well point, are useful for determining the directions of groundwater flow in an aquifer system. In Fig. 2.15a, three piezometers installed to the same depth enable the determination of the direction of groundwater flow and, with the application of Darcy's law (eq. 2.5), the calculation of the horizontal component of flow. In Fig. 2.15b, two examples of piezometer nests are shown that allow the measurement of hydraulic head and the direction of groundwater flow in the vertical direction to be determined either at different levels in the same aquifer formation or in different formations.

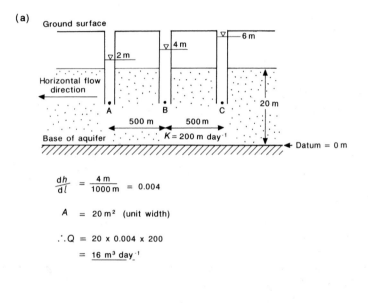

(a)

$$\frac{dh}{dl} = \frac{4\,m}{1000\,m} = 0.004$$

$$A = 20\,m^2 \;\;(unit\;width)$$

$$\therefore Q = 20 \times 0.004 \times 200$$
$$= 16\,m^3\,day^{-1}$$

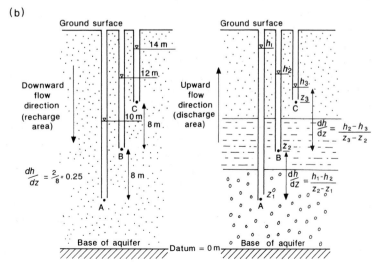

Fig. 2.15 Determination of groundwater flow direction and hydraulic head gradient from piezometer measurements for (a) horizontal flow and (b) vertical flow. The elevation of the water level indicating the hydraulic head at each of the points A, B and C is noted adjacent to each piezometer.

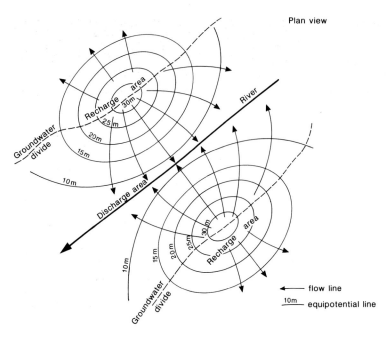

Plan view

flow line
10m equipotential line

Fig. 2.16 Sketch map of the surface of the water table in an unconfined aquifer showing recharge and discharge areas and the position of groundwater divides.

2.9.2 Water table and potentiometric surface maps

Observation boreholes and piezometers located within a district provide a picture of the three-dimensional distribution of hydraulic head throughout an aquifer system. Lines drawn joining points of equal groundwater head, or groundwater potential, are termed equipotential lines. Lines perpendicular to the equipotential lines are flow lines and can be used in the construction of a flow net (Box 2.3). In plan view, the construction of equipotential contours results in a map of the potentiometric surface. In an unconfined aquifer, the potentiometric surface is a map of the water table, where the groundwater is by definition at atmospheric pressure. In a confined aquifer the potentiometric surface predicts the position that the water level would rise to in a borehole that penetrates the buried aquifer. As shown in Fig. 2.16, areas of high hydraulic head may be interpreted as groundwater recharge zones while areas of low hydraulic head are typically in groundwater discharge zones. Box 2.4 provides an example of an actual potentiometric surface map for the Chalk aquifer underlying the London Basin.

2.9.3 Types of groundwater conditions

Groundwater conditions are strongly influenced by the juxtaposition of lithological units and by geological structure. The nature of aquifer geometry can give rise to four basic types of groundwater conditions, as depicted in Fig. 2.17, and also determines the occurrence of springs. An unconfined aquifer exists when a water table is developed that separates the unsaturated zone above from the saturated zone below. It is possible for an unconfined aquifer to develop below the lower surface of an aquitard layer. In this case, a concealed unconfined aquifer is recognized.

In heterogeneous material, for example sedimentary units containing intercalated lenses or layers of clay, perched water table conditions can develop. As shown in Fig. 2.18, above the regional water table and within the unsaturated zone, a clay layer within a sand matrix causes water to be held above the lower permeability material creating a perched water table. Because water table conditions occur where groundwater is at atmospheric pressure, inverted water tables occur at the base of the perched lens of water in the clay layer and also below the ground

BOX
2.3

Flow nets and the tangent law

The construction of a flow net, for example a water table or potentiometric surface map, and the interpretation of groundwater flow lines, requires the implicit assumption that flow is perpendicular to the lines of equal hydraulic head (i.e. the porous material is isotropic), with flow in the direction of decreasing head. All flow nets, however simple or advanced, can be drawn using a set of basic rules. When attempting to draw a two-dimensional flow net for isotropic porous material by trial and error, the following rules must be observed (Fig. 1):

1 Flow lines and equipotential lines should intersect at right angles throughout the groundwater flow system.

2 Equipotential lines should meet an impermeable boundary at right angles resulting in groundwater flow parallel to the boundary.

3 Equipotential lines should be parallel to a boundary that has a constant hydraulic head resulting in groundwater flow perpendicular to the boundary.

4 In a layered, heterogeneous groundwater flow system, the tangent law must be satisfied at geological boundaries.

5 If squares are created in one portion of one formation, then squares must exist throughout that formation and throughout all formations with the same hydraulic conductivity with the possible exception of partial stream tubes at the edge. Rectangles will be created in formations with different hydraulic conductivity.

The last two rules are particularly difficult to observe when drawing a flow net by hand but even a qualitative flow net, in which orthogonality is preserved but with no attempt to create squares, can help provide a first understanding of a groundwater flow system. In simple flow nets, the squares are actually 'curvilinear squares' that have equal central dimensions able to enclose a circle that is tangent to all four bounding lines (Fig. 2).

In Fig. 2, a flow net is constructed for groundwater flow beneath a dam structure that is partially buried in an isotropic and homogeneous sand aquifer. To calculate the flow beneath the dam, consider the mass balance for box ABCD for an incompressible fluid. Under steady-state conditions, and assuming unit depth into the page, the flow into the box across face AB with width, Δw, will equal

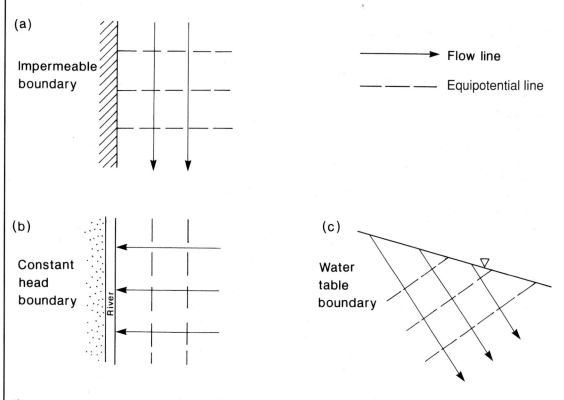

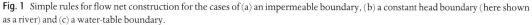

Fig. 1 Simple rules for flow net construction for the cases of (a) an impermeable boundary, (b) a constant head boundary (here shown as a river) and (c) a water-table boundary.

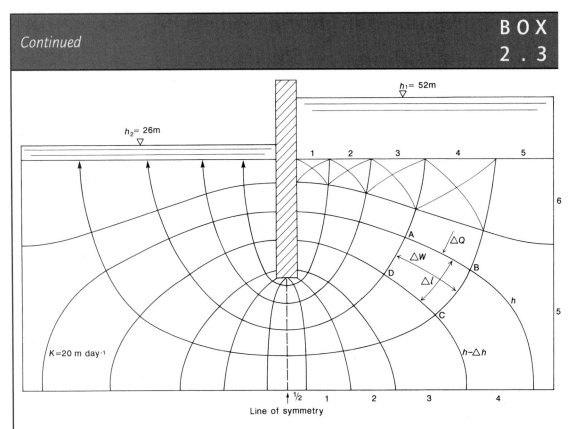

Fig. 2 Flow net constructed for groundwater flow beneath a dam in a homogeneous, isotropic aquifer.

the flow out of the box across face DC. From Darcy's law (eq. 2.5) the best estimate of flow through box ABCD, ΔQ, is equal to:

$$\Delta Q = \Delta w \cdot K \frac{\Delta h}{\Delta l}$$
eq. 1

or, on rearrangement:

$$\Delta Q = K \cdot \Delta h \frac{\Delta w}{\Delta l}$$
eq. 2

If the flow net is equi-dimensional (curvilinear squares), then $\Delta w / \Delta l$ is about equal to unity and equation 2 becomes:

$$\Delta Q = K \cdot \Delta h$$
eq. 3

Δh is found from the total head drop $(h_1 - h_2)$ along the stream tube divided by the number of head divisions, n, in the flow net:

$$\Delta h = \frac{(h_1 - h_2)}{n}$$
eq. 4

If the number of stream tubes in the region of flow is m, then the total flow below the dam is:

$$Q = \frac{m}{n} \cdot K(h_1 - h_2)$$
eq. 5

For the example of flow beneath a dam shown in Fig. 2, $m = 5$, $n = 13$ and $(h_1 - h_2) = 26$ m. If the hydraulic conductivity, K, is 20 m day^{-1}, then the total flow is found from:

$$Q = \frac{5}{13} \times 20 \times 26$$

$$= 200 \text{ m}^3 \text{ day}^{-1}$$

In homogeneous but anisotropic porous material, flow net construction is complicated by the fact that flow lines and equipotential lines are not orthogonal. To overcome this problem, a transformed section is prepared through the application of a hydraulic conductivity ellipse (Fig. 3a). Considering a two-dimensional region in a homogeneous, anisotropic aquifer with principal hydraulic conductivities K_x and K_z, the hydraulic conductivity ellipse will have

Continued

(a)
Hydraulic conductivity ellipse

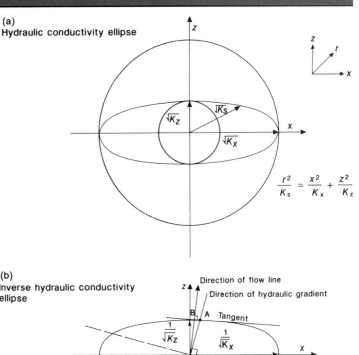

$$\frac{r^2}{K_s} = \frac{x^2}{K_x} + \frac{z^2}{K_z}$$

Fig. 3 In (a) the hydraulic conductivity ellipse for a homogeneous, anisotropic material is shown with principal hydraulic conductivities K_x and K_z. The hydraulic conductivity value K_s for any direction of flow in an anisotropic material can be found graphically if K_x and K_z are known. Also shown are two circles representing the possible isotropic transformations for flow net construction (see text for explanation). In (b) the method for determining the direction of flow in an anisotropic material at a specified point is shown. A line drawn in the direction of the hydraulic gradient intersects the ellipse at point A. If a tangent is drawn to the ellipse at A, the direction of flow is then found perpendicular to this tangent (point B).

(b)
Inverse hydraulic conductivity ellipse

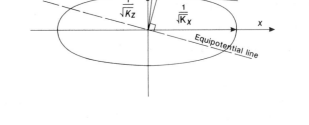

semi-axes $\sqrt{K_x}$ and $\sqrt{K_z}$. The co-ordinates in the transformed region, X–Z, are related to the original x–z system by:

$$X = x$$

$$Z = \frac{z\sqrt{K_x}}{\sqrt{K_z}} \qquad \text{eq. 6}$$

For $K_x > K_z$, this transformation will expand the vertical scale of the region of flow and also expand the hydraulic conductivity ellipse into a circle of radius $\sqrt{K_x}$. The fictitious, expanded region of flow will then act as if it were homogeneous with hydraulic conductivity $\sqrt{K_x}$. The graphical construction of the flow net follows from the transformation of the co-ordinates and using the above rules for homogeneous, isotropic material. The final step is to redraw the flow net by inverting the scaling ratio to the original dimensions. If

discharge quantities or flow velocities are required, it is easiest to make these calculations in the transformed section and applying the hydraulic conductivity value K', found from:

$$K' = \sqrt{K_x \cdot K_z} \qquad \text{eq. 7}$$

In the absence of a transformation of the co-ordinate system, the direction of groundwater flow at a point in an anisotropic material can be found using the construction shown in Fig. 3b. A line drawn in the direction of the hydraulic gradient intersects the ellipse at point A. If a tangent is drawn to the ellipse at A, then the direction of flow is perpendicular to this tangent line. For a further treatment of the topic of flow net construction, refer to Cedergren (1967) and Freeze and Cherry (1979).

When groundwater flows across a geological boundary between two formations with different values of hydraulic conductivity, the flow lines refract in an analogous way to light passing between two

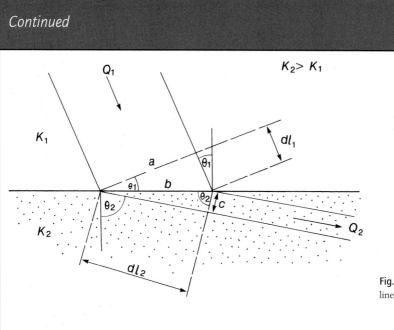

$K_2 > K_1$

Fig. 4 Refraction of groundwater flow lines at a geological boundary.

materials. Unlike in the case of light that obeys a sine law, groundwater refraction obeys a tangent law, as explained below.

In Fig. 4, a stream tube is shown with flow from a region with hydraulic conductivity K_1 to a region with hydraulic conductivity K_2, where $K_2 > K_1$. Considering a stream tube of unit depth perpendicular to the page, for steady flow, the inflow Q_1 must equal the outflow Q_2; then, from Darcy's law (eq. 2.5):

$$K_1 a \frac{dh_1}{dl_1} = K_2 c \frac{dh_2}{dl_2} \qquad \text{eq. 8}$$

where dh_1 is the decrease in head across distance dl_1 and dh_2 is the decrease in head across distance dl_2. In that dl_1 and dl_2 bound the same two equipotential lines, then dh_1 equals dh_2; and from a consideration of the geometry of Fig. 4, $a = b \cdot \cos \theta_1$ and $c = b \cdot \cos \theta_2$. Noting that $b/dl_1 = 1/\sin \theta_1$ and $b/dl_2 = 1/\sin \theta_2$, equation 8 now becomes:

$$K_1 \frac{\cos \theta_1}{\sin \theta_1} = K_2 \frac{\cos \theta_2}{\sin \theta_2} \qquad \text{eq. 9}$$

or

$$\frac{K_1}{K_2} = \frac{\tan \theta_1}{\tan \theta_2} \qquad \text{eq. 10}$$

Equation 10 is the tangent law for the refraction of groundwater flow lines at a geological boundary in heterogeneous material. In

layered aquifer systems, as shown in Fig. 5, the outcome of the tangent law is that flow lines have longer, horizontal components of flow in aquifer layers and shorter, vertical components of flow across intervening aquitards. The aquifer layers act as conduits for groundwater flow. If the ratio of the aquifer to aquitard hydraulic conductivities is greater than 100, then flow lines are almost horizontal in aquifer layers and close to vertical across aquitards. This is commonly the case, as the values of hydraulic conductivity of natural geological materials range over many orders of magnitude (Table 2.1).

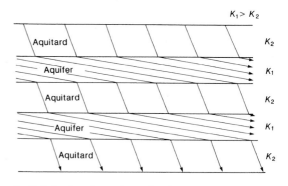

Fig. 5 Refraction of groundwater flow lines across a layered aquifer system.

B O X
2 . 4

Potentiometric surface map of the London Basin

The recording of groundwater levels in wells and boreholes and their reference to a common datum such as sea level is a basic requirement in hydrogeology. Maps of the water table or, more correctly, the potentiometric surface assist in the management of groundwater resources by enabling the identification of recharge and discharge areas and groundwater conditions. The repeat mapping of an area enables the storage properties of the aquifer to be understood from an examination of observed fluctuations in groundwater levels (see Section 5.2). Potentiometric surface maps can also be used as the basis for constructing a regional flow net (Box 2.3) and can provide useful information on the hydraulic conductivity of the aquifer from inspection of the gradients of the equipotential contour lines. A further important application is in groundwater modelling where a high quality potentiometric surface map and associated observation borehole hydrographs are necessary in developing a well-calibrated groundwater flow model.

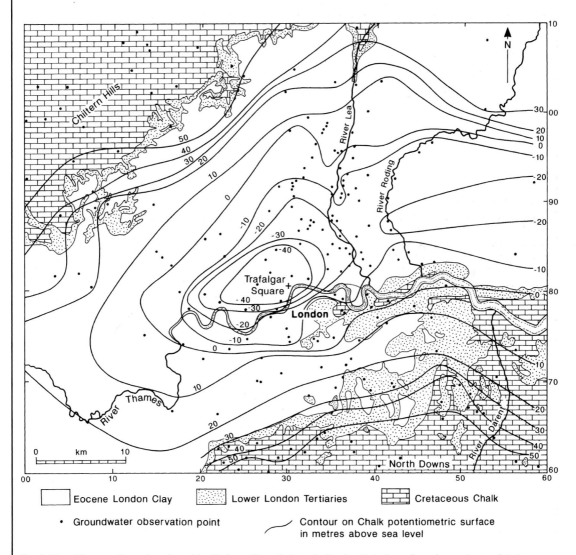

	Eocene London Clay		Lower London Tertiaries		Cretaceous Chalk

• Groundwater observation point   ⌒ Contour on Chalk potentiometric surface in metres above sea level

Fig. 1 Map of the potentiometric surface of the Chalk aquifer underlying the London Basin drawn from observations made in January 1994. After Lucas and Robinson (1995).

A map of the potentiometric surface of the Chalk aquifer below the London Basin is shown in Fig. 1 and illustrates a number of the above points. Areas of high groundwater level in excess of 50 m above sea level are present in unconfined areas where the Chalk is exposed on the northern and southern rims of the synclinal basin. Here, the Chiltern Hills and North Downs are the recharge areas for the London Basin. In the centre of the Basin, the residual drawdown in Chalk groundwater levels due to earlier over-exploitation of the aquifer (Box 2.5) is clearly visible in the wide area where the Chalk

potentiometric surface is less than 10 m below sea level. Additional disturbance of the regional groundwater level is noticeable along the River Lea valley to the north of London where large abstractions have disturbed the equipotential contours. To the east of Central London, the Chalk potentiometric surface is at about sea level along the estuary of the River Thames and here saline water can intrude the aquifer where the overlying Lower London Tertiaries and more recent deposits are thin or absent.

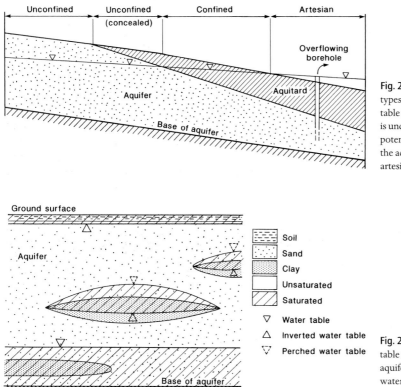

Fig. 2.17 Cross-section showing four types of groundwater conditions. A water table is developed where the aquifer is unconfined or concealed and a potentiometric surface is present where the aquifer experiences confined or artesian conditions.

Fig. 2.18 Perched and inverted water-table conditions developed within a sand aquifer containing clay lenses. An inverted water table is also shown below the wetted ground surface in the soil zone.

surface following a rainfall event in which water infiltrates the soil zone.

A confined aquifer is contained between two aquitards or aquicludes. Water held in a confined aquifer is under pressure, such that groundwater in a borehole penetrating a confined unit will rise to a level above the top of the aquifer. If the groundwater

level rises to the top of the borehole above ground level and overflows, then an overflowing artesian groundwater condition is encountered. If over-abstraction of groundwater occurs from boreholes exploiting a confined aquifer, the groundwater level can be drawn down below the top of the aquifer such that it becomes unconfined. As shown in

Box 2.5, this situation developed in the Chalk aquifer of the London Basin as a result of over-abstraction of groundwater to sustain the growth of London's industry and population from between the early 1800s and the early 1960s.

If the overlying geological unit behaves as an aquitard then leakage of water to the underlying aquifer can occur if a vertical hydraulic gradient is developed across the aquitard–aquifer boundary. This situation is commonly encountered where fluvial or glacio-fluvial silts and sandy clays overlie an aquifer and results in a semiconfined aquifer condition. A more complete regional hydrogeological interpretation requires the combination and analysis of mapped geological and geomorphological information, surveyed groundwater-level data and hydrogeological field observations. As an example of such an integrated approach, groundwater conditions prevailing in the Qu'Appelle Valley of Saskatchewan are explained in Box 2.6. Interpretation of the relationship between geology, groundwater occurrence and potentiometric head distribution demonstrates the connection between groundwater and spring-fed lakes in this large glacial meltwater channel feature.

2.10 Classification of springs and intermittent streams

The classification of springs has been discussed from as early as Bryan (1919) who recognized the following types: volcanic, fissure, depression, contact, artesian and springs in impervious rock. Simply defined, springs represent the termination of underground flow systems and mark the point at which fluvial processes become dominant. The vertical position of the spring marks the elevation of the water table or a minimum elevation of the potentiometric surface at the point of discharge from the aquifer. The influence which springs exert on the aquifers they drain depends principally upon the topographic and structural context of the spring. Ford and Williams (1989) discuss hydrogeological controls on springs and recognize three principle types of springs (free draining, dammed and confined), principally in relation to karst aquifers in which some of the world's largest springs occur (Table 2.2).

With reference to Fig. 2.19, free draining springs experience groundwater discharge under the influence of gravity and are entirely or dominantly in the unsaturated (vadose) zone. Dammed springs are a

History of groundwater exploitation in the Chalk aquifer of the London Basin

BOX 2.5

The industrial revolution and associated population growth of Greater London resulted in a large demand for water. With surface water resources becoming polluted, increasing use was made of the substantial storage in the Cretaceous Chalk aquifer that underlies the London Basin. The Chalk forms a gentle syncline with extensive outcrops to the south and north-west of London (Fig. 1) and is confined by Tertiary strata, mainly Eocene London Clay. The aquifer is recharged at outcrop by rainfall on the Chiltern Hills and North Downs. A component of this water flows through the aquifer towards and along the easterly dipping axis of the syncline. Prior to

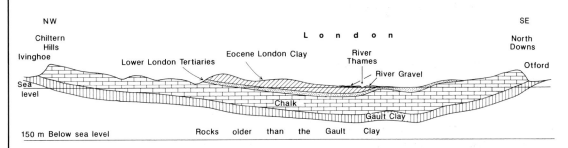

Fig. 1 Schematic geological section across the London Basin. The vertical scale is greatly exaggerated. After Sherlock (1962).

exploitation, this groundwater eventually discharged into the Thames estuary in the Woolwich area via springs (Marsh & Davies 1983).

The first deep boreholes were sunk in the middle of the eighteenth century, and there was rapid development from about 1820. At this time, as shown in Fig. 2a, an artesian situation existed throughout much of the London Basin. The potentiometric surface of the Chalk aquifer was typically at shallow depth and in low-lying areas, such as the Hackney Marshes, water seeped upwards to the surface. In Central London, the natural potentiometric surface was a few metres above sea level (+7.5 m; Fig. 2a) creating artesian groundwater conditions. At the start of the twentieth century, rest

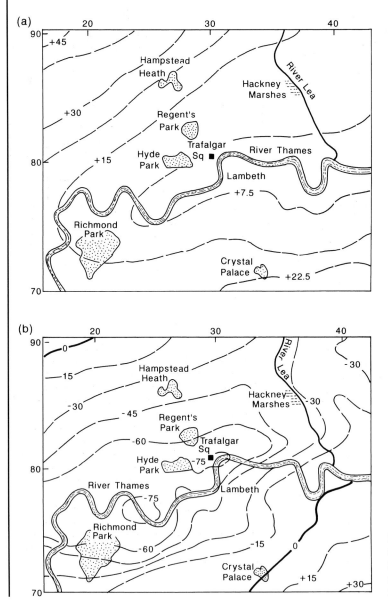

Fig. 2 Maps of the Chalk potentiometric surface in London showing groundwater levels (a) before groundwater exploitation commenced in the early 1800s and (b) in 1950. After Marsh and Davies (1983).

water levels in the borehole that at first supplied the fountains in Trafalgar Square under an artesian condition (Kirkaldy 1954) had declined to 60 m below sea level (Fig. 3), a drop of 40 m since the borehole was drilled in 1844, causing the groundwater condition in the Chalk aquifer to change from confined to unconfined. The decrease in hydraulic head caused a loss of yield from 2600 m³ day⁻¹ to 36 m³ day⁻¹ by 1900.

By 1950, and over an area extending from the centre of London south-west to Richmond (Fig. 2b), groundwater levels were more than 60 m below sea level over an area of 200 km². In the 100 years up to 1950, groundwater abstractions are estimated to have increased from 9.0×10^6 m³ a⁻¹ to 73.0×10^6 m³ a⁻¹ (Marsh & Davies 1983).

The general decline in groundwater levels had a number of impacts on groundwater resources in the London Basin, including a loss of yield and eventual failure of supply boreholes, and saline water intrusion into the Chalk aquifer. Saline water was recorded as far west as Lambeth in a zone 5–8 km wide centred on the River Thames. As discussed in Box 2.8, falling groundwater levels in a confined aquifer system can cause land subsidence, although the well-consolidated nature of the marine London Clay formation prevented substantial compaction in this case. Nevertheless, the overall settlement approached 0.2 m in Central London over the period 1865–1931 (Wilson & Grace 1942) as a consequence of the combined effects of dewatering the aquifer system, heavy building development and the removal of fine sediment during pumping from sand and gravel horizons.

An increasing awareness of the problems of over-abstraction from the Chalk aquifer together with a reduction in industrial abstractions since 1945 have led to a virtual end to further aquifer exploitation in London. Water supplies switched to piped supplies drawn predominantly from reservoirs in the Thames and Lea valleys, or from major abstraction boreholes located outside of Greater London. Since 1965, the reduction in groundwater abstraction has resulted in commencement of a recovery of groundwater levels at a rate of about 1 m a⁻¹ in the centre of the cone of depression (Lucas & Robinson 1995; see Fig. 5.3b). The re-saturation of the Tertiary strata has implications for building structures because of increased hydrostatic pressure on deep foundations and the flooding of tunnels constructed during the main period of depressed groundwater levels between 1870 and 1970. To combat this problem it is recognized that pumping will be required to control groundwater levels. Regional groundwater modelling of the London Basin indicated that an additional 70×10^3 m³ day⁻¹ of pumping would be necessary to control the rise in groundwater to an acceptable level assuming that existing abstractions continued at their current level (Lucas & Robinson 1995). To be effective, much of this abstraction would need to be concentrated in an area incorporating Central London together with an extension eastwards along the River Thames where structural engineering problems due to groundwater level rise are predicted (Simpson et al. 1989).

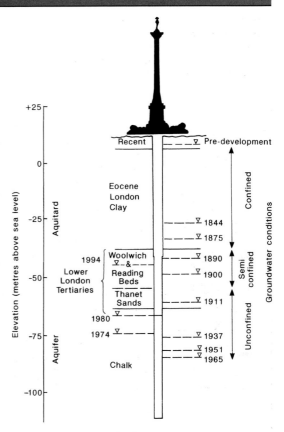

Fig. 3 Variation in groundwater level at Trafalgar Square (borehole TQ28/119) showing the changes in groundwater conditions in the Chalk aquifer below London. In the period before groundwater exploitation, the natural potentiometric surface was near to ground surface at about 7.5 m above sea level. At this time, the Chalk aquifer was confined by the Eocene London Clay. By 1965, the groundwater level had declined to 83 m below sea level as a result of over-exploitation of the groundwater resource. The lower section of the Tertiary strata includes sands of the Lower London Tertiaries which are in hydraulic contact with the Chalk and these extend the aquifer vertically. By 1900, the Chalk potentiometric surface was within these strata and the groundwater condition had become semiconfined. For most of the twentieth century until the mid-1960s when groundwater levels began to rise, the Chalk potentiometric surface was below the base of the Tertiary strata and the groundwater condition was unconfined. After Marsh and Davies (1983).

Relationship between geology, geomorphology and groundwater in the Qu'Appelle Valley, Saskatchewan

BOX 2.6

The Qu'Appelle Valley in southern Saskatchewan in Canada is a major landscape feature that owes its origin to the continental glaciations of the Quaternary Period. During the advance of the first glacier, the Hatfield Valley was cut into bedrock by glacial meltwater to a width of 20 km from near the Manitoba border to Alberta. When the glacier advanced to the vicinity of the Hatfield Valley, sands of the Empress Group were deposited and these now form a major aquifer. A similar sequence of events occurred during the advance of the third glacier which deposited the Floral Formation till. Meltwaters from this glacier cut the Muscow Valley, which was then filled with silt, sand and gravel of the Echo Lake Gravel. The fourth and final ice advance finally retreated from the Qu'Appelle area about 14,000 years ago, leaving its own distinctive till, the Battleford Formation, and other major landscape features.

The Qu'Appelle Valley was carved by meltwater issuing from the last retreating ice sheet, draining eastwards to glacial Lake Agassiz through the ice marginal Qu'Appelle Spillway. As the water continued eastwards through the spillway it cut a wide valley to a depth of 180 m into the underlying glacial deposits and bedrock. Where it crossed the buried Hatfield Valley it cut into the sand deposits. Since the retreat of the glacier, the present Qu'Appelle Valley has continued to fill with alluvial material derived from down-valley transport of sediment and from the erosion of valley sides and adjacent uplands.

Presently, a number of freshwater lakes and the existence of tributary valleys owe their existence to groundwater discharge and demonstrate the relationship between geology, geomorphology and groundwater. To illustrate this relationship, the cross-section

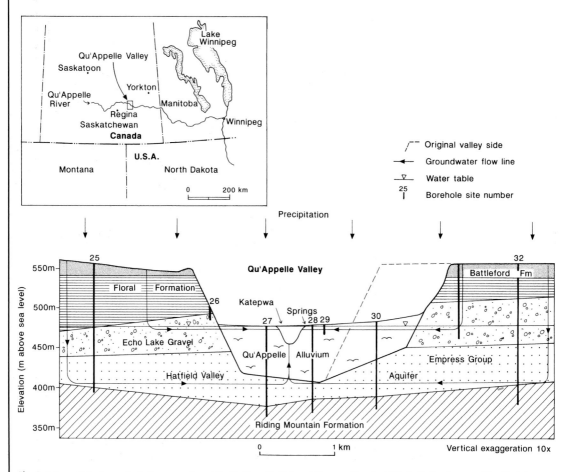

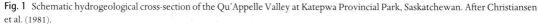

Fig. 1 Schematic hydrogeological cross-section of the Qu'Appelle Valley at Katepwa Provincial Park, Saskatchewan. After Christiansen et al. (1981).

BOX
2.6

Continued

through the Qu'Appelle Valley at Katepwa Provincial Park (Fig. 1) shows recharge from rainfall and snowmelt on the adjacent prairie moving vertically downwards to the Echo Lake Gravel and Empress Group, then horizontally through these more permeable deposits, before finally moving vertically upwards to discharge into Katepwa Lake as underwater springs.

The evolution of the tributary valleys, as shown in Fig. 2, is linked to past and present groundwater flow regimes. At the time the Qu'Appelle Valley was cut by glacial meltwater, the water-bearing Echo Lake Gravel was penetrated, and large quantities of groundwater discharged from this aquifer into the Qu'Appelle Valley in the form of major springs. It is conjectured that discharging groundwater carried sand and gravel from the Echo Lake Gravel and, to a lesser extent, from the Empress Group into the Qu'Appelle Valley, where part of it was swept away by meltwater flowing through the Qu'Appelle Spillway. This loss of sand and gravel by 'spring sapping' caused the overlying till to collapse, forming a tributary valley, which developed headwards along the path of maximum groundwater flow. Spring sapping forms short, wide tributaries and accounts for the short, well-developed gullies observed in the valley sides that deliver material to build the alluvial fans that today project into Katepwa Lake.

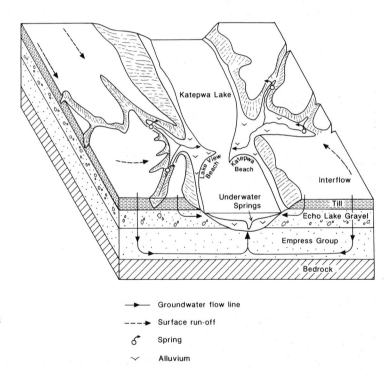

Fig. 2 Block diagram showing the final stage of evolution of tributary valleys and alluvial fans by spring sapping in the Qu'Appelle Valley, Saskatchewan. After Christiansen et al. (1981).

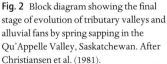

→   Groundwater flow line

---→   Surface run-off

⚲   Spring

⌄   Alluvium

common type and result from the location of a major barrier in the path of the underground drainage. The barrier may be caused by another lithology, either faulted or in conformable contact, or be caused by valley aggradation, such as by the deposition of glacial deposits. The denser salt water of the sea also forms a barrier to submarine groundwater discharge of freshwater. In each case, temporary overflow springs may form at times of higher water table elevation. Confined springs arise where artesian conditions are caused by an overlying impervious formation. Fault planes occasionally provide a discharge route for the confined groundwater as in the case of the Bath thermal springs (Box 2.7) and the example of the Permian Magnesian Limestones in South Yorkshire (Fig. 2.20). Elsewhere, the groundwater escapes

Table 2.2 Discharges of ten of the world's largest karst springs. After Ford and Williams (1989).

Spring	Discharge (m³ s⁻¹)			Basin area (km²)
	Mean	Max.	Min.	
Matali, Papua New Guinea	90	>240	20	350
Bussento, Italy	–	117	76	–
Dumanli*, Turkey	50	–	25	2800
Trebišnjca, Bosnia-Herzegovina	50	250	3	–
Chingshui, China	33	390	4	1040
Vaucluse, France	29	200	4.5	2100
Frió, Mexico	28	515	6	>1000?
Silver, USA	23.25	36.5	15.3	1900
Waikoropupu, New Zealand	15	21	5.3	450
Maligne, Canada	13.5	45	1	730

* Dumanli spring is the largest of a group of springs that collectively yield a mean flow of 125–130 m³ s⁻¹ at the surface of the Manavgat River.

where the overlying strata are removed by erosion. Since the emerging water is usually rapidly equilibrating to atmospheric pressure, dissolved gases can create a 'boiling' appearance within the spring pool.

In rivers that flow over an aquifer outcrop, both influent and effluent conditions can develop depending on the position of the water table in relation to the elevation of the river bed. With the seasonal fluctuation of the water table, the sections of river that receive groundwater discharge in addition to surface runoff will also vary. A good example of this type of river is the intermittent streams that appear over areas of Chalk outcrop in southern England. In these areas, the low specific yield of the Chalk aquifer causes large fluctuations in the position of the water table between the summer and winter and, therefore, in the length of the intermittent streams. The intermittent streams, or Chalk bournes or winterbournes as they are known, flow for part of the year, usually during or after the season of most precipitation. An example is the River Bourne located on the north-east of Salisbury Plain as shown in Fig. 2.21a. In this area of undulating Chalk downland in central southern England, the intermittent section of the River Bourne is 10 km in length until the point of the perennial stream head is met below which the Chalk water table permanently intersects the river bed. In drought years, the intermittent section may remain dry while in wet years, as shown in Fig. 2.21b, the upper reaches sustain a bank full discharge.

2.11 Groundwater flow theory

In this section, the mathematical derivation of the steady-state and transient groundwater flow equations will be presented followed by a demonstration of simple analytical solutions to groundwater flow problems. Following from this, different scales of flow systems are shown to exist in regional aquifer systems. Before starting, it is necessary to define the two aquifer properties of transmissivity and storativity for both unconfined and confined aquifers.

2.11.1 Transmissivity and storativity of confined aquifers

For a confined aquifer of thickness, b, the transmissivity, T, is defined as:

$$T = Kb \qquad \text{eq. 2.23}$$

and represents the rate at which water of a given density and viscosity is transmitted through a unit width of aquifer or aquitard under a unit hydraulic gradient. Transmissivity has the units of $L^2 \, T^{-1}$.

The storativity (or storage coefficient), S, of a confined aquifer is defined as:

$$S = S_s b \qquad \text{eq. 2.24}$$

where S_s is the specific storage term, and represents the volume of water that an aquifer releases from storage per unit surface area of aquifer per unit decline in the component of hydraulic head normal to that surface (Fig. 2.22a). Storativity values are dimensionless and range in value from 0.005 to 0.00005, such that large head changes over extensive areas are required to produce significant yields from confined aquifers.

2.11.2 Release of water from confined aquifers

At the beginning of the last century, Meinzer and Hard (1925) observed in a study of the Dakota sandstone that more water was pumped from the region than could be accounted for (as water was pumped, a cone of depression developed and the rate of

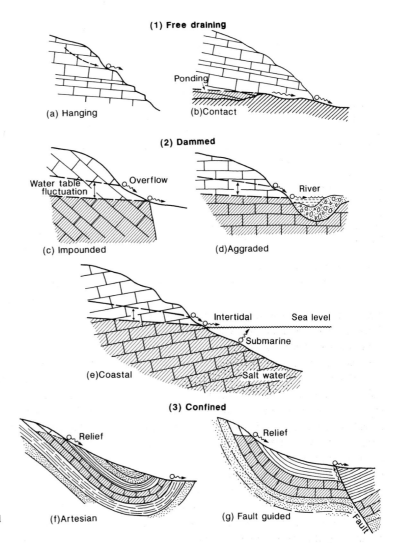

Fig. 2.19 Types of springs. After Ford and Williams (1989).

abstraction decreased, but with no apparent effect on groundwater levels in the recharge zone), such that the water-bearing formation was demonstrating elastic behaviour in releasing water from storage. Later, in deriving the general partial differential equation describing transient groundwater flow, Jacob (1940) formally described the elastic behaviour of porous rocks. There are two mechanisms that explain how water is produced by confined aquifers: the porosity of the aquifer is reduced by compaction and groundwater is released; and the water itself expands since water is slightly compressible.

As shown in Fig. 2.23, the total downward stress, σ_T, applied at the top of a confined aquifer is supported by an upward effective stress, σ_e, on the aquifer material and the water pressure contained in the pore space, P_w, such that:

$$\sigma_T = \sigma_e + P_w \qquad\qquad eq.\ 2.25$$

If the porewater pressure is decreased by groundwater pumping or by natural groundwater outflow, the stress on the aquifer material will increase causing it to undergo compression.

Compressibility is a material property that describes the change in volume (strain) induced in a material under an applied stress and is the inverse of the modulus of elasticity (equal to a change in stress divided by a change in strain). The compressibility of water, β, is defined as:

$$\beta = \frac{-dV_w/V_w}{dP_w} \qquad \text{eq. 2.26}$$

where P_w is porewater pressure, V_w is volume of a given mass of water and dV_w/V_w is volumetric strain for an induced stress dP_w. For practical purposes, β can be taken as a constant equal to 4.4×10^{-10} m^2 N^{-1} (or Pa^{-1}).

The compressibility of aquifer material, α, is defined as:

$$\alpha = \frac{-dV_T/V_T}{d\sigma_e} \qquad \text{eq. 2.27}$$

where V_T is total volume of aquifer material and dV_T/V_T is volumetric strain for an induced change in effective stress $d\sigma_e$.

Now, with reference to equation 2.27, for a reduction in the total volume of aquifer material, dV_T, the amount of water produced by compaction of the aquifer, dV_w, is:

$$dV_w = -dV_T = \alpha V_T d\sigma_e \qquad \text{eq. 2.28}$$

The thermal springs of Bath, England

BOX 2.7

The thermal springs at Bath in the west of England are the principal occurrence of thermal springs in the British Isles and have been exploited for at least the past 2000 years. The springs have temperatures of 44–47°C with an apparently constant flow of 15 L s^{-1}. Three springs, the King's (Fig. 1), Cross and Hetling Springs, issue from what were probably once pools on a floodplain terrace on the River Avon, in the centre of Bath. A succession of buildings has been constructed over the springs, beginning with the Roman Baths and temple of the first century AD (Fig. 2). Further details and an account of the hydrogeology of the thermal springs are given by Atkinson and Davison (2002).

The origin of the thermal waters has been subject to various investigations. Andrews et al. (1982) examined the geochemistry of the hot springs and other groundwaters in the region, and demonstrated that they are of meteoric origin. The silica content indicates

Fig. 1 The King's Spring (or Sacred Spring) emerging into the King's Bath at the Roman Baths, Bath, England.

Fig. 2 The overflow of thermal spring water adjacent to the King's Bath shown in Fig. 1.

BOX
2 . 7

Continued

that the thermal waters attain a maximum temperature between 64°C and 96°C, the uncertainty depending on whether chalcedony or quartz controls the silica solubility. Using an estimated geothermal gradient of 20°C km⁻¹, Andrews et al. (1982) calculated a circulation depth for the water of between 2.7 and 4.3 km from these temperatures. The natural groundwater head beneath central Bath is about 27–28 m above sea level, compared with normal spring pool levels at about 20 m. For this head to develop, Burgess et al. (1980) argued that the recharge area is most likely the

Carboniferous limestone outcrop in the Mendip Hills (see Section 2.7), 15–20 km south and south-west of Bath, in order to drive recharge down along a permeable pathway and then up a possible thrust fault to the springs themselves. The structural basin containing the Carboniferous limestone lies at depths exceeding 2.7 km at the centre of the basin, sufficient for groundwater to acquire the necessary temperature indicated by its silica content. This 'Mendips Model' for the origin of the Bath hot springs is summarized by Andrews et al. (1982) and shown in Fig. 3.

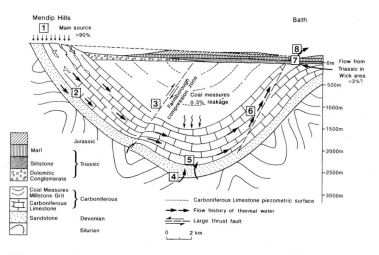

Fig. 3 Conceptual model for the origin of the Bath thermal springs. The numbers shown in squares indicate: (1) recharge (9–10°C) at the Carboniferous limestone / Devonian sandstone outcrop on the Mendip Hills; (2) flow down-dip and downgradient; (3) possible downward leakage from Upper Carboniferous Coal Measures; (4) possible leakage of very old ⁴He-bearing groundwater from Devonian sandstone and Lower Palaeozoic strata; (5) storage and chemical equilibration within the Carboniferous limestone at 64–96°C; (6) rapid ascent, probably along Variscan thrust faults re-activated by Mesozoic tectonic extension; (7) lateral spread of thermal water into Permo-Triassic strata at Bath; and (8) discharge of the thermal springs at Bath (46.5°C). After Andrews et al. (1982).

If the total stress does not change ($d\sigma_T = 0$), then from a knowledge that $P_w = \rho g\psi$ and $\psi = h - z$ (eq. 2.22), with z remaining constant, then, using equation 2.25:

$$d\sigma_e = 0 - \rho g d\psi = -\rho g dh \qquad \text{eq. 2.29}$$

For a unit decline in head, $dh = -1$, and if unit volume is assumed ($V_T = 1$), then equation 2.28 becomes:

$$dV_w = \alpha(1)(-\rho g)(-1) = \alpha\rho g \qquad \text{eq. 2.30}$$

The water produced by the expansion of water is found from equation 2.26 thus:

$$dV_w = -\beta V_w dP_w \qquad \text{eq. 2.31}$$

Recognizing that the volume of water, V_w, in the total unit volume of aquifer material, V_T, is nV_T where n is porosity, and that $dP = \rho g d\psi$ or $-\rho g$ for a unit decline in hydraulic head (where $\psi = h - z$ (eq. 2.22), with z remaining constant), then for unit volume, $V_T = 1$, equation 2.31 gives:

$$dV_w = -\beta n(1)(-\rho g) = \beta n\rho g \qquad \text{eq. 2.32}$$

Finally, the volume of water that a unit volume of aquifer releases from storage under a unit decline in

(a)

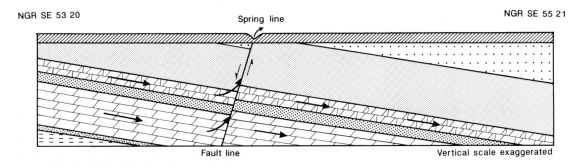

Quaternary		Undifferentiated peat, silt, clay and sand
Triassic		Sherwood Sandstone
Permian		Upper Permian Marl
		Upper Magnesian Limestone
		Middle Permian Marl
		Lower Magnesian Limestone
		Basal Permian Sands
Carboniferous		Upper Coal measures
	→	Direction of groundwater flow

0   200m

NGR SE 53 20   Spring line   NGR SE 55 21

Fault line   Vertical scale exaggerated

(b)

Fig. 2.20 (a) Geological section at Gale Common, South Yorkshire showing a spring line at the position of a normal fault that acts as a barrier to groundwater movement. The existence of the fault causes groundwater flow through the overlying Upper Permian Marl via a zone of enhanced permeability. (b) Springs from Permian Magnesian Limestones appearing through Quaternary deposits at Gale Common, South Yorkshire. For scale, a camera lens cap is shown to the right of the spring pool.

hydraulic head (the specific storage, S_s) is the sum of the volumes of water produced by the two mechanisms of compaction of the aquifer (eq. 2.30) and expansion of the water (eq. 2.32) thus:

$$S_s = \alpha\rho g + \beta n\rho g = \rho g(\alpha + n\beta) \qquad \text{eq. 2.33}$$

In other words, groundwater pumped from a confined aquifer does not represent a dewatering of the physical pore space in the aquifer but, instead, results from the secondary effects of aquifer compaction and water expansion. As a consequence, for an equivalent unit decline in hydraulic head, yields from confined aquifers are much less than from unconfined aquifers. Hence, storage coefficient values of confined aquifers are much smaller than for unconfined aquifers.

Values of material compressibility, α, range from 10^{-6} to 10^{-9} m^2 N^{-1} for clay and sand and from 10^{-8} to 10^{-10} m^2 N^{-1} for gravel and jointed rock (Table 2.3). These values indicate that a greater, largely irrecoverable compaction is expected in a previously unconsolidated clay aquitard, while smaller, elastic

(a)

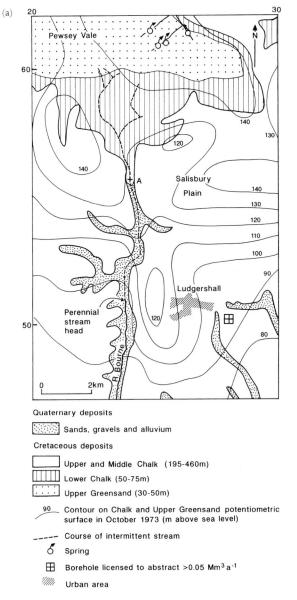

(b)

Quaternary deposits

Sands, gravels and alluvium

Cretaceous deposits

Upper and Middle Chalk (195–460m)

Lower Chalk (50–75m)

Upper Greensand (30–50m)

_90_ Contour on Chalk and Upper Greensand potentiometric surface in October 1973 (m above sea level)

------ Course of intermittent stream

⚲ Spring

⊞ Borehole licensed to abstract >0.05 Mm^{3}a^{-1}

▨ Urban area

Fig. 2.21 (a) Hydrogeological map of the north-east of Salisbury Plain showing the intermittent and perennial sections of the River Bourne. Position A is the site of the photograph shown in (b) located at Collingbourne Kingston and looking north in December 2002. The river bed is covered by lesser water parsnip *Berula erecta*, a plant that proliferates in still to medium flows of base-rich water. Hydrogeological information based on the British Geological Survey (1978); plant species identification using Haslam et al. (1975).

deformations are likely in gravel or indurated sedimentary aquifers. A possible consequence of groundwater abstraction from confined aquifers is ground subsidence following aquifer compaction, especially in sand–clay aquifer–aquitard systems. A notable example is the Central Valley in California (Box 2.8).

2.11.3 Transmissivity and specific yield of unconfined aquifers

For an unconfined aquifer, the transmissivity is not as well defined as in a confined aquifer, but equation 2.23 can be applied with b now representing the

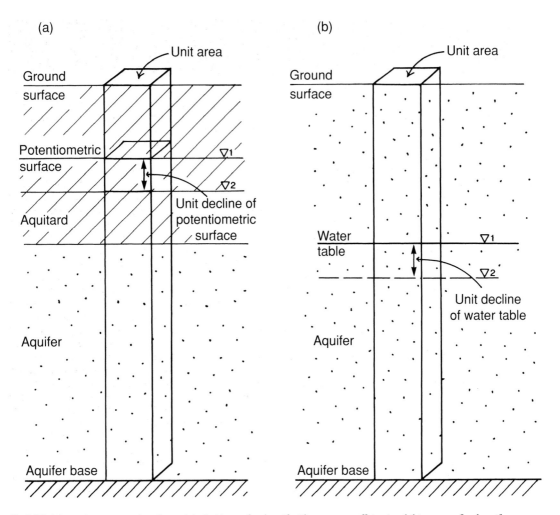

Fig. 2.22 Schematic representation of storativity in (a) a confined aquifer (the storage coefficient) and (b) an unconfined aquifer (the specific yield).

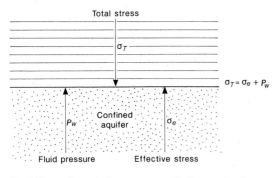

Fig. 2.23 Total stress, effective stress and fluid pressure at the top of a confined aquifer.

Table 2.3 Range of values of compressibility. After Freeze and Cherry (1979).

	Compressibility, α ($m^2\ N^{-1}$ or Pa^{-1})
Clay	$10^{-6}–10^{-8}$
Sand	$10^{-7}–10^{-9}$
Gravel	$10^{-8}–10^{-10}$
Jointed rock	$10^{-8}–10^{-10}$
Sound rock	$10^{-9}–10^{-11}$
Water (β)	4.4×10^{-10}

Agricultural production in the Central Valley of California is dependent on the availability of water for irrigation. One-half of this irrigation water is supplied by groundwater and accounts for 74% of California's total abstractions and about 20% of the irrigation abstractions in the United States (Williamson et al. 1989). Groundwater abstraction is especially important in dry years when it supplements highly variable surface water supplies. In 1975, about 57% of the total land area (5.2×10^6 ha) in the Central Valley was irrigated. The intensive agricultural development during the past 100 years has had major impacts on the aquifer system.

The Central Valley is a large structural trough filled with marine sediments overlain by continental deposits with an average total thickness of about 730 m (Fig. 1). More than half of the thickness of the continental deposits is composed of fine-grained sediments. When development began in the 1880s, flowing wells and marshes were found throughout most of the Central Valley. The northern one-third of the valley, the Sacramento Valley, is considered to be an unconfined aquifer with a water table and the southern two-thirds, the San Joaquin Valley, as a two-layer aquifer system separated by a regional confining clay layer, the Pleistocene Corcoran Clay. This clay layer is highly susceptible to compaction. Figure 1 is a conceptual model of the hydrogeology of the Central Valley based on Williamson et al. (1989) who considered the entire thickness of continental deposits to be one aquifer system that has varying vertical leakance (ratio of vertical hydraulic conductivity to bed thickness) and confinement depending on the proportion of fine-grained sediments encountered.

During 1961–1977, an average of 27×10^9 m^3 a^{-1} of water was used for irrigation with about one-half derived from groundwater. This amount of groundwater abstraction has caused water levels to decline by in excess of 120 m in places (Fig. 2) resulting in the largest volume of land subsidence in the world due to groundwater abstraction. Land subsidence has caused problems such as cracks in roads and canal linings, changing slopes of water channels and ruptured well casings.

From pre-development until 1977, the volume of water in aquifer storage declined by about 74×10^9 m^3, with 49×10^9 m^3 from the water table zone, 21×10^9 m^3 from inelastic compaction of fine-grained sediments and 4×10^9 m^3 from elastic storage. Elastic storage is a result of the expansion of water and the compression of sediments resulting from a change in fluid pressure (see Section 2.11.2). The estimated average elastic specific storage, S_s (eq. 2.33), is 1×10^{-5} m^{-1}. The inelastic compaction of fine-grained sediments in the aquifer system caused by a decline in the hydraulic head results in a reorientation of the grains and a reduction in pore space within the compacted beds, thus releasing water. The volume of water released by compaction is approximately equal to the volume of land subsidence observed at the surface. The loss of pore space represents a permanent loss of storage capacity in the aquifer system. Even if water levels were to recover to their previous highest position, the amount of water stored in the aquifer system would be less than the amount stored prior to compaction. Inelastic compaction means permanent compaction. This type of land subsidence represents a once-only abstraction of water from storage.

The cumulative volume of subsidence in the San Joaquin Valley is shown in Fig. 3. By 1970, the total volume of subsidence was 19×10^9 m^3. Also included in Fig. 3 are cumulative volumes of subsidence for each of the three major subsiding areas. The volume of subsidence in the Los Banos–Kettleman City area accounted for nearly two-thirds of the total volume of subsidence as of 1970. From 1970 to 1975 there was little further subsidence in this area because surface water imports from the California Aqueduct greatly

Fig. 1 Conceptual model of the hydrogeology of the Central Valley of California. Before groundwater development, water that recharged the aquifer at the valley margins moved downwards and laterally into the aquifer system and then moved upwards to discharge at rivers and marshes along the valley axis. The entire aquifer system is considered to be a single heterogeneous system in which vertically and horizontally scattered lenses of fine-grained materials provide increasing confinement with depth. After Williamson et al. (1989).

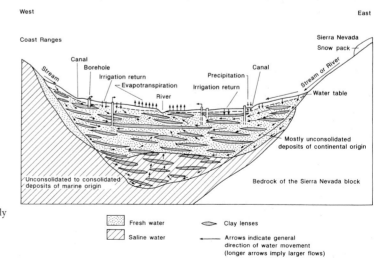

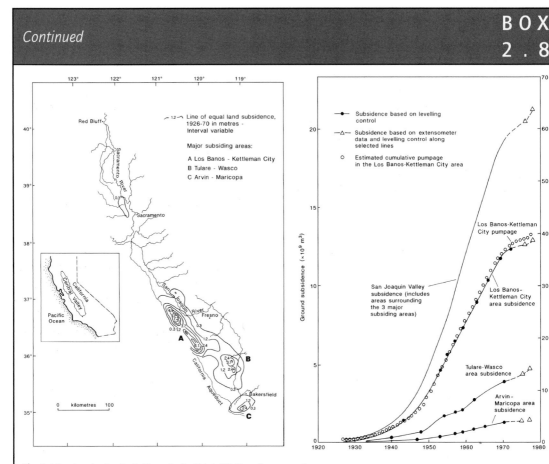

Fig. 2 Map showing land subsidence in the Central Valley of California. After Williamson et al. (1989).

Fig. 3 Volumes of land subsidence in the major subsiding areas of the San Joaquin Valley and groundwater abstraction in the Los Banos–Kettleman City area, 1925–1977. After Williamson et al. (1989).

reduced the demand for groundwater. However, subsidence recurred during the drought of 1976–1977 owing to an increase in groundwater abstraction. The correlation between groundwater abstraction and the volume of subsidence in the Los Banos–Kettleman City area is good, indicating that about 43% of the water pumped from the lower pumped zone (at least 75–80% of the total) was derived from compaction of the fine-grained sediments in the aquifer system.

Land subsidence continues to be a problem in some areas of the Sacramento and San Joaquin Valleys, although the areas of greater

subsidence have been controlled by importing surface water and decreasing groundwater abstractions. In these areas, the recovery of the lower pumped zone water levels to nearly their predevelopment elevation may lead to future overestimation of the available groundwater resources. If groundwater pumping increases again, water levels will drop rapidly towards the previous lows because of the loss of aquifer storage capacity that resulted from the previous compaction of fine-grained sediments.

saturated thickness of the aquifer or the height of the water table above the top of a lower aquitard boundary. The transmissivity will, therefore, vary if there are large seasonal fluctuations in the elevation of the water table or if the saturated thickness of the aquifer

shows lateral variation as a result of an irregular lower aquitard boundary or differences between recharge and discharge areas in the same aquifer.

The storage term for an unconfined aquifer is known as the specific yield, S_y, (or the unconfined storativity) and is that volume of water that an unconfined aquifer releases from storage per unit surface area of aquifer per unit decline in the water table (Fig. 2.22b), and is approximately equivalent to the total porosity of a soil or rock (see Section 2.2). Specific yield is a dimensionless term and the normal range is from 0.01 to 0.30. Relative to confined aquifers, the higher values reflect the actual dewatering of pore space as the water table is lowered. Consequently, the same yield can be obtained from an unconfined aquifer with smaller head changes over less extensive areas than can be produced from a confined aquifer. Although not commonly used, by combining the aquifer properties of transmissivity (T or K) and storativity (S or S_s) it is possible to define a single formation parameter, the hydraulic diffusivity, D, defined as either:

$$D = \frac{T}{S} \text{ or } \frac{K}{S_s} \qquad \text{eq. 2.34}$$

Aquifer formations with a large hydraulic diffusivity respond quickly in transmitting changed hydraulic conditions at one location to other regions in an aquifer, for example in response to groundwater abstraction.

2.11.4 Equations of groundwater flow

Equations of groundwater flow are derived from a consideration of the basic flow law, Darcy's law (eq. 2.5), and an equation of continuity that describes the conservation of fluid mass during flow through a porous material. In the following treatment, which derives from the classic paper by Jacob (1950), steady-state and transient saturated flow conditions are considered in turn. Under steady-state conditions, the magnitude and direction of the flow velocity at any point are constant with time. For transient conditions, either the magnitude or direction of the flow velocity at any point may change with time, or the potentiometric conditions may change as groundwater either enters into or is released from storage.

Steady-state saturated flow

First, consider the unit volume of a porous material (the elemental control volume) depicted in Fig. 2.24.

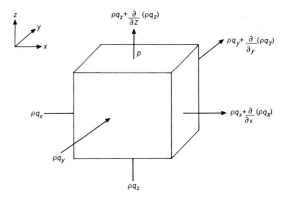

Fig. 2.24 Unit volume (elemental control volume) for flow through porous material.

The law of conservation of mass for steady-state flow requires that the rate of fluid mass flow into the control volume, ρq (fluid density multiplied by specific discharge across a unit cross-sectional area), will be equal to the rate of fluid mass flow out of the control volume, such that the incremental differences in fluid mass flow, in each of the directions x, y, z, sum to zero, thus:

$$\left(\rho q_x + \frac{\partial(\rho q_x)}{\partial x} - \rho q_x \right) + \left(\rho q_y + \frac{\partial(\rho q_y)}{\partial y} - \rho q_y \right)$$

$$+ \left(\rho q_z + \frac{\partial(\rho q_z)}{\partial z} - \rho q_z \right) = 0 \qquad \text{eq. 2.35}$$

From equation 2.35, the resulting equation of continuity is:

$$\frac{\partial(\rho q_x)}{\partial x} + \frac{\partial(\rho q_y)}{\partial y} + \frac{\partial(\rho q_z)}{\partial z} = 0 \qquad \text{eq. 2.36}$$

If the fluid is incompressible, then density, $\rho(x, y, z)$, is constant and equation 2.36 becomes:

$$\frac{\partial q_x}{\partial x} + \frac{\partial q_y}{\partial y} + \frac{\partial q_z}{\partial z} = 0 \qquad \text{eq. 2.37}$$

From Darcy's law, each of the specific discharge terms can be expressed as:

$$q_x = -K_x \frac{\partial h}{\partial x}, \, q_y = -K_y \frac{\partial h}{\partial y}, \, q_z = -K_z \frac{\partial h}{\partial z} \qquad \text{eq. 2.38}$$

and upon substitution in equation 2.37:

$$\frac{\partial}{\partial x}\left(K_x \frac{\partial h}{\partial x}\right) + \frac{\partial}{\partial y}\left(K_y \frac{\partial h}{\partial y}\right) + \frac{\partial}{\partial z}\left(K_z \frac{\partial h}{\partial z}\right) = 0$$

eq. 2.39

For an isotropic and homogeneous porous material, $K_x = K_y = K_z$ and $K(x, y, z) = $ constant, respectively. By substituting these two conditions in equation 2.39 it can be shown that:

$$\frac{\partial^2 h}{\partial x^2} + \frac{\partial^2 h}{\partial y^2} + \frac{\partial^2 h}{\partial z^2} = 0$$

eq. 2.40

Thus, the steady-state groundwater flow equation is the Laplace equation and the solution $h(x, y, z)$ describes the value of the hydraulic head at any point in a three-dimensional flow field. By solving equation 2.40, either in one, two or three dimensions depending on the geometry of the groundwater flow problem under consideration, a contoured equipotential map can be produced and, with the addition of flow lines, a flow net drawn (Box 2.3).

Transient saturated flow

The law of conservation of mass for transient flow in a saturated porous material requires that the net rate of fluid mass flow into the control volume (Fig. 2.24) is equal to the time rate of change of fluid mass storage within the control volume. The equation of continuity is now:

$$\frac{\partial(\rho q_x)}{\partial x} + \frac{\partial(\rho q_y)}{\partial y} + \frac{\partial(\rho q_z)}{\partial z} = \frac{\partial(\rho n)}{\partial t} = n\frac{\partial \rho}{\partial t} + \rho\frac{\partial n}{\partial t}$$

eq. 2.41

The first term on the right-hand side of equation 2.41 describes the mass rate of water produced by expansion of the water under a change in its density, ρ, and is controlled by the compressibility of the fluid, β. The second term is the mass rate of water produced by the compaction of the porous material as influenced by the change in its porosity, n, and is determined by the compressibility of the aquifer, α.

Changes in fluid density and formation porosity are both produced by a change in hydraulic head and the volume of water produced by the two mechanisms for a unit decline in head is the specific storage, S_s. Hence, the time rate of change of fluid mass storage within the control volume is:

$$\rho S_s \frac{\partial h}{\partial t}$$

and equation 2.41 becomes:

$$\frac{\partial(\rho q_x)}{\partial x} + \frac{\partial(\rho q_y)}{\partial y} + \frac{\partial(\rho q_z)}{\partial z} = \rho S_s \frac{\partial h}{\partial t}$$

eq. 2.42

By expanding the terms on the left-hand side of equation 2.42 using the chain rule (eliminating the smaller density gradient terms compared with the larger specific discharge gradient terms) and, at the same time, inserting Darcy's law to define the specific discharge terms, then:

$$\frac{\partial}{\partial x}\left(K_x \frac{\partial h}{\partial x}\right) + \frac{\partial}{\partial y}\left(K_y \frac{\partial h}{\partial y}\right) + \frac{\partial}{\partial z}\left(K_z \frac{\partial h}{\partial z}\right) = S_s \frac{\partial h}{\partial t}$$

eq. 2.43

If the porous material is isotropic and homogeneous, equation 2.43 reduces to:

$$\frac{\partial^2 h}{\partial x^2} + \frac{\partial^2 h}{\partial y^2} + \frac{\partial^2 h}{\partial z^2} = \frac{S_s}{K}\frac{\partial h}{\partial t}$$

eq. 2.44

or, expanding the specific storage term, S_s (eq. 2.33):

$$\frac{\partial^2 h}{\partial x^2} + \frac{\partial^2 h}{\partial y^2} + \frac{\partial^2 h}{\partial z^2} = \frac{\rho g(\alpha + n\beta)}{K}\frac{\partial h}{\partial t}$$

eq. 2.45

Equations 2.43, 2.44 and 2.45 are all transient groundwater flow equations for saturated anisotropic (eq. 2.43) and homogeneous and isotropic (eqs 2.44 and 2.45) porous material. The solution $h(x, y, z, t)$ describes the value of hydraulic head at any point in a three-dimensional flow field at any time. A solution requires knowledge of the three hydrogeological parameters, K, α and n, and the fluid parameters, ρ and β. A simplification is to take the special case of a horizontal confined aquifer of thickness, b, storativity, S ($= S_s b$), and transmissivity, T ($= Kb$), and substitute in equation 2.44, thus:

$$\frac{\partial^2 h}{\partial x^2} + \frac{\partial^2 h}{\partial y^2} = \frac{S}{T}\frac{\partial h}{\partial t} \qquad \text{eq. 2.46}$$

The solution of this equation, $h(x, y, t)$, describes the hydraulic head at any point on a horizontal plane through the horizontal aquifer at any time. A solution requires knowledge of the aquifer parameters T and S, both of which are measurable from field pumping tests (see Section 5.8.2).

Transient unsaturated flow

A treatment of groundwater flow in unsaturated porous material must incorporate the presence of an air phase. The air phase will affect the degree of connectivity between water-filled pores, and will therefore influence the hydraulic conductivity. Unlike in saturated material where the pore space is completely water filled, in unsaturated material the partial saturation of pore space, or moisture content (θ), means that the hydraulic conductivity is a function of the degree of saturation, $K(\theta)$. Alternatively, since the degree of moisture content will influence the pressure head (ψ), the hydraulic conductivity is also a function of the pressure head, $K(\psi)$. In soil physics, the degree of change in moisture content for a change in pressure head ($d\theta/d\psi$) is referred to as the specific moisture capacity, C (the unsaturated storage property of a soil), and can be empirically derived from the slope of a soil characteristic curve (see Section 5.4.1).

Returning to Fig. 2.24, for flow in an elemental control volume that is partially saturated, the equation of continuity must now express the time rate of change of moisture content as well as the time rate of change of storage due to water expansion and aquifer compaction. The fluid mass storage term (ρn) in equation 2.41 now becomes $\rho\theta$ and:

$$\frac{\partial(\rho q_x)}{\partial x} + \frac{\partial(\rho q_y)}{\partial y} + \frac{\partial(\rho q_z)}{\partial z} = \frac{\partial(\rho\theta)}{\partial t} = \theta\frac{\partial\rho}{\partial t} + \rho\frac{\partial\theta}{\partial t}$$

$$\text{eq. 2.47}$$

The first term on the right-hand side of equation 2.47 is insignificantly small and by inserting the unsaturated form of Darcy's law, in which the hydraulic conductivity is a function of the pressure head, $K(\psi)$, then equation 2.47 becomes, upon cancelling the ρ terms:

$$\frac{\partial}{\partial x}\left(K(\psi)\frac{\partial h}{\partial x}\right) + \frac{\partial}{\partial y}\left(K(\psi)\frac{\partial h}{\partial y}\right)$$

$$+ \frac{\partial}{\partial z}\left(K(\psi)\frac{\partial h}{\partial z}\right) = \frac{\partial\theta}{\partial t} \qquad \text{eq. 2.48}$$

It is usual to quote equation 2.48 in a form where the independent variable is either θ or ψ. Hence, noting that $h = z + \psi$ (eq. 2.22) and defining the specific moisture capacity, C, as $d\theta/d\psi$, then:

$$\frac{\partial}{\partial x}\left(K(\psi)\frac{\partial\psi}{\partial x}\right) + \frac{\partial}{\partial y}\left(K(\psi)\frac{\partial\psi}{\partial y}\right)$$

$$+ \frac{\partial}{\partial z}\left(K(\psi)\left(\frac{\partial\psi}{\partial z} + 1\right)\right) = C(\psi)\frac{\partial\psi}{\partial t} \qquad \text{eq. 2.49}$$

This equation (eq. 2.49) is the ψ-based transient unsaturated flow equation for porous material and is known as the Richards equation. The solution $\psi(x, y, z, t)$ describes the pressure head at any point in a flow field at any time. It can be easily converted into a hydraulic head solution $h(x, y, z, t)$ through the relation $h = z + \psi$ (eq. 2.22). To be able to provide a solution to the Richards equation it is necessary to know the characteristic curves $K(\psi)$ and $C(\psi)$ or $\theta(\psi)$ (see Section 5.4.1).

2.12 Analytical solution of one-dimensional groundwater flow problems

The three basic steps involved in the mathematical analysis of groundwater flow problems are the same whatever the level of mathematical difficulty and are: (i) conceptualizing the problem; (ii) finding a solution; and (iii) evaluating the solution (Rushton 2003).

Simple one-dimensional problems can be solved using ordinary differential and integral calculus. Two-dimensional problems or transient (time-variant) flow problems require the use of partial derivatives and more advanced calculus.

In conceptualizing the groundwater flow problem, the basic geometry should be sketched and the aquifers, aquitards and aquicludes defined. Simplifying assumptions, for example concerning isotropic and homogeneous hydraulic conductivity, should be stated and, if possible, the number of dimensions reduced (for example, consider only the horizontal component of flow or look for radial symmetry or approximately parallel flow). If the groundwater flow is confined, then a mathematically linear solution results which can be combined to represent more complex situations. Unconfined situations produce higher order equations.

As a first step, an equation of continuity is written to express conservation of fluid mass. For incompressible fluids this is equivalent to conservation of fluid volume. Water is only very slightly compressible, so conservation of volume is a reasonable approximation. By combining the equation of continuity with a flow law, normally Darcy's law, and writing down equations that specify the known conditions at the boundaries of the aquifer, or at specified points (for example, a well), provides a general differential equation for the specified system. Solving the problem consists of finding an equation (or equations) which describes the system and satisfies both the differential equation and the boundary conditions. By integrating the differential equation, the resulting equation is the general solution. If the constants of integration are found by applying the boundary conditions, then a specific solution to the problem is obtained (for examples, see Box 2.9).

The final step in the mathematical analysis is the evaluation of the solution. The specific solution is normally an equation that relates groundwater head to position and to parameters contained in the problem such as hydraulic conductivity, or such factors as the discharge rate of wells. By inserting numerical values for the parameters, the solution can be used to evaluate groundwater head in terms of position in the co-ordinate system. The results might be expressed as a graph or contour diagram, or as a predicted value of head for a specified point.

2.13 Groundwater flow patterns

Preceding sections in this chapter have introduced the fundamental principles governing the existence and movement of groundwater, culminating in the derivation of the governing groundwater flow equations for steady-state, transient and unsaturated flow conditions. Within the water or hydrological cycle, groundwater flow patterns are influenced by geological factors such as differences in aquifer lithologies and structure of confining strata. A further influence on groundwater flow, other than aquifer heterogeneity, is the topography of the ground surface. Topography is a major influence on groundwater flow at local, intermediate and regional scales. The elevation of recharge areas in regions of aquifer outcrop, the degree to which river systems incise the landscape and the location and extent of lowland areas experiencing groundwater discharge determine the overall configuration of groundwater flow. As shown by Freeze and Witherspoon (1967), the relative positions and difference in elevation of recharge and discharge areas determine the hydraulic gradients and the length of groundwater flowpaths.

To understand the influence of topography, consider the groundwater flow net shown in Fig. 2.25 for a two-dimensional vertical cross-section through a homogeneous, isotropic aquifer. The section shows a single valley bounded by groundwater divides and an impermeable aquifer base. The water table is a sub-dued replica of the topography of the valley sides. The steady-state equipotential and groundwater flow lines are drawn using the rules for flow net analysis introduced in Box 2.3. It is obvious from the flow net that groundwater flow occurs from the recharge areas on the valley sides to the discharge area in the valley bottom. The hinge line separating the recharge from the discharge areas is also marked in Fig. 2.25. For most common topographic profiles, hinge lines are positioned closer to valley bottoms than to catchment divides with discharge areas commonly comprising only 5–30% of the catchment area (Freeze & Cherry 1979). Tòth (1963) determined a solution to the boundary-value problem represented by the flow net shown in Fig. 2.25. The solution provides an analytical expression for the hydraulic head in the flow field for simple situations of an inclined water table of constant slope and cases in which a sine curve

Examples of analytical solutions to one-dimensional groundwater flow problems

BOX 2.9

To illustrate the basic steps involved in the mathematical analysis of groundwater flow problems, consider the one-dimensional flow problem shown in Fig. 1 for a confined aquifer with thickness, b. The total flow at any point in the horizontal (x) direction is given by the equation of continuity of flow:

$$Q = q \cdot b \qquad \text{eq. 1}$$

where q, the flow per unit width (specific discharge), is found from Darcy's law:

$$q = -K \frac{dh}{dx} \qquad \text{eq. 2}$$

where x increases in the direction of flow. Combining equations 1 and 2 gives the general differential equation:

$$Q = -Kb \frac{dh}{dx} \qquad \text{eq. 3}$$

By integrating equation 3, it is possible to express the groundwater head, h, in terms of x and Q:

$$\int dh = -\int \frac{Q}{Kb} dx$$

$$\therefore h = -\frac{Q}{Kb} x + c \qquad \text{eq. 4}$$

c is the constant of integration and can be determined by use of a supplementary equation expressing a known combination of h and x. For example, for the boundary condition $x = 0$, $h = h_0$ and by applying this condition to equation 4 gives:

$$h_0 = -\frac{Q}{Kb} \cdot 0 + c$$

$$\therefore c = h_0$$

which gives the solution:

$$h = h_0 - \frac{Q}{Kb} x \qquad \text{eq. 5}$$

The specific solution given in equation 5 relates h to location, x, in terms of two parameters, Q and Kb (transmissivity), and one boundary value, h_0. The solution is the equation of a straight line and predicts the position of the potentiometric surface as shown in Fig. 1. Also note, by introducing a new pair of values of x where h is known, for example $x = D$, $h = h_D$, we can use the following equation to evaluate the parameter combination Q/Kb since:

$$\frac{Q}{Kb} = \frac{h_0 - h_D}{D} \qquad \text{eq. 6}$$

If we know Kb, we can find Q, or vice versa.

As a further example, the following groundwater flow problem provides an analytical solution to the situation of a confined aquifer receiving constant recharge, or leakage. A conceptualization of the problem is shown in Fig. 2 and it should be noted that recharge, W, at the upper boundary of the aquifer is assumed to be constant everywhere.

Between $x = 0$ and $x = L$, continuity and flow equations can be written as:

$$Q = W(L - x) \qquad \text{eq. 7}$$

Section (x,z plane)

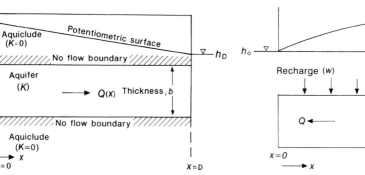

Fig. 1 Definition sketch of steady flow through a uniform thickness, homogeneous confined aquifer.

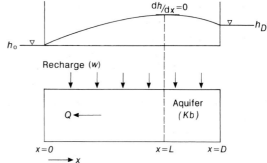

Fig. 2 Definition sketch of steady flow through a uniform thickness, homogeneous confined aquifer with constant recharge.

$$Q = Kb\frac{dh}{dx} \qquad \text{eq. 8}$$

and between $x = L$ and $x = D$:

$$Q = W(x - L) \qquad \text{eq. 9}$$

$$Q = -Kb\frac{dh}{dx} \qquad \text{eq. 10}$$

By combining the first pair of equations (eqs 7 and 8) between $x = 0$ and $x = L$ and integrating:

$$Kb\frac{dh}{dx} = W(L - x)$$

$$\int dh = \int \frac{W}{Kb}(L - x)dx$$

$$\therefore h = \frac{W}{Kb}\left(Lx - \frac{x^2}{2}\right) + c \qquad \text{eq. 11}$$

for the boundary condition $x = 0$, $h = h_0$, then the constant of integration $c = h_0$ and the following partial solution is found:

$$h - h_0 = \frac{W}{Kb}\left(Lx - \frac{x^2}{2}\right) \qquad \text{eq. 12}$$

Note that in equation 12 the position of the divide, L, appears as a parameter. We can eliminate L by invoking a second boundary condition, $x = D$, $h = h_D$ and rearranging equation 12 such that:

$$L = \frac{Kb}{W}\frac{(h_D - h_0)}{D} + \frac{D}{2} \qquad \text{eq. 13}$$

By substituting L found from equation 13 into equation 12 we obtain the specific solution:

$$h - h_0 = x\frac{(h_D - h_0)}{D} + \frac{W}{2Kb}(Dx - x^2) \qquad \text{eq. 14}$$

Thus, the potentiometric surface in a confined aquifer with constant transmissivity and constant recharge is described by a parabola as shown in Fig. 2.

To obtain a solution to simple one-dimensional groundwater flow problems in unconfined aquifers, it is necessary to adopt the Dupuit assumptions that in any vertical section the flow is horizontal, the flow is uniform over the depth of flow and the flow velocities are proportional to the slope of the water table and the saturated depth. The last assumption is reasonable for small slopes of the

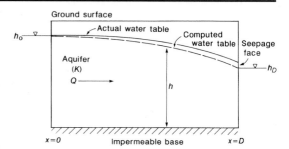

Fig. 3 Definition sketch of steady flow in a homogeneous unconfined aquifer between two water bodies with vertical boundaries.

water table. From a consideration of flow and continuity, and with reference to Fig. 3, the discharge per unit width, Q, for any given vertical section is found from:

$$Q = -Kh\frac{dh}{dx} \qquad \text{eq. 15}$$

and integrating:

$$Qx = -\frac{K}{2}h^2 + c \qquad \text{eq. 16}$$

For the boundary condition $x = 0$, $h = h_0$, we obtain the following specific solution known as the Dupuit equation:

$$Q = \frac{K}{2x}(h_0^2 - h^2) \qquad \text{eq. 17}$$

The Dupuit equation therefore predicts that the water table is a parabolic shape. In the direction of flow, the curvature of the water table, as predicted by equation 17, increases. As a consequence, the two Dupuit assumptions become poor approximations to the actual groundwater flow and the actual water table increasingly deviates from the computed position as shown in Fig. 3. The reason for this difference is that the Dupuit flows are all assumed horizontal whereas the actual velocities of the same magnitude have a vertical downward component so that a greater saturated thickness is required for the same discharge. As indicated in Fig. 3, the water table actually approaches the right-hand boundary tangentially above the water body surface and forms a seepage face. However, this discrepancy aside, the Dupuit equation accurately determines heads for given values of boundary heads, Q and K.

As a final example, Fig. 4 shows steady flow to two parallel stream channels from an unconfined aquifer with continuous recharge applied uniformly over the aquifer. The stream channels are idealized as two long parallel streams completely penetrating

BOX
2 . 9

Continued

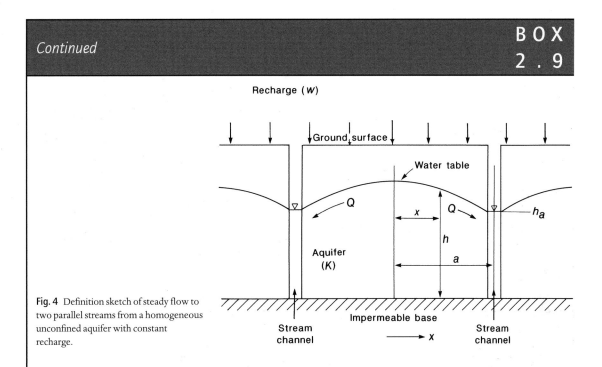

Recharge (*W*)

Fig. 4 Definition sketch of steady flow to two parallel streams from a homogeneous unconfined aquifer with constant recharge.

the aquifer. Adopting the Dupuit assumptions, the flow per unit width is given by equation 15 and from continuity:

$$Q = Wx \qquad \text{eq. 18}$$

By combining equations 15 and 18, then integrating and setting the boundary condition $x = a$, $h = h_a$, the following specific solution is obtained:

$$h^2 = h_a^2 + \frac{W}{K}(a^2 - x^2) \qquad \text{eq. 19}$$

From symmetry and continuity, the baseflow entering each stream per unit width of channel is equal to $2aW$. If h is known at any point, then the baseflow or recharge rate can be computed provided the hydraulic conductivity is known. This type of analysis has been applied to the design of parallel drains on agricultural soils to calculate the necessary spacing for specified soil, crop and irrigation conditions. A reappraisal of techniques used to analyse the horizontal flow of water to tile drains that also separate the external boundary of the water table from the internal boundary of the tile drain is given by Khan and Rushton (1996).

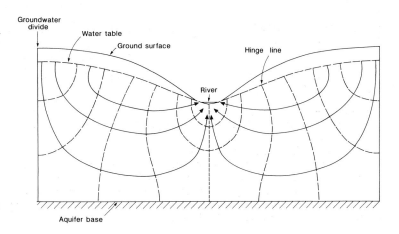

Fig. 2.25 Groundwater flow net for a two-dimensional vertical cross-section through a homogeneous, isotropic aquifer with a symmetrical valley topography.

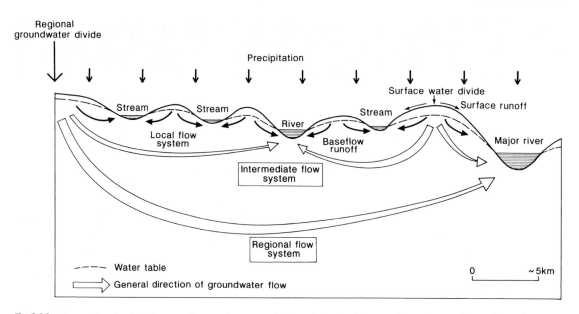

Fig. 2.26 Diagram showing the influence of hummocky topography in producing local, intermediate and regional groundwater flow systems. After Tòth (1963).

is superimposed on the incline to represent hummocky topography. Freeze and Witherspoon (1967) developed this mathematical approach further by employing numerical simulations to examine the effects of topography and geology on the nature of regional groundwater flow patterns.

As illustrated in Fig. 2.26, Tòth (1963) proposed that it is possible to differentiate between local, intermediate and regional groundwater flow systems. Tòth (1963) further showed that as the ratio of depth to lateral extent of the entire aquifer system becomes smaller and as the amplitude of the hummocks becomes larger, the local flow systems are more likely to reach the aquifer base, thus creating a series of small groundwater flow cells. In general, where the local topography is subdued, only regional systems develop compared to areas of pronounced local relief where only local flow systems develop. Groundwater flow in the local and intermediate systems moves relatively quickly along short flowpaths and discharges as baseflow to streams at the local scale and rivers at the intermediate scale. The regional component of groundwater flow has a relatively long residence time and follows long flowpaths before discharging to major rivers. Field examples

of regional-scale and continental-scale groundwater flow systems are given in Boxes 2.10 and 2.11 and illustrate the combined influence of geology and topography on groundwater flow.

2.14 Groundwater and geology

The occurrence of groundwater and the extent and distribution of aquifers and aquitards in a region are determined by the lithology, stratigraphy and structure of the geological strata present. The lithology refers to the general characteristics of the geological strata in terms of mineral composition and texture of the formations present. The stratigraphy describes the character of the rocks and their sequence in time as well as the relationship between various deposits in different localities. Structural features, such as folds and faults, determine the geometric properties of the formations that are produced by deformation and fracturing after deposition or crystallization. In unconsolidated strata, the lithology and stratigraphy comprise the most important controls.

In any hydrogeological investigation, a clear understanding of the geology of an area is essential if

BOX
2.10

Regional groundwater flow in the Lincolnshire Limestone aquifer, England

The Lincolnshire Limestone aquifer in eastern England is part of a relatively uniform stratigraphical succession of Jurassic strata that dip 1–2° to the east. The pattern of recharge and groundwater flow in the aquifer is influenced by various geological and hydrological controls as shown in the schematic cross-section of Fig. 1. The Upper Lias Clay forms an effective impermeable base to the limestone aquifer. In the west of the section, the limestone is 6–8 km wide at outcrop. Formations overlying the limestone have a significant effect on surface water and groundwater flows since they form an alternating sequence of limestone aquifer units and confining beds of low permeability consisting of clays, shales and marls. The uppermost deposits of the confining strata include the Oxford Clay that extends to the east of the area where the continuation of the limestone below the Lincolnshire Fens is uncertain. Quaternary glacial deposits consisting mainly of boulder clay (glacial till) and sands and gravels occur in the west of the area and also in west–east drainage channels.

From a consideration of catchment water balances (see Section 8.2.1) and the use of numerical modelling, Rushton and Tomlinson (1999) showed that recharge to the Lincolnshire Limestone aquifer occurs as direct recharge in the western outcrop area and as runoff recharge and downward leakage from the low permeability boulder clay and confining strata. The boulder clay produces large volumes of runoff which commonly recharges the limestone through the many swallow holes that occur at the edge of the boulder clay. In places, the West Glen River and East Glen River incise the confining strata and provide a mechanism for localized groundwater discharge from the limestone aquifer. Further east, the groundwater potentiometric surface is above the top of the aquifer, with overflowing artesian conditions developed in the Fens. In this region of the aquifer, slow upward leakage of the regionally extensive groundwater flow system is expected, although the large volumes of groundwater abstracted in recent years for public water supply has disrupted the natural flow patterns and drawn modern recharge water further into the confined aquifer.

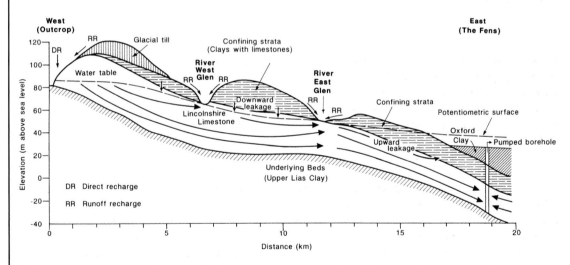

Fig. 1 Schematic cross-section of the Lincolnshire Limestone aquifer in eastern England showing mechanisms of groundwater recharge and directions of local and regional groundwater flow. After Rushton and Tomlinson (1999).

the identification of aquifers and aquitards and the mechanisms of groundwater flow are to be properly understood. To illustrate the range of geological conditions in which groundwater resources can occur, the following description is based on the wide variation in groundwater occurrence in the United Kingdom.

2.14.1 Hydrogeological environments of the United Kingdom

The United Kingdom is fortunate in the variety of its rocks, structures and natural resources. As shown in Fig. 2.27, the principal groundwater resources are located in the Midlands and south-east of England where the Cretaceous Chalk, Permo-Triassic sandstones and the Jurassic limestones contribute one-third of abstracted water supplies, of which half is reliant on the Chalk aquifer. Unlike many other countries, these important water supplies are dependent on fissure flow in making the limestones and sandstones permeable. Aquifers also exist in the older rocks such as the Carboniferous limestones and in more recent formations such as the Pleistocene sands and gravels, but these aquifers are not of such regional significance. Although less important than surface water sources, the Precambrian and Palaeozoic rocks in the remoter areas of Ireland, Scotland and Wales have sufficient storage to be of local importance for domestic supplies and in supporting baseflows to minor rivers.

The hydrogeology of England and Wales, Scotland and Northern Ireland is presented in a number of maps produced by the Institute of Geological Sciences (1977) and the British Geological Survey (1988, 1994), respectively, with the hydrogeology of Scotland documented in a memoir (Robins 1990). In

---

Large-scale groundwater flow in the Great Artesian Basin, Australia

BOX 2.11

The Great Artesian Basin of Australia covers an area of 1.7×10^6 km^2 or about one-fifth of the continent (Fig. 1). The Basin underlies arid and semi-arid regions where surface water is sparse and unreliable. Discovery of the Basin's artesian groundwater resources in 1878 made settlement possible and led to the development of an important pastoral industry. Farming and public water supplies, and increasingly the mining and petroleum industries, are largely or totally dependent on the Basin's artesian groundwater (Habermehl 1980; Habermehl & Lau 1997).

The Basin consists of a multilayered confined aquifer system, with aquifers occurring in continental quartzose sandstones of Triassic, Jurassic and Cretaceous age and ranging in thickness from several metres to several hundreds of metres. The intervening confining beds consist of siltstone and mudstone, with the main confining unit formed by a sequence of argillaceous Cretaceous sediments of marine origin (Fig. 2). Basin sediments are up to 3000 m thick and form a large synclinal structure, uplifted and exposed along its eastern margin and tilted towards the south-west (Fig. 3).

Fig. 1 Map of the location and extent of the Great Artesian Basin, Australia. After Habermehl (1980).

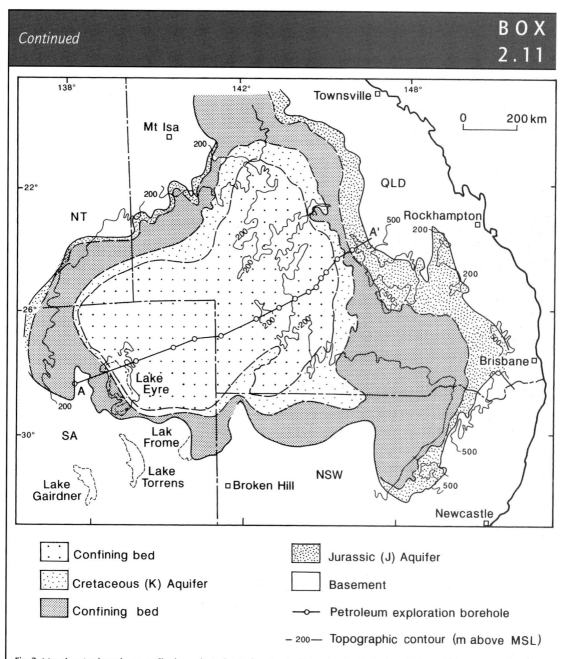

Fig. 2 Map showing lateral extent of hydrogeological units forming the Great Artesian Basin. After Habermehl (1980).

Groundwater recharge occurs mainly in the eastern marginal zone on the western slope of the Great Dividing Range in an area of relatively high rainfall (median annual rainfall >500 mm) and large-scale regional groundwater movement is predominantly towards the southwestern, western and southern discharge margins, as well as a component northwards (Fig. 4). Recharge also occurs along the western margin of the Basin and in the arid centre of the continent with groundwater flow towards the southwestern discharge margin. Natural groundwater discharge occurs as groups of springs (Fig. 4) that are commonly characterized by conical mounds

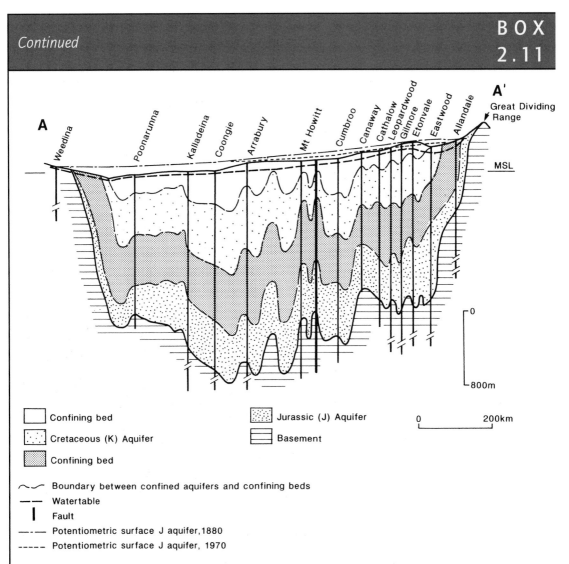

Fig. 3 Simplified cross-section of the Great Artesian Basin showing the position of major aquifer and confining units (see Fig. 2 for location of section A–A'). After Habermehl (1980).

of sand and silt-size sediment and carbonate which range in height and diameter from a few metres to tens of metres, some with water-filled craters (Habermehl 1980). The location of many springs appears to be fault-controlled with others present where aquifers abut low permeability basement rocks or where only thin confining beds are present. Diffuse discharge also occurs from the artesian aquifers near the Basin margins where the overlying confining beds are thin.

The potentiometric surface of the Triassic, Jurassic and Lower Cretaceous confined aquifers is still above ground level in most areas (Fig. 3) despite considerable lowering of heads due to extensive

groundwater development since the 1880s. Groundwater levels in the confined aquifer in the upper part of the Cretaceous sequence have always been below ground level throughout most of the Basin area and development of these non-flowing artesian aquifers requires the installation of pumping equipment. Transmissivity values of the main aquifers in the Lower Cretaceous–Jurassic sequence, from which most overflowing artesian groundwater is obtained, are typically several tens to several hundreds of m^2 day^{-1}. Average groundwater flow rates in the eastern and western parts of the Basin range from 1 to 5 m a^{-1} based on hydraulic data and radiometric dating using carbon-14 and chlorine-36 radioisotopes,

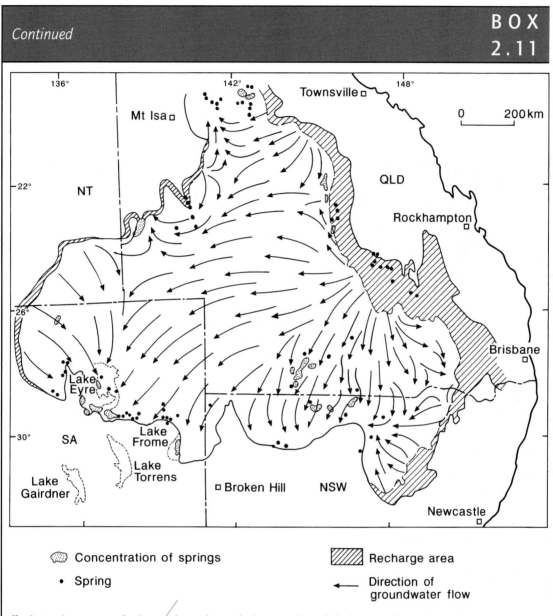

Fig. 4 Map showing areas of recharge and natural spring discharge together with the directions of regional groundwater flow. After Habermehl (1980).

and yield residence times from several thousands of years near the marginal recharge areas to more than one million years near the centre of the Basin (for further discussion see Section 4.4.3 and Bentley et al. 1986; Love et al. 2000).

Groundwater in the most widely exploited artesian aquifers in the Lower Cretaceous–Jurassic sequence generally contains between 500 and 1000 mg L^{-1} of total dissolved solids, predominantly

as sodium and bicarbonate ions. Water quality improves with increasing depth of aquifers in the sequence and on the whole the groundwater is suitable for domestic and stock use, although it is generally unsuitable for irrigation use due to the high sodium concentration and its chemical incompatibility with the montmorillonite clay soils. Groundwater from aquifers in the upper part of the Cretaceous sequence has a higher salinity.

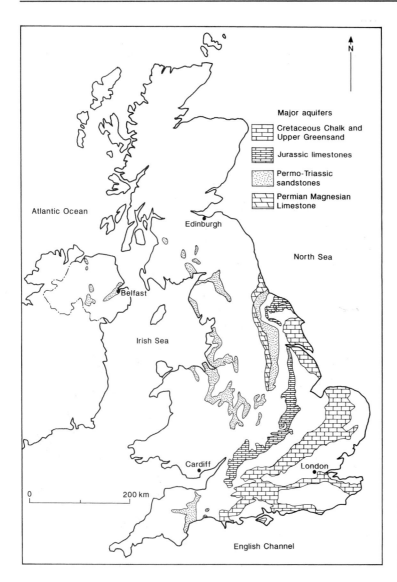

Fig. 2.27 Hydrogeological map of the United Kingdom showing the location of major aquifers. After CEH (1998) http://www.nwl.ac.uk/ih/nrfa/groundwater/figure11_1998.htm.

addition, Maps B3 Edinburgh and B4 London from the International Hydrogeological Map of Europe series cover the British Isles (UNESCO 1976, 1980). In the following sections, the hydrogeological environments of the British Isles are described in terms of the three major rock types (sedimentary, metamorphic and igneous) and their associated aquifer properties. To help in locating rock types, Fig. 2.28 shows the geological map of Britain and Ireland and a geological timescale is provided in Appendix 3.

Sedimentary rocks

The extensive sedimentary rocks range from unconsolidated Quaternary deposits to ancient, highly indurated Late Precambrian sandstones and siltstones. The most prolific aquifers are associated with the Mesozoic sandstones and limestones. The groundwater resources of the principal Mesozoic aquifers in England and Wales derived from the infiltration of rainfall amount to 7300 km^3 a^{-1} (Table 2.4) of which

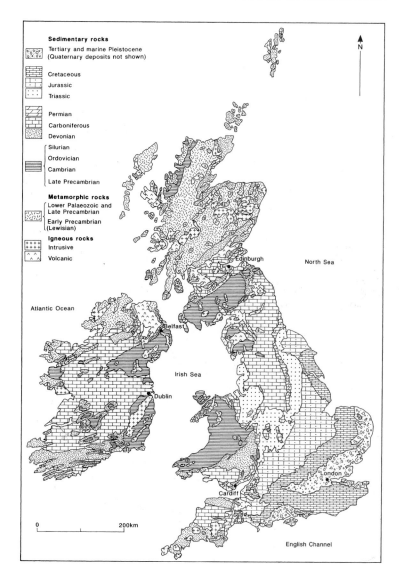

Fig. 2.28 Geological map of Britain and Ireland. After Geological Museum (1978), South Kensington, London.

63% is from the Cretaceous Chalk and 20% from the Permo-Triassic sandstones.

Recent coastal dune sands and raised beach deposits are restricted in distribution but provide limited supplies to individual farm and domestic users. Riverine alluvium occurs along many valley bottoms and includes alluvial fans, deltas, lake and estuarine deposits. Alluvial deposits include fine-grained sands, silts and clays with the presence of lenses of gravels and cobbles. Together with glacial sands and gravels, these superficial deposits are of significance as locally

important aquifers in the hard rock areas of Ireland, Scotland and Wales (Fig. 2.29). In Scotland, glacial sands and gravels form terraced and gently sloping hillocky ground with the groundwater potential dependent on the extent and thickness of the saturated material. Borehole yields of up to 10^{-2} m^3 s^{-1} can typically be achieved in these deposits.

The fine, largely unconsolidated Pleistocene shelly marine sands and silts of eastern England form the regionally important Crag aquifer. The Crag can attain a thickness of 80 m representing a single water-bearing

Table 2.4 Groundwater resources of the principal Mesozoic aquifers in England and Wales (in km^3 a^{-1}). Abstraction data are for the year 1977. After Downing (1993).

Aquifer	Infiltration (I)	Abstraction (A)	Balance	A/I
Cretaceous Chalk	4631	1255	3376	0.27
Cretaceous Lower Greensand	275	86	189	0.31
Jurassic Lincolnshire Limestone	86	43	43	0.50
Middle Jurassic limestones	627	65	562	0.10
Permo-Triassic sandstones	1443	587	856	0.41
Permian Magnesian Limestone	247	41	206	0.17
Total	7309	2077	5232	0.28

Fig. 2.29 Site of a groundwater source developed in alluvial deposits in the Rheidol Valley, west Wales. In general, the well-sorted fluvial and glaciofluvial sands and gravels reach a thickness of 30 m and are exploited for locally important groundwater supplies. At this site, Lovesgrove, transmissivities are 0.05–0.07 m^2 s^{-1} (4000–6000 m^2 day^{-1}), specific yield about 5% and sustainable yield 0.05 m^3 s^{-1}. The number 1 borehole is positioned below the top of the concrete chamber shown at left and reaches a depth of 30 m through a sequence of river gravels (Hiscock & Paci 2000).

unit with overlying glacial sands and can yield supplies of 10^{-2} m^3 s^{-1}. Elsewhere in south-east England, Tertiary strata form a variable series of clays, marls and sands ranging in thickness from 30 to 300 m. The Eocene London Clay, up to 150 m thick, is an important confining unit in the London Basin. The underlying Lower London Tertiaries include clays, fine sands and pebble beds and where permeable sands rest on the underlying Chalk, these Basal Sands are generally in hydraulic continuity and can yield small supplies.

The Cretaceous Period resulted in the transgression of shallow, warm tropical seas and the deposition of

the Chalk Formation, the most important source of groundwater in the south and east of England (Fig. 2.30) (Downing et al. 1993). The Chalk is a pure, white, microporous limestone made up of minute calcareous shells and shell fragments of plankton together with bands of harder nodular chalk and flints, marly in the lower part. In total the Chalk is up to 500 m in thickness. The intrinsic permeability of the Chalk matrix is low, so that good yields of typically 10^{-1} m^3 s^{-1} depend on the intersection of fissures and fractures, solutionally developed along bedding plains and joints. The permeability is best developed in the upper 80–100 m in the zone of greatest secondary permeability development. Below this level the fissures are infrequent and closed by the overburden pressure and the groundwater becomes increasingly saline. In Northern Ireland, a hard microporous and fissured Chalk is found with recrystallized calcite partly infilling pore spaces. The Chalk attains a maximum thickness of only 150 m and is largely covered by Tertiary basalt lavas. Recharge via the lavas supports numerous springs at the base of the outcrop along the Antrim coast. Borehole yields from the Chalk beneath the lavas are typically less than 10^{-3} m^3 s^{-1} and the number of boreholes is few. Beneath the Chalk, the Lower Cretaceous glauconitic and ferruginous sands and sandstones of the Upper and Lower Greensand Formations and the alternating sequence of sandstones and clays of the Hastings Beds occur and form locally important aquifers in southern England.

The Jurassic Period also resulted in the formation of important limestone aquifer units, namely the Corallian and Lincolnshire Limestone aquifers. The Corallian is well developed in Yorkshire where well-

Fig. 2.30 Cretaceous Chalk outcrop at Hunstanton in north-west Norfolk. Unusually at this location, the hard, well-fissured lower Chalk passes into the highly fossiliferous, iron-rich, red-stained Chalk known as the Red Chalk (now Hunstanton Formation). Below is the Lower Cretaceous Carstone Formation of medium to coarse, pebbly, glauconitic, quartz sand stained brown by a limonitic cement.

Fig. 2.31 The Lincolnshire Limestone at Clipsham Quarry, Lincolnshire. In this exposure, which is approximately 12 m high, solutional weathering of bedding surfaces and vertical joints imparts a high secondary permeability. The pale centre in the block at the top right-hand side is an example of the heart stone of unoxidized limestone. Above is the grey-green clay of the Rutland Formation that acts as an aquitard above the limestone aquifer.

Fig. 2.32 An exposed joint surface on a block of Lincolnshire Limestone showing a scallop-like feature caused by groundwater flow. The dark area to the right is a large vertical conduit formed by solutional weathering of a joint normal to the exposed face.

jointed oolitic limestones and grits are found up to 110 m thick. Southwards the formation thins to 20 m and is replaced by clay before reappearing in the Cotswolds, up to 40 m thick, giving yields of 10^{-3} to 10^{-2} m^3 s^{-1}. The Great Oolite and Inferior Oolite limestones of Central England and the Cotswolds are part of a variable group of limestones, clays and sands, up to 60 m thick that can yield copious supplies. The Inferior Oolite Lincolnshire Limestone aquifer is partly karstic in nature with rapid groundwater flow through conduits (Figs 2.31 & 2.32). The highly developed secondary permeability supports yields of 10^{-1} m^3 s^{-1}. Beneath the limestones, the

Upper Lias sands form locally important aquifers such as the Midford Sands close to Bath in the west of England.

The Permo-Triassic sandstones, marls and conglomerates comprise an extensive sequence up to 600 m thick and form deep sedimentary basins in the Midlands and North West of England and a smaller basin typically over 100 m thick in south-west England (Fig. 2.33). The red sandstones originated in

Fig. 2.33 Exposure of Triassic Sherwood Sandstone (Otter Sandstone Formation) at Ladram Bay, South Devon. The Otter Sandstone Formation comprises predominantly fine- to medium-grained red-brown, micaceous, variably cemented, ferruginous sands and sandstones, with occasional thin silt and conglomerate lenses. Frequent, well-developed bedding planes show a gentle 2–4° east-south-east dip, with, as shown in this cliff face, cross-bedding also typically present. Cementation along bedding planes exerts a small but potentially significant control in promoting horizontal groundwater movement (Walton 1981). Two prominent fissure openings along cemented bedding planes are visible to the left, in the lower one-third and upper two-thirds, of the face shown.

a desert environment and much of the fine- to medium-grained cross-bedded sandstones are soft, compact rock that is only weakly cemented. Groundwater can flow through the intergranular matrix but the presence of fractures enhances the permeability considerably giving good yields of up to 10^{-1} m^3 s^{-1} of good quality. Above the Triassic sandstones, the Mercia Mudstone confines the underlying aquifer although local supplies are possible from minor sandstone intercalations. Up to 300 m of Triassic sandstone occurs in the Lagan Valley and around Newtownards in Northern Ireland where it is intruded by many basalt dykes and sills with yields of up to 10^{-2} m^3 s^{-1} obtained from a fine- to medium-grained sandstone. Smaller isolated Permian sands and sandstones are found in north-west England and south-west Scotland. At the base of the Permian, and overlain by red marls, the Magnesian Limestone forms a sequence of massive dolomitic and reef limestones that are important for water supply in the north-east of England where typical yields range up to 10^{-2} m^3 s^{-1}.

The Carboniferous strata include the massive, well-fissured karstic limestones that give large supplies of up to 10^{-1} m^3 s^{-1} from springs in the Mendip Hills (see Section 2.7), South Wales and, to a lesser extent, Northern Ireland. Later, rhythmic sequences of massive grits, sandstone, limestone, shale and coals produce minor supplies from fissured horizons in sandstone and limestone and form aquifers of local significance, particularly from the Yoredale Series in northern England and in the Midland Valley of Scotland.

The Old Red Sandstone is the principal aquifer unit of the Devonian Period and includes sandstones, marls and conglomerates that yield small supplies from sandstones in the Welsh borders. In Scotland, the Upper Old Red Sandstone is of much greater significance and around Fife and the southern flank of the Moray Firth consists of fine- to medium-grained sandstones, subordinate mudstones and conglomerates with good intergranular permeability yielding supplies of up to 10^{-2} m^3 s^{-1}. Widespread outcrops of fine- to medium-grained Lower and Middle Old Red Sandstones, in places flaggy, with siltstones, mudstones and conglomerates as well as interbedded lavas give borehole yields ranging from 10^{-3} m^3 s^{-1} in the Borders to 10^{-2} m^3 s^{-1} in Ayrshire and parts of Strathmore. In Northern Ireland the principal Devonian lithology is conglomerate with some sandstone, subordinate mudstone and volcanic rocks but these are indurated and poorly jointed with small borehole yields of 10^{-3} m^3 s^{-1} where secondary permeability is present.

Silurian, Ordovician and other Lower Palaeozoic and Late Precambrian sedimentary facies predominate in southern Scotland, the north-east and north-west of Ireland and Wales and consist of great thicknesses of highly indurated and tectonically deformed shales, mudstones, slates and some limestones and sandstones (Fig. 2.34). Some groundwater may occur in shallow cracks and joints that produce a subsurface permeable zone in which perched water tables may occur that support occasional springs and shallow boreholes providing small yields.

Metamorphic rocks

The Lower Palaeozoic and Precambrian crystalline basement rocks of the Highlands and Islands of

Fig. 2.34 Silurian shales at outcrop in the Rheidol Valley, west Wales showing the grey mudstones. Secondary permeability is developed in the weathered upper horizon from which shallow well supplies can be obtained as well as from occasional fracture openings (as seen at lower left) that intersect drilled boreholes.

Scotland and north-west Ireland offer little potential for groundwater storage and flow other than in cracks and joints that may be associated with tectonic features or near-surface weathering. Available groundwater can support a yield up to 10^{-3} m^3 s^{-1}.

Igneous rocks

Groundwater flow in well-indurated igneous intrusive and extrusive rocks occurs in shallow cracks and joints opened by weathering. Yields of occasional springs are small except where tectonic influences have enhanced the secondary porosity (isolated yields of up to 10^{-3} m^3 s^{-1} are typical). Tertiary basalts up to 800 m thick occur in the Antrim and Lough Neagh areas of Northern Ireland. The basalts have some primary permeability in weathered zones but the principal, secondary permeability is developed in joints and fissures which provide sustainable borehole yields of only 10^{-4} to 10^{-3} m^3 s^{-1}.

2.15 FURTHER READING

Albu, M., Banks, D. & Nash, H. (1997) *Mineral and Thermal Groundwater Resources*. Chapman & Hall, London.

Back, W., Rosenshein, J.S. & Seaber, P.R. (eds) (1988) *Hydrogeology. The geology of North America*, vol. O-2. The Geological Society of North America, Boulder, Colorado.

Cripps, J.C., Bell, F.G. & Culshaw, M.G. (eds) (1986) Groundwater in engineering geology. In: *Proceedings of the 21st Annual Conference of the Engineering Group of the Geological Society*. The Geological Society of London.

Domenico, P.A. & Schwartz, F.W. (1998) *Physical and Chemical Hydrogeology*, 2nd edn. John Wiley, New York.

Downing, R.A., Price, M. & Jones, G.P. (1993) *The Hydrogeology of the Chalk of north-west Europe*. Clarendon Press, Oxford.

Fetter, C.W. (2001) *Applied Hydrogeology*, 4th edn. Pearson Higher Education, New Jersey.

Freeze, R.A. & Cherry, J.A. (1979) *Groundwater*. Prentice-Hall, Englewood Cliffs, New Jersey.

Robins, N.S. & Misstear, B.D.R. (2000) *Groundwater in the Celtic Regions: studies in hard rock and quaternary hydrogeology*. Geological Society, London, Special Publications **182**.

Rushton, K.R. (2003) *Groundwater Hydrology: conceptual and computational models*. John Wiley, Chichester.

Chemical hydrogeology

3

3.1 Introduction

The study of groundwater chemistry, or hydrochemistry, is useful in hydrogeology in a number of ways. Interpretation of the distribution of hydrochemical parameters in groundwater can help in the understanding of hydrogeological conditions and can also aid decisions relating to the quality of water intended for drinking water. Hydrochemical processes are also significant in attenuating groundwater contaminants. In this chapter, the major hydrochemical processes of importance in groundwater are introduced. Interpretation techniques for combining data and defining hydrochemical types are also discussed as part of an integrated approach to understanding groundwater flow mechanisms.

3.2 Properties of water

The chemical structure of water is illustrated in Fig. 3.1, which shows one oxygen atom bonded asymmetrically to two hydrogen ions with a bond angle of 105°. The shape results from the geometry of the electron orbits involved in the bonding. Oxygen has a much higher electronegativity (a measure of the tendency of an atom to attract an additional electron) than hydrogen and pulls the bonding electrons towards itself and away from the hydrogen atom. The oxygen thus carries a partial negative charge (usually expressed as δ−), and each hydrogen a partial positive charge (δ+), creating a dipole, or electrical charges of equal magnitude and opposite sign a small distance apart. As a consequence, the opposite charges of water molecules attract each other to form clusters of molecules, through a type of interaction

known as hydrogen bonding. The size of the clusters increases with decreasing temperature reaching a maximum at 4°C. When water is cooled from 4°C to 0°C the size of the clusters creates a more open structure and the water becomes less dense, with further expansion on freezing. Hence, ice has a lower density than liquid water. Values for water density, viscosity, vapour pressure and surface tension over a temperature range of 0–100°C are given in Appendix 2.

As illustrated in Table 3.1, water is not simply H_2O but rather a mixture of six molecules depending on the hydrogen and oxygen isotopes that combine to form the water molecule. Eighteen combinations are possible, the most common of which is $^1H_2{}^{16}O$. Pure water contains hydrogen and oxygen in ionic form as well as in the combined molecular form. The ions are formed when water dissociates as follows:

$$H_2O \rightleftharpoons H^+ + OH^- \qquad \text{eq. 3.1}$$

The H^+ ion is normally in the form H_3O^+ (the hydronium ion) and in rock–water interactions, the transfer of protons (H^+ ions) between the liquid and solid

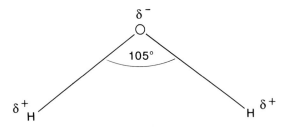

Fig. 3.1 The structure of the water molecule showing the dipole created by the partial negative charge on the oxygen atom and partial positive charge on each hydrogen atom.

Table 3.1 The relative abundance of hydrogen and oxygen isotopes in the water molecule.

Isotope		Relative abundance	Type
^1H	Proteum	99.984	Stable
^2H	Deuterium	0.016	Stable
^3H	Tritium	$0-10^{-15}$	Radioactive*
^{16}O	Oxygen	99.76	Stable
^{17}O	Oxygen	0.04	Stable
^{18}O	Oxygen	0.20	Stable

* Half-life = 12.3 years.

phases is known as proton transfer, an important consideration in carbonate chemistry.

The polarity of the water molecule makes water an effective solvent for ions; the water molecules are attracted to the ions by electrostatic forces to form a cluster with either oxygen or hydrogen oriented towards the ions as shown in Fig. 3.2. This phenomenon is known as hydration and acts to stabilize the solution. Polar solvents such as water easily dissolve crystalline solids like sodium chloride (NaCl) and break down the ionic crystal into a solution of separately charged ions:

$$Na^+Cl^- \overset{H_2O}{\rightleftharpoons} Na^+_{(aq)} + Cl^-_{(aq)} \qquad \text{eq. 3.2}$$

Positively charged atoms like sodium are known as cations, while negatively charged ions like chloride are called anions. When writing chemical equations, the sum of charges on one side of the equation must balance the sum of charges on the other side. On the left-hand side of equation 3.2, NaCl is an electrically neutral compound, while on the right-hand side, the aqueous sodium and chloride ions each carry a single but opposite charge so that the charges cancel, or balance, each other.

The degree of hydration increases with increasing electrical charge of the dissociated ion and also with decreasing ionic radius. Because cations are generally smaller than anions, cations are usually more strongly hydrated. The effect of hydration is to increase the size of an ion and so reduce its mobility and affect the rates of chemical reaction. In addition, the complexation of cations, particularly the transition metals such as iron, copper, zinc, cobalt and chromium, where the ions chemically bond with water molecules, leads to the creation of complex ions of fixed composition and great chemical stability.

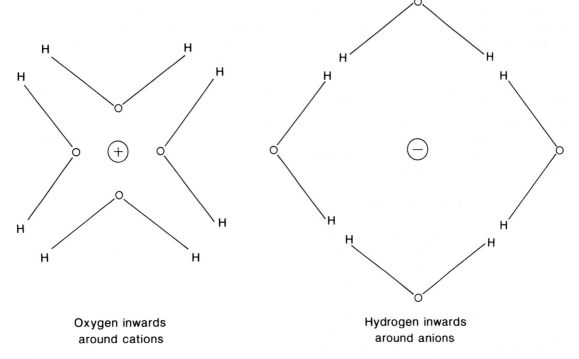

Oxygen inwards
around cations

Hydrogen inwards
around anions

Fig. 3.2 Water molecules surrounding a positively charged cation and a negatively charged anion to form hydrated ions.

3.3 Chemical composition of groundwater

The chemical and biochemical interactions between groundwater and the geological materials of soils and rocks provide a wide variety of dissolved inorganic and organic constituents. Other important considerations include the varying composition of rainfall and atmospheric dry deposition over groundwater recharge areas, the modification of atmospheric inputs by evapotranspiration, differential uptake by biological processes in the soil zone and mixing with seawater in coastal areas. As shown in Table 3.2, and in common with freshwaters in the terrestrial aquatic environment, the principal dissolved components of groundwater are the six major ions sodium (Na^+), calcium (Ca^{2+}), magnesium (Mg^{2+}), chloride (Cl^-), bicarbonate (HCO_3^-) and sulphate (SO_4^{2-}). These cations and anions normally comprise over 90% of the total dissolved solids content, regardless of whether the water is dilute rainwater or has a salinity greater than seawater (typical analyses of rainwater and seawater are given in Appendix 5). Minor ions include potassium (K^+), dissolved iron (Fe^{2+}), strontium (Sr^{2+}) and fluoride (F^-) while aqueous solutions commonly also contain amounts of trace elements and metal species. The introduction of contaminants into groundwater from human activities can result in some normally minor ions reaching concentrations equivalent to major ions. An example is nitrate, as excessive application of nitrogenous fertilizers can raise nitrate concentrations in

---

Concentration units used in hydrochemistry

BOX 3.1

Concentration is a measure of the relative amount of *solute* (the dissolved inorganic or organic constituent) to the *solvent* (water). A list of atomic weights is supplied in Appendix 4.

There are various types of concentration unit as follows:

Molarity (M): number of moles of solute dissolved in 1 L of solution (mol L^{-1}). For example, if we have 10 grammes of potassium nitrate (molar mass of $KNO_3 = 101$ grammes per mole) then this is $(10\,g)/(101\,g\,mol^{-1}) = 0.10$ moles of KNO_3. If we place this in a flask and add water until the total volume = 1 L we would then have a 0.1 molar solution. Molarity is usually denoted by a capital M, for example a 0.10 M solution. It is important to recognize that molarity is moles of solute per litre of solution, not per litre of solvent, and that molarity changes slightly with temperature because the volume of a solution changes with temperature.

Molality (m): number of moles of solute dissolved in 1 kg of solvent (mol kg^{-1}). Notice that, compared with molarity, molality uses mass rather than volume and uses solvent instead of solution. Unlike molarity, molality is independent of temperature because mass does not change with temperature. If we were to place 10 g of KNO_3 (0.10 moles) in a flask and then add one kilogramme of water we would have a 0.50 molal solution. Molality is usually denoted with a small m, for example a 0.10 m solution.

Mass concentration: this unit of concentration is often used to express the concentration of very dilute solutions in units of parts per million (ppm) or, more commonly, mg L^{-1}. Since the amount of solute relative to the amount of solvent is typically very small, the density of the solution is approximately the same as the density of the solvent. For this reason, parts per million may be expressed in the following two ways:

ppm = mg of solute/L of solution

ppm = mg of solute/kg solution

Chemical equivalence: the concept of chemical equivalence takes into account ionic charge and is useful when investigating the proportions in which substances react. This aspect of chemistry is called **stoichiometry**.

Equivalents per litre (eq L^{-1}) = number of moles of solute multiplied by the valence of the solute in 1 L of solution. From this it follows that meq L^{-1} = mg L^{-1} × (valence/atomic weight).

As an example of the application of chemical equivalence, take the effect of ion exchange when a fresh groundwater in contact with a rock is able to exchange a chemically equivalent amount of calcium (a divalent cation with atomic weight = 40 g) with sodium (a monovalent cation with atomic weight = 23 g) contained within the aquifer. If the groundwater has an initial calcium concentration of 125 mg L^{-1} and sodium concentration = 15 mg L^{-1}, what will be the new groundwater sodium concentration if all the calcium were exchanged with the clay material?

Initial calcium concentration = 125 mg L^{-1} × (2/40)
= 6.25 meq L^{-1}

Initial sodium concentration = 12 mg L^{-1} × (1/23)
= 0.52 meq L^{-1}

New sodium concentration after ion exchange = 6.25 + 0.52
= 6.77 meq L^{-1}

= 6.77 × (23/1) = 155.71 mg L^{-1}

Therefore, the extra sodium contributed to the groundwater by ion exchange is ~141 mg L^{-1}.

Table 3.2 Chemical composition of groundwater divided into major and minor ions, trace constituents and dissolved gases. After Freeze and Cherry (1979).

Groundwater composition

Major ions (>5 mg L^{-1})

Bicarbonate	Sodium
Chloride	Calcium
Sulphate	Magnesium

Minor ions (0.01–10.0 mg L^{-1})

Nitrate	Potassium
Carbonate	Strontium
Fluoride	Iron
Phosphate	Boron

Trace constituents (<0.1 mg L^{-1})

Aluminium	Manganese
Arsenic	Nickel
Barium	Radium
Bromide	Selenium
Cadmium	Silica
Caesium	Silver
Chromium	Thorium
Cobalt	Tin
Gold	Titanium
Iodide	Uranium
Lead	Vanadium
Lithium	Zinc

Dissolved gases (trace to 10 mg L^{-1})

Nitrogen	Methane
Oxygen	Hydrogen sulphide
Carbon dioxide	Nitrous oxide

soil water and groundwater to levels in excess of 50 mg L^{-1} (see Box 3.1 for a discussion of the different concentration units used in hydrochemistry).

Organic compounds are usually present in groundwater at very low concentrations of less than 0.1 mg L^{-1} as a result of oxidation of organic matter to carbon dioxide during infiltration through the soil zone (see Section 3.7). In environments rich in organic carbon such as river floodplains and wetlands, biogeochemical processes can generate anaerobic groundwater conditions and the production of dissolved gases such as nitrogen (N_2), hydrogen sulphide (H_2S) and methane (CH_4). Other dissolved gases include oxygen (O_2) and carbon dioxide (CO_2) mostly of an atmospheric source, and nitrous oxide (N_2O) from biogeochemical processes in soils and groundwater. Radon (^{222}Rn) gas, a decay product of uranium (U) and thorium (Th), is common in groundwater and can accumulate to undesirable concentrations in unventilated homes, mines and caves. Uranium is present in crustal rocks (e.g. in the mineral uranite, UO_2), silicates (e.g. in the mineral zircon, $ZrSiO_4$) and phosphates (e.g. in the mineral apatite, $Ca_5(PO_4)_3(OH, F, Cl)$) and is common in granitic rocks but also in other rock types, sediments and soil. Radon and its decay products such as polonium (^{218}Po and ^{216}Po) and ultimately isotopes of lead (Pb) are harmful when inhaled by humans.

Minor concentrations of the inert gases argon (Ar), helium (He), krypton (Kr), neon (Ne) and xenon (Xe) are found dissolved in groundwater and these can provide useful information on the age and temperature of groundwater recharge and therefore help in the interpretation of hydrochemical and hydrogeological conditions in aquifers (see Section 4.5).

The degree of salinization of groundwater expressed as the total dissolved solids (TDS) content is a widely used method for categorizing groundwaters (Table 3.3). In the absence of any specialist analytical equipment for measuring individual dissolved components, a simple determination of TDS by weighing the solid inorganic and organic residue remaining after evaporating a measured volume of filtered sample to dryness can help determine the hydrochemical characteristics of a regional aquifer. Equally, a measurement of the electrical conductivity (EC) of a solution will also give a relative indication of the amount of dissolved salts, made possible by the fact that groundwater is an electrolytic solution with the dissolved components present in ionic form. For any investigation, it is possible to relate the TDS value to electrical conductivity (usually expressed in units of micro-siemens centimetre^{-1}) as follows:

$$\text{TDS (mg L}^{-1}) = k_e \cdot \text{EC (μS cm}^{-1}) \qquad \text{eq. 3.3}$$

where the correlation factor, k_e, is typically between 0.5 and 0.8 and can be determined for each field investigation. The electrical conductivity for fresh groundwater is of the order of 100s μS cm^{-1} while rainwater is of the order of 10s μS cm^{-1} and brines 100,000s μS cm^{-1}. Given that ionic activity, and therefore electrical conductivity, increases with temperature at a rate of about a 2% per °C, measurements are usually normalized to a specific temperature of 25°C and recorded as SEC$_{25}$.

Table 3.3 Simple classification of groundwater based on total dissolved solids (TDS) content. After Freeze and Cherry (1979).

Category	TDS (mg L^{-1})
Freshwater	0–1000
Brackish water	1000–10,000
Saline water (seawater)	10,000–100,000 (35,000)
Brine water	>100,000

Note: TDS > 2000–3000 mg L^{-1} is too salty to drink.

3.4 Sequence of hydrochemical evolution of groundwater

In a series of three landmark papers, and based on nearly ten thousand chemical analyses of natural waters, Chebotarev (1955) put forward the concept that the salinity distribution of groundwaters obeys a definite hydrological and geochemical law which can be formulated as the cycle of metamorphism of natural waters in the crust of weathering. Chebotarev (1955) recognized that the distribution of groundwaters with different hydrochemical facies depended on rock–water interaction in relation to hydrogeological environment, with groundwaters evolving from bicarbonate waters at outcrop to saline waters at depth in the Earth's crust.

In a later paper, Hanshaw and Back (1979) described the chemistry of groundwater as a result of the intimate relationship between mineralogy and flow regime since these determine the occurrence, sequence, rates and progress of reactions. With reference to carbonate aquifers, Hanshaw and Back (1979) presented the conceptual model shown in Fig. 3.3 to depict the changes in groundwater chemistry from the time of formation of a carbonate aquifer through to the development of the aquifer system. When carbonate sediments first emerge from the marine environment, they undergo flushing of seawater by freshwater during which time the salinity decreases and the hydrochemical facies becomes dominated by Ca-HCO$_3$. At this time, the carbonate sediments are selectively dissolved, recrystallized, cemented and perhaps dolomitized to form the rock aquifer. Gradually, as recharge moves downgradient (R → D in Fig. 3.3), Mg^{2+} increases due to dissolution of dolomite and high-magnesium calcite while Ca^{2+} remains relatively constant. With this chemical evolution, SO$_4^{2-}$ increases as gypsum dissolves and HCO$_3^-$ remains relatively constant. For coastal situations or where extensive accumulations of evaporite minerals

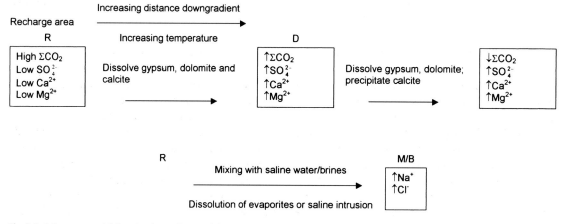

Fig. 3.3 Schematic model showing the evolution of chemical character of groundwater in carbonate aquifers. In areas of recharge (R), the high concentrations of CO$_2$ and low dissolved solids content cause solution of calcite, dolomite and gypsum. As the concentrations of ions increase and their ratios change downgradient (D), groundwater becomes saturated with respect to calcite which begins to precipitate. Dedolomitization (dissolution of dolomite to form calcite with a crystalline structure similar to dolomite) occurs in response to gypsum solution with calcite precipitation. Where extensive accumulations of evaporite minerals occur, their dissolution results in highly saline brines (B). Another common pathway is caused by mixing with seawater (M) that has intruded the deeper parts of coastal aquifers. After Hanshaw and Back (1979).

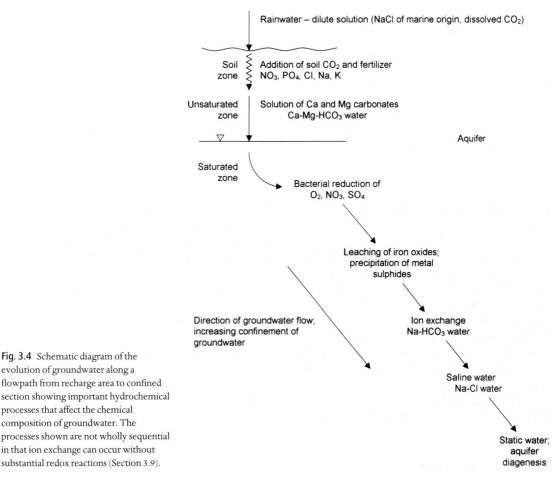

Fig. 3.4 Schematic diagram of the evolution of groundwater along a flowpath from recharge area to confined section showing important hydrochemical processes that affect the chemical composition of groundwater. The processes shown are not wholly sequential in that ion exchange can occur without substantial redox reactions (Section 3.9).

occur, highly saline waters or brines result (R → M/B in Fig. 3.3).

In interpreting groundwater chemistry and identifying hydrochemical processes it is useful to adopt the concept of hydrochemical facies, or water type, introduced by Chebotarev (1955). A hydrochemical facies is a distinct zone of groundwater that can be described as having cation and anion concentrations within definite limits. A pictorial representation of the typical changes in hydrochemical facies along a groundwater flowpath is shown in Fig. 3.4. Dilute rainwater with a Na-Cl water type and containing CO_2 enters the soil zone, whereupon further CO_2, formed from the decay of organic matter, dissolves in the infiltrating water. Where relevant, the application of agricultural chemicals such as fertilizers add further Na^+, K^+, Cl^-, NO_3^- and PO_4^{3-}. Within the soil and unsaturated zone, the dissolved CO_2 produces a weakly acidic solution of carbonic acid, H_2CO_3, which itself dissociates and promotes the dissolution of calcium and magnesium carbonates giving a Ca-Mg-HCO_3 water type. Away from the reservoir of oxygen in the soil and unsaturated zone, the groundwater becomes increasingly anoxic below the water table with progressive reduction of oxygen, nitrate and sulphate linked to bacterial respiration and mineralization of organic matter. Under increasing reducing conditions, Fe and Mn become mobilized and then later precipitated as metal sulphides (see Section 3.9). In the presence of disseminated clay material within the aquifer, ion exchange causes Ca^{2+} to be replaced by Na^+ in solution and the water evolves to a Na-HCO_3 water type. In the deeper, confined section of the aquifer, mixing with saline water occurs to produce a Na-Cl water type before a region of static water and aquifer diagenesis is reached. Either part or all of this classic sequence of hydrochemical change is identified in a number of aquifers including the Great Artesian Basin of Australia (see Box 2.11), the Floridan aquifer system (Box 3.2) and the Chalk aquifer of the London Basin (Ineson & Downing 1963; Mühlherr et al. 1998).

3.5 Groundwater sampling and graphical presentation of hydrochemical data

The aim of field sampling is to collect a raw water sample that is representative of the hydrochemical conditions in the aquifer. To obtain a groundwater sample for analysis some type of sampling device is required, for example a bailer, depth sampler, gas lift sampler or pump (inertial, suction or submersible) in either an open hole or borehole, or from purpose-designed piezometers or nested piezometers. Care must be taken to first flush the water standing in the well or borehole by removing up to three or four times the well or borehole volume prior to sampling. Abstraction boreholes that are regularly pumped provide a sample that is representative of a large volume of aquifer, whereas depth-specific sampling from a multilevel system, or from a section of borehole column isolated by inflatable packers, provides discrete samples that are representative of a small volume of aquifer. For a comparison of the performances of various depth sampling methods, the reader is referred to Price and Williams (1993) and Lerner and Teutsch (1995).

Groundwater samples can be conveniently collected in new or cleaned screw-cap, high-density polyethylene bottles. When filling sampling bottles, the bottle should first be rinsed two to three times with the water sample and, for samples containing suspended sediment or particles, filtered using a 0.45 μm membrane filter. A 0.1 μm filter is recommended where trace metals are to be analysed. Typically, several filled sample bottles are collected at one site, with one sample preserved to stabilize the dissolved metals in solution. The conventional method of preservation prior to analysis is to add a few drops of concentrated nitric or acetic acid after the sample has been collected and filtered to lower the pH to 2. A test using pH paper can be applied to indicate if the pH after acidification is adequate. Once collected, samples should be stored in a cool box and shipped as soon as possible, ideally the same day, to the laboratory for storage at 4°C in order to limit bacterial activity and degradation of nutrient species (NO_3^-, SO_4^{2-} and PO_4^{3-}). Freezing of samples should be avoided as this can lead to precipitation of some elements.

While at a sampling location, and as part of the sampling methodology, the well-head chemistry should be measured, again with the aim of collecting data representative of the hydrochemical conditions in the aquifer. Parameters such as temperature, pH, redox potential (Eh) and dissolved oxygen (DO) content will all change once the groundwater sample is exposed to ambient conditions at the ground surface, and during storage and laboratory analysis. On-site measurement of electrical conductivity (EC) and alkalinity should also be conducted. To limit the exsolution of gases from the groundwater, for example the loss of CO_2 which will cause the pH value of the sample to increase, and in order to prevent mixing of atmospheric oxygen with the sample and so affecting the Eh and DO values, measurements should be made on a flowing sample within an isolation cell or flow cell. The flow cell is designed with a plastic tube leading from the sampling tap into the base of the cell and an overflow tube at the top of the cell. The lid of the flow cell has access holes into which the various measurement probes can be inserted. A typical time to allow for the Eh and DO electrodes to stabilize in the flow cell prior to recording the final values is between 20 and 30 minutes. For further information on hydrochemical parameter measurement and sample collection, including sampling of springs and pore waters, the reader is referred to Lloyd and Heathcote (1985) and Appelo and Postma (1994).

Before attempting any hydrochemical interpretation, it is first necessary to check the quality of laboratory chemical analyses and that the condition of electroneutrality has been met whereby the sum of the equivalent weight of cations equates to the sum of the equivalent weight of anions. This check is commonly carried out by calculating the ionic balance error of the major ions where:

Ion balance error (%) = {Σcations − Σanions)/ (Σcations + Σanions)} × 100   eq. 3.4

Hydrochemical evolution in the Floridan aquifer system

The Floridan aquifer system occurs in the south-east of the United States (Fig. 1) and is one of the most productive aquifers in the world. The aquifer system is a vertically continuous sequence of Tertiary carbonate rocks of generally high permeability. Limestones and dolomites are the principal rock types, although in south-western and north-eastern Georgia and in South Carolina, the limestones grade into lime-rich sands and clays. The Floridan aquifer is composed primarily of calcite and dolomite with minor gypsum, apatite, glauconite, quartz, clay minerals and trace amounts of metallic oxides and sulphides.

The aquifer system generally consists of an Upper and Lower Floridan aquifer separated by a less permeable confining unit having highly variable hydraulic properties (Fig. 1). The Upper Floridan aquifer is present throughout but the Lower Floridan is absent in most of northern Florida and Georgia. Recharge occurs primarily in outcrop areas of Alabama, Georgia and north-central Florida. Most discharge is to rivers and springs with only a small fraction (<5%) discharged directly into the sea. Where the system is unconfined, recharge is rapid and groundwater circulation and discharge rates are high, and secondary permeability is developed by mineral dissolution. Where confining units are thick, the carbonate chemistry of the groundwater evolves in a closed system (Section 3.7) and the development of secondary permeability and flushing of residual saline water within the aquifer system is slow (Sprinkle 1989).

In an extensive study of the hydrochemistry of the Floridan aquifer system, Sprinkle (1989) identified the following major hydrochemical processes:

* dissolution of aquifer minerals towards equilibrium;
* mixing of groundwater with seawater, recharge or leakage;
* sulphate reduction;
* cation exchange between water and rock minerals.

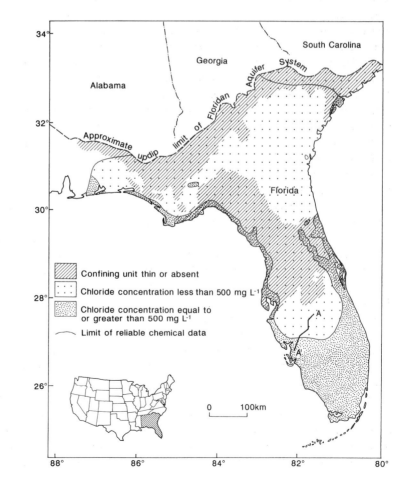

Fig. 1 Map of the extent of the Upper Floridan aquifer showing the relation between the spatial distribution of chloride concentrations and areas of confined–unconfined groundwater conditions. After Sprinkle (1989).

A sequence of hydrochemical evolution is observed starting with calcite dissolution in recharge areas that produces a Ca-HCO$_3$ dominated water type with a TDS concentration of generally less than 250 mg L^{-1}. Downgradient, dissolution of dolomite leads to a Ca-Mg-HCO$_3$ hydrochemical facies. Where gypsum is abundant, sulphate becomes the predominant anion. In coastal areas, as shown in Fig. 1, seawater increases the TDS concentrations and the hydrochemical facies changes to Na-Cl. Leakage from underlying or adjacent sand aquifers in south-central Georgia enters the Floridan aquifer and lowers TDS concentrations but does not change the hydrochemical facies. In the western panhandle of Florida, cation exchange leads to the development of a Na-HCO$_3$ water type.

The hydrogeological and hydrochemical sections shown in Figs 2 and 3 illustrate the main features of the hydrochemical environment in south Florida. The confining unit of the Upper Floridan aquifer is thick along this section and there are evident changes in the concentrations of the major ions in the general direction of groundwater flow. Chemical stratification of the Upper Floridan aquifer may be indicated by the fact that concentrations of Ca^{2+},

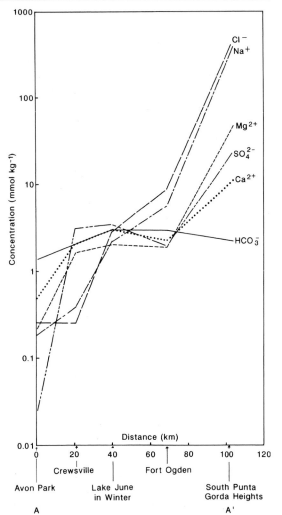

Fig. 3 Hydrochemical section along line A–A' (see Fig. 2) showing the variation in the concentrations of major ions downgradient in the direction of groundwater flow. After Sprinkle (1989).

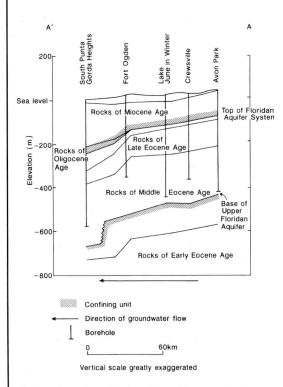

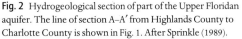

Fig. 2 Hydrogeological section of part of the Upper Floridan aquifer. The line of section A–A' from Highlands County to Charlotte County is shown in Fig. 1. After Sprinkle (1989).

Mg^{2+} and SO$_4^{2-}$ in water from the Lake June in Winter (LJIW) well are greater than in water from the downgradient Fort Ogden well, which is about 100 m shallower than the LJIW well. However, the Fort Ogden well is nearer the coast and Na$^+$ and Cl$^-$ concentrations increase by about three times compared with concentrations at the LJIW well.

An ionic balance error of less than 5% should be achievable with modern analytical equipment and certainly less than 10%. Larger errors are unacceptable and suggest that one or more analyses are in error.

Various methods have been developed for the visual inspection of hydrochemical data in order to look for discernible patterns and trends. By grouping chemical analyses it becomes possible to identify hydrochemical facies and begin to understand the hydrogeological processes that influence the groundwater chemistry. The simplest methods include plotting distribution diagrams, bar charts, pie charts, radial diagrams and pattern diagrams (as presented by Stiff 1951). Although these are easy to construct, they are not convenient for graphical presentation of large numbers of analyses and for this reason other techniques are used including Schoeller (named after Schoeller 1962), trilinear (Piper 1944) and Durov (Durov 1948) diagrams.

The distribution diagram shown in Fig. 3.5 represents concentrations of Na^+ and Cl^- in groundwaters of the Milk River aquifer system located in the southern part of the Western Canadian Sedimentary Basin. The Milk River aquifer is an artesian aquifer, one of several sandstone units developed within a Tertiary–Cretaceous section comprising mainly shale and mudstones. The aquifer crops out in the southern part where recharge occurs but in the northern part is covered by up to 400 m of younger rocks. Groundwater flow is generally northward with a significant component of upward leakage through confining beds. Concentrations of Na^+ and Cl^- are characterized by marked spatial variability and, as shown in Fig. 3.5, are lowest in the south where most of the freshwater recharge occurs. An important feature is the marked northward tongue of fresh groundwater (delimited by the 100 mg L^{-1} Cl^- contour) that coincides with a well-developed zone of aquifer permeability. This hydrochemical pattern is interpreted as a broad zone of mixing that forms as meteoric recharge water flushes pre-existing more saline formation water (Schwartz & Muehlenbachs 1979).

The Schoeller diagram visualizes concentrations as meq L^{-1} (see Box 3.1 for a discussion of concentration units). The example shown in Fig. 3.6 is for the chemistry of groundwater in crystalline rocks in which the near-surface groundwater is typically recharged by rain and snowmelt. The rock-forming minerals in

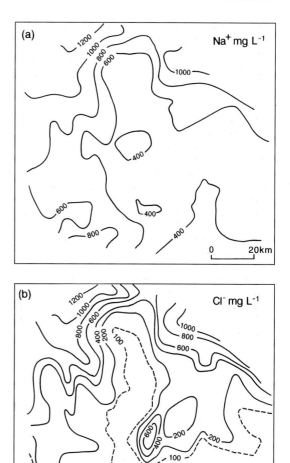

Fig. 3.5 Distribution diagrams of (a) Na^+ and (b) Cl^- concentrations in groundwaters from the Milk River aquifer system, Alberta. After Schwartz and Muehlenbachs (1979).

crystalline rocks are silicates, and their chief cations are Ca^{2+}, Mg^{2+}, Na^+ and K^+. For North America, the principal weathering agent is carbonic acid, chiefly from CO_2 dissolved in the soil zone. Under these conditions the water typically attains a Ca- or Na-HCO_3 composition. Variants of this general water type are due to major differences in the composition of the aquifer rock (Trainer 1988).

To illustrate simple methods for presenting hydrochemical data, Fig. 3.7 shows bar charts, pie charts, a radial diagram and pattern diagrams for the major ion analyses given in Table 3.4. These analyses are for the

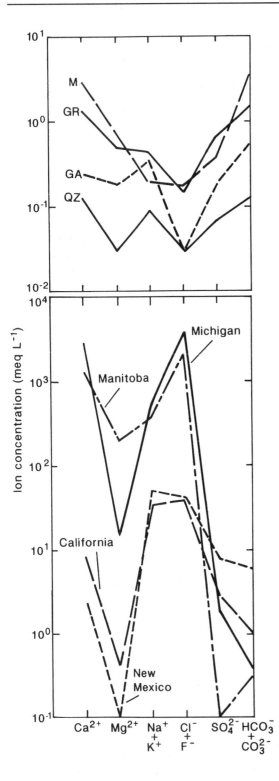

Milligan Canyon area, south-west Montana, in the eastern region of the Northern Rocky Mountain Province. The area is a broad synclinal basin with folded carbonate rocks of Palaeozoic and Mesozoic age on the southern rim, and volcanic breccia and andesitic lava overlying older, deformed rocks on the northern rim. The basin is infilled with unconsolidated alluvial and aeolian deposits and Tertiary sediments of siltstones, limestones and sandstones that contain deposits of gypsum ($CaSO_4 \cdot 2H_2O$) and anhydrite ($CaSO_4$). A number of Upper Cretaceous and early Tertiary igneous intrusives are also present. Groundwater flow is predominantly from west to east but with a contribution of upward groundwater flow from the Madison limestone aquifer underlying the basin. High groundwater yields are obtained from the Tertiary basin.

The same major ion analyses for the Milligan Canyon area are also presented as trilinear and Durov diagrams in Figs 3.8 and 3.9. With these methods of graphical presentation, the concentrations of individual samples are plotted as percentages of the total cation and/or anion concentrations, such that samples with very different total ionic concentrations can occupy the same position in the diagrams. Also, with the trilinear diagram (Fig. 3.8), samples that plot on a straight line within the central diamond field represent mixing of groundwaters between two end-member solutions, for example freshwater and saline water. Further hydrochemical interpretations can be obtained from the Durov diagram (Fig. 3.9). Lines from the central square field can be extended to the adjacent scaled rectangles to allow for representation in terms of two further parameters.

The next step in the hydrochemical interpretation after plotting the chemical data in a variety of ways is to identify the hydrochemical facies present and to prepare maps and cross-sections to show the regional distribution of water types. In the example of groundwaters in the Milligan Canyon area, Fig. 3.10 shows

Fig. 3.6 (*left*) Schoeller (semi-logarithmic) diagram illustrating near-surface groundwater chemistry in crystalline rocks. Symbols on the plot indicate rock type: QZ, quartzite; GA, gabbro; GR, granite; M, marble. Analyses shown are for Houghton County, Michigan; Thompson, Manitoba; New Mexico (mixture of native and injected surface waters) and California. After compilation by Trainer (1988).

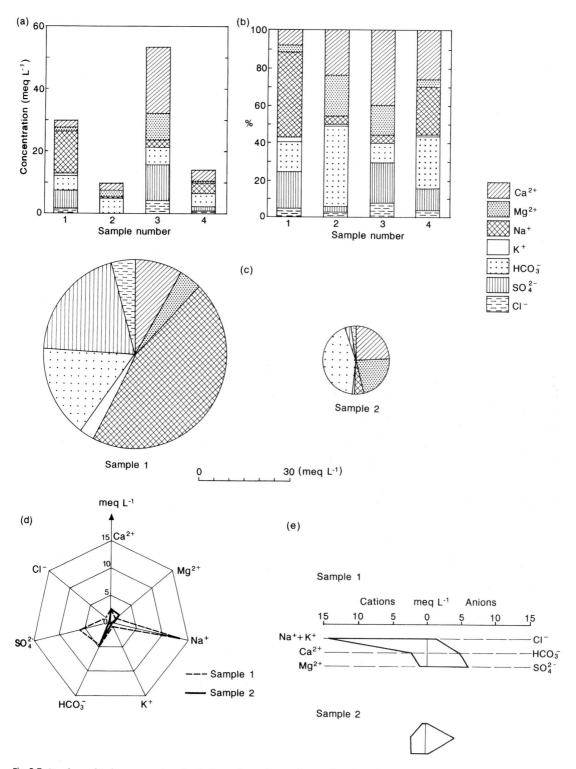

Fig. 3.7 Simple graphical representation of major ion analyses of groundwaters from the Milligan Canyon area, south-west Montana, given in Table 3.4. A stacked bar chart is shown in (a) with major ion concentration values given in meq L^{-1}. The same information is shown in (b) but with the individual major ions shown as a percentage of the total ion concentration for each sample. The two pie charts in (c) represent samples 1 and 2 with the radii scaled to the total ion concentration of each sample. Samples 1 and 2 are also shown in the radial diagram (d) and the pattern, or Stiff, diagram (e).

Table 3.4 Major ion analyses of four groundwater samples from the Milligan Canyon area, south-west Montana, used to plot Figs 3.7–3.9. Data from Krothe and Bergeron (1981).

Water type	Sample number	Ca^{2+}	Mg^{2+}	Na^+	K^+	HCO_3^-	SO_4^{2-}	Cl^-
Ion concentration (mg L^{-1})								
Ca-HCO$_3$	2	47	25.2	10	3.1	251	29	7
Ca-SO$_4$	3	426	103	56	5.4	328	1114	143
Na-HCO$_3$	4	73	6.4	83	4.5	236	158	17
Na-SO$_4$	1	49	13.7	312	29.2	286	566	43
Ion concentration (meq L^{-1})								
Ca-HCO$_3$	2	2.35	2.10	0.43	0.08	4.11	0.30	0.20
Ca-SO$_4$	3	21.30	8.58	2.43	0.14	5.38	11.60	4.09
Na-HCO$_3$	4	3.65	0.53	3.61	0.12	3.87	1.65	0.49
Na-SO$_4$	1	2.45	1.14	13.57	0.75	4.69	5.90	1.23

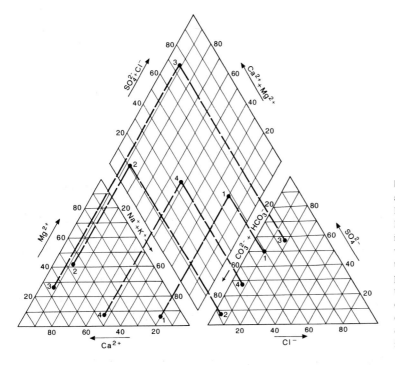

Fig. 3.8 Trilinear diagram of major ion analyses of groundwaters from the Milligan Canyon area, south-west Montana, given in Table 3.4. The individual cation and anion concentration values are expressed as percentages of the total cations and total anions and then plotted within the two triangular fields at the lower left and lower right of the diagram. The two points representing each sample are then projected to the central diamond field and the point of intersection found.

the overall hydrochemical and hydrogeological interpretation. In Fig. 3.10a, the central diamond field of the trilinear diagram is shown and the samples grouped depending on the hydrochemical facies present (Ca-HCO$_3$, Ca-SO$_4$, Na-HCO$_3$ and Na-SO$_4$). When the spatial distribution of the hydrochemical facies are plotted in Fig. 3.10b and compared with the regional geology and groundwater flow direction obtained from a potentiometric map, it is clear that an elongate groundwater body with a distinct Ca-SO$_4$ water type is found in the centre of the basin. The hydrochemical interpretation given by Krothe and

Bergeron (1981) is that recharge occurring in the structurally high areas around the rim of the basin, and formed by older carbonate rocks, results in groundwater of a Ca-HCO$_3$ character. As groundwater flows through the Tertiary deposits, the recharge water undergoes a change in chemical character with increasing residence time and possible solution of anhydrite and/or gypsum resulting in the Ca-SO$_4$ water type and high concentrations of SO$_4^{2-}$ and TDS. Groundwaters with high Na$^+$ concentrations are localized in the north-east and east-central portions of the basin at the point of groundwater discharge out of

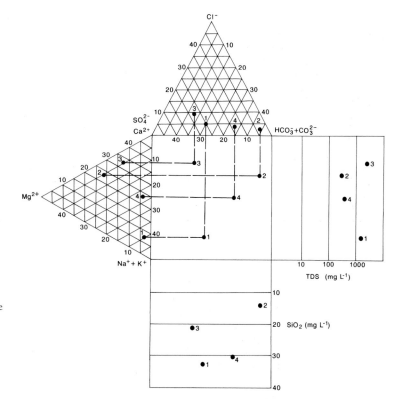

Fig. 3.9 Durov diagram of major ion analyses of groundwaters from the Milligan Canyon area, south-west Montana, given in Table 3.4 with additional SiO$_2$ and TDS data from Krothe and Bergeron (1981). The method of plotting is similar to that for the trilinear diagram shown in Fig. 3.8 but with the additional projection of points from the central square to the two adjacent scaled rectangles.

the basin where the HCO$_3^-$ ion is again dominant. A combination of three factors accounts for these chemical changes, including: sulphate reduction; mixing between Ca-HCO$_3$ and Ca-SO$_4$ waters at the exit from the basin; and possible ion exchange between Ca^{2+} and Na$^+$ associated with montmorillonite clay in the Tertiary sediments and the weathering of the Elkhorn Mountain Volcanics that are rich in Na-plagioclases.

It might be concluded from Fig. 3.10 that sharp boundaries exist between adjacent water types. In reality, the groundwater chemistry is evolving along a flowpath and this can be illustrated by constructing a hydrochemical section such as the example shown for the Floridan aquifer system described in Box 3.2 (Fig. 3). Another interpretation technique for understanding regional hydrochemistry is to prepare a series of X-Y plots and dilution diagrams on either linear or semilogarithmic paper that can demonstrate hydrochemical processes such as simple mixing, ion exchange and chemical reactions. Examples are shown in Fig. 3.11.

All the graphical techniques described above have been applied principally in regional hydrochemical

studies. For contaminated groundwater investigations these techniques are not always appropriate, except for the simpler diagrams such as bar charts, due to the wide spatial variation in concentrations of contaminant species between background and contaminated groundwaters. In this case, plotting the contaminant concentrations as pie charts on a site map can give a visual indication of the 'hot spots' of contamination. An example is shown in Fig. 3.12.

3.6 Concept of chemical equilibrium

Hydrochemical processes in groundwater can be viewed as proceeding slowly towards chemical equilibrium, a concept that is common to aqueous chemistry. Shifts in a system's equilibrium can be qualitatively described by the Le Chatelier Principle that states that, if a system at equilibrium is perturbed, the system will react in such a way as to minimize the imposed change. For example, consider groundwater flowing through a limestone aquifer composed of calcite:

$$CaCO_3 + H_2CO_3 \rightleftharpoons Ca^{2+} + 2HCO_3^- \qquad \text{eq. 3.5}$$

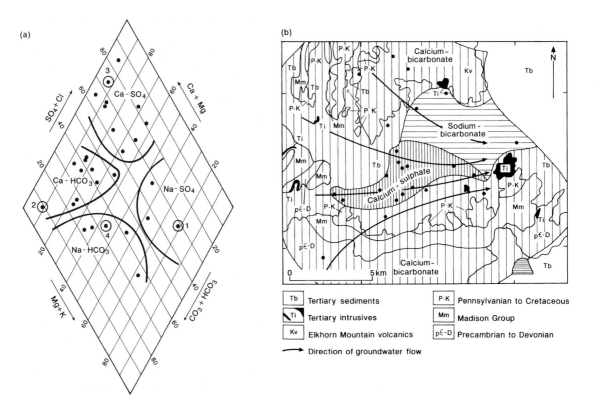

Fig. 3.10 (a) Identification using a trilinear diagram and (b) mapping of the distribution of hydrochemical facies in the Milligan Canyon area, south-west Montana. The four numbered points refer to the major ion analyses given in Table 3.4. After Krothe and Bergeron (1981).

The chemical reaction described by equation 3.5 will proceed to the right (mineral dissolution) or to the left (mineral precipitation) until equilibrium is reached. Looking at this in another way, any increase in Ca^{2+} or HCO_3^- in solution would be lessened by a tendency for the reaction to shift to the left, precipitating calcite. The concept of chemical equilibrium establishes boundary conditions towards which chemical processes will proceed, and can be discussed from either a kinetic or an energetic viewpoint. The kinetic approach is described first.

3.6.1 Kinetic approach to chemical equilibrium

The equilibrium relationship is often called the law of mass action which describes the equilibrium chemical mass activities of a reversible reaction. The rate of reaction is proportional to the effective concentration of the reacting substances. For the reaction:

$$aA + bB \rightleftharpoons cC + dD \qquad \text{eq. 3.6}$$

the law of mass action expresses the relation between the reactants and the products when the chemical reaction is at equilibrium such that:

$$K = \frac{[C]^c[D]^d}{[A]^a[B]^b} \qquad \text{eq. 3.7}$$

where the square brackets indicate the thermodynamically effective concentration, or activity (Box 3.3).

Equation 3.7 is a statement of chemical equilibrium where K is the thermodynamic equilibrium constant (or stability constant). Values of K depend on temperature with solute concentrations expressed in terms of activities. An equilibrium constant greater than unity suggests that equilibrium lies to the right-hand side of the equation describing the chemical reaction and that the forward reaction is favoured.

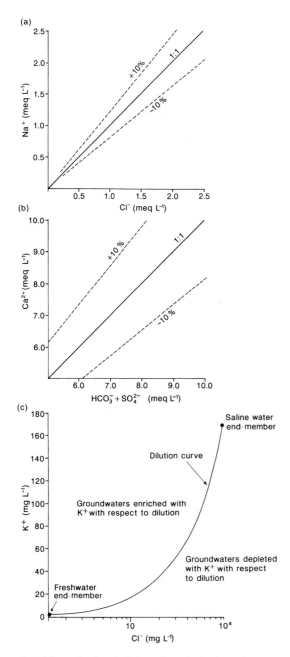

Fig. 3.11 Graphical methods for exploring hydrochemical processes. The X-Y plots in (a) and (b) indicate simple mixing within confidence limits of ±10% between groundwaters that plot close to the lines with a 1 : 1 ratio. Processes such as ion exchange and sulphate reduction would cause samples to plot above the line in (a) and deviate from the line in (b). The semi-logarithmic plot of a dilution diagram in (c) shows a line representing mixing between fresh and saline end-member groundwaters. Points plotting above and below the dilution line represent enrichment and depletion, respectively, of the ionic concentration with respect to the conservative chloride ion.

3.6.2 Energetic approach to chemical equilibrium

In this approach to chemical equilibrium, the most stable composition of a mixture of reactants is the composition having the lowest energy. This more rigorous thermodynamic treatment, compared with the kinetic approach, involves enthalpy, the heat content, H, at constant pressure, and entropy, S, a measure of the disorder of a system. The change in enthalpy (ΔH, measured in J mol^{-1}) in a reaction is a direct measure of the energy emitted or absorbed. The change in entropy in most reactions (ΔS, measured in J mol^{-1} K^{-1}) proceeds to increase disorder, for example by splitting a compound into its constituent ions. For a reversible process, the change in entropy is equal to the amount of heat taken up by a reaction divided by the absolute temperature, T, at which the heat is absorbed.

The total energy released, or the energy change in going from reactants to products, is termed the Gibbs free energy, G (measured in kJ mol^{-1}). If energy is released, in which case the products have lower free energy than the reactants, G is considered negative. The change in Gibbs free energy is defined as:

$$\Delta G = \Delta H - T\Delta S \qquad \text{eq. 3.8}$$

By convention, elements in their standard state (25°C and 1 atmosphere pressure) are assigned enthalpy and free energy values of zero. Standard state thermodynamic data, indicated by the superscript °, and tabulated as values of standard free energies, enthalpies and entropies, are given in most geochemistry and aqueous chemistry textbooks, for example Krauskopf and Bird (1995) and Stumm and Morgan (1981). Values of $\Delta G°$ for different reactions can be calculated by simple arithmetic combination of the tabulated values. Any reaction with a negative $\Delta G°$ value will, in theory, proceed spontaneously (the chemical equivalent of water flowing down a hydraulic gradient), releasing energy. The reverse reaction requires an input of energy. For example, consider the reaction of aqueous carbon dioxide (H_2CO_3) with calcite (eq. 3.5). Relevant data for this reaction at standard state are $\Delta G°$ (H_2CO_3) = −623.1 kJ mol^{-1}, $\Delta G°$ ($CaCO_3$) = −1128.8 kJ mol^{-1}, $\Delta G°$ (Ca^{2+}) = −553.6 kJ mol^{-1} and $\Delta G°$ (HCO_3^-) = −586.8 kJ mol^{-1}. Therefore, the change in standard free energy for this

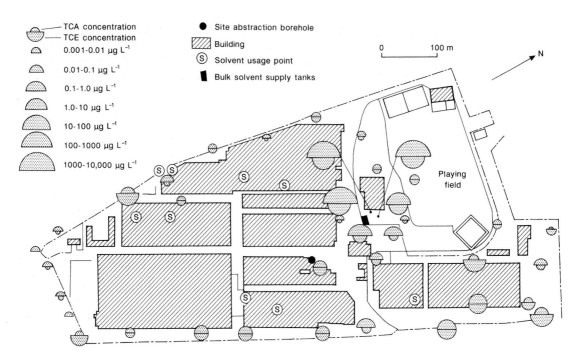

Fig. 3.12 Site plan showing points of solvent use and soil gas concentrations of TCE (trichloroethene) and TCA (1,1,1-trichloroethane) at an industrial site in the English Midlands. After Bishop et al. (1990).

reaction, $\Delta G^{\circ} = \Sigma \Delta G^{\circ}_{products} - \Sigma \Delta G^{\circ}_{reactants} = -1727.2 - (-1751.9) = 24.7$ kJ mol^{-1}.

Since an energetically favoured reaction proceeds from reactants to products, the relationship between ΔG and the equilibrium constant, K, for a reaction is given by:

$$\Delta G = -RT \log_e K \qquad \text{eq. 3.9}$$

where R is the universal gas constant relating pressure, volume and temperature for an ideal gas (8.314 J mol^{-1} K^{-1}).

Hence, one useful application of the energetic approach to chemical equilibrium is the use of thermodynamic data to derive equilibrium constants, K, using equation 3.9. Now, for the reaction given in equation 3.5, and using equation 3.9 at standard conditions ($T = 298$ K), then $\log_e K = -\Delta G^{\circ}/RT = -(24.7 \times 10^3)/(8.314 \times 298) = -9.97$. Hence, K, the thermodynamic equilibrium constant for the reaction of dissolved carbon dioxide with calcite at 25°C is equal to 4.68×10^{-5} or, expressed as the negative logarithm to base 10 of K (pK) = 4.33 and $K = 10^{-4.33}$.

3.7 Carbonate chemistry of groundwater

Acids and bases exert significant control over the chemical composition of water (Box 3.4). The most important acid–base system with respect to the hydrochemistry of most natural waters is the carbonate system. The fate of many types of contaminants, for example metal species, can depend on rock–water interactions involving groundwater and carbonate minerals. Later, in Section 4.4.2, the interpretation of groundwater ages based on the carbon-14 dating method will require knowledge of carbonate chemistry and how the water has interacted with carbonate minerals in an aquifer.

The fundamental control on the reaction rates in a carbonate system is the effective concentration of dissolved CO_2 contained in water. The proportion of CO_2 in the atmosphere is about 0.03% but this increases in the soil zone due to the production of CO_2 during the decay of organic matter, such that the amount of CO_2 increases to several per cent of the soil atmosphere. As groundwater infiltrates the soil zone and recharges the aquifer, reactions can occur

BOX 3.3

Active concentration

The active concentration or **activity** of an ion is an important consideration in concentrated and complex solutions such as seawater, but also for groundwaters and surface waters that contain dissolved ions from many sources. Ions in a concentrated solution are sufficiently close to one another for electrostatic interactions to occur. These interactions reduce the effective concentration of ions available to participate in chemical reactions and, if two salts share a common ion, they mutually reduce each other's solubility and exhibit the **common ion effect**. In order to predict accurately chemical reactions in a concentrated solution, it is necessary to account for the reduction in concentration as follows:

$$a = \gamma m \qquad \text{eq. 1}$$

where a is the solute activity (dimensionless), γ is the constant of proportionality known as the activity coefficient (kg mol^{-1}) and m is the molality. In most cases, it is convenient to visualize the activities of aqueous species as modified molalities in order to take account of the influence on the concentration of a given solute species of other ions in solution.

The activity coefficient of an ion is a function of the ionic strength, I, of a solution given by:

$$I = \tfrac{1}{2}\sum_i c_i z_i^2 \qquad \text{eq. 2}$$

where c_i is the concentration of ion, i, in mol L^{-1}, z_i is the charge of ion, i, and Σ represents the sum of all ions in the solution. As a measure of the concentration of a complex electrolyte solution, ionic strength is better than a simple sum of molar concentrations since it accounts for the effect of the charge of multivalent ions. Freshwaters typically have ionic strengths between 10^{-3} and 10^{-4} mol L^{-1}, whereas seawater has a fairly constant ionic strength of 0.7 mol L^{-1}.

For dilute solutions such as rainwater, γ is about equal to unity. Activity coefficients can be calculated by the extended Debye–Hückel equation, examples of which for a number of charged and uncharged species are shown graphically in Fig. 1. In most practical applications involving dilute or fresh groundwaters, it is adequate to assume that the activity of a dissolved species is equal to the concentration, although measured concentrations of any chemical species should strictly be converted to activities before comparison with thermodynamic data.

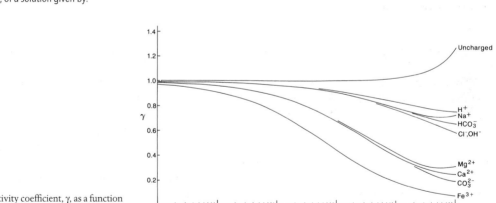

Fig. 1 Activity coefficient, γ, as a function of ionic strength for common ionic constituents in groundwater.

with carbonate minerals, typically calcite and dolomite, which are present.

CO$_2$ dissolves in water forming small quantities of weak carbonic acid, as follows:

$$CO_{2(g)} + H_2O \rightleftharpoons H_2CO_3 \qquad \text{eq. 3.10}$$

From the law of mass action, the equilibrium constant for this reaction is:

$$K_{CO_2} = \frac{[H_2CO_3]}{[H_2O][CO_{2(g)}]} \qquad \text{eq. 3.11}$$

According to Henry's law, in dilute solutions the partial pressure of a dissolved gas, expressed in atmospheres, is equal to its molality (or activity for dilute solutions). Also, given that the activity of water is unity except for very saline solutions, then equation 3.11 becomes:

Acid–base reactions

Acids and bases are important chemical compounds that exert particular control over reactions in water. Acids are commonly considered as compounds that dissociate to yield hydrogen ions (protons) in water:

$$HCl_{(aq)} \rightleftharpoons H^+_{(aq)} + Cl^-_{(aq)} \qquad \text{eq. 1}$$

Bases (or alkalis) can be considered as those substances which yield hydroxide (OH^-) ions in aqueous solutions:

$$NaOH_{(aq)} \rightleftharpoons Na^+_{(aq)} + OH^-_{(aq)} \qquad \text{eq. 2}$$

Acids and bases react to neutralize each other, producing a dissolved salt plus water:

$$HCl_{(aq)} + NaOH_{(aq)} \rightleftharpoons Cl^-_{(aq)} + Na^+_{(aq)} + H_2O_{(l)} \qquad \text{eq. 3}$$

Hydrochloric acid (HCl) and sodium hydroxide (NaOH) are recognized, respectively, as strong acids and bases that dissociate completely in solution to form ions. Weak acids and bases dissociate only partly.

The acidity of aqueous solutions is often described in terms of the pH scale. The pH of a solution is defined as:

$$pH = -\log_{10}[H^+] \qquad \text{eq. 4}$$

Water undergoes dissociation into two ionic species as follows:

$$H_2O \rightleftharpoons H^+ + OH^- \qquad \text{eq. 5}$$

In reality, H^+ cannot exist, and H_3O^+ (hydronium) is formed by the interaction of water and H^+. However, it is convenient to use H^+ in chemical equations. From the law of mass action, the equilibrium constant for this dissociation is:

$$K_{H_2O} = \frac{[H^+][OH^-]}{[H_2O]} = 10^{-14} \qquad \text{eq. 6}$$

For water that is neutral, there are exactly the same concentrations (10^{-7}) of H^+ and OH^- ions such that $pH = -\log_{10}[10^{-7}] = 7$. If pH < 7, there are more H^+ ions than OH^- ions and the solution is acidic. If pH > 7, there are more OH^- ions than H^+ ions and the solution is basic. It is important to notice that pH is a logarithmic scale and so it is not appropriate to average pH values of solutions. Instead, it is better to average H^+ concentrations.

Acid–base pairs commonly present in groundwater are those associated with carbonic acid and water itself. Boric, orthophosphoric and humic acids are minor constituents of groundwater but are relatively unimportant in controlling acid–base chemistry. Many aquifers of sedimentary origin contain significant amounts of solid carbonate such as calcite ($CaCO_3$), a fairly strong base that contributes a carbonate ion, thus rendering the solution more alkaline, and dolomite ($CaMg(CO_3)_2$) which participate in equilibrium reactions involving carbonic acid. All acid–base reactions encountered in natural aqueous chemistry are fast such that acid–base systems are always in equilibrium in solution.

$$K_{CO_2} = \frac{[H_2CO_3]}{P_{CO_2}} \qquad \text{eq. 3.12}$$

Carbonic acid is polyprotic (i.e. it has more than one H^+ ion) and dissociates in two steps:

$$H_2CO_3 \rightleftharpoons H^+ + HCO_3^- \qquad \text{eq. 3.13}$$

$$HCO_3^- \rightleftharpoons H^+ + CO_3^{2-} \qquad \text{eq. 3.14}$$

From the law of mass action, dissociation constants can be expressed as follows:

$$K_{H_2CO_3} = \frac{[H^+][HCO_3^-]}{[H_2CO_3]} \qquad \text{eq. 3.15}$$

$$K_{HCO_3^-} = \frac{[H^+][CO_3^{2-}]}{HCO_3^-} \qquad \text{eq. 3.16}$$

Using a mass balance expression for the dissolved inorganic carbon (DIC) in the acid and its dissociated anionic species, expressed in terms of molality, then:

$$DIC = (H_2CO_3) + (HCO_3^-) + (CO_3^{2-}) \qquad \text{eq. 3.17}$$

Rearranging equations 3.13–3.17, and taking an arbitrary value of unity (1) for DIC, equations for the relative concentrations of H_2CO_3, HCO_3^- and CO_3^{2-} as a function of pH are obtained as shown graphically in Fig. 3.13. It can be seen from Fig. 3.13 that over most of the normal pH range of groundwater (6–9), HCO_3^- is the dominant carbonate species and this explains why HCO_3^- is one of the major dissolved inorganic species in groundwater.

To calculate actual concentrations of inorganic carbon species in groundwater, first consider the dissolution of calcite by carbonic acid (eq. 3.5). With reference to equation 3.10, if the partial pressure of

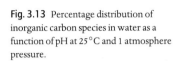

Fig. 3.13 Percentage distribution of inorganic carbon species in water as a function of pH at 25°C and 1 atmosphere pressure.

carbon dioxide (Pco_2) in the infiltrating groundwater increases, then reaction 3.5 proceeds further to the right to achieve equilibrium. Now, at 25°C, substitution of equations 3.12, 3.15 and 3.16 into the equation expressing the equilibrium constant for the dissociation of calcite:

$$K_{calcite} = [Ca^{2+}][CO_3^{2-}] \qquad \text{eq. 3.18}$$

yields:

$$[H^+] = 10^{-4.9}\{[Ca^{2+}]Pco_2\}^{1/2} \qquad \text{eq. 3.19}$$

To obtain the solubility of calcite for a specified Pco_2, an equation of electroneutrality is required for the condition $\Sigma zm_c = \Sigma zm_a$ for calcite dissolution in pure water where z is the ionic valence and m_c and m_a are, respectively, the molalities of the cation and anion species involved:

$$2(Ca^{2+}) + (H^+) = (HCO_3^-) + 2(CO_3^{2-}) + (OH^-) \quad \text{eq. 3.20}$$

The concentrations of H^+ and OH^- are negligible compared with the other terms in equation 3.20 with respect to the groundwater Pco_2 values and by combining equations 3.19 and 3.20 with equations 3.12, 3.15 and 3.16 gives a polynomial expression in terms of two of the variables and the activities. For a specified Pco_2 value, iterative solutions by computer can be obtained. The results of these calculations for equilibrium calcite dissolution in water for a condition with no limit on the supply of carbon dioxide are shown graphically in Fig. 3.14.

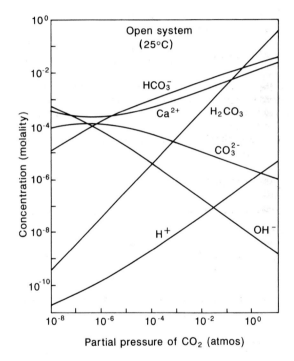

Fig. 3.14 Dissolved species in water in equilibrium with calcite as a function of Pco_2 at 25°C. Note the parallel lines at the top right side for Ca^{2+} and HCO_3^- (just 0.301 unit apart) that demonstrate that at high Pco_2 values, such as found in groundwater, these are the major species formed by dissolving calcite. After Guenther (1975).

It can be seen from Fig. 3.14, that the solubility of calcite is strongly dependent on the Pco_2 and that the equilibrium value of H^+ (i.e. pH) also varies strongly with Pco_2. Hence, the accurate calculation of the inorganic carbon species in groundwater requires the careful measurement of pH in the field with no loss of carbon dioxide by exsolution from the sample. In the groundwater environment, Pco_2 is invariably $>10^{-4}$ atmospheres and again it is noticed that HCO_3^- rather than CO_3^{2-} is the dominant ionic species of DIC in groundwater. In Box 3.5, the carbonate chemistry of a limestone aquifer in the west of England is described to illustrate the changes in the distribution of carbonate species along a groundwater flowpath.

The chemical evolution of the carbonate system can be considered as occurring under 'open' or 'closed' conditions. As shown in Fig. 3.15, under open conditions of carbonate dissolution, a constant Pco_2 is maintained while under 'closed' conditions, carbonate dissolution occurs without replenishment of CO_2. Using the theory outlined above, it is possible to model paths of chemical evolution for groundwater dissolving carbonate. Steps along the paths are computed by hypothetically dissolving small amounts of carbonate material (calcite or dolomite) for a given temperature and starting condition for Pco_2 until the water becomes saturated (lines 2 and 3 at 15 °C in Fig. 3.16). Lower temperatures will shift the saturation line to higher solubilities, while higher temperatures will result in saturation at lower solubilities.

It is noticeable in Fig. 3.16 that, for closed-system dissolution, the pH values at saturation are higher and the HCO_3^- concentrations lower. In reality, very small quantities of calcite and dolomite exert a strong influence on the carbonate chemistry of groundwater flowing through the soil zone such that open-system conditions typically dominate, resulting in a pH invariably between 7 and 8.

To illustrate carbonate dissolution pathways in a limestone aquifer, Fig. 3.17 shows the results of chemical modelling of recharge to the Chalk aquifer in north Norfolk, eastern England. In this area, the Chalk aquifer is covered by glacial deposits including outwash sands and gravel and two types of till deposits that are distinguished by the mixture of contained clay, sand and carbonate fractions. The evolution of the carbonate chemistry for this aquifer system can be modelled for both open- and closed-systems as demonstrated by Hiscock (1993). Two shallow well waters in the glacial deposits that are both under-saturated with respect to calcite represent starting conditions for the two models. Using the computer program MIX2 (Plummer et al. 1975) small increments of carbonate are added to the groundwaters until saturation is reached. The initial and final chemical compositions of the well waters are given in Table 3.5 and the evolution paths followed in each case are shown in Fig. 3.17. The agreement between the modelled chemistry and the actual Chalk groundwaters sampled from beneath the glacial deposits demonstrates that calcite saturation is achieved under open-system conditions. The distribution of points about the phase boundary in Fig. 3.17 is, as explained by Langmuir (1971), the result of either variations in the soil Pco_2 at the time of groundwater recharge or changes in groundwater chemistry that affect Pco_2. For the Chalk aquifer in north Norfolk, the main reason is variations in soil Pco_2. In areas of sandy till cover, recharge entering soils depleted in carbonate attain calcite saturation for lower values of soil Pco_2, in the range $10^{-2.5}$ to $10^{-2.0}$ atmospheres. By contrast, the

---

Carbonate chemistry of the Jurassic limestones of the Cotswolds, England

BOX 3.5

The Jurassic limestones of the Cotswolds, England, form two major hydrogeological units in the Upper Thames catchment: (i) limestones in the Inferior Oolite Series; and (ii) limestones in the Great Oolite Series. These aquifer units are separated by the Fullers Earth clay which, in this area, acts as an aquitard. The groundwater level contours shown in Fig. 1 indicate a regional groundwater flow direction to the south-east although with many local variations superimposed upon it. In the upper reaches, river valleys are often eroded to expose the underlying Lias clays. The presence of these clays together with the Fullers Earth clay, intercalated clay and marl

bands within the limestones, as well as the presence of several faults, gives rise to numerous springs. Flow through the limestones is considered to be dominantly through a small number of fissures. At certain horizons a component of intergranular movement may be significant, but on the whole the limestone matrix has a very low intrinsic intergranular permeability due to its normally well-cemented and massive lithology (Morgan-Jones & Eggboro 1981).

A survey of the hydrochemistry of boreholes and springs between 1976 and 1979 is reported by Morgan-Jones and Eggboro (1981). The hydrochemical profiles shown in Fig. 2 are for the selected

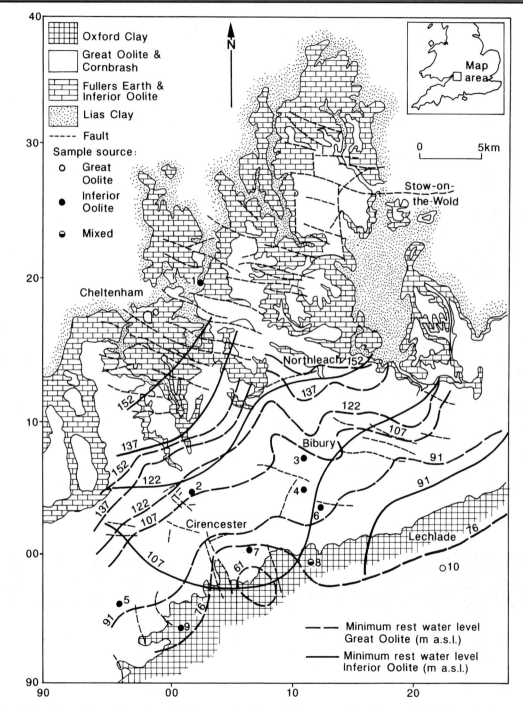

Fig. 1 Map of the geology and groundwater level contours for the Jurassic limestone aquifers of the Cotswolds. The sample points shown are from the survey of Morgan-Jones and Eggboro (1981) with the numbered samples used to construct the hydrochemical sections shown in Fig. 2. After Morgan-Jones and Eggboro (1981).

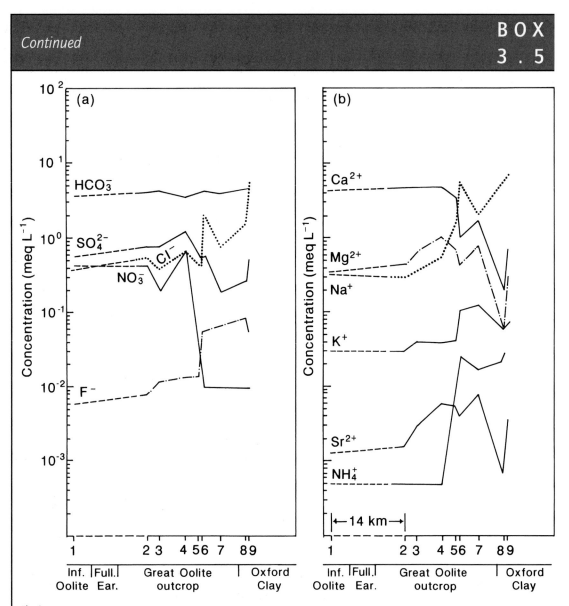

Fig. 2 Hydrochemical sections showing major and minor anion (a) and cation (b) data for the sample locations shown in Fig. 1. Data after Morgan-Jones and Eggboro (1981).

locations shown in Fig. 1. The dominant major ions in the limestone groundwaters are Ca^{2+} and HCO_3^- in the unconfined areas but in areas where the limestones become increasingly confined beneath clays, the groundwater changes to a Na-HCO₃ type. Groundwaters from deep within the confined limestones contain a Na-Cl component as a result of mixing with increasingly saline groundwater.

The degree of groundwater saturation with respect to calcite and dolomite is shown diagrammatically in Fig. 3. A confidence limit of $\pm 10\%$ is shown to account for any errors in the chemical analyses or in the basic thermodynamic data used to compute the saturation indices (Box 3.6). Samples from locations in the unconfined limestones where high Pco_2 values are maintained under open-system conditions are supersaturated with respect to calcite. The consistently high Ca^{2+}/Mg^{2+} ratios and low degrees of dolomite saturation in the samples analysed suggest that both limestone series contain predominantly low magnesium calcite.

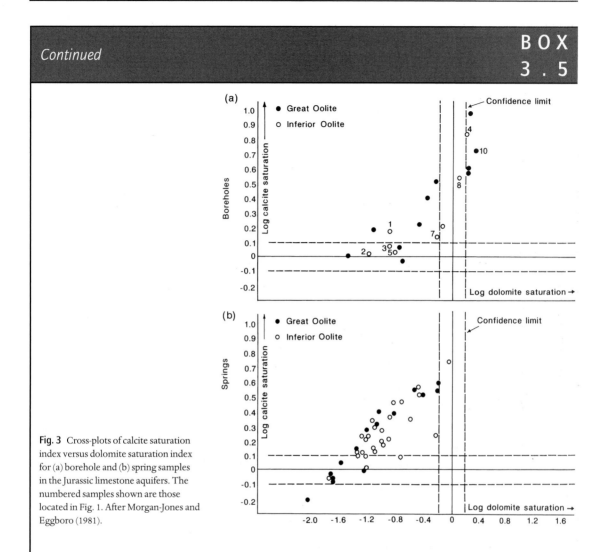

Fig. 3 Cross-plots of calcite saturation index versus dolomite saturation index for (a) borehole and (b) spring samples in the Jurassic limestone aquifers. The numbered samples shown are those located in Fig. 1. After Morgan-Jones and Eggboro (1981).

Values of pH of about 7.5 were recorded for all unconfined groundwater samples. With the onset of confined conditions, pH values increase as a result of changes in HCO_3^- concentration and ion exchange reactions lead to an increase in Na^+ and decrease in Ca^{2+} concentrations (Fig. 2). The removal of Ca^{2+} causes the groundwater to dissolve more carbonate material, but without the replenishment of CO_2 in the closed-system conditions that prevail, the pH begins to rise as H^+ ions are consumed in the calcite dissolution reaction (eq. 3.5). For example, the confined groundwater of sample 8 has a pH of 9.23 and Ca^{2+}, HCO_3^- and Na^+ concentrations of 4, 295 and 137 mg L^{-1}, respectively, compared with the unconfined sample 2 which has corresponding values of 7.31 (pH) and 100 (Ca^{2+}), 246 (HCO_3^-) and 6.9 (Na^+) mg L^{-1}.

Of the minor ions, fluoride shows considerable enrichment with increasing confinement of the groundwaters. The majority of spring samples have F^- concentrations of <100 μg L^{-1} (Fig. 2). Most

limestones contain small amounts of fluorite (CaF_2) but this mineral has a very low solubility (K_{sp} at 25°C = $10^{-10.4}$) and the majority of Ca-rich waters have low F^- values. Morgan-Jones and Eggboro (1981) considered that rainwater is the principal source of F^- in the unconfined aquifer areas. A maximum value of 9.8 mg L^{-1} was recorded at sample location 10 in the confined Great Oolite aquifer. Here, the increase is related to the onset of ion exchange and to the availability and solubility of fluorite within the limestone. The equilibrium activity of F^- is dependent on the activity of Ca^{2+} as defined by the equilibrium constant:

$$K_{CaF_2} = \frac{[Ca^{2+}][F^-]^2}{[CaF_2]} \qquad \text{eq. 1}$$

Hence, the decline in the activity of Ca^{2+} with increasing ion exchange allows the F^- activity to increase.

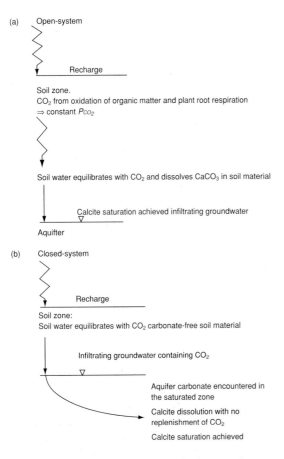

(a) Open-system

Recharge

Soil zone.
CO₂ from oxidation of organic matter and plant root respiration
⇒ constant P_{CO_2}

Soil water equilibrates with CO₂ and dissolves CaCO₃ in soil material

Calcite saturation achieved infiltrating groundwater

Aquifter

(b) Closed-system

Recharge

Soil zone:
Soil water equilibrates with CO₂ carbonate-free soil material

Infiltrating groundwater containing CO₂

Aquifer carbonate encountered in
the saturated zone

Calcite dissolution with no
replenishment of CO₂

Calcite saturation achieved

Fig. 3.15 Schematic representation of the development of open and closed systems of calcite dissolution in soil–aquifer systems.

carbonate-rich soils developed in areas of chalky, clay-rich till experience calcite saturation for higher soil P_{CO_2} values, in the range $10^{-2.1}$ to $10^{-2.0}$ (Hiscock 1993).

Although a useful framework for considering the chemical evolution of the carbonate system, several factors have not been considered, including: seasonal variation in soil temperature and P_{CO_2}; processes such as adsorption, cation exchange and gas diffusion and dispersion that influence the concentrations of Ca^{2+} and P_{CO_2}; and the process of incongruent dissolution whereby the dissolution is not stoichiometric, with one of the dissolution products being a mineral phase sharing a common ionic component with the dissolving phase.

The above treatment of calcite dissolution assumed independent dissolution of calcite and dolomite.

However, if both minerals occur in a hydrogeological system they may both dissolve simultaneously or sequentially leading to different equilibrium relations compared to those shown in Fig. 3.16. In this situation, a comparison of equilibrium constants for calcite and dolomite for the particular groundwater temperature is necessary to define which mineral is dissolving incongruently.

For example, considering the thermodynamic data shown in Table 3.6, at about 20°C, $K_{calcite} = K_{dolomite}^{1/2}$. Under these conditions, since the solubility product (Box 3.6) of calcite is equal to $[Ca^{2+}][CO_3^{2-}]$ and for dolomite is equal to $[Ca^{2+}][Mg^{2+}][CO_3^{2-}][CO_3^{2-}]$, if groundwater saturated with dolomite flows into a zone that contains calcite, no calcite dissolution will occur because the water is already saturated with respect to calcite. At temperatures lower than 20°C, $K_{calcite} < K_{dolomite}^{1/2}$ and if groundwater dissolves dolomite to equilibrium the water becomes supersaturated with respect to calcite which can then precipitate. In a system where the rate of dolomite dissolution is equal to the rate of calcite precipitation, this is the condition of incongruent dissolution of dolomite. At temperatures higher than 20°C, $K_{calcite} > K_{dolomite}^{1/2}$ and if dolomite saturation is achieved with the groundwater then entering a region containing calcite, calcite dissolution will occur leading to an increase in Ca^{2+} and CO_3^{2-} concentrations. The water will now be supersaturated with respect to dolomite and dolomite precipitation, although sluggish, will occur, to achieve a condition of incongruent dissolution of calcite. In cases where groundwater first dissolves calcite to equilibrium and then encounters dolomite, dolomite dissolves regardless of the temperature because the water must acquire appreciable Mg^{2+} before dolomite equilibrium is achieved. However, as the water becomes supersaturated with respect to calcite due to the influx of Ca^{2+} and CO_3^{2-} ions from dolomite dissolution, calcite precipitates and the dolomite dissolution becomes incongruent (Freeze & Cherry 1979).

Over long periods of time, incongruent calcite and dolomite dissolution may exert an important influence on the chemical evolution of the groundwater and on the mineralogical evolution, or diagenesis, of the aquifer rock; for example, dolomitization of calcareous sediments. Dedolomitization, the process whereby a dolomite-bearing rock is converted to

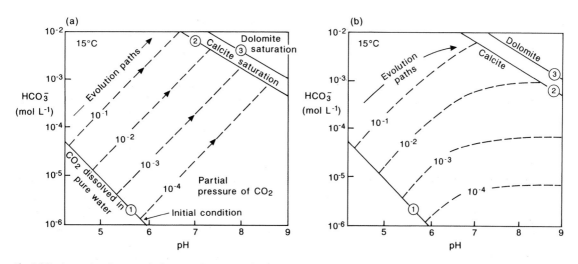

Fig. 3.16 Chemical evolution paths for water dissolving calcite at 15°C for (a) open-system dissolution and (b) closed-system dissolution. Line 1 represents the initial condition for the water containing dissolved CO_2; line 2 represents calcite saturation; and line 3 represents dolomite saturation if dolomite is dissolved under similar conditions. After Freeze and Cherry (1979).

a calcite-bearing rock has been reported, for example in the case of the Floridan aquifer system described in Box 3.2. Concentrations of Mg^{2+} in the Upper Floridan aquifer generally range from 1 to 1000 mg L^{-1}, with the highest concentrations occurring where the aquifer contains seawater. Where the aquifer contains freshwater, Mg^{2+} concentrations generally increase in downgradient directions because of dedolomitization of the aquifer, although data are insufficient to prove that formation of dolomite limits Mg^{2+} concentrations in the Upper Floridan aquifer (Sprinkle 1989).

3.8 Adsorption and ion exchange

Adsorption and ion exchange reactions in aquifers can significantly influence the natural groundwater chemistry and are an important consideration in predicting the migration of contaminants such as heavy metals and polar organic chemicals (see Section 6.3.2 for further discussion). Major ion exchange reactions affect not only the exchanging ions but also other species, especially via dissolution and precipitation reactions. The attenuation of some pollutants, for example NH_4^+, is mainly by the process of ion exchange (Carlyle et al. 2004). Ion exchange reactions

can also lead to changes in the hydraulic conductivity of natural materials (Zhang & Norton 2002).

Ionic species present in groundwater can react with solid surfaces. As shown in Fig. 3.18a, adsorption occurs when a positively charged ion in solution is attracted to and retained by a solid with a negatively charged surface. Depending on the point of zero charge (PCZ) of the rock-forming mineral, where PCZ is the pH at which the mineral has zero charge (at pH values less than the PCZ the mineral has a net positive charge, at values greater than the PCZ a net negative charge), different minerals will attract anions or cations to their surfaces depending on the pH of the solution. Clay minerals have negative surface charges in all but the most acidic solutions and therefore attract cations to their surfaces to neutralize the negative charge. The surface charge for oxides and hydroxides of Fe can be either negative or positive in the pH range of most groundwaters (PCZ values for haematite and goethite are between 5–9 and 7.3–7.8, respectively; Krauskopf & Bird 1995), giving the potential for anion adsorption (for example, arsenic oxyanions) at low pH, and release at high pH.

In general, the degree of adsorption increases with an increasing surface area and with decreasing grain size. Hence, clays are typically most reactive since

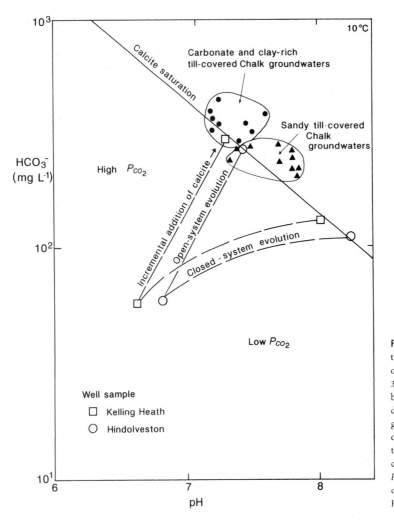

Fig. 3.17 Graphical representation of the modelled open- and closed-system carbonate dissolution paths given in Table 3.5. Starting conditions are represented by well samples contained in glacial till deposits. The two fields showing Chalk groundwaters represent saturated conditions with respect to calcite with the differences explained by the greater dissolution of carbonate material at higher P_{CO_2} values in soils developed on the carbonate and clay-rich till deposits. After Hiscock (1993).

Table 3.5 Chemical models for the formation of Chalk groundwaters beneath glacial deposits in north Norfolk, eastern England, under the processes of open- and closed-system dissolution of calcite. From Hiscock (1993).

Well sample name	$CaCO_3$ added ($\times 10^{-3}$ mol L^{-1})	Ca^{2+} (mg L^{-1})	HCO_3^- (mg L^{-1})	pH	P_{CO_2} (atmos)	$\Omega_{calcite}$
Open-system conditions						
Kelling Heath	0	10.0	56.7	6.62	1.94	−2.47
	2.00	90.2	299.5	7.30	1.94	−0.03
Hindolveston	0	9.4	59.2	6.81	2.12	−2.09
	1.75	79.5	271.5	7.44	2.12	0.02
Closed-system conditions						
Kelling Heath	0	10.0	56.7	6.62	1.94	−2.47
	0.70	36.0	130.5	8.02	3.00	−0.05
Hindolveston	0	9.4	59.2	6.81	2.12	−2.09
	0.44	28.4	111.6	8.24	3.29	0.02

Solubility product and saturation index

BOX 3.6

The dynamic equilibrium between a mineral and its saturated solution when no further dissolution occurs is quantified by the thermodynamic equilibrium constant. For example, the reaction:

$$CaCO_{3(calcite)} \rightleftharpoons Ca^{2+}_{(aq)} + CO^{2-}_{3(aq)}$$ eq. 1

is quantified by the equilibrium constant, K, found from:

$$K_{calcite} = \frac{[Ca^{2+}][CO_3^{2-}]}{[CaCO_3]}$$ eq. 2

Since $CaCO_3$ is a solid crystal of calcite, its activity is effectively constant and by convention is assigned a value of one or unity.

The equilibrium constant for a reaction between a solid and its saturated solution is known as the solubility product, K_{sp}. Solubility products have been calculated for many minerals, usually using pure water under standard conditions of 25°C and 1 atmosphere pressure and are tabulated in many textbooks, for example data for major components in groundwater are given by Appelo and Postma (1994). The solubility product for calcite (eq. 2) is $10^{-8.48}$ and equation 2 now becomes:

$$K_{sp} = \frac{[Ca^{2+}][CO_3^{2-}]}{1} = [Ca^{2+}][CO_3^{2-}] = 10^{-8.48} \, mol^2 \, L^{-2}$$ eq. 3

The solubility product can be used to calculate the solubility (mol L^{-1}) of a mineral in pure water. The case for calcite is straightforward since each mole of $CaCO_3$ that dissolves produces one mole of Ca^{2+} and one mole of CO_3^{2-}. Thus, the calcite solubility = $[Ca^{2+}] = [CO_3^{2-}]$ and therefore calcite solubility = $(10^{-8.48})^{1/2} = 10^{-4.24} = 5.75 \times 10^{-5}$ mol L^{-1}.

The state of saturation of a mineral in aqueous solution can be expressed using a saturation index, where:

$$\Omega = \frac{IAP}{K_{sp}}$$ eq. 4

in which IAP is the ion activity product of the ions in solution obtained from analysis. A Ω value of 1 indicates that mineral saturation (equilibrium) has been reached. Values greater than 1 represent oversaturation or supersaturation and the mineral is likely to be precipitated from solution. Values less than 1 indicate undersaturation and further mineral dissolution can occur. An alternative to equation 4 is to define $\Omega = \log_{10}(IAP/K_{sp})$ in which case the value

of Ω is 0 at equilibrium with positive values indicating supersaturation and negative values undersaturation.

By calculating saturation indices, it is possible to determine from hydrochemical data the equilibrium condition of groundwater with respect to a given mineral. For example, a groundwater from the unconfined Chalk in Croydon, South London, gave the following results: temperature = 12°C, pH = 7.06, Ca^{2+} concentration = 121.6 mg L^{-1} ($10^{-2.52}$ mol L^{-1}) and $HCO_3^- = 217$ mg L^{-1} ($10^{-2.45}$ mol L^{-1}). By making the assumption that the concentrations are equal to activities for this dilute groundwater sample, the first step in calculating a calcite saturation index is to find the CO_3^{2-} concentration. The dissociation of HCO_3^- can be expressed as:

$$HCO_3^- \rightleftharpoons H^+ + CO_3^{2-}$$ eq. 5

for which the approximate equilibrium constant at 10°C (Table 3.6) is:

$$K_{HCO_3^-} = \frac{[H^+][CO_3^{2-}]}{HCO_3^-} = 10^{-10.49}$$ eq. 6

Rearranging equation 6 for the unknown CO_3^{2-} concentration and substituting the measured values for H^+ and HCO_3^- gives:

$$[CO_3^{2-}] = \frac{K_{HCO_3^-}[HCO_3^-]}{[H^+]} = \frac{10^{-10.49} \cdot 10^{-2.45}}{10^{-7.06}} = 10^{-5.88}$$ eq. 7

Now, using the result of equation 7, the calcite saturation index is found from:

$$\Omega_{calcite} = \log_{10}\frac{[Ca^{2+}][CO_3^{2-}]}{K_{sp}}$$ eq. 8

If K_{sp} for calcite = $10^{-8.41}$ at 10°C, then:

$$\Omega_{calcite} = \log_{10}\frac{[10^{-2.52}][10^{-5.88}]}{10^{-8.41}} = \log_{10}10^{0.01} = 0.01$$ eq. 9

Hence, the Chalk groundwater is marginally supersaturated and, given the assumptions used in the calculation of $\Omega_{calcite}$, can be regarded as at equilibrium. A more accurate calculation using the chemical program WATEQ (Truesdell & Jones 1973) that accounts for the chemical activity and speciation of the sample as well as the actual sample temperature of 12.0°C gives a calcite saturation index of −0.12, again indicating equilibrium with respect to calcite for practical purposes (Mühlherr et al. 1998).

they have a small grain size and therefore have a large surface area on which sorption reactions can occur. In addition, clays tend to be strong adsorbers since they have an excess of negative charge at the surface due to crystal lattice defects on to which cations can

adsorb. The adsorption may be weak, essentially a physical process caused by Van der Waals' force, or strong, if chemical bonding occurs. Divalent cations are usually more strongly adsorbed than monovalent ions as a result of their greater charge density, a

Table 3.6 Thermodynamic equilibrium constants for calcite, dolomite and major aqueous carbonate species in pure water for a temperature range of 0–30°C and 1 atmosphere total pressure. Note that pK = $-\log_{10}K$. Thermodynamic data from Langmuir (1971) and Plummer and Busenberg (1982).

Temperature (°C)	pK_{CO_2}	p$K_{H_2CO_3}$	p$K_{HCO_3^-}$	p$K_{calcite}$	p$K_{dolomite}$	$K_{calcite}/(K_{dolomite})^{1/2}$
0	1.11	6.58	10.63	8.38	16.18	0.51
5	1.19	6.52	10.56	8.39	16.39	0.63
10	1.27	6.47	10.49	8.41	16.57	0.75
15	1.34	6.42	10.43	8.43	16.74	0.87
20	1.41	6.38	10.38	8.45	16.88	0.97
25	1.47	6.35	10.33	8.48	17.00	1.05
30	1.52	6.33	10.29	8.51	17.11	1.11

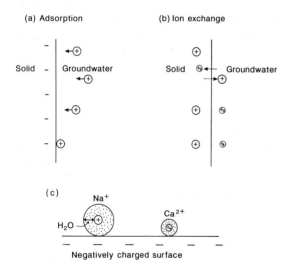

Fig. 3.18 Pictorial representation of (a) adsorption and (b) cation exchange reactions in groundwater. In (c) the divalent Ca^{2+} ions with a smaller hydrated radius are more strongly adsorbed by the negatively charged surface (e.g. clay) than are the monovalent Na^+ ions with a larger hydrated radius.

consequence of valence and smaller hydrated radius (Fig. 3.18c). The adsorptive capacity of specific soils or sediments is usually determined experimentally by batch or column experiments.

Ion exchange occurs when ions within the mineral lattice of a solid are replaced by ions in the aqueous solution (Fig. 3.18b). Ion exchange sites are found primarily on clays, soil organic matter and metal oxides and hydroxides which all have a measurable cation exchange capacity (CEC). In cation exchange the divalent ions are more strongly bonded to a solid surface such that the divalent ions tend to replace monovalent ions. The amounts and types of cations exchanged are the result of the interaction of the con-

centration of cations in solution and the energy of adsorption of the cations at the exchange surface. The monovalent ions have a smaller energy of adsorption and are therefore more likely to remain in solution. As a result of a larger energy of adsorption, divalent ions are more abundant as exchangeable cations. Ca^{2+} is typically more abundant as an exchangeable cation than is Mg^{2+}, K^+ or Na^+. The energy absorption sequence is: $Ca^{2+} > Mg^{2+} > K^+ > Na^+$ and this provides a general ordering of cation exchangeability for common ions in groundwater.

Values of CEC are found experimentally with laboratory results reported in terms of meq $(100\ g)^{-1}$. CEC is commonly determined by extraction of the cations from soils or aquifer materials with a solution containing a known cation, normally NH_4^+. Ammonium acetate (CH_3COONH_4), the salt of a weak acid and weak base, is usually used for this purpose, its pH being adjusted to the value most suited to the investigation. A review of methods for determining exchangeable cations is provided by Talibudeen (1981). The surface area and CEC of various clays and Fe and Al oxy-hydroxides are given in Table 3.7.

An example of cation exchange occurring in groundwater is found in the Jurassic Lincolnshire Limestone aquifer in eastern England. The Lincolnshire Limestone aquifer comprises 10–30 m of oolitic limestone with a variable content of finely disseminated iron minerals and dispersed clay and organic matter acting as reactive exchange sites for Ca^{2+} and Na^+. The aquifer is confined down-dip by thick marine clays. A hydrochemical survey (Fig. 3.19) showed that cation exchange occurs at around 12 km from the aquifer outcrop and that within a further 12 km downgradient the Ca^{2+} decreases to a minimum of <4 mg L^{-1} as a result of exchange with Na^+ (Edmunds & Walton

Table 3.7 Surface area and cation exchange capacity (CEC) values for clays and Fe and Al oxyhydroxides. After Talibudeen (1981) and Drever (1988).

	Surface area $(m^2 g^{-1})$	CEC $(meq\ 100\ g^{-1})$
Fe and Al oxy-hydroxides (pH ~8.0)	25–42	0.5–1
Smectite	750–800	60–150
Vermiculite	750–800	120–200
Bentonite	750	100
Illite	90–130	10–40
Kaolinite	10–20	1–10
Chlorite	–	<10

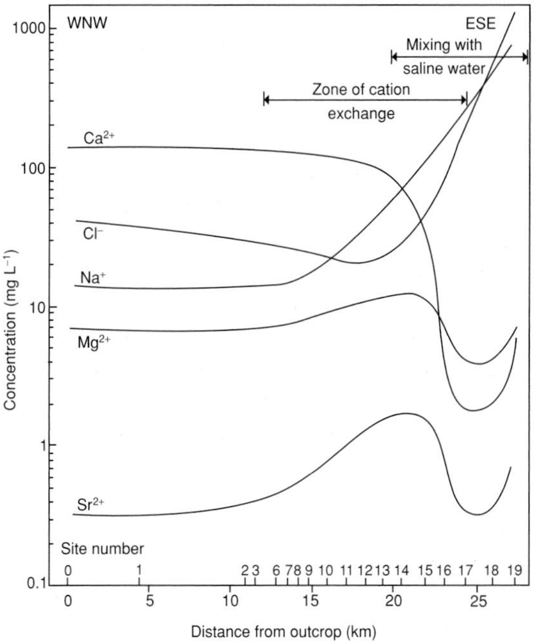

Fig. 3.19 Hydrogeochemical trends in the Lincolnshire Limestone for Ca^{2+}, Mg^{2+}, Na^+, Sr^{2+} and Cl^-. The trend lines are for 1979 and illustrate the effect of cation exchange between Ca^{2+} and Na^+ and the onset of mixing with saline water in the deeper aquifer. After Edmunds and Walton (1983).

1983). The lack of cation exchange closer to the aquifer outcrop is explained by the exhaustion of the limited cation exchange capacity of the limestone. The concentrations of Sr^{2+} and, to a lesser extent, Mg^{2+} continue to increase for around 22 km from outcrop as a result of incongruent dissolution. The removal of Ca^{2+} by cation exchange causes calcite dissolution

to occur to restore carbonate equilibrium. However, once the Ca^{2+}/Sr^{2+} and Ca^{2+}/Mg^{2+} equivalents ratios fall below a certain critical level (~20 : 1 and 1 : 1, respectively) cation exchange reactions become dominant and both Sr^{2+} and Mg^{2+} concentrations begin to decrease (Edmunds & Walton 1983).

Cation exchange reactions are a feature of saline water intrusion in coastal areas. Freshwater in coastal areas is typically dominated by Ca^{2+} and HCO_3^- ions from the dissolution of calcite such that cation exchangers present in the aquifer have mostly Ca^{2+} adsorbed on their surfaces. In seawater, Na^+ and Cl^- are the dominant ions and aquifer materials in contact with seawater will have Na^+ attached to the exchange surfaces. When seawater intrudes a coastal freshwater aquifer, the following cation exchange reaction can occur:

$$Na^+ + {}^1\!/_2Ca\text{-}X \rightarrow Na\text{-}X + {}^1\!/_2Ca^{2+} \qquad eq.\ 3.21$$

where X indicates the exchange material. As the exchanger takes up Na^+, Ca^{2+} is released, and the hydrochemical water type evolves from Na-Cl to Ca-Cl. The reverse reaction can occur when freshwater flushes a saline aquifer:

$$^1\!/_2Ca^{2+} + Na\text{-}X \rightarrow {}^1\!/_2Ca\text{-}X + Na^+ \qquad eq.\ 3.22$$

where Ca^{2+} is taken up from water in return for Na^+ resulting in a Na-HCO$_3$ water type. An example of this reaction is given in Box 3.7 for the Lower Mersey Basin Permo-Triassic sandstone aquifer of north-west England.

The chemical reactions that occur during freshwater and saline water displacements in aquifers can be identified from a consideration of conservative mixing of fresh and saline water end-member solutions and comparing with individual water analyses. For conservative mixing:

$$c_{i,mix} = f_{saline} \cdot c_{i,saline} + (1 - f_{saline})c_{i,fresh} \qquad eq.\ 3.23$$

where c_i is the concentration of ion i; $_{mix}$, $_{fresh}$ and $_{saline}$ indicate the conservative mixture and end-member fresh and saline waters; and f_{saline} is the fraction of saline water. Any change in the sample composition as a result of reactions, for example cation exchange, other than by simple mixing ($c_{i,react}$) is then simply found from:

$$c_{i,react} = c_{i,sample} - c_{i,mix} \qquad \text{eq. 3.24}$$

As an example calculation, the data shown in Table 3.8 are for samples 0, 14 and 19 of Fig. 3.19 representing fresh, mixed and saline groundwaters present in the Lincolnshire Limestone aquifer. To calculate how much Na^+ has been added to the mixed groundwater sample by cation exchange, equation 3.23 can be re-written as:

$$c_{Na,mix} = f_{saline} \cdot c_{Na,saline} + (1 - f_{saline})c_{Na,fresh} \qquad \text{eq. 3.25}$$

Now, using the mixed and saline Cl^- concentration values (samples 14 and 19) to indicate the fraction of the saline water ($f_{saline} = 114/1100$), and assuming a freshwater end-member Cl^- concentration value equal to zero, then substituting the values from Table 3.8 into equation 3.25:

Cation exchange in the Lower Mersey Basin Permo-Triassic sandstone aquifer, England

BOX 3.7

The Lower Mersey Basin Permo-Triassic sandstone aquifer of north-west England demonstrates the effect of very long-term natural flushing of a saline aquifer. The aquifer comprises two main units: the Permian Collyhurst Sandstone Formation and the Triassic Sherwood Sandstone Formation that dip southwards at about 5° and are up to 500 m thick. To the south, the aquifer unit is overlain by the Triassic Mercia Mudstone Group, a formation which contains evaporites. Underlying the aquifer is a Permian sequence, the upper formations of which are of low permeability that rest unconformably on Carboniferous mudstones. The sequence is extensively faulted with throws frequently in excess of 100 m. Overlying the older formations are highly heterogeneous, vertically variable Quaternary deposits dominated by glacial till. Pumping of the aquifer system has caused a decline in water levels such that much of the sandstone aquifer is no longer confined by the till.

Typical compositions of the sandstones are quartz 60–70%, feldspar 3–6%, lithic clasts 8%, calcite 0–10% and clays (including smectite) <15%. Haematite imparts a red colour to most of the sequence. The sandstones contain thin mudstone beds often less than 10 cm in thickness. Cation exchange capacities are of the order of 1 meq $(100\,g)^{-1}$. Underlying the fresh groundwaters present in the area are saline groundwaters attaining a Cl^- concentration of up to 100 g L^{-1}, which appear to be derived from dissolution of evaporites in the overlying Mercia Mudstone Group (Tellam 1995). This saline water is present within 50 m of ground level immediately up-flow of the Warburton Fault Block and along the Mersey Valley but the freshwater–saline water interface is found deeper both to the north and south (Tellam et al. 1986).

A hydrochemical survey of around 180 boreholes across the Lower Mersey Basin was conducted in the period 1979–1980. The results were presented by Tellam (1994) and five water types were identified (Table 1, Fig. 1). Salinities ranged from 100 mg L^{-1} up to brackish water concentrations. Ion proportions varied widely, with Ca-HCO_3, Ca-SO_4 and Na-HCO_3 being dominant water types in various locations. The large storativity of the aquifer means that the groundwater chemistry does not substantially change seasonally. The spatial distribution of water types is shown in Fig. 2. In general, the hydrochemical distribution of water types correlates with the broad pattern of groundwater flow in the aquifer. Type 1 water is found in areas of recent groundwater recharge, in contrast to older fresh groundwaters (Types 4 and 5) and saline groundwater

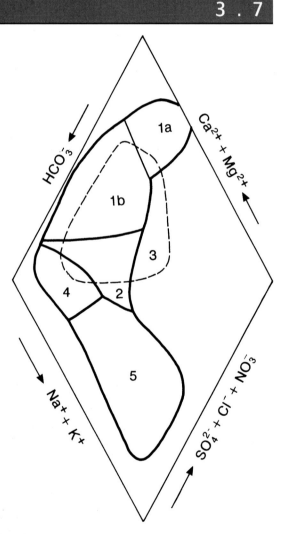

Fig. 1 The central area of a Piper diagram showing the distribution of hydrochemical water types in the Lower Mersey Basin Permo-Triassic sandstone aquifer. After Tellam (1994).

BOX
3.7

Continued

(Type S2) located in areas of the aquifer with low hydraulic gradients. The effects of modern groundwater abstractions are reflected by saline intrusion (Type S1) and recently induced recharge through the Quaternary deposits (Type 2).

Example analyses of water Types 4 and 5 are shown in Table 2. Type 5 water that occurs in the eastern confined areas and in the

Mersey Valley area are very similar to Type 4, except it has a high Na^+/Cl^- equivalents ratios (Table 2) and a range of Cl^- concentrations from <10 to 1000 mg L^{-1}. In both areas the saline–water interface is at a high level and groundwater flow rates are low. The hydrochemistry of Type 5 water is interpreted as having experienced Na^+ release by cation exchange following invasion by Type 4 water

Table 1 Hydrochemical water types for the Lower Mersey Basin Permo-Triassic sandstone aquifer as determined from borehole water chemistry. Table provided courtesy of J.H. Tellam.

Water type	General characteristics	Occurrence	Interpretation
1A [1B]	High NO_3, SO_4, Cl, tritium Low HCO_3, Ca, Mg, pH SIC << 0 [As for 1A but SIC ~ 0]	Predominantly where aquifer outcrops [As above]	Samples dominated by recent recharge, affected by agricultural and industrial activity [Slightly older water, or water which has been induced to flow through carbonate-containing parts of the aquifer near outcrop]
2	NO_3 often low, but not always Variable SO_4 and Cl pH ~7 SIC ~0	Predominantly under glacial till	Mixing in borehole of water recharged locally through Quaternary deposits and water recharged at outcrop. Through-Quaternary recharge occurs because of leakage induced by abstraction. Approximately a mixture of Types 1 and 4, though modified during travel through the Quaternary deposits
4	NO_3 < d.l., low SO_4, low Cl High HCO_3, Ca, Mg pH ~8 SIC ~0; SID ~ 0	Predominantly under glacial till	Samples dominated by pre-industrially recharged water with low NO_3, SO_4, Cl. Has encountered enough mineral carbonate to become saturated with respect to calcite and dolomite
5	NO_3 < d.l., low SO_4, low to very high Cl Very high HCO_3 pH 8–9 Very low to low Ca, Mg Na/Cl > to >> 1 SIC ~0; SID ~0	Below glacial till, adjacent to saline groundwater	Type 4 water, but in a part of the aquifer previously occupied by saline groundwaters. The Na/Cl ratio indicates ion exchange, the exchangers releasing Na sorbed when the saline groundwater was present. As Ca and Mg are taken up, more carbonate is dissolved, and pH and HCO_3 rise. Cl concentration depends on whether there is any saline water left in the system
3	Generally low and variable concentrations of major determinands SIC < 0	Collyhurst Sandstone Formation and in an isolated fault block in west of the area	Post-industrial recharge, but with limited pollutants. Limited carbonate in aquifer results in carbonate undersaturation, even for waters with considerable residence time
Saline groundwater S1	Cl to 5 g L^{-1}, high I, SO_4 < expected from Mersey Estuary High Ca/Cl, low Na/Cl, SIC ~0	Bordering the Mersey Estuary (average Cl concentration ~5 g L^{-1})	Intrusion from Mersey Estuary (Carlyle et al. 2004), accompanied by various reactions including SO_4 reduction, CO_2 degassing and ion exchange
Saline groundwater S2	Cl to 105 g L^{-1} Low Br/Cl compared with S1 Low SO_4/Cl compared with S1 Very light $\delta^{18}O$ and δ^2H	Below fresh groundwater in Mersey Valley inland from Mersey Estuary and adjacent areas; interface with freshwater up to 250 m below ground level	Brines resulting from dissolution of evaporites in the Mercia Mudstone Group, and subsequent migration. Upper part of the saline groundwater was diluted by freshwater recharged under climatic conditions significantly cooler than at present

SIC, saturation index for calcite; SID, saturation index for dolomite [$\log_{10}IAP/K$]; d.l., detection limit.

Continued

Table 2 Example of hydrochemical analyses of Types 4 and 5 groundwaters from the Lower Mersey Basin Permo-Triassic sandstone aquifer. After Tellam (1994).

Water type	Site*	Na^+ (mg L^{-1})	K^+ (mg L^{-1})	Ca^{2+} (mg L^{-1})	Mg^{2+} (mg L^{-1})	Cl^- (mg L^{-1})	SO_4^{2-} (mg L^{-1})	HCO_3^- (mg L^{-1})	NO_3^- (mg L^{-1})	Na^+/Cl^-	pH	SIC
4	20 (100)	15	4.5	48	32	20	2	322	0.0	1.14	7.7	+0.1
	88 (P)	46	4.0	108	35	24	17	540	1.1	2.92	7.8	+0.8
	104 (180)	17	4.0	83	48	28	8	513	0.4	0.92	7.3†	+0.3
5	109 (47)	192	1.2	3.7	1.7	25	9	487	0.4	11.69	6.8	−1.7
	123 (95)	947	2.7	2.3	1.7	1196	76	454	0.4	1.20	8.8	−0.1
	140 (P)	170	3.7	37	14	102	8	438	0.0	2.54	7.9	−0.2

SIC, saturation index for calcite [$\log_{10}$IAP/K].
* Figures in parentheses indicate depth in metres; (P), pumped.
† Laboratory measured pH value.

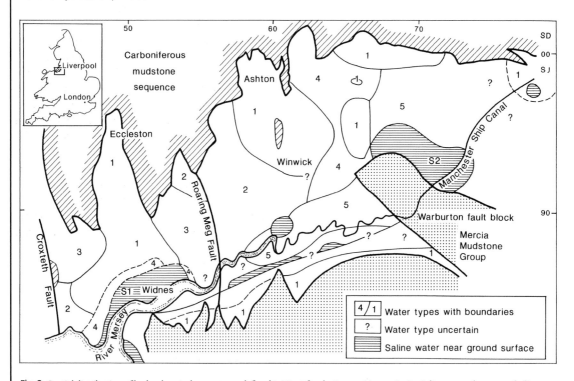

Fig. 2 Spatial distribution of hydrochemical water types defined in Fig. 1 for the Lower Mersey Basin. Saline groundwater underlies much of the fresh groundwater in the Permo-Triassic sandstone aquifer. After Tellam (1994).

into regions originally occupied by saline groundwater. Where flushing has been less complete, higher Cl^- concentrations occur. The Na^+ release is accompanied by Ca^{2+} uptake by the aquifer as described by equation 3.22. An extreme example is provided by sample 123 in Table 2 where the water contains greater than 1000 mg L^{-1} Cl^-, 2 mg L^{-1} Ca^{2+} and 2 mg L^{-1} Mg^{2+}. The removal of Ca^{2+} promotes calcite dissolution with HCO_3^- concentrations able to reach 500–600 mg L^{-1} (Tellam 1994).

Table 3.8 Selected hydrochemical data for the Lincolnshire Limestone aquifer to illustrate cation exchange. Data from Edmunds and Walton (1983).

	Na^+ (mg L^{-1})	Ca^{2+} (mg L^{-1})	Cl^- (mg L^{-1})	Na^+/Cl^-
Freshwater (Sample 0, Ropsley)	14	135	42	0.51
Mixed water (Sample 14, Pepper Hill)	280	17	114	3.74
Saline water (Sample 19, Deeping St Nicholas)	920	9.5	1100	1.27

$$c_{Na,mix} = (114/1100)920 + (1 - 114/1100)14$$
$$= 107.9 \text{ mg L}^{-1}$$

Similarly, equation 3.24 can be re-written to calculate the amount of Na^+ released in the cation exchange reaction:

$$c_{Na,react} = c_{Na,sample} - c_{Na,mix} \qquad \text{eq. 3.26}$$

Now, using the result for $c_{Na,mix}$ found above:

$$c_{Na,react} = 280 - 107.9 = 172.1 \text{ mg L}^{-1}$$
$$(\text{or } 7.5 \text{ meq L}^{-1})$$

A similar calculation for Ca^{2+} removed from the mixed groundwater sample results in:

$$c_{Ca,mix} = (114/1100)9.5 + (1 - 114/1100)135$$
$$= 122.0 \text{ mg L}^{-1}$$

and

$$c_{Ca,react} = 17 - 122.0 = -105.0 \text{ mg L}^{-1}$$
$$(\text{or } 5.3 \text{ meq L}^{-1})$$

3.9 Redox chemistry

Reactions involving a change in oxidation state are referred to as oxidation-reduction or redox reactions. Redox reactions have a controlling influence on the solubility and transport of some minor elements in groundwater such as Fe and Mn and also on redox sensitive species such as NO_3^- and SO_4^{2-}. The extent to which redox reactions occur in groundwater systems is therefore significant with respect to many practical problems, for example issues of groundwater quality for drinking water, the attenuation of landfill leachate plumes and the remediation of sites contaminated by organic pollutants. The major redox sensitive components of groundwaters and aquifers are O_2, $NO_3^-/N_2/NH_4^+$, SO_4^{2-}/HS^-, $Mn(II)/Mn(IV)$ and $Fe(II)/Fe(III)$. Redox sensitive trace elements include As, Se, U and Cr in addition to Fe and Mn. The toxic effects of these elements differ greatly for various redox species, for example $Cr(III)/Cr(VI)$, and so it is important that the behaviour of these elements can be predicted on the basis of the groundwater redox conditions.

During redox reactions, electrons are transferred between dissolved, gaseous or solid constituents and result in changes in the oxidation states of the reactants and products. The oxidation state (or oxidation number) represents the hypothetical charge that an atom would have if the ion or molecule were to dissociate. The oxidation states that can be achieved by the most important multi-oxidation state elements that occur in groundwater are listed in Table 3.9. By definition, oxidation is the loss of electrons and reduction is the gain of electrons. Every oxidation is accompanied by a reduction and vice versa, so that an electron balance is always maintained (Freeze & Cherry 1979).

For every redox half-reaction, the following form of an equation can be written:

$$\text{oxidized state} + ne^- = \text{reduced state} \qquad \text{eq. 3.27}$$

As an example, the redox reaction for the oxidation of Fe can be expressed by two half-reactions:

$$^1/_2O_2 + 2H^+ + 2e^- = H_2O \quad (\text{reduction}) \qquad \text{eq. 3.28}$$

$$2Fe^{2+} = 2Fe^{3+} + 2e^- \quad (\text{oxidation}) \qquad \text{eq. 3.29}$$

The complete redox reaction for the oxidation of Fe is found from the addition of equations 3.28 and 3.29

Table 3.9 Examples of oxidation states for various compounds that occur in groundwater. The oxidation state of free elements, whether in atomic or molecular form, is zero. Other rules for assigning oxidation states include: the oxidation state of an element in simple ionic form is equal to the charge on the ion; the sum of oxidation states is zero for molecules; and for ion pairs or complexes it is equal to the formal charge on the species. After Freeze and Cherry (1979).

Carbon compounds		Sulphur compounds		Nitrogen compounds		Iron compounds	
Substance	C state	Substance	S state	Substance	N state	Substance	Fe state
HCO_3^-	+IV	S	0	N_2	0	Fe	0
CO_3^{2-}	+IV	H_2S	–II	SCN^-	+II	FeO	+II
CO_2	+IV	HS^-	–II	N_2O	–III	$Fe(OH)_2$	+II
CH_2O	0	FeS_2	–I	NH_4^+	+III	$FeCO_3$	+II
$C_6H_{12}O_6$	0	FeS	–II	NO_2^-	+V	FeO_3	+III
CH_4	–IV	SO_3^{2-}	+IV	NO_3^-	–III	$Fe(OH)_3$	+III
CH_3OH	–II	SO_4^{2-}	+VI	HCN	–I	FeOOH	+III

and expresses the net effect of the electron transfer with the absence of free electrons, thus:

$$\tfrac{1}{2}O_2 + 2Fe^{2+} + 2H^+ = 2Fe^{3+} + H_2O \qquad \text{eq. 3.30}$$

By expressing redox reactions as half-reactions, the concept of pe is used to describe the relative electron activity where:

$$pe = -\log_{10}[e^-] \qquad \text{eq. 3.31}$$

pe is a dimensionless quantity and is a measure of the oxidizing or reducing tendency of the solution where pe and pH are functions of the free energy involved in the transfer of 1 mole of electrons or protons, respectively, during a redox reaction.

For the general half-reaction:

$$\text{oxidants} + ne^- = \text{reductants} \qquad \text{eq. 3.32}$$

where n is the number of electrons transferred, then, from the law of mass action:

$$K = \frac{[\text{reductants}]}{[\text{oxidants}][e^-]^n} \qquad \text{eq. 3.33}$$

A numerical value for such an equilibrium constant can be computed using Gibbs free energy data for conditions at 25°C and 1 atmosphere pressure. By convention, the equilibrium constant for a half-reaction is always expressed in the reduction form. The oxidized forms and electrons are written on the left and the reduced products on the right.

Rearrangement of equation 3.33 gives the electron activity $[e^-]$ for a half-reaction as:

$$[e^-] = \left\{ \frac{[\text{reductants}]}{[\text{oxidants}]K} \right\}^{1/n} \qquad \text{eq. 3.34}$$

Rewriting equation 3.34 by taking the negative logarithm of both sides yields:

$$-\log_{10}[e^-] = pe$$

$$= \frac{1}{n} \left\{ \log_{10} K - \log_{10} \frac{[\text{reductants}]}{[\text{oxidants}]} \right\} \qquad \text{eq. 3.35}$$

When a half-reaction is written in terms of a single electron transfer, or $n = 1$, the $\log_{10} K$ term is written as pe^o such that:

$$pe = pe^o - \log_{10} \frac{[\text{reductants}]}{[\text{oxidants}]} \qquad \text{eq. 3.36}$$

Tabulations of thermodynamic data for redox reactions are commonly expressed as pe^o values. A set of reduction reactions of importance in groundwater, together with their respective pe^o values, is listed in Table 3.10. The reactions are listed on the basis

Table 3.10 Table of reduction reactions of importance in groundwater. After Champ et al. (1979).

Reaction	$pe^{o} = \log_{10}K$
(1) $^{1}/_{4}O_{2(g)} + H^{+} + e^{-} = ^{1}/_{2}H_{2}O$	+20.75
(2) $^{1}/_{5}NO_{3}^{-} + ^{6}/_{5}H^{+} + e^{-} = ^{1}/_{10}N_{2(g)} + ^{3}/_{5}H_{2}O$	+21.05
(3) $^{1}/_{2}MnO_{2(s)} + 2H^{+} + e^{-} = ^{1}/_{2}Mn^{2+} + H_{2}O$	+20.8
(4) $^{1}/_{8}NO_{3}^{-} + ^{5}/_{4}H^{+} + e^{-} = ^{1}/_{8}NH_{4}^{+} + ^{3}/_{8}H_{2}O$	+14.9
(5) $Fe(OH)_{3(s)} + 3H^{+} + e^{-} = Fe^{2+} + 3H_{2}O$	+17.1
(6) $^{1}/_{8}SO_{4}^{2-} + ^{9}/_{8}H^{+} + e^{-} = ^{1}/_{8}HS^{-} + ^{1}/_{2}H_{2}O$	+4.25
(7) $^{1}/_{8}CO_{2(g)} + H^{+} + e^{-} = ^{1}/_{8}CH_{4(g)} + ^{1}/_{4}H_{2}O$	+2.87
(8) $^{1}/_{6}N_{2(g)} + ^{4}/_{3}H^{+} + e^{-} = ^{1}/_{3}NH_{4}^{+}$	+4.68
(9) $^{1}/_{4}CO_{2(g)} + H^{+} + e^{-} = ^{1}/_{4}CH_{2}O + ^{1}/_{4}H_{2}O$	−1.20

of decreasing oxidizing ability, such that species associated with a reaction of more positive pe^{o} act as electron acceptors or oxidizing agents in the oxidation of species associated with reactions of significantly more negative pe^{o}.

As an example of this law of mass action approach to redox reactions, the following equilibrium constants can be written for the two half-reactions describing the oxidation of Fe^{2+} to Fe^{3+} by free oxygen (eqs 3.28 and 3.29):

$$K = \frac{1}{Po_{2}^{1/4}[H^{+}][e^{-}]} = 10^{20.75} \qquad \text{eq. 3.37}$$

$$K = \frac{[Fe^{2+}]}{[Fe^{3+}][e^{-}]} = 10^{13.05} \qquad \text{eq. 3.38}$$

Rewriting equations 3.37 and 3.38 in logarithmic form produces:

$$pe = 20.75 + ^{1}/_{4}\log_{10}Po_{2} - pH \qquad \text{eq. 3.39}$$

$$pe = 13.05 + \log_{10}\left(\frac{[Fe^{3+}]}{[Fe^{2+}]}\right) \qquad \text{eq. 3.40}$$

If the complete redox reaction (eq. 3.30) is at equilibrium, and if the concentrations of Fe^{2+} and Fe^{3+}, Po_{2} and pH are known, then the pe obtained from both these relations (eqs 3.39 and 3.40) is the same.

In groundwater systems there is an interdependency of pe and pH which can be conveniently represented

as pe–pH stability diagrams. Methods for the construction of pe–pH diagrams are presented by Stumm and Morgan (1981). As an example, a Fe stability diagram is shown in Fig. 3.20. The equilibrium equations required to construct this type of diagram provide boundary conditions towards which a redox

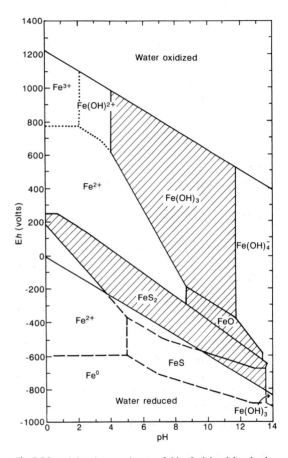

Fig. 3.20 Stability diagram showing fields of solid and dissolved forms of Fe as a function of Eh and pH at $25\,^{\circ}C$ and 1 atmosphere pressure. The diagram represents a system containing activities of total sulphur species of 96 mg L^{-1} as SO_{4}^{2-}, total carbon dioxide species of 61 mg L^{-1} as HCO_{3}^{-}, and dissolved Fe of 56 µg L^{-1}. Solids indicated by the shaded areas would be thermodynamically stable in their designated domains. Boundaries between solute species are not sensitive to specific dissolved Fe activity, but the domains of solid species will increase in area if more dissolved Fe is present. The boundaries for sulphides and elemental Fe extend below the water stability boundary and show the conditions required for thermodynamic stability of Fe^{o}. Under the conditions specified, siderite ($FeCO_{3}$) saturation is not reached. Therefore, $FeCO_{3}$ is not a stable phase and does not have a stability domain in the diagram. After Hem (1985).

system is proceeding. In practical terms, stability diagrams can be used to predict the likely dissolved ion or mineral phase that may be present in a groundwater for a measured pe–pH condition.

As an alternative to pe, the redox condition for equilibrium processes can be expressed in terms of Eh. Eh is commonly referred to as the (platinum electrode) redox potential and is defined as the energy gained in the transfer of 1 mole of electrons from an oxidant to H_2. Eh is defined by the Nernst equation:

$$Eh(\text{volts}) = Eh^\circ + \frac{2.303RT}{nF} \log_{10}\left(\frac{[\text{oxidants}]}{[\text{reductants}]}\right)$$

$$\text{eq. 3.41}$$

where F is the Faraday constant (9.65×10^4 C mol^{-1}), R the gas constant and T the absolute temperature (degrees K). Eh° is a standard or reference condition at which all substances involved in the redox reaction are at a hypothetical unit activity. Tabulated values of Eh° are readily available, for example in Krauskopf and Bird (1995).

The equation relating Eh° to the thermodynamic equilibrium constant is:

$$Eh^\circ = \frac{RT}{nF} \log_e K$$

$$\text{eq. 3.42}$$

Finally, the relationship between Eh and pe is:

$$Eh = \frac{2.303RT}{nF} pe$$

$$\text{eq. 3.43}$$

which at 25°C becomes:

$$Eh = \frac{0.059}{n} pe$$

$$\text{eq. 3.44}$$

The Nernst equation (eq. 3.41) assumes that the species participating in the redox reactions are at equilibrium and that the redox reactions are reversible, both in solution and at the electrode–solution interface. Since the redox reactions involving most of the dissolved species in groundwater are not reversible, such correlations between Eh, pe and pH can be limited in practice. Instead, and as proposed by Champ et al. (1979), it is valuable to consider redox processes from a qualitative point of view in which overall changes in redox conditions in an aquifer are described. As part of the concept of redox sequences, Champ et al. (1979) suggested that three redox zones exist in aquifer systems: the oxygen-nitrate, iron-manganese and sulphide zones (Fig. 3.21).

Two general types of hydrochemical systems are recognized by Champ et al. (1979): closed and open oxidant systems. In the closed system, the groundwater

Table 3.11 Sequence of redox processes in a closed system. In this example, the simplest carbohydrate, CH_2O, represents the dissolved organic carbon (DOC) that acts as a reducing agent to reduce the various oxidized species initially present in a recharging groundwater. For a confined aquifer containing excess DOC and some solid phase Mn(IV) and Fe(III), it is predicted, on the basis of decreasing negative values of free energy change, that the oxidized species will be reduced in the sequence O_2, NO_3^-, Mn(IV), Fe(III), SO_4^{2-}, HCO_3^- and N_2. As the reactions proceed, and in the absence of other chemical reactions such as ion exchange, the equations show that the sum of dissolved inorganic carbon species (H_2CO_3, HCO_3^-, CO_3^{2-} and complexes) rises as DOC is consumed. The pH of the groundwater may also increase depending on the relative importance of the Fe and Mn reduction processes. After Champ et al. (1979).

Reaction	Equation
Aerobic respiration	$CH_2O + O_2 = CO_2 + H_2O$
Denitrification	$CH_2O + \frac{4}{5}NO_3^- + \frac{4}{5}H^+ = CO_2 + \frac{2}{5}N_2 + \frac{7}{5}H_2O$
Mn(IV) reduction	$CH_2O + 2MnO_2 + 4H^+ = 2Mn^{2+} + 3H_2O + CO_2$
Fe(III) reduction	$CH_2O + 8H^+ + 4Fe(OH)_3 = 4Fe^{2+} + 11H_2O + CO_2$
Sulphate reduction	$CH_2O + \frac{1}{2}SO_4^{2-} + \frac{1}{2}H^+ = \frac{1}{2}HS^- + H_2O + CO_2$
Methane fermentation	$CH_2O + \frac{1}{2}CO_2 = \frac{1}{2}CH_4 + CO_2$
Nitrogen fixation	$CH_2O + H_2O + \frac{2}{3}N_2 + \frac{4}{3}H^+ = \frac{4}{3}NH_4^+ + CO_2$

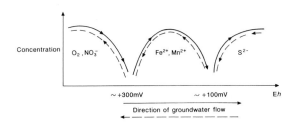

Fig. 3.21 Variation in the concentration of the major dissolved species affected by redox processes within a groundwater flow system. After Champ et al. (1979).

Table 3.12 Sequence of redox processes in an open system in which an excess of dissolved oxygen reacts with reduced species. Under these conditions, the reduced species will be oxidized in the sequence: dissolved organic carbon (CH_2O), HS^-, Fe^{2+}, NH_4^+ and Mn^{2+}. As each of these reactions proceeds, the Eh will become more positive, while the pH of the groundwater should decrease. After Champ et al. (1979).

Reaction	Equation
Aerobic respiration	$O_2 + CH_2O = CO_2 + H_2O$
Sulphide oxidation	$O_2 + \frac{1}{2}HS^- = \frac{1}{2}SO_4^{2-} + \frac{1}{2}H^+$
Fe(II) oxidation	$O_2 + 4Fe^{2+} + 10H_2O = 4Fe(OH)_3 + 8H^+$
Nitrification	$O_2 + \frac{1}{2}NH_4^+ = \frac{1}{2}NO_3^- + H^+ + \frac{1}{2}H_2O$
Mn(II) oxidation	$O_2 + 2Mn^{2+} + 2H_2O = 2MnO_2 + 4H^+$

initially contains dissolved oxidized species such as O_2, NO_3^-, SO_4^{2-} and CO_2 and also excess reduced dissolved organic carbon (DOC). After entering the groundwater flow system via the recharge area, the groundwater is then closed to the input of further oxidants or oxidized species. In the open system, excess dissolved oxygen is present which may react with reduced species such as HS^- and NH_4^+, for example in situations where landfill leachate is in contact with groundwater. Champ et al. (1979) recognized that in closed systems containing excess reducing agent (DOC) and open systems containing an excess of dissolved oxygen, sequences of redox reactions can be identified as summarized in Tables 3.11 and 3.12, respectively. An example of the identification of a groundwater redox sequence is provided in Box 3.8 and further discussion of the specific redox reaction of microbially mediated denitrification is presented in Box 3.9.

3.10 Groundwater chemistry of crystalline rocks

Weathering processes participate in controlling the hydrogeochemical cycles of many elements. In soluble carbonate and evaporite deposits, solution processes are rapid and, generally, congruent but in lithologies composed of silicates and quartz, the solution processes are very slow and incongruent. In many sedimentary sandstone aquifers, the solution of traces of carbonate present either as cement or detrital grains may predominate over any chemical contribution from the silicate minerals. Crystalline rocks of igneous or metamorphic origin, on the other hand, contain appreciable amounts of quartz and aluminosilicate minerals such as feldspars and micas. Although the dissolution of most silicate minerals

---

BOX 3.8

Redox processes in the Lincolnshire Limestone aquifer, England

The Jurassic Lincolnshire Limestone aquifer in eastern England provides an example of the zonation of redox conditions in an aquifer system. In this fissured, oolitic limestone aquifer (Fig. 1), an eastward decline in NO_3^- concentration accompanies progressive removal of dissolved oxygen and lowering of the redox potential (Eh) from an initial value of about +400 mV. Over a distance of 10 km, the groundwater NO_3^- concentration declines from a maximum of 25 mg L^{-1} as N at outcrop to less than 5 mg L^{-1} as N when confined. Four cored boreholes were drilled along a groundwater flow line and the material recovered examined for organic carbon content and the presence of denitrifying bacteria. The results

showed that the DOC content of the pore water ranged between 13 and 28 mg L^{-1} and that of the mobile groundwater between 1.6 and 3.4 mg L^{-1}. Denitrifying bacteria were cultured from samples scraped from fissure walls, but not from samples incubated with pore water. Thus, it appears that the source of DOC supporting denitrification is contained in the limestone matrix, and that the very small pore size of the matrix restricts denitrification to short distances from fissure walls (Lawrence & Foster 1986).

Further into the confined aquifer, at about 12 km from outcrop, the Eh falls to less than +100 mV and the groundwater environment remains anaerobic. In this region, SO_4^{2-} reduction is noticeable by the presence of H_2S and lowered SO_4^{2-} concentrations, although the process is sluggish with SO_4^{2-} concentrations persisting for at least 5 km beyond the decline in redox potential. According to Edmunds and Walton (1983), Fe is initially quite soluble in this reducing environment ($Fe^{2+} < 0.5$ mg L^{-1}) but once sulphide is pro-

duced by SO_4^{2-} reduction (see reaction equation in Table 3.11) much of the Fe^{2+} is removed as ferrous sulphide (FeS) as follows:

$$Fe^{2+} + HS^- \rightleftharpoons FeS_{(s)} + H^+ \qquad \text{eq. 1}$$

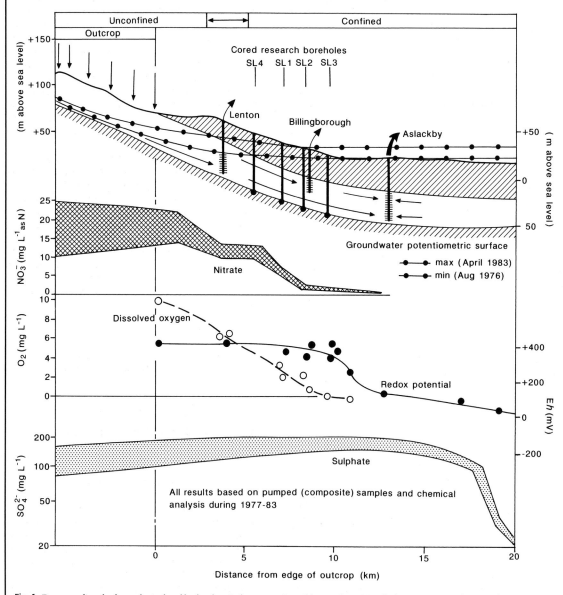

Fig. 1 Downgradient hydrogeological and hydrochemical cross-section of the southern Lincolnshire Limestone showing the sequential changes in the redox-sensitive species O_2, NO_3^- and SO_4^{2-}. After Lawrence and Foster (1986).

BOX
3.9

Microbially mediated denitrification

The sequences of redox reactions shown in Tables 3.11 and 3.12 are thermodynamically favoured, but their reaction rates are slow in the absence of catalysts. In natural waters, the most important electron-transfer mechanism is catalysis associated with microbially produced enzymes. As shown in Fig. 1, the most stable nitrogen species within the Eh-pH range encountered in the majority of groundwaters is erroneously predicted to be gaseous nitrogen (N_2). The observed departure from equilibrium is explained by the catalysing effect of bacteria in accelerating the biological reduction of NO_3^- at lower redox potentials (Hiscock et al. 1991).

The viability of micro-organisms in groundwater is dependent on two important factors that limit enzymatic function and cell growth, namely temperature and nutrient availability. At low temperatures, microbial activity decreases markedly but is measurable between 0 and 5°C. For the process of denitrification, a general doubling is observed with every 10°C increase in temperature (Gauntlett & Craft 1979). The nutrients required for biosynthesis include those

elements required in large amounts (C, H, O, N, P, S), minor amounts of minerals (K, Na, Mg, Ca, Fe) and trace amounts of certain metals (Mn, Zn, Cu, Co and Mo). On the basis of average cellular composition, the favourable ratio of C : N : P : S is about 100 : 20 : 4 : 1 (Spector 1956). For energy generation, electron donors (dissolved organic carbon (DOC), H_2S, NH_3, Fe^{2+}) and electron acceptors (DO, NO_3^-, Fe(III), Mn(IV), SO_4^{2-}, CO_2) are required. It is normally considered that most groundwaters should be capable of supplying the very low concentrations of minerals and trace metals required by microbes, as well as the electron donors, particularly DOC (Box 3.8). From a survey of 100 groundwaters, Thurman (1985) reported a median DOC content of 0.7 mg L^{-1} for sandstone, limestone and sand and gravel aquifers which should meet microbial requirements which have been reported to be less than 0.1 mg L^{-1} (Zobell & Grant 1942). In aquifer situations where the availability of DOC is limited, then other electron donors such as reduced sulphur species become important.

Denitrification is observed to proceed at reduced oxygen levels via a number of microbially mediated steps, the end product of which is normally gaseous nitrogen (N_2) (Korom 1992). The denitrification process requires a suitable electron donor or donors to complete the dissimilatory reduction of NO_3^- to N_2, the most likely of which are organic carbon (heterotrophic denitrification) and reduced Fe and S species (autotrophic denitrification). The stoichiometry of denitrification reactions can be expressed by the following simple equations which describe a generalized progression of reactions with depth below the water table:

1 Heterotrophic denitrification in which an arbitrary organic compound (CH_2O) is oxidized:

$$5CH_2O + 4NO_3^- + 4H^+ = 5CO_2 + 2N_{2(g)} + 7H_2O \qquad \text{eq. 1}$$

In this reaction, 1.1 g of C are required to reduce 1.0 g of N.
2 Autotrophic denitrification by reduced iron in which ferrous iron is oxidized:

$$5Fe^{2+} + NO_3^- + 12H_2O = 5Fe(OH)_3 + {}^1\!/_2N_{2(g)} + 9H^+ \qquad \text{eq. 2}$$

In this reaction, 20.0 g of Fe^{2+} are required to reduce 1.0 g of N.
3 Autotrophic denitrification by reduced sulphur in which pyrite is oxidized:

$$5FeS_{2(s)} + 14NO_3^- + 4H^+ = 5Fe^{2+} + 10SO_4^{2-} + 7N_{2(g)} + 2H_2O \qquad \text{eq. 3}$$

In this reaction, 1.6 g of S are required to reduce 1.0 g of N.
In some highly reducing, carbon-rich environments, NO_3^- may be converted to NH_4^+ by dissimilatory nitrate reduction to ammonium (DNRA). DNRA may be important in some marine sediments but in less reducing groundwater environments, denitrification is generally favoured over DNRA (Korom 1992).

Although denitrification by organic matter is thermodynamically favourable, the reactivity of organic matter in denitrification is much lower than the reactivity of pyrite. To demonstrate, Fig. 2

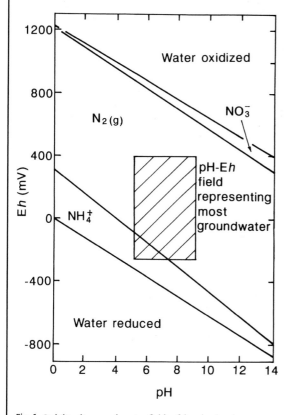

Fig. 1 Stability diagram showing fields of dissolved and gaseous forms of N as a function of Eh and pH at 25°C and 1 atmosphere pressure. The diagram represents a system containing an activity of total N species of 14 mg L^{-1}. After Stumm and Morgan (1981).

shows depth profiles for pH and the redox-sensitive species NO_3^-, SO_4^{2-} and Fe^{2+} for a multilevel observation well installed near to the Vierlingsbeek wellfield in the south-east of the Netherlands. The unconfined aquifer at this site is composed of unconsolidated fluvial sands containing no calcite and low amounts of organic matter (0.1–2%) and pyrite (<0.01–0.2%), the amounts of which decrease with depth (Fig. 3). An aquitard composed of fine-grained cemented deposits exists at 30 m below ground level (m bgl) and the water table varies between 2 and 4 m bgl. Very intensive cattle farming in the vicinity of the wellfield with the application of liquid manure has provided a source of high NO_3^- concentrations in the shallow groundwater (van Beek 2000).

At Vierlingsbeek, at 10 m bgl, Fig. 2 shows that measured NO_3^- concentrations are above 200 mg L^{-1} but that below 21 m bgl NO_3^- is absent. The decrease in NO_3^- coincides with an increase in SO_4^{2-} and Fe^{2+} between 20 and 21 m bgl that can be explained by autotrophic denitrification in which pyrite is oxidized (eq. 3). In this reaction, and in the absence of carbonate to act as a buffer, protons are consumed and a steady rise in pH is observed below 20 m bgl.

The weight percentage of pyrite in the fluvial sands is about a factor of 10 lower than the content of organic matter (Fig. 3) and it would be expected that the reduction capacity of organic matter per unit weight of solid material would be far greater than that of pyrite. Hence, the measured hydrochemical profiles demonstrate the higher reactivity of pyrite compared with organic matter in the denitrification process at this site (van Beek 2000).

This example illustrates how, with consumption of the source of electron donor, in this case pyrite, the denitrification front will migrate downwards such that the aquifer will gradually loose the ability to attenuate NO_3^-. For a similar hydrogeological situation to Vierlingsbeek, and also for denitrification in the presence of reduced sulphur with the oxidation of pyrite, Robertson et al. (1996) measured a downward rate of movement of a denitrification front in silt-rich sediments of 1 mm a^{-1}.

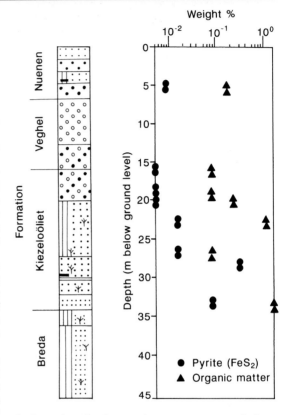

Fig. 3 Depth profile of pyrite and organic matter content in the unconsolidated fluvial sands aquifer recorded at multilevel observation well NP1 located near to Vierlingsbeek, south-east Netherlands. After van Beek (2000).

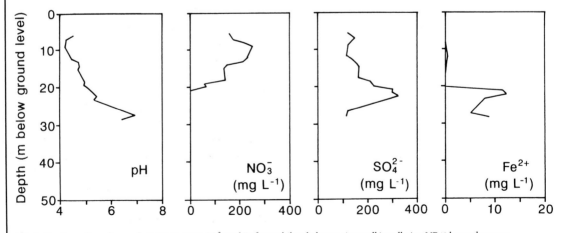

Fig. 2 Depth profiles of groundwater pH, NO_3^-, SO_4^{2-} and Fe for multilevel observation well installation NP40 located near to Vierlingsbeek, south-east Netherlands. After van Beek (2000).

Table 3.13 Estimated average mineralogical composition of the upper continental crust (volume %). The composition of the exposed crust differs from the upper continental crustal estimate primarily in the content of volcanic glass. After Nesbitt and Young (1984).

	Upper continental crust	Exposed continental crust surface
Plagioclase feldspar	39.9	34.9
Potassium feldspar	12.9	11.3
Quartz	23.2	20.3
Volcanic glass	0.0	12.5
Amphibole	2.1	1.8
Biotite mica	8.7	7.6
Muscovite mica	5.0	4.4
Chlorite	2.2	1.9
Pyroxene	1.4	1.2
Olivine	0.2	0.2
Oxides	1.6	1.4
Others	3.0	2.6

results in very dilute solutions, the weathering of silicate minerals is estimated to contribute about 45% to the total dissolved load of the world's rivers, underlining the significant role of these processes in the overall chemical denudation on the Earth's surface (Stumm & Wollast 1990).

Upper crustal rocks have an average composition similar to the rock granodiorite (Table 3.13) that is composed of the framework silicates plagioclase feldspar, potassium feldspar and quartz, with plagioclase feldspar being the most abundant. Depending on the nature of the parent rocks, various secondary minerals such as gibbsite, kaolinite, smectite and illite are formed as reaction products. In all cases, water and carbonic acid (H_2CO_3), which is the source of H^+, are the main reactants. The net result of the reactions is the release of cations (Ca^{2+}, Mg^{2+}, K^+, Na^+) and the production of alkalinity via HCO_3^-.

The two following reactions (eqs 3.45 and 3.46) provide examples of important silicate weathering processes. Firstly, taking Ca-rich plagioclase feldspar (anorthite) the incongruent weathering reaction resulting in the aluminosilicate residue kaolinite is written (Andrews et al. 2004):

$$CaAl_2Si_2O_{8(s)} + 2H_2CO_{3(aq)} + H_2O \rightarrow$$
$$Ca^{2+}_{(aq)} + 2HCO_{3(aq)} + Al_2Si_2O_5(OH)_{4(s)} \qquad \text{eq. 3.45}$$

In this weathering reaction, H^+ ions dissociated from H_2CO_3 hydrate the silicate surface and naturally buffer the infiltrating soil water or groundwater. The ionic bonds between Ca^{2+} and the SiO_4 tetrahedra are easily broken, releasing Ca^{2+} into solution resulting in a Ca-HCO_3 water type.

Secondly, for the Na-rich plagioclase feldspar (albite) the incongruent reaction producing kaolinite and releasing Na^+ and HCO_3^- ions is:

$$2NaAlSi_3O_{8(s)} + 9H_2O + 2H_2CO_{3(aq)} \rightarrow$$
$$Al_2Si_2O_5(OH)_{4(s)} + 2Na^+_{(aq)} + 2HCO^-_{3(aq)} +$$
$$4Si(OH)_{4(aq)} \qquad \text{eq. 3.46}$$

Further examples of weathering reactions for some common primary minerals are listed in Table 3.14. When ferrous iron (Fe^{2+}) is present in the lattice, as in the case of biotite mica, oxygen consumption may become an important factor affecting the rate of dissolution that results in Fe-oxide as an insoluble weathering product. For example, in the following equation, biotite weathers to gibbsite and goethite:

$$KMgFe_2AlSi_3O_{10}(OH)_{2(s)} + \tfrac{1}{2}O_{2(aq)} + 3CO_{2(aq)} +$$
$$11H_2O \rightarrow Al(OH)_{3(s)} + 2Fe(OH)_{3(s)} + K^+_{(aq)} +$$
$$Mg^{2+}_{(aq)} + 3HCO^-_{3(aq)} + 3Si(OH)_{4(aq)} \qquad \text{eq. 3.47}$$

Differences in solution rates between the silicate minerals lead to their successive disappearance as weathering proceeds. This kinetic control on the distribution of primary silicates is known as the Goldich weathering sequence (Fig. 3.22). As shown in Fig. 3.22, olivine and Ca-plagioclase are the most easily weathered minerals while quartz is the most resistant to weathering.

The leaching of different weathering products depends not just on the rate of mineral weathering, but also on the hydrological conditions. Montmorillonite ($Na_{0.5}Al_{1.5}Mg_{0.5}Si_4O_{10}(OH)_2$) is formed preferentially in relatively dry climates, where the flushing rate in the soil is low, and its formation is favoured when rapidly dissolving material such as volcanic rock is available. By contrast, gibbsite ($Al(OH)_3 \cdot 3H_2O$) forms typically in tropical areas with intense rainfall and under well-drained conditions. Here, gibbsite and other aluminium hydroxides may form a thick weathering residue of bauxite. Because of differences in residence times, water in areas where

Table 3.14 Reactions for incongruent dissolution of some aluminosilicate minerals (solid phases are underlined). After Freeze and Cherry (1979).

Aluminosilicate mineral	Reaction
Gibbsite-kaolinite	$\underline{Al_2O_3 \cdot 3H_2O} + 2Si(OH)_4 = \underline{Al_2Si_2O_5(OH)_4} + 5H_2O$
Na-montmorillonite-kaolinite	$\underline{Na_{0.33}Al_{2.33}Si_{3.67}O_{10}(OH)_2} + ^1/_3H^+ + ^{23}/_6H_2O = ^7/_6\underline{Al_2Si_2O_5(OH)_4} + ^1/_3Na^+ + ^4/_3Si(OH)_4$
Ca-montmorillonite-kaolinite	$\underline{Ca_{0.33}Al_{4.67}Si_{7.33}O_{20}(OH)_4} + ^2/_3H^+ + ^{23}/_2H_2O = ^7/_3\underline{Al_2SiO_2O_5(OH)_4} + ^1/_3Ca^{2+} + ^8/_3Si(OH)_4$
Illite-kaolinite	$\underline{K_{0.6}Mg_{0.25}Al_{2.30}Si_{3.5}O_{10}(OH)_2} + ^{11}/_{10}H^+ + ^{63}/_{60}H_2O = ^{23}/_{30}\underline{Al_2Si_2O_5(OH)_4} + ^3/_5K^+ + ^1/_4Mg^{2+} + ^6/_5Si(OH)_4$
Biotite-kaolinite	$\underline{KMg_3AlSi_3O_{10}(OH)_2} + 7H^+ + ^1/_2H_2O = ^1/_2\underline{Al_2Si_2O_5(OH)_4} + K^+ + 3Mg^{2+} + 2Si(OH)_4$
Albite-kaolinite	$\underline{NaAlSi_3O_8} + H^+ + ^9/_2H_2O = ^1/_2\underline{Al_2Si_2O_5(OH)_4} + Na^+ + 2Si(OH)_4$
Albite-Na-montmorillonite	$\underline{NaAlSi_3O_8} + ^6/_7H^+ + ^{20}/_7H_2O = ^3/_7\underline{Na_{0.33}Al_{2.33}Si_{3.67}O_{10}(OH)_2} + ^6/_7Na^+ + ^{10}/_7Si(OH)_4$
Microcline-kaolinite	$\underline{KAlSi_3O_8} + H^+ + ^9/_2H_2O = ^1/_2\underline{Al_2Si_2O_5(OH)_4} + K^+ + 2Si(OH)_4$
Anorthite-kaolinite	$\underline{CaAl_2Si_2O_8} + 2H^+ + H_2O = \underline{Al_2Si_2O_5(OH)_4} + Ca^{2+}$
Andesine-kaolinite	$\underline{Na_{0.5}Ca_{0.5}Al_{1.5}Si_{2.5}O_8} + ^3/_2H^+ + ^{11}/_4H_2O = ^3/_4\underline{Al_2Si_2O_5(OH)_4} + ^1/_2Na^+ + ^1/_2Ca^{2+} + Si(OH)_4$

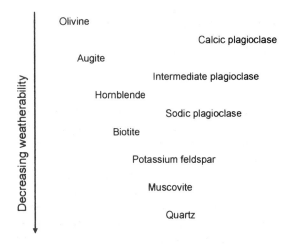

Decreasing weatherability

Olivine

Calcic plagioclase

Augite

Intermediate plagioclase

Hornblende

Sodic plagioclase

Biotite

Potassium feldspar

Muscovite

Quartz

Fig. 3.22 The Goldich weathering sequence based on observations of the sequence of disappearance of primary silicate minerals in soils. After Goldich (1938).

$$K_{albite-kaolinite} = \frac{[Na^+][Si(OH)_4]^2}{[H^+]} \qquad \text{eq. 3.48}$$

where $K_{albite-kaolinite}$ is the equilibrium constant. In this approach, the activities of the mineral phases and water are taken as unity. Expressing equation 3.48 in logarithmic form gives:

$$\log_{10}K_{albite-kaolinite} = \log_{10}[Na^+] + 2\log_{10}[Si(OH)_4] - pH$$
$$\text{eq. 3.49}$$

or

$$\log_{10}K_{albite-kaolinite} = \log_{10}\left(\frac{[Na^+]}{[H^+]}\right) + 2\log_{10}[Si(OH)_4]$$
$$\text{eq. 3.50}$$

which indicates that the equilibrium condition for the albite-kaolinite reaction can be expressed in terms of pH and activities of Na^+ and $Si(OH)_4$. Equilibrium relations such as this are the basis for the construction of stability diagrams, examples of which are shown in Fig. 3.23. These types of diagrams represent minerals with ideal chemical compositions, which may not accurately represent real systems, but, nevertheless, are useful in the interpretation of chemical data from hydrogeological systems. In igneous terrain, nearly all groundwaters within several hundred metres of the ground surface plot in the kaolinite

montmorillonite is forming is high in dissolved ions while in areas with high rainfall with the formation of gibbsite, the dissolved ion concentrations are low (Appelo & Postma 1994).

Now, by taking a thermodynamic equilibrium approach, it is possible to gain further insight into some of the more specific results of groundwater interactions with feldspars and clays. For example, considering the albite ($NaAlSi_3O_8$) dissolution reaction given in Table 3.14, then from the law of mass action:

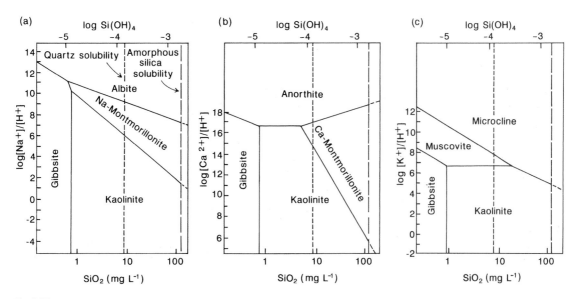

Fig. 3.23 Stability relations for gibbsite, kaolinite, montmorillonite, muscovite and feldspar at $25\,^{\circ}C$ and 1 bar pressure as functions of pH and the activities of Na^+, Ca^{2+}, K^+ and $Si(OH)_4$. (a) Gibbsite, $Al_2O_3 \cdot H_2O$; kaolinite, $Al_2Si_2O_5(OH)_4$; Na-montmorillonite, $Na_{0.33}Al_{2.33}Si_{3.67}O_{10}(OH)_2$; and albite, $NaAlSi_3O_8$. (b) Gibbsite; kaolinite; Ca-montmorillonite, $Ca_{0.33}Al_{4.67}Si_{7.33}O_{20}(OH)_4$; and anorthite, $CaAl_2Si_2O_8$. (c) Gibbsite; kaolinite; muscovite, $KAl_2(AlSi_3O_{10})(OH)_2$; and microcline (feldspar), $KAlSi_3O_8$. After Tardy (1971).

fields of Fig. 3.23. A small percentage of samples plot in the montmorillonite fields and hardly any occur in the gibbsite, mica or feldspar fields or exceed the solubility limit of amorphous silica. This observation suggests that alteration of feldspars and micas to kaolinite is a common process in groundwater flow systems in igneous rocks (Freeze & Cherry 1979).

The chemical composition of groundwaters in crystalline rocks is characterized by very low major ion concentrations. Normally, without exception, HCO_3^- is the dominant anion (see eqs 3.45 and 3.46) with silicon present in major concentrations relative to the cations. In the pH range that includes nearly all groundwaters (pH = 6–9), the dominant dissolved silicon released by weathering is the extremely stable and non-ionic monosilicic acid, $Si(OH)_4$.

$Si(OH)_4$ is usually expressed as SiO_2 in water analyses. Representative chemical analyses of groundwaters from granitic massifs are shown in Table 3.15 in which the SiO_2 concentration ranges from 1 to 85 mg L^{-1}. For comparison, Haines and Lloyd (1985) presented data for the major British sedimentary limestone and sandstone aquifers that demonstrate

a groundwater SiO_2 concentration in the range 6–27 mg L^{-1}, with the higher values associated with weathering of Mg-smectites found in carbonate aquifers.

In crystalline rocks, the groundwater concentrations of the anions Cl^- and SO_4^{2-} usually occur in only minor or trace amounts and typically have an atmospheric source or occur as trace impurities in rocks and minerals. Saline groundwaters can occur in granitic rocks, with the example of the Carnmenellis Granite in Cornwall, south-west England illustrated in Box 3.10.

3.11 FURTHER READING

Andrews, J.E., Brimblecombe, P., Jickells, T.D., Liss, P.S. & Reid, B.J. (2004) *An Introduction to Environmental Chemistry*, 2nd edn. Blackwell Science, Oxford.

Appelo, C.A.J. & Postma, D. (1994) *Geochemistry, Groundwater and Pollution*. A.A. Balkema, Rotterdam.

Chapelle, F.H. (1993) *Ground-water Microbiology and Geochemistry*. John Wiley, New York.

Domenico, P.A. & Schwartz, F.W. (1998) *Physical and Chemical Hydrogeology*, 2nd edn. John Wiley, New York.

Fetter, C.W. (2001) *Applied Hydrogeology*, 4th edn. Pearson Higher Education, New Jersey.

Freeze, R.A. & Cherry, J.A. (1979) *Groundwater*. Prentice-Hall, Englewood Cliffs, New Jersey.

Krauskopf, K.B. & Bird, D.K. (1995) *Introduction to Geochemistry*, 3rd edn. McGraw-Hill, New York.

Lloyd, J.W. & Heathcote, J.A. (1985) *Natural Inorganic Hydrochemistry in Relation to Groundwater: an introduction*. Clarendon Press, Oxford.

Mazor, E. (1997) *Chemical and Isotopic Groundwater Hydrology: the applied approach*, 2nd edn. Marcel Dekker, New York.

Stumm, W. & Morgan, J.J. (1996) *Aquatic Chemistry: chemical equilibria and rates in natural waters*, 3rd edn. John Wiley, New York.

Table 3.15 Mean chemical composition of groundwaters from European and African granitic massifs as presented by Tardy (1971).

Location	Number of samples	pH	HCO_3^- (mg L^{-1})	Cl^- (mg L^{-1})	SO_4^{2-} (mg L^{-1})	SiO_2 (mg L^{-1})	Na^+ (mg L^{-1})	K^+ (mg L^{-1})	Ca^{2+} (mg L^{-1})	Mg^{2+} (mg L^{-1})
Norway	28	5.4	4.9	5.0	4.6	3.0	2.6	0.4	1.7	0.6
Vosges	51	6.1	15.9	3.4	10.9	11.5	3.3	1.2	5.8	2.4
Brittany	7	6.5	13.4	16.2	3.9	15.0	13.3	1.3	4.4	2.6
Central Massif	10	7.7	12.2	2.6	3.7	15.1	4.2	1.2	4.6	1.3
Alrance Spring F	77	5.9	6.9	<3	1.15	5.9	2.3	0.6	1.0	0.4
Alrance Spring A	47	6.0	8.1	<3	1.1	11.5	2.6	0.6	0.7	0.3
Corsica	25	6.7	40.3	22.0	8.6	13.2	16.5	1.4	8.1	4.0
Sahara	8	6.9	30.4	4.0	20	9	30	1.8	40	–
Senegal	7	7.1	43.9	4.2	0.8	46.2	8.4	2.2	8.3	3.7
Chad	2	7.9	54.4	<3	1.4	85	15.7	3.4	8.0	2.5
Ivory Coast (dry season)	54	5.5	6.1	<3	0.4	10.8	0.8	1.0	1.0	0.10
Ivory Coast (wet season)	59	5.5	6.1	<3	0.5	8.0	0.2	0.6	<1	<0.1
Malagasy (High Plateaux)	2	5.7	6.1	1	0.7	10.6	0.95	0.62	0.40	0.12

Hydrogeochemical characteristics of the Carnmenellis Granite, Cornwall, England

BOX 3.10

The Carnmenellis Granite and its aureole in south-west England contain the only recorded thermal groundwaters in British granites and occur as springs in tin mines. Most of the groundwaters are saline with a maximum mineralization of 19,310 mg L^{-1} (Edmunds et al. 1984). The Carnmenellis Granite forms a near-circular outcrop of the Cornubian batholith (Fig. 1) which was intruded about 290 Ma into Devonian argillaceous sediments. The rock is highly fractured. Beneath a weathered zone of variable thickness, the granite is characterized by secondary permeability. Most groundwater flow and storage occurs in open horizontal fractures and is shallow in depth (commonly above 50 m) and localized.

The granite is composed of coarse- and fine-grained porphyritic muscovite-biotite granite, the former being more common, and has undergone a long history of alteration. The granite is enriched in volatile elements (B, Cl, F, Li) compared with many other granite terrains. There is extensive hydrothermal mineralization of Variscan age which has produced economic vein deposits of Sn, Cu, Pb and Zn. The principal mineral lodes occur in a mineralized belt north of the Carnmenellis Granite (Edmunds et al. 1985).

Saline groundwaters are encountered in four accessible mines (Fig. 1 for locations) in the granite or its thermal aureole, as well as in several disused mines, all at the northern margin. The saline waters generally issue from cross-courses with discharges between 1 and 10 L s^{-1} at depths between 200 and 700 m below surface. The discharge temperatures of up to 52°C are typically in excess of the average regional thermal gradients of 30°C km^{-1} in the granite and 50°C km^{-1} in the aureole. As proposed by Edmunds et al. (1985) and shown schematically in Fig. 2, the temperature anomaly implies that ancient, warmer saline fluids are upwelling by convective circulation and mixing with recent, fresh, shallow groundwaters. The driving force for the current circulation system is the hydraulic sink created by the former mining operations. Also, the existence of old, flooded mine workings locally increases the secondary porosity of the rocks.

Chemical analyses of the four mine waters and two shallow groundwater samples are given in Table 1. The fresh, shallow groundwater is generally of good quality, has a low total dissolved solids content and may be acidic (pH < 5.5). The most important features of the hydrochemistry of the Carnmenellis Granite, in addition to the high Cl^- concentrations, are the depletion of Na^+ relative to Cl^-, the enhanced Ca^{2+} levels and especially the significantly enriched Li^+, with Li^+ values as high as 125 mg L^{-1}. The unusual chemistry combined with stable isotope data demonstrates a meteoric origin for all the groundwaters that excludes seawater as the source of the salinity.

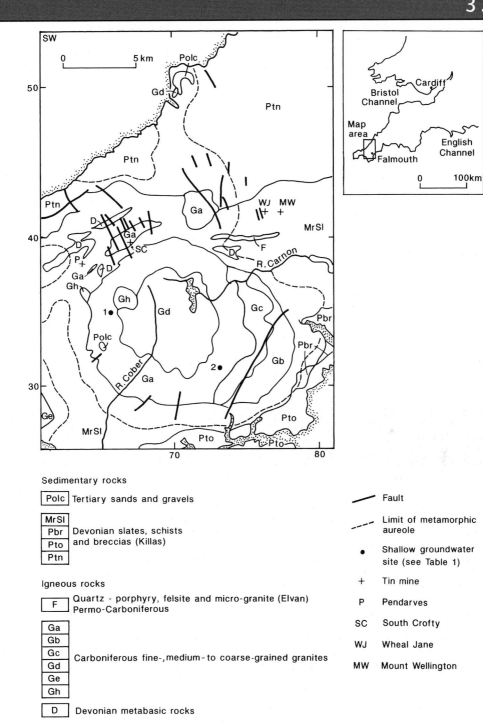

Fig. 1 Map showing the location and solid geology of the Carnmenellis Granite in Cornwall, south-west England. The sites of tin mines are shown as: P, Pendarves Mine; SC, South Crofty Mine; WJ, Wheal Jane Mine; MW, Mount Wellington Mine. After British Geological Survey (1990).

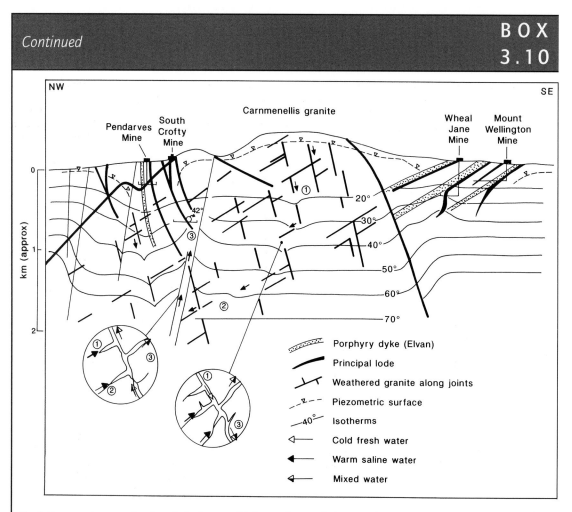

Fig. 2 Conceptual cross-section through the Carnmenellis Granite showing the location of tin mines and saline groundwaters. Isotherms in the granite and its aureole show convective distortion due to groundwater circulation. The suggested circulated route of groundwater flow is shown as follows: (1) Rapid percolation of recent meteoric water with vertical drainage enhanced by mining. Local distortion of isotherms by groundwater flow; (2) Ancient meteoric water with high salinity derived from granite–water reactions and stored in the fracture system; (3) Circulation of mixed groundwater which discharges as springs in the mines. Local enhancement of thermal gradient by upward groundwater flow. After Edmunds et al. (1985).

In their hydrogeochemical interpretation, Edmunds et al. (1985) concluded that hydrolysis of biotite $(K_2(Mg,Fe)_4(Fe,Al,Li)_2$ $[Si_6Al_2O_{20}](OH)_2(F,Cl)_2)$ could account for the increase in Cl^-, Li^+, K^+ and other species in the groundwaters. The Li^+ and Cl^- are considered as conservative products of biotite alteration, unlikely to be assimilated by reaction products. The fact that the Li/Cl ratio is only 1 : 100 compared to 1 : 2 for unaltered granite suggests that Cl^- release is a relatively easy and rapid process that accounts for the observed salinities over very long timescales, while Li^+ and Mg^{2+} and other ions are only released upon a breakdown of the trioctahedral biotite structure. The depletion of Mg^{2+} in most of the saline groundwaters is consistent with the observed biotite alteration to chlorite.

A second important silicate weathering process affecting the hydrogeochemistry is the acid hydrolysis of plagioclase feldspars (eqs 3.45 and 3.46) that contributes the principal sources of Na^+ and Ca^{2+} to the groundwaters. Both biotite and plagioclase alteration by weathering produces significant amounts of silica as well as kaolinite and other clay minerals; in some areas, kaolinization along fractures reduces the available secondary porosity. Concentrations of SiO_2 (Table 1) in these groundwaters represent saturation or supersaturation with respect to chalcedony, and this tendency for silica deposition is argued to rule out fluid inclusions acting as a potential source of salinity (Edmunds et al. 1985).

BOX
3.10

Continued

Table 1 Representative chemical analyses of saline and thermal groundwaters from four tin mines together with two fresh shallow groundwater sites in the Carnmenellis Granite or its aureole. Locations are shown in Fig. 1. After Edmunds et al. (1985).

Site	Mount Wellington Mine	Pendarves Mine	Wheal Jane Mine	South Crofty Mine	Shallow groundwater	
					1	2
Depth (m)	240	260	300	690	30	41
Flow rate (L s^{-1})	15	0.5	10	3.5	–	–
Temperature (°C)	21.6	21.4	39.5	41.5	10.8	10.5
pH	5.6	6.9	6.4	6.5	4.92	–
Na (mg L^{-1})	125	29	1250	4300	14	12
K	12	3.1	72.0	180	2.8	4.0
Li	3.55	0.06	26.0	125	<0.01	<0.01
Ca	93	18	835	2470	7	13
Mg	11.9	5.0	22.0	73.0	2.7	2.4
Sr	1.43	0.06	12.8	40.0	<0.06	0.23
HCO$_3$ (mg L^{-1})	9	67	21	68	4	9
NO$_3$	11.2	8.3	10	<0.2	2.1	26.0
SO$_4$	275	38	148	145	15	17
Cl	287	32	3300	11,500	25	21
F	0.29	2.90	3.30	2.70	<0.1	0.11
Br	0.9	<0.3	–	43.0	–	
B	0.80	<0.01	3.3	11.0	–	
SiO$_2$	19.2	34.2	28.4	34.2		
Fe (mg L^{-1})	43.0	0.62	22.4	4.75	0.280	0.014
Mn	2.90	0.30	4.00	4.50	0.031	0.008
Cu	0.024	0.002	0.005	0.023	0.004	0.027
Ni	0.134	0.007	0.027	0.190	0.002	0.0015
Total mineralization (mg L^{-1})	885	230	5747	19,002	73	105
δ^{2}H (‰ SMOW)*	−35	−38	−31	−29		
δ^{18}O (‰ SMOW)*	−5.4	−5.2	−5.7	−5.2		

* δ^{2}H and δ^{18}O are measured in per mil relative to Standard Mean Ocean Water (SMOW; see Section 4.2).

Environmental isotope hydrogeology

4

4.1 Introduction

The stable and radioactive isotopes of the common elements of oxygen, hydrogen, carbon, sulphur and nitrogen have a wide range of applications in hydrogeology. The stable isotopes of water (^{16}O, ^{18}O, ^{1}H, ^{2}H) can be used as tracers of the origin of groundwater recharge and, together with noble (inert) gases (Ne, Ar, Kr, Xe), used to provide information on aquifer evolution. Sulphur and nitrogen isotopes have applications in contaminant hydrogeology in the identification of pollution sources and their fate in the groundwater environment. For further discussion of sulphur and nitrogen isotopes refer to Bottrell et al. (2000), Heaton (1986) and Kendall (1998). The radioactive isotopes of water, ^{3}H (tritium), carbon (^{14}C) and chlorine (^{36}Cl) are useful in providing estimates of aquifer residence times that can assist in managing groundwater resources.

The following sections describe, with examples, the basis for the application of environmental isotopes in groundwater investigations with emphasis given to groundwater source identification and age dating using the stable and radioactive isotopes of water, ^{14}C and ^{36}Cl. In discussing the origin of groundwater recharge, a section is also included to demonstrate the application of noble gas concentration data in reconstructing present and past groundwater recharge temperatures. The combined interpretation of environmental isotope and noble gas data can therefore enable reconstruction of palaeoenvironmental conditions and give insight into the history of aquifer evolution.

4.2 Stable isotope chemistry and nomenclature

Modern double inlet, double collector mass spectrometers are capable of detecting small changes in relative isotopic abundances with the results expressed using the δ notation. In general, the δ notation, normally expressed in parts per thousand (per mil or ‰) with respect to a known standard, is written as follows:

$$\delta = \frac{R_{sample} - R_{standard}}{R_{standard}} \times 1000 \qquad \text{eq. 4.1}$$

where R_{sample} and $R_{standard}$ are the isotopic ratios (for example $^{18}O/^{16}O$ and $^{2}H/^{1}H$) of the sample and standard, respectively. With this notation, an increasing value of δ means an increasing proportion of the rare, heavy isotope. In this case, the sample is said to have a heavier, more positive or enriched isotope composition compared with another, isotopically lighter sample. For water, the accepted international standard is V_{SMOW} (Vienna Standard Mean Ocean Water) with values of $\delta^{18}O$ and $\delta^{2}H$ equal to zero. Measurements of $\delta^{18}O$ and $\delta^{2}H$ can usually be determined to an accuracy of better than ±0.2‰ and ±2‰, respectively.

The mechanisms of isotope separation that lead to enrichment and depletion of isotopic ratios between phases or species can be divided into three types (Krauskopf & Bird 1995):

1 mechanisms depending on physical properties, for example evaporation or precipitation;

2 exchange reactions resulting in isotopic equilibrium between two or more substances;

3 separation depending on reaction rate.

An example of mechanism 1 is the evaporation of water which leads to the concentration of the light isotopes ^{16}O and ^{1}H in the vapour phase and the heavy isotopes in the liquid phase. This is because water molecules containing the light isotopes move more rapidly and thus have a higher vapour pressure. Most samples of freshwater have negative values of $\delta^{18}O$ (ranging down to −60‰) in that the light ^{16}O isotope is concentrated in the vapour evaporating from the sea surface. Oxygen in air has a high positive isotopic signature of +23.5‰.

As an example of mechanism 2, if CO_2 containing only ^{16}O is mixed with water containing only ^{18}O, exchange will occur according to the following reaction, until equilibrium is reached among the four species:

$$^{1}/_{2}C^{16}O_2 + H_2^{18}O \rightleftharpoons {}^{1}/_{2}C^{18}O_2 + H_2^{16}O \qquad \text{eq. 4.2}$$

Although the bond strengths in the two compounds are different, at equilibrium, the ratio $^{18}O/^{16}O$ will be nearly the same in the CO_2 and H_2O.

The variable separation of isotopes depending on reaction rates (mechanism 3 above) is particularly associated with reactions catalysed by bacterial activity. For example, in the bacterial reduction of SO_4^{2-}, the production of sulphide (S^{2-}, HS^- and H_2S) is faster for the light isotope, ^{32}S, than for the heavy isotope, ^{34}S, such that the light isotope becomes concentrated in the sulphide species and the heavy isotope enriched in residual SO_4^{2-}.

Regardless of mechanism, the extent of isotope separation between two phases A and B can be represented by a fractionation factor, α, where:

$$\alpha_{A-B} = \frac{R_A}{R_B} \qquad \text{eq. 4.3}$$

where R_A is the ratio of concentrations of heavy to light isotope in phase A ($^{18}O/^{16}O$ in liquid water, for example) and R_B is the same ratio in phase B ($^{18}O/^{16}O$ in water vapour). If equilibrium is established between liquid water and vapour at 25°C, the value of α is about 1.0092. Similar fractionation factors, very slightly greater or less than 1, are obtained for other examples of isotope separation and it is for this reason that the descriptive δ notation is adopted.

The relationship between δ and α is given by the expression:

$$\alpha_{A-B} = \frac{R_A}{R_B} = \frac{1000 + \delta_A}{1000 + \delta_B} \qquad \text{eq. 4.4}$$

For example, in the condensation of water vapour (phase v) to liquid (phase l), equation 4.4 becomes:

$$\alpha_v^l = \frac{1000 + \delta^{18}O_l}{1000 + \delta^{18}O_v} \qquad \text{eq. 4.5}$$

and if $\delta^{18}O_l = -5‰$ and $\delta^{18}O_v = -14‰$, then the fractionation factor can be calculated:

$$\alpha_v^l = \frac{(1000 + (-5))}{(1000 + (-14))} = 1.0092 \qquad \text{eq. 4.6}$$

4.3 Stable isotopes of water

The relative abundances of hydrogen and oxygen isotopes found naturally in the water molecule are given in Table 3.1. Meteoric water shows a wide range of $\delta^{18}O$ and δ^2H values reflecting the extent of isotope fractionation during successive cycles of evaporation and condensation of water originally evaporated from the sea. When condensation occurs to form precipitation, the isotopic concentration changes according to a Rayleigh distillation process for which the isotopic ratio, R, in a diminishing reservoir of reactant is a function of its initial ratio, R_o, the remaining reservoir fraction, f, and the fractionation factor, α, such that $R = R_o f^{(\alpha - 1)}$. The $^2H/^1H$ fractionation is proportional to, and about eight times as large as, the $^{18}O/^{16}O$ fractionation. Both fractionations change proportionally as temperature changes. Craig (1961) showed that δ values for meteoric water samples of global distribution, for the most part, define a straight line on a cross-plot of δ^2H against $\delta^{18}O$, represented by the approximate equation, know as the World Meteoric Water Line (WMWL):

$$\delta^2H = 8\delta^{18}O + 10 \qquad \text{eq. 4.7}$$

In general, samples with $\delta^{18}O$ and δ^2H lighter than −22‰ and −160‰, respectively, represent snow and

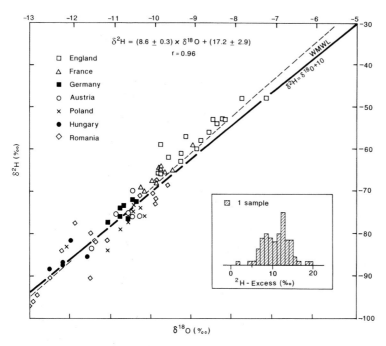

$$\delta^2H = (8.6 \pm 0.3) \times \delta^{18}O + (17.2 \pm 2.9)$$

$$r = 0.96$$

□ England
△ France
■ Germany
○ Austria
× Poland
● Hungary
◇ Romania

☑ 1 sample

0   10   20
^2H - Excess (‰)

WMWL

$\delta^2H = \delta^{18}O + 10$

δ^2H (‰)

$\delta^{18}O$ (‰)

Fig. 4.1 The δ^2H versus $\delta^{18}O$ relationship for western and central European palaeowaters (groundwaters having a radiocarbon age of between approximately 15 and 30 ka). Old European groundwaters are depleted in deuterium (^2H) by about 12‰ compared with modern recharge waters (as shown in the frequency histogram of deuterium excess values, d, where $d = \delta^2H - 8\delta^{18}O$). The continental gradient in deuterium is very similar to modern meteoric water as indicated by the essentially identical comparison with the World Meteoric Water Line (WMWL) and strongly suggests a constant atmospheric circulation regime over Europe during the past 35 ka years. After Rozanski (1985).

ice from high latitudes while tropical samples show very small depletions relative to ocean water. This distribution is expected for an atmospheric Rayleigh process as vapour is removed from tropospheric air that is moving poleward. Linear correlations with coefficients only slightly different to equation 4.7 are obtained from studies of local precipitation. For example, in the British Isles, a 20-year monthly dataset from 1982 to 2001 for Wallingford, a station in central, southern England, gave the following regression: $\delta^2H = 7.00\delta^{18}O + 0.98$, with a slope a little less than the WMWL but consistent with those from other long-term stations in north-west Europe (Darling & Talbot 2003).

In addition to latitudinal and, similarly, altitudinal effects of temperature, the location of a site in relation to the proximity of the evaporating water mass is also important. As water vapour moves inland across continental areas, and as the process of condensation and evaporation is repeated many times, rain or snow becomes increasingly isotopically depleted. This continental 'rain-out' effect of the heavy isotopes has been shown for Europe by Rozanski (1985), as illustrated in Fig. 4.1, and is also apparent in the stable isotope composition of recent groundwaters measured for the British Isles (Fig. 4.2).

Since both condensation and isotope separation are temperature dependent, the isotope composition of meteoric water displays a strong seasonal variation at a given location. In interpreting groundwater isotopic compositions, individual recharge events are mixed in the region of water-table fluctuation such that isotopic variations over short timescales become obscured. Thus, it is possible to use a weighted mean isotopic composition to represent the isotopic signature of the seasonal recharge. In Fig. 4.3, a 1-year record of monthly precipitation amount and composite $\delta^{18}O$ values is shown for a rain gauge situated in north Norfolk, eastern England. The data show the general trend of isotopically enriched precipitation in the warmer summer months and isotopically depleted precipitation in the colder winter months. The winter rainfall provides a representative volume-weighted mean isotopic composition for groundwater recharge of $-7.20‰$ and $-47.6‰$ for $\delta^{18}O$ and δ^2H, respectively.

Recognition of the effects of season, latitude, altitude and continentality is the basis for using $\delta^{18}O$ and δ^2H isotope values as non-reactive, naturally occurring tracers to identify the climatic and palaeo-geographic conditions of groundwater recharge. For example, modern recharge waters in the Chalk

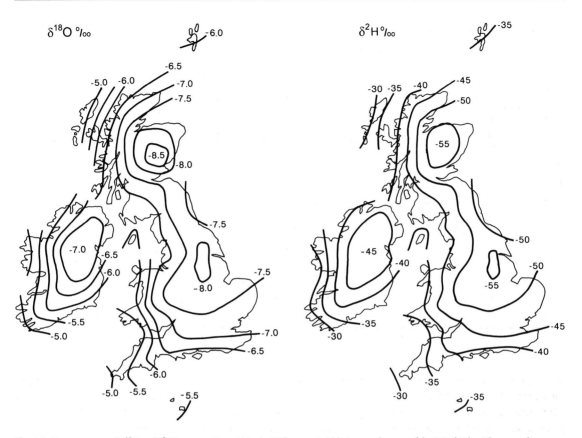

Fig. 4.2 Contour maps of $\delta^{18}O$ and δ^2H in recent (i.e. within the Holocene, 0–10 ka) groundwaters of the British Isles. The maps show similar features with relatively large variations in isotopic composition covering ranges of almost 4‰ in $\delta^{18}O$ and 30‰ in δ^2H. The areal isotopic composition is controlled mainly by the predominant source of rainfall from the south-west of the British Isles with some topographic variation noticeable over the Highlands of Scotland and the Pennines of England where isotopic depletion occurs between the west and east of the country due to the orographic patterns of rainfall distribution. After Darling et al. (2003).

aquifer of Norfolk, with a $\delta^{18}O$ composition of about −7.0‰, are found in the major unconfined river valleys (compare with the isotopic composition of rainfall; Fig. 4.3). Palaeogroundwaters with a $\delta^{18}O$ composition of <−7.5‰ are found trapped below extensive low-permeability glacial till deposits in the interfluvial areas. The existence of the palaeogroundwater, and an isotopic shift of greater than 0.5‰ between the modern water and palaeogroundwater, is evidence for groundwater recharge during the late Pleistocene when the mean surface air temperature is estimated to have been at least 1.7°C cooler than at present. This estimate is based on the slope of the best-fit line of $\delta^{18}O$ data for global precipitation in the temperature range 0–20°C and equal to 0.58‰°C^{-1}, in close agreement with theoretical predictions based

on the Rayleigh condensation model (Rozanski et al. 1993). In the Norfolk Chalk aquifer, the isotopic data confirm the conceptual hydrogeological model (Fig. 4.4) by which most active groundwater recharge and flow is restricted to relatively limited areas where the overlying glacial till deposits are thin or absent (Hiscock et al. 1996).

Other examples of the application of the stable isotopes of water in the interpretation of palaeogroundwaters include the identification of freshwaters at depth in European coastal aquifers (Fig. 4.5) and the existence of fossil freshwater bodies in arid and semiarid areas (Fig. 4.6). A compilation and synthesis of stable isotope data ($\delta^{18}O$, δ^2H) for palaeogroundwaters in the British Isles is given by Darling et al. (1997). Evidence from the major British sandstone

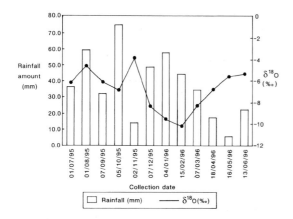

Fig. 4.3 Record of monthly precipitation amount and composite $\delta^{18}O$ values for a rain gauge situated at Salle, Norfolk (NGR TG 6126 3243), eastern England. The established Local Meteoric Water Line (LMWL) for precipitation in north Norfolk is $\delta^2H = 6.48\delta^{18}O - 0.62$. The volume-weighted mean winter rainfall (recharge) values are $\delta^2H = -47.6‰$ and $\delta^{18}O = -7.20‰$. After George (1998).

and limestone aquifers shows that atmospheric circulation patterns over Britain have probably remained the same since the late Pleistocene. However, additional ^{14}C groundwater age data highlight a hiatus in recharge occurrence under periglacial conditions between the late Pleistocene and early Holocene at about the time of the last glacial maximum (LGM).

In deep aquifers where temperatures can exceed $50–100°C$, the ^{18}O content of groundwater emerging as hot springs can be significantly altered by chemical interactions with the host rock. Measurements in such areas fall close to a horizontal line, indicating that the hot water contains an excess of ^{18}O over the meteoric water of the same region, but with approximately the same ^2H content (Fig. 4.7). This suggests that the infiltrating water has exchanged some of its ^{16}O for ^{18}O from the silicate minerals of the host rock. A similar exchange of hydrogen isotopes is insignificant because most minerals only contain a small amount of this element (Krauskopf & Bird 1995). Further discussion of the isotopic composition of thermal waters can be found in Albu et al. (1997).

Another isotopic effect is observed for water evaporating from shallow soil or surface water bodies. Under natural conditions, the surface water becomes enriched in the heavy isotopes as evaporation occurs and provides a means for identifying surface water

inputs to groundwater, and also for separating stream hydrographs into components of event (rainfall) and pre-event (soil) water (see Section 5.7.1).

4.4 Age dating of groundwater

The age of a groundwater relates to the time when an aquifer experiences recharge and is a measure of the groundwater residence time. The exploitation of groundwater resources at a rate in excess of the time to replenish the aquifer storage will risk mining the groundwater. Hence, knowledge of the age of groundwater is useful in aquifer management. Given that groundwater velocities are typically small and variable, a wide range of residence times are encountered in natural systems from a few days in karst aquifers to millennia in unfractured mudstones. Qualitative indicators of the age of a groundwater body include whether the groundwater is chemically oxidizing (aerobic, modern water) or reducing (anoxic, older water) in chemical character. Quantitative measures of the age of groundwater use radioisotopes as a dating method. To demonstrate, the next section defines the law of radioactive decay. The following sections present applications of the ^{14}C and tritium dating techniques together with an introduction to the more advanced methods of ^{36}Cl dating and $^3H/^3He$ dating.

4.4.1 Law of radioactive decay

The activity of a radioisotope at a given time can be calculated using the basic radioactive decay law:

$$\frac{A}{A_0} = 2^{-t/t_{1/2}} \qquad \text{eq. 4.8}$$

where A_0 is the radioactivity at time $t = 0$, A is the measured radioactivity at time t and $t_{1/2}$ is the half-life of the radionuclide found from:

$$t_{1/2} = \frac{\log_e 2}{\lambda} \qquad \text{eq. 4.9}$$

where λ is the decay constant of the radionuclide.

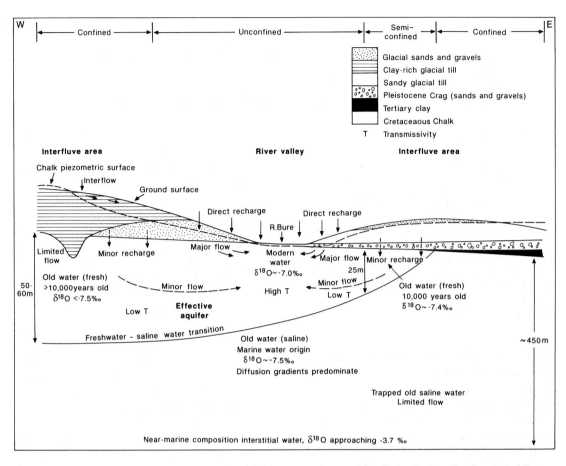

Fig. 4.4 Schematic cross-section showing a conceptual model of the extent and nature of the effective Chalk aquifer of north Norfolk, eastern England. After Hiscock et al. (1996).

Hence, with knowledge of A/A_0, the fraction or percentage of a radionuclide remaining at time t, and the decay constant λ, then it is possible to calculate the apparent age of the groundwater. The age is considered an apparent age due to interpretation difficulties that arise from the general problem of mixing of groundwater bodies with different ages and, in the case of ^{14}C dating, from reactions between groundwater and aquifer carbonate material. To overcome these problems, corrections are required to the apparent groundwater age in order to obtain a corrected age.

4.4.2 ^{14}C dating

The radioisotope ^{14}C is produced by cosmic ray bombardment of nitrogen in the upper atmosphere. For every 10^{12} atoms of the stable isotopes of carbon (^{12}C and ^{13}C) in the atmosphere and oceans there is an abundance of one atom of ^{14}C. ^{14}C decays back to nitrogen together with the emission of a β particle. The half-life of ^{14}C is measured as 5730 years and provides a useful dating tool in the age range up to 40,000 years for the most accurate determinations. Using the radioactive decay law (eq. 4.8) and substituting $t_{1/2} = 5730$ years, the dating equation becomes:

$$t = -8267 \log_e \frac{A}{A_0} \qquad \text{eq. 4.10}$$

The ^{14}C activity measured in the laboratory by accelerator mass spectrometry is given in terms of per cent modern carbon (pmc), with a counting statistics error of about ±0.6 pmc. Calculated groundwater ages are

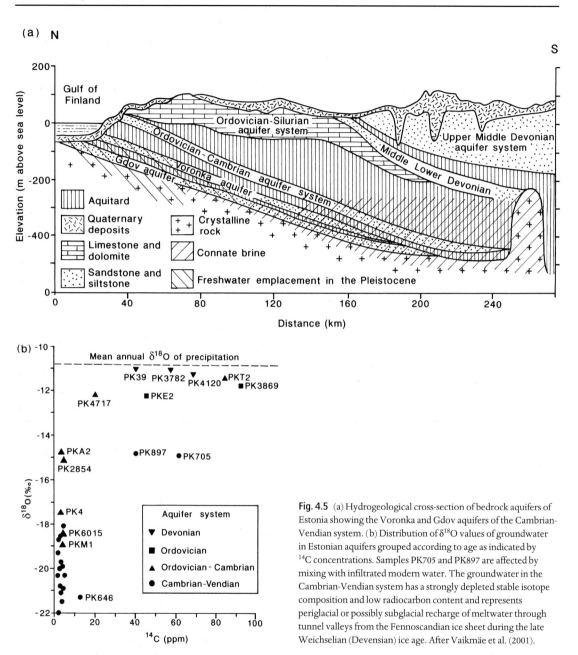

(a)

Fig. 4.5 (a) Hydrogeological cross-section of bedrock aquifers of Estonia showing the Voronka and Gdov aquifers of the Cambrian-Vendian system. (b) Distribution of $\delta^{18}O$ values of groundwater in Estonian aquifers grouped according to age as indicated by ^{14}C concentrations. Samples PK705 and PK897 are affected by mixing with infiltrated modern water. The groundwater in the Cambrian-Vendian system has a strongly depleted stable isotope composition and low radiocarbon content and represents periglacial or possibly subglacial recharge of meltwater through tunnel valleys from the Fennoscandian ice sheet during the late Weichselian (Devensian) ice age. After Vaikmäe et al. (2001).

quoted in years Before Present (pre-1950) using the notation a (for example, 1 ka indicates 1000 years Before Present).

Allowance must be made in the interpretation of the apparent age provided by measurement of the ^{14}C activity of a sample for reaction with sources of inorganic carbon encountered along the groundwater flowpath. The two principal sources are:

1 'biogenic' carbon with a source in the CO_2 of the atmosphere and also respired by the decay of organic carbon in the soil zone, which is usually assumed to contain 100 pmc;

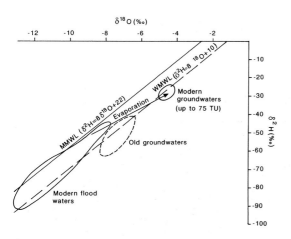

Fig. 4.6 Cross-plot of $\delta^2 H$ versus $\delta^{18}O$ for groundwaters in southern Jordan. Old groundwaters with ^{14}C ages of 5–30 ka plot as a distinct group compared with modern, tritiated groundwaters that have a flood water source. The flood waters plot close to the Mediterranean Meteoric Water Line (MMWL) while the modern (evaporated) and old groundwaters relate more closely to the World Meteoric Water Line (WMWL). The fact that the old groundwaters conform to the WMWL suggests that their recharge was associated with storms tracking through the area from either the Atlantic or the Indian Oceans during the late Pleistocene and early Holocene. The limited occurrence of groundwaters showing a modern isotope composition also suggests that current recharge in this arid region is limited. After Lloyd and Heathcote (1985).

2 'lithic' carbon with a source in the soil and rock carbonate and containing 'dead' carbon in which all the ^{14}C has decayed away (0 pmc).

The distribution of biogenic and lithic sources of carbon in groundwater can be described by the solution of calcite by weak carbonic acid (eq. 3.5) as follows:

$$CaCO_3 \text{ (lithic carbon)} + H_2CO_3 \text{ (biogenic carbon)}$$
$$\rightarrow Ca^{2+} + 2HCO_3^- \text{ (sample carbon)} \qquad \text{eq. 4.11}$$

If this reaction predominates, then the stoichiometry of equation 4.11 predicts that the ^{14}C activity of the groundwater bicarbonate will be 50% of the modern activity. In other words, because of dilution of the sample with 'dead' carbon from soil and rock carbonate then, without correction, the apparent age of the groundwater will appear older than it actually is. To correct for this effect, one method is to use the $^{13}C/^{12}C$ ratio ($\delta^{13}C$) of the groundwater bicarbonate as a chemical tracer of the ^{14}C activity. The interpretation of the ^{14}C data from the sample $\delta^{13}C$ value is based on the fact that the principal sources of carbon (lithic and biogenic) contributing to the carbonate system in the water have different $\delta^{13}C$ values.

Analyses of $\delta^{13}C$ values are relatively easy to obtain by conversion of the sample carbonate to CO_2 followed by measurement of the isotope ratio on a mass spectrometer. Results are quoted relative to the standard Pee Dee Belemnite (PDB), a rock unit from the Cretaceous Period, with an accuracy of about ±0.2‰. The CO_2 associated with organic carbon in temperate soils has a $\delta^{13}C$ value of about −26‰ for Calvin photosynthetic cycle plants (−12‰ for Hatch–Slack cycle plants in hot, arid climates). Values for limestone rock are usually between 0 and +2‰ (Schoelle & Arthur 1980).

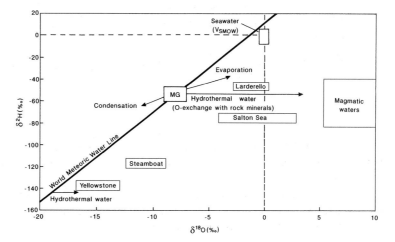

Fig. 4.7 Diagram illustrating the possible isotopic shifts from the World Meteoric Water Line (WMWL) on a plot of $\delta^2 H$ versus $\delta^{18}O$. The arrows indicate isotope separation effects away from a typical groundwater of modern meteoric origin (MG). The diagram includes the plotting positions of the range of isotopic composition of hydrothermal waters (Larderello, Salton Sea, Steamboat Springs and Yellowstone), illustrating a shift to the right (enrichment of ^{18}O) relative to the meteoric water line. After compilation of Albu et al. (1997).

The result of calcite dissolution (eq. 4.11) is that the groundwater would be expected to have a $\delta^{13}C$ value of about $-12‰$. Away from the soil zone, $\delta^{13}C$ values are less negative than those near to the aquifer outcrop as a result of the continuous precipitation and dissolution of carbonate as the water flows through the aquifer. The isotopic composition of the precipitated carbonate mineral differs from that of the carbonate species in solution. As a result there is isotopic fractionation between the aqueous phase and solid carbonate-containing mineral phase.

To account for the two sources of inorganic and organic carbon in the groundwater, let Q equal the proportion of biogenic carbon in the sample. To determine the proportions of lithic and biogenic carbon, then the following relationship can be used with the $\delta^{13}C$ data:

$$\delta^{13}C_{sample} = \delta^{13}C_{biogenic}(Q) + \delta^{13}C_{lithic}(1 - Q) \qquad \text{eq. 4. 12}$$

or, rearranged:

$$Q = \frac{\delta^{13}C_{sample} - \delta^{13}C_{lithic}}{\delta^{13}C_{biogenic} - \delta^{13}C_{lithic}} \qquad \text{eq. 4.13}$$

Using the calculated value of Q, the corrected groundwater age, t_c, can be found from the dating equation (eq. 4.10) by applying the fraction of biogenic carbon to the initial ^{14}C activity, A_0, such that:

$$t_c = -8267 \log_e \frac{A}{A_0 Q} \qquad \text{eq. 4.14}$$

or, separating terms:

$$t_c = -8267 \log_e \frac{A}{A_0} + 8267 \log_e Q \qquad \text{eq. 4.15}$$

As an example, the following information was obtained for a groundwater sample from Waltham Abbey PS (sample 12, Table 4.1) in the Chalk aquifer of the London Basin:

$$A_0 = {}^{14}C_{atmospheric} = 100 \text{ pmc}$$

$$A = {}^{14}C_{sample} = 13 \text{ pmc}$$

$$\delta^{13}C_{sample} = -6.3‰$$

$$\delta^{13}C_{lithic} = +2‰$$

$$\delta^{13}C_{biogenic} = -26‰.$$

From equation 4.13, the fraction of biogenic carbon, $Q = (-6.3 - (+2))/(-26 - (+2)) = 0.2964$ and substitution in equation 4.15 gives the following corrected groundwater age:

$$t_c = -8267 \log_e(13/100) + 8267 \log_e(0.2964) = 6813.5 \text{ a}$$

or, in practical terms, about 7 ka.

Further corrected groundwater ages for the Chalk aquifer of the London Basin are given in Table 4.1 and the results shown as age contours in Fig. 4.8c. The oldest groundwaters are found below the Eocene London Clay in the confined aquifer at the centre of the Basin. Modern groundwaters are present in the recharge area on the northern limb of the basin. The groundwater becomes progressively older towards the centre of the basin with flow induced by pumping of the aquifer (compare with the map of the Chalk potentiometric surface shown in Box 2.4). The upper limit for the age measurement is about 25 ka with groundwater at the centre of the Basin known to be older than this limit for the ^{14}C dating method (Smith et al. 1976a).

In situations where a negative corrected groundwater age is obtained, then the sample has become swamped by the modern atmospheric ^{14}C content which is in excess of 100 pmc. The extra ^{14}C was contributed by the detonation of thermonuclear devices in the 1950s and early 1960s. In such cases, the groundwater age can be assumed to be modern. This effect and other inherent shortcomings, for example contamination with atmospheric ^{14}C during field sampling and laboratory errors in the measurement of $\delta^{13}C$ and ^{14}C, mean that any groundwater sample with an age of up to 0.5 ka can be regarded as modern.

More advanced approaches to correcting ^{14}C ages that account for carbon isotope exchange between soil gas, dissolved carbonate species and mineral carbonate in the unsaturated and saturated zones under open- and closed-system conditions (Section 3.7) are included in Mook (1980). However, the extra effort expended may not be justified, especially if mixing between two groundwaters is suspected. For example, in the case of the Chalk aquifer discussed above,

Table 4.1 Tritium, carbon-14 and stable isotope measurements for Chalk groundwaters in the London Basin. The location of numbered sampling sites are shown in Fig. 4.8a. From Smith et al. (1976a).

Site	Tritium (TU)	^{14}C age (ka)	^{14}C (pmc)	$\delta^{13}C$ (‰)	$\delta^{18}O$ (‰)	δ^2H (‰)
1. Burnham PS	57	modern	60.8	−13.2	−7.1	−46
2. Duffield House	13	3.9	31.4	−11.9	−7.1	−45
3. Iver PS	19	9.1	6.8	−3.5	−7.4	−48
4. West Drayton	0	22	1.0	−1.8	−7.9	−50
5. Crown Cork	0	19	1.1	−0.8	−7.7	−50
6. Callard & Bowser	0	>20	0.8*	−0.5	−7.9	−51
7. Polak's Frutal	3	20	1.1	−1.1	−7.8	−50
8. Southall AEC	1	>20	0.5*	−0.9	−7.8	−51
9. Morganite Carbon	4	15	4.1	−4.5	−7.4	−48
10. White City Stadium	0	14	2.2	−0.9	−7.9	−51
11. Broxbourne PS	37	0.8	49.1	−13.0	−7.2	−47
12. Waltham Abbey PS	3	7.0	13.0	−6.3	−7.1	−46
13. Hadley Road PS	2	20	1.7	−3.3	−7.2	−48
14. Hoe Lane PS	2	8.2	14.9	−9.1	−7.1	−46
15. Chingford Mill PS	12	4.3	21.5	−7.9	−7.2	−46
16. Initial Services	53	modern	41.4	−8.3	−6.6	−44
17. Berrygrove	45	modern	58.5	−13.1		
18. Kodak	1	11	3.8	−1.7	−7.8	−47
19. New Barnet	1	17	3.3	−4.9	−7.8	−48
20. Schwepps	0	>25	0.7*	−1.5		
21. Kentish Town	1	>25	1.2*	−4.0		
22. Bouverie House	1	>25	0.6*	−1.7		

PS, Pumping station.
* Samples near the limit of detection. In these cases, values given represent the lowest possible value.

diffusive mixing of old water contained in the pores of the rock matrix with modern water moving through the fissured component will dilute the ^{14}C content of the pumped groundwater sample, thus increasing the apparent groundwater age. Hence, even after correction for isotopic exchange with mineral carbonate, the calculated age may not be the true age.

4.4.3 ^{36}Cl dating

Chlorine-36 (^{36}Cl), with a half-life of $301{,}000 \pm 4000$ years, is produced primarily in the atmosphere via cosmic ray bombardment of ^{40}Ar. ^{36}Cl is potentially an ideal tracer for age dating on timescales of up to 1.5 Ma in large groundwater systems, provided that sources and sinks of ^{36}Cl and Cl can be accounted for. The advantage of using ^{36}Cl is that the Cl anion behaves conservatively and, in the absence of Cl-bearing minerals such as halite, it is neither added nor removed from solution via rock–water interactions

and moves at approximately the same velocity as the groundwater.

Although ^{36}Cl and Cl concentrations in groundwater may be modified after recharge by mixing between aquifers or by diffusion from adjacent aquitards, these problems can be overcome by incorporating supplementary chemical and isotopic data to account for these added contributions. Cosmogenic production of ^{36}Cl in the near-surface environment by interaction of cosmic rays with minerals in surface rocks and soils, and nucleogenic production via neutrons generated within the aquifer matrix through decay of U and Th, can be reasonably estimated and are relatively small compared with atmospheric production. The most difficult parameter to estimate in the age determination is the initial $^{36}Cl/Cl$ ratio at the time of recharge. A further issue is the secular variation of ^{36}Cl production over long timescales (Love et al. 2000).

For ^{36}Cl determinations, about 20 mg of Cl is precipitated as AgCl and analysed by accelerator mass spectrometry. A decrease in the measured $^{36}Cl/Cl$

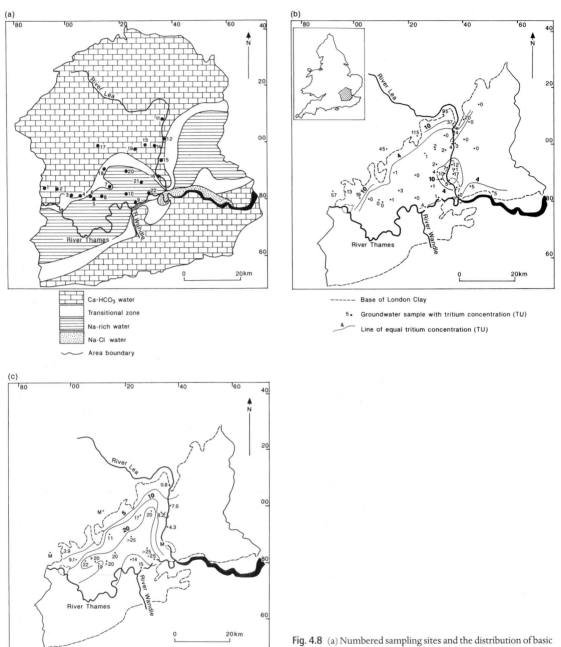

Fig. 4.8 (a) Numbered sampling sites and the distribution of basic hydrochemical water types, (b) tritium concentrations (in TU) and (c) ^{14}C ages (in ka) of Chalk groundwaters in the central part of the London Basin for the data presented in Table 4.1. After Smith et al. (1976a).

ratio along a groundwater hydraulic gradient represents ^{36}Cl decay, and after correcting for different sources of Cl, ^{36}Cl ages can be estimated. In an ideal situation, if Cl and ^{36}Cl are solely derived from atmospheric sources with no internal sources or sinks, except for ^{36}Cl decay and nucleogenic production, and if the initial $^{36}Cl/Cl$ ratio (R_o) and the secular equilibrium $^{36}Cl/Cl$ ratio (R_{se}) can be estimated, then groundwater age estimates can be determined by the following equation (Bentley et al. 1986):

$$t = -\frac{1}{\lambda_{36}} \log_e \frac{R - R_{se}}{R_o - R_{se}} \qquad \text{eq. 4.16}$$

where λ_{36} is the decay constant for ^{36}Cl, and R is the $^{36}Cl/Cl$ measured in groundwater.

Love et al. (2000) presented two modifications of equation 4.16 to allow for the addition of Cl via leakage or diffusion from an adjacent aquitard. In their study, Love et al. (2000) applied the ^{36}Cl dating technique to groundwaters of the south-west flow system of the Great Artesian Basin (see Box 2.11) in north-east South Australia. The main aquifer system comprises Cretaceous gravels, sands and silts of the Jurassic-Cretaceous aquifer, with mean flow velocities, calculated from the rate of decrease of absolute ^{36}Cl concentrations, of 0.24 ± 0.03 m a^{-1}. Calculated ^{36}Cl ages of the confined aquifers, although complicated by addition of Cl via diffusion from the overlying aquitard, range from 200 to 600 ka (Table 4.2; Love et al. 2000). Groundwater trends, from the eastern margin of the Basin in north-east Queensland towards the centre of the Basin in South Australia, show a decrease in the $^{36}Cl/Cl$ ratio in the direction of the hydraulic gradient from approximately 100×10^{-15} to about 10×10^{-15} over a distance of 1000 km indicating ages approaching 1.1 Ma near the end of the flowpath (Bentley et al. 1986).

4.4.4 Tritium dating

Tritium (^3H or T), the radioisotope of hydrogen, has a relative abundance of about $0-10^{-15}\%$. The unit of measurement is the tritium unit (TU) defined as 1 atom of tritium occurring in 10^{18} atoms of H and equal to 3.19 pCi L^{-1} or 0.118 Bq L^{-1}. The half-life of tritium is 12.38 years for ^3H decaying to ^3He with the

emission of a β particle. In a similar way to ^{14}C production, tritium is produced naturally mainly in the upper atmosphere by interaction of cosmic ray-produced neutrons with nitrogen. After oxidation to $^1H^3HO$, tritium becomes part of the hydrological cycle. Analysis of tritium requires distillation followed by electrolytic enrichment of the tritium content. The enriched sample is converted to ethane and gas scintillation techniques are used to measure the tritium content. Combined field and laboratory errors yield an accuracy of ±2 TU or better.

Natural levels of tritium in precipitation are estimated to be between 0.5 and 20 TU. The tritium concentrations of four long-term precipitation records are shown in Fig. 4.9. The records illustrate the effect of atmospheric testing of thermonuclear devices between 1952, prior to which there were no measurements of natural tritium levels in the Earth's atmosphere, and the test ban treaty of 1963 when tritium concentrations, as a result of nuclear fusion reactions, reached a peak of over 2000 TU in the northern hemisphere. A clear seasonal variation is also evident with measured tritium concentrations less in the winter compared with the summer when tritium is rained out of the atmosphere. In the southern hemisphere (and at coastal sites in general), the greater influence of the oceans leads to a greater dilution with water vapour resulting in lower tritium concentrations overall. Currently, atmospheric background levels in the northern hemisphere are between about 5 and 30 TU and in the southern hemisphere between 2 and 10 TU (IAEA/WMO 1998).

The application of the dating equation (eq. 4.8) with tritium data to obtain groundwater ages is problematic given the variation in the initial activity, A_0. Ambiguity arises in knowing whether the input concentration relates to the time before or after the 1963 bomb peak. Nevertheless, tritium concentrations provide a relative dating tool with the presence of tritium concentrations above background concentrations indicating the existence of modern, post-1952 water. Again, as with ^{14}C dating, mixing between different water types with different recharge histories complicates the interpretation. Some authors (for example, Downing et al. 1977) have attempted to use the amount of tritium in groundwater to correct the ^{14}C content of a sample for the effects of dilution with modern water.

Table 4.2 Environmental isotope data and modelled ^{36}Cl ages for groundwaters from the south-west margin of the Great Artesian Basin (26°30′–28°30′S; 133°30′–136°30′E). After Love et al. (2000).

Unit number	Sample name	Distance along transect (km)	δ²H (‰ V$_{SMOW}$)	δ¹⁸O (‰ V$_{SMOW}$)	δ¹³C (‰ PDB)	¹⁴C (pmc)	^{36}Cl/Cl × 10^{-15}	^{36}Cl × 10^6 (atoms L^{-1})	^{36}Cl age (ka)*
Northern Transect (NW to SE)									
574400015	Lambina Homestead	47	−46.3	−5.84	−11.0	6.6 ± 1.2	129 ± 8	1888 ± 140	
574400003	Warrungadinna	61	−42.6	−5.97	−10.6	11 ± 1.3	132 ± 10	2520 ± 235	
574400004	Lambina Soak	71	−49.7	−6.33	−11.7	2.4 ± 1.2	108 ± 7	2026 ± 160	
584300025	Marys Well 3	129	−47.6	−6.71	−11.1		102 ± 8	1162 ± 110	58–94
584300026	Murdarinna 2	145	−51.3	−6.66	−11.7	2.5 ± 2.5	109 ± 10	1186 ± 125	47–190
594300017	Midway Bore	154	−45.7	−6.56			89 ± 4	934 ± 60	160–290
594200001	Oodnadatta Town Bore 1	203	−48.2	−6.71	−11.6	2.0 ± 2.0	52 ± 5	594 ± 65	380–490
604200021	Watson Creek 2	246	−42.7	−6.35			25 ± 3	439 ± 60	550–790
614200004	Duckhole 2 (24701)	261	−49.5	−6.61	−11.0	1.7 ± 1.2	31 ± 4	444 ± 65	540–670
574300005	Appatinna Bore†	96	−35.7	−4.85	−10.4	49.1 ± 5.1	134 ± 13	780 ± 85	
Southern Transect (W to E)									
574100007	C.B. Bore	6	−33.0	−3.81	−8.9	41.2 ± 4.5	115 ± 8	1030 ± 85	
574100014	Ross Bore	33	−39.4	−5.02			54 ± 6	1478 ± 175	
574100049	Evelyn Downs Homestead	43	−40.0	−5.16	−10.9		54 ± 5	1296 ± 140	
584100053	Woodys Bore Windmill	67	−38.5	−4.89			47 ± 5	1914 ± 250	460
584100050	Robyns Bore 2	80	−39.4	−5.10	−11.4	1.8 ± 1.2	41 ± 4	1691 ± 210	30–530
584100011	Ricky Bore 2	89	−41.4	−5.47			37 ± 3	1295 ± 85	110–580
594100003	Paulines Bore	95	−43.3	−5.29	−11.9	3.3 ± 1.2	40 ± 4	1149 ± 130	200–540
594100017	Nicks Bore	114	−42.8	−5.73			46 ± 5	1135 ± 145	200–470
594100013	Leos Bore	136	−42.0	−5.68			53 ± 5	1012 ± 115	250–400
594100006	Fergys Bore	141	−44.1	−6.07	−11.5		61 ± 1.2	969 ± 210	270–340
604100037	Lagoon Hill Drill Hole 15 (New Peake)	183	−46.4	−6.56	−10.7	3.8 ± 1.2	29 ± 3	461 ± 60	600–710

pmc, % modern carbon.

* Age range calculated using equation 4.16 and two variants to allow for the addition of Cl via leakage or diffusion from an adjacent aquitard.

† Data for the unconfined Appatinna Bore have been used to determine the initial ^{36}Cl/Cl ratio.

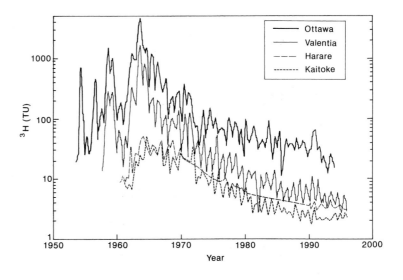

Fig. 4.9 Variations in mean monthly tritium concentrations in precipitation since 1953 at four IAEA stations (IAEA/WMO 1998): Ottawa, Canada (northern hemisphere, continental); Valencia, Ireland (northern hemisphere, marine); Harare, Zimbabwe (southern hemisphere, continental); Kaitoke, New Zealand (southern hemisphere, marine).

Although the bomb peak has decayed away, making tritium less useful as a dating tool, earlier studies in the 1970s and 1980s were able to relate large quantities of tritium in a sample to groundwater recharge in the 1960s. For example, in Fig. 4.8b, modern water with a tritium concentration up to 57 TU (Table 4.1) was sampled in the recharge area of the Chalk outcrop on the northern rim of the London Basin. Away from the Chalk outcrop, tritium values decline rapidly below the confining Eocene London Clay. Another area of interest is the River Lea valley where a tongue of water containing measurable tritium (53 TU maximum) extends south from the Chalk outcrop. This area coincides with the region of the North London Artificial Recharge Scheme (see Box 8.2) and it appears likely that the tritium anomaly is due to artificial recharge of treated mains water containing a component of modern water.

Another application for tritium is providing a tracer for modern pollutant inputs to aquifers such as agricultural nitrate and landfill leachate. For example, sequential profiling of tritium in porewater contained in the unsaturated zone of the English Chalk has helped in the understanding of the transport properties of this dual-porosity aquifer. As shown in Fig. 4.10a for a site in west Norfolk, sequential redrilling and profiling of tritium concentrations in the Chalk matrix revealed the preservation of the tritium bomb peak. The results are interpreted as indicating slow downward migration of water through the low-permeability, saturated Chalk matrix. The tritium appears to migrate with a 'piston-like' displacement with only limited dispersion of the peak concentrations. Detailed analysis of the profiles in terms of their mass balance and peak movement revealed the possibility of rapid water movement, or 'by-pass' flow. It is estimated that up to 15% of the total water movement occurs through the numerous fissures that comprise the secondary porosity of the Chalk once the fluid pressure in the unsaturated zone (Section 5.4.1) increases to the range −5 to 0 kPa (Gardner et al. 1991). At other locations, a 'forward tailing' and broadening of the tritium peaks are observed in sequential tritium profiles providing evidence of greater dispersion (Parker et al. 1991).

Laboratory measurements have shown that tritiated water and NO_3^- have similar diffusion coefficients and so similar concentration profiles might be expected for NO_3^- (Fig. 4.10b). Dispersion of NO_3^- is suggested where there has been less downward movement than would be expected if piston flow were occurring. An apparent flattening of the original NO_3^- peaks leads to higher concentrations than expected in the deeper part of the profiles. The rate of solute movement in the Chalk implied by the tritium and NO_3^- profiles is between 0.4 and 1.1 m a^{-1}. The actual rate of movement at a specific site will depend on many factors, including the physical properties

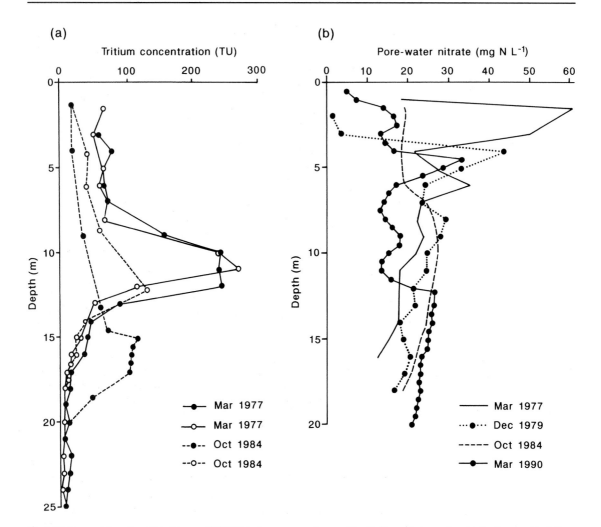

Fig. 4.10 Sequential profiles of (a) tritium and (b) NO$_3^-$ in the unsaturated zone of the Chalk aquifer beneath arable land in west Norfolk, Eastern England. After Parker et al. (1991).

of the Chalk, temporal variations in rainfall and antecedent moisture content, and seasonal variations in solute input (Parker et al. 1991).

4.4.5 ^{3}H/^{3}He dating

The limitations of using tritium alone to estimate groundwater residence times that arise from its short radioactive half-life and the decay of the bomb peak can be addressed with the additional measurement of ^{3}He. As tritium in groundwater enters the unsaturated

zone and is isolated from the atmosphere below the water table, dissolved ^{3}He concentrations from the decay of tritium will increase as the groundwater becomes older. Although groundwaters contain helium from several sources other than tritium decay, including atmospheric and radiogenic ^{3}He and ^{4}He, determination of both tritium and tritiogenic ^{3}He (^{3}He*) can be used as a dating tool (Solomon et al. 1993). The tritium-helium (^{3}H/^{3}He) age is defined as:

$$t_{^3H/^3He} = \frac{1}{\lambda_T} \log_e \left(\frac{^3He^\star}{^3H} + 1 \right)$$ eq. 4.17

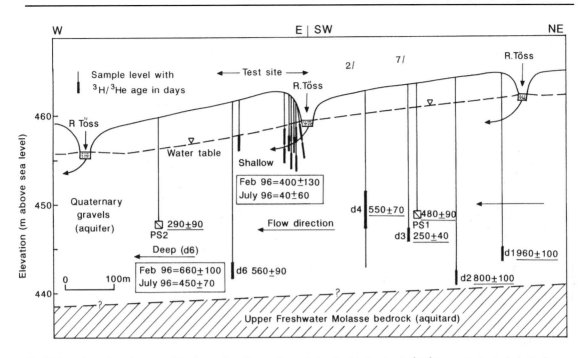

Fig. 4.11 Hydrogeological section of the Linsental aquifer, north-east Switzerland, indicating the ^{3}H/^{3}He ages (in days) of samples from boreholes and two pumping stations, Sennschür (PS1) and Obere Au (PS2), for various sampling dates. The ages of samples collected from shallow and deep (point d6) sections of the aquifer in February and July 1996 are also shown. In February, older groundwater dominated in the shallow aquifer during a time of decreased river water infiltration. In July, heavy rain and flood events were common and locally infiltrated river water with a younger age dominated in the shallow aquifer. After Beyerle et al. (1999).

where λ_T is the tritium decay constant. Analytical techniques for tritium and ^{3}He determinations are such that under ideal conditions, ages between 0 and about 30 years can be determined with typical analytical uncertainties of <10% (Solomon et al. 1993). Further confidence in deriving an apparent age is obtained when multiple techniques are applied to the same sample, for example combining measurements of ^{3}H/^{3}He and chlorofluorocarbons (CFCs) (Section 5.8.3). Applications for ^{3}H/^{3}He dating are found in the accurate dating of groundwater residence times of a few tens of years (Plummer et al. 2001), in identifying groundwater flowpaths (Beyerle et al. 1999) and in tracing sources and rates of contaminant movement (Shapiro et al. 1999).

As an example, Beyerle et al. (1999) applied ^{3}H/^{3}He dating at a bank filtration site at Linsental in north-east Switzerland. The Linsental aquifer is in the lower reach of the River Töss valley and has a maximum thickness of 25 m and an average width of 200 m. The aquifer is composed of heterogeneous

Quaternary gravels reworked by fluvial erosion. The lower aquifer boundary is formed by a low permeability molasse bedrock. Hydraulically, the Linsental aquifer is predominantly influenced by groundwater abstraction from two pumping stations, Sennschür and Obere Au (PS1 and PS2, respectively in Fig. 4.11), and infiltration from the River Töss.

Groundwater and river water samples were collected between 1995 and 1997 and analysed for their tritium and helium contents and the resulting ^{3}H/^{3}He ages interpreted along a hydrogeological section. As shown in Fig. 4.11, the ^{3}H/^{3}He groundwater ages from the deep part of the aquifer are similar (about 600 days), whereas water ages of the shallow aquifer depend on the time of sampling. As river infiltration is reduced at times of low river stage, only older groundwater remains and the ^{3}H/^{3}He ages close to the river become comparable to the age of groundwater in the deeper aquifer. The ages of groundwater abstracted from the two pumping stations are about 1 year old, much higher than initially considered. An estimated mixing ratio of 50% between the younger infiltrated river water and older deep groundwater explains the age of the water abstracted from the production boreholes. Hence, the ^{3}H/^{3}He ages have provided useful information on the recharge dynamics and residence times of the bank filtration system (Beyerle et al. 1999).

4.5 Noble gases

The temperatures at which recharge water is equilibrated with air in the soil zone can be determined from the noble (inert) gas contents of groundwater. The concentrations of Ar, Kr, Ne and Xe dissolved during equilibration of recharge water with air are controlled by their solubility relationship with temperature, generally decreasing in concentration with increasing temperature. In temperate latitudes, the potential effects of solar insolation at the ground surface are minor and the recharge temperature generally reflects the mean annual temperature of the soil zone. Any increase in groundwater temperature as the water moves from the soil zone to greater depths in the aquifer will not result in exsolution of noble gases, since the increase in hydrostatic pressure maintains the groundwater undersaturated with respect to the gases. It is therefore valid to derive recharge temperatures from noble gas contents (Andrews & Lee 1979). The quantity of a dissolved noble gas, for example Ar, is given by:

$$[Ar] = s_T P_{Ar} (cm^3 \text{ STP cm}^{-3} H_2O) \qquad \text{eq. 4.18}$$

where s_T is the solubility of Ar at 1 atmosphere pressure and the temperature of recharge, T; and P_{Ar} is the partial pressure of Ar in the atmosphere. Similar relationships apply to the other noble gases.

Noble gas concentrations of groundwater commonly exceed those calculated for thermodynamic solubility equilibrium with air in the unsaturated zone (eq. 4.18). This additional component, termed 'excess air', is most likely the result of fluctuations in the water-table trapping and partially or entirely dissolving small bubbles under increased hydrostatic pressure or surface tension. Corrections for the

excess air component are possible, normalized to dissolved Ne contents (Elliot et al. 1999). Some of this excess air may be lost by gas exchange across the water table. The rates of gas exchange of the individual noble gases decrease with molecular weight such that fractionation of the residual excess air occurs with significant gas loss. Typically, noble gas recharge temperatures are determined by an iterative procedure. Initially, unfractionated air is subtracted successively from the measured concentrations, and the remaining noble gas concentrations are converted into temperatures on the basis of solubility data and the atmospheric pressure at the elevation of the water table. This procedure is repeated until optimum agreement among the four calculated noble gas temperatures is achieved (Stute et al. 1995).

When combined with stable isotope data and ^{14}C ages, noble gas recharge temperatures can further elucidate the history of aquifer evolution and provide a proxy indicator of palaeoenvironmental conditions. Noble gas data are presented by Dennis et al. (1997) and Elliot et al. (1999) for the Chalk aquifer of the London Basin and provide evidence for Late Pleistocene (cold stage interstadial) groundwaters recharged at temperatures 5–7°C cooler than present in confined zones north of the River Thames (Table 4.3). Recharge conditions at this time were probably controlled by the occurrence of areas of unfrozen ground within the summer permafrost (talik zones), and so the noble gas recharge temperature may be more representative of the mean summer air temperature. Mean annual air temperatures spanning the frozen winter period were probably lower than the noble gas recharge temperatures.

As shown in Fig. 4.12, similar mean recharge temperatures during the last glacial period of about 5°C cooler than modern Holocene waters are reported for the East Midlands Triassic sandstone aquifer in the United Kingdom (Andrews & Lee 1979) and Devonian sandstones in the semi-arid Piaui Province in north-east Brazil (Stute et al. 1995). Hence, terrestrial records of palaeogroundwaters provide convincing evidence for surface temperature cooling at both high and low latitudes during the last (Devensian or Weichselian) ice age.

Table 4.3 Isotope and noble gas recharge temperatures for Chalk groundwaters in the London Basin along a N–S transect from the outcrop on the North Downs to the confined centre of the Basin in the City of London. After Elliot et al. (1999).

Sample location	Distance N along transect (km)	δ^2H*	$\delta^{18}O*$	$\delta^{13}C\dagger$	^{14}C (pmc)	RT (°C)	±1 s.e.
1. Paynes u	0	−51	−7.4	−13.6	66.2	12.5	0.7
2. Philips 2 u	1.7	−48	−7.2	−13.4		11.4	0.9
3. BXL Plastics c	3.7			−3.6	4.3	9.0	0.5
4. Modeluxe c	6.3	−50	−7.4	−3.8	4.5	9.2	0.8
5. Sunlight c	10.9	−49		−8.7	14.6	10.2	0.6
6. Unigate c	13.4	−47	−7.2	−12.8	12.6	10.9	1.0
7. Harrods c	15.1	−51	−7.4	−4.8	17.1	9.5	0.8
8. Buchanan House c	16.3	−52	−7.3	−3.5		8.7	0.7
11. Unilever c	16.3	−53	−7.6	−2.5	1.0	6.8	0.6
10. Sainsburys c	16.3	−51	−7.4	−1.7		7.1	0.5
9. Dorset House c	17.4	−53	−7.8	−2.4	0.8	7.1	0.5
13. Kentish Town c	20.0	−53	−7.8	−2.6	0.8	5.4	0.6
12. Hornsey Road c	21.1	−54	−7.7	−2.6	1.4	5.8	0.5

pmc, % modern carbon; u, unconfined; c, confined; RT, noble gas recharge temperature; 1 s.e., 1 standard error.
* ±1‰ and ±0.1‰ relative to V_{SMOW}.
† ±0.1‰ relative to PDB.

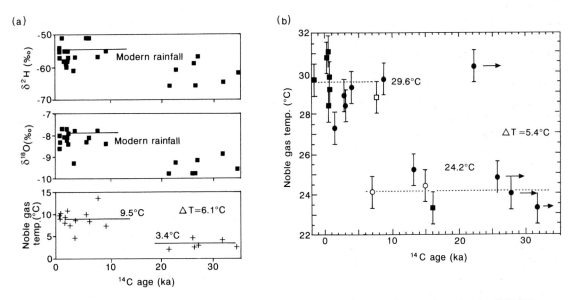

Fig. 4.12 Noble gas palaeotemperature records for (a) the East Midlands Triassic sandstone aquifer, United Kingdom (53.5°N, 1°W) after Edmunds (2001), and (b) Devonian sandstones of the Cabeças (circles) and Serra Grande (squares) aquifers of Piaui Province, Brazil (7°S, 41.5°W) after Stute et al. (1995). Noble gas recharge temperatures are plotted as a function of ^{14}C age. Relative to modern waters, the climate of both tropical and mid-latitudes was between 5–6°C cooler during the last ice age. At mid-latitudes the mean annual temperatures may have been cooler given that recharge was probably restricted to talik zones within the summer permafrost. The ^{14}C data for the East Midlands aquifer demonstrate that groundwater recharge occurred during the Holocene as well as during the Devensian (the oldest waters are beyond the limits of radiocarbon dating). An age gap is evident between 10 and 20 ka when the recharge areas were frozen at the time of the last glacial maximum (LGM). In lowland Brazil, the arrows shown indicate those samples with a ^{14}C content below the detection limit and the open symbols represent samples obtained by Verhagen et al. (1991). The cooling of surface temperatures in Brazil by 5.4°C during the period between 10 and 35 ka is consistent with about a 1000-m lowering of the snowline altitudes in the tropical mountain ranges for the LGM. After Edmunds (2001) and Stute et al. (1995).

4.6 FURTHER READING

Attendorn, H.-G. & Bowen, R.N.C. (1997) *Radioactive and Stable Isotope Geology*. Chapman & Hall, London.

Clark, I.D. & Fritz, P. (1997) *Environmental Isotopes in Hydrogeology*. Lewis Publishers, Boca Raton, Florida.

Faure, G. (1986) *Principles of Isotope Geology*, 2nd edn. John Wiley, New York.

Fritz, P. & Fontes, J.Ch. (1980) *Handbook of Environmental Isotope Chemistry*, vol. 1: *the terrestrial environment, A.* Elsevier, Amsterdam.

Kendall, C. & McDonnell, J.J. (eds) (1998) *Isotope Tracers in Catchment Hydrology*. Elsevier, Amsterdam.

Lloyd, J.W. & Heathcote, J.A. (1985) *Natural Inorganic Hydrochemistry in Relation to Groundwater: an introduction*. Clarendon Press, Oxford.

Mazor, E. (1997) *Chemical and Isotopic Groundwater Hydrology: the applied approach*, 2nd edn. Marcel Dekker, New York.

Groundwater investigation techniques 5

5.1 Introduction

The purpose of this chapter is to introduce basic field methods and techniques used in hydrogeological investigations. The assessment and management of groundwater resources requires measurement of water flows into and out of an aquifer. In that groundwater is intimately linked to surface water within a river basin, it is useful to adopt the concept of a catchment water balance, in which all known inflows and outflows are accounted for, when discussing measurement requirements. Discussion of groundwater development schemes, including a catchment water balance equation, is provided in Section 8.2, but the measurements needed to complete a water balance, such as precipitation, evaporation, streamflow and groundwater recharge, are presented in this chapter. Pumping test methods for the assessment of the aquifer properties of transmissivity and storativity and the application of tracer techniques in groundwater investigations are also discussed, together with an introduction to the application of geophysical techniques and numerical groundwater flow and solute transport modelling in hydrogeology.

5.2 Measurement and interpretation of groundwater level data

The measurement and collection of groundwater level data are of fundamental importance in hydrogeology. Groundwater level data for an aquifer unit can be used for several purposes including plotting a hydrograph, determining the direction of groundwater flow by constructing a map of the potentiometric surface, and in completing a flow net (see Box 2.3). Values of hydraulic head are also essential in the process of designing and testing a numerical groundwater flow model for the purpose of making predictions of aquifer behaviour under future conditions

5.2.1 Water level measurement

Measurement of groundwater level in the field is undertaken using either a water level dipper or by use of a submersible pressure transducer positioned just below the lowest expected groundwater level. The required field measurement is the depth from a convenient measurement datum, for example the well top or borehole flange, to the position of the groundwater level. If the elevation of the measurement datum is known from levelling techniques (Pugh 1975) then the elevation of the groundwater level (or groundwater head, h) can be recorded as the height above or below a local base level, typically mean sea level.

A water level dipper for use in the field comprises a length of twin-core cable, graduated in centimetres and metres, wound on to a drum and with a pair of electrodes attached to the end (Fig. 5.1). When the electrodes touch the water surface, a circuit is completed which activates either a light or a buzzer or both. Water levels can be measured to a precision of ± 0.005 m (Brassington 1998).

A pressure transducer consists of a solid-state pressure sensor encapsulated in a stainless steel, submersible housing. A waterproof cable, moulded to the transducer, connects the water pressure sensor to a monitoring device such as a data logger from which

(a) General assembly

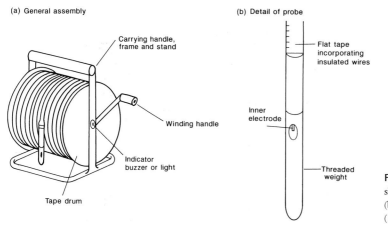

Carrying handle,
frame and stand

Winding handle

Indicator
buzzer or light

Tape drum

(b) Detail of probe

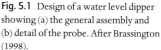

Flat tape
incorporating
insulated wires

Inner
electrode

Threaded
weight

Fig. 5.1 Design of a water level dipper showing (a) the general assembly and (b) detail of the probe. After Brassington (1998).

a temporal record of groundwater level fluctuations can be obtained (Fig. 5.5). When employing a pressure transducer, and in order to convert to a value of groundwater head, h, the measurement of fluid pressure, $P (= \rho g \psi)$, recorded at a depth elevation, z, is converted to a value of groundwater head using the relationship given in equation 2.22.

Groundwater levels are measured in either a well (Fig. 5.2a) or a purpose-built observation borehole (Fig. 5.2b). Wells are typically shallow in depth, lined with unmortared bricks and penetrate the top of the local water table in an unconfined aquifer. Observation boreholes can either be uncased (open) or cased, depending on the strength of the aquifer rock, and are drilled with either a percussion or rotary rig (Brassington 1998). Further guidance on well and borehole design and construction methods are contained in useful textbooks by Driscoll (1986) and Clark (1988).

Observation boreholes record the groundwater level in unconfined aquifers (for which there is a water table) or confined aquifers (for which there exists a potentiometric surface). A special type of installation known as a piezometer (Fig. 5.2c) is designed to provide a measurement of the hydraulic head at a given depth in an aquifer. A bundle of piezometers nested in a single borehole installation can provide information on hydraulic heads at several depths in an aquifer from which the vertical component of groundwater flow, either downwards in a recharge area or upwards in a discharge zone, can be ascertained.

5.2.2 Borehole hydrographs and barometric efficiency

Well or borehole hydrographs typically display data collected at monthly intervals and provide a record of fluctuations in groundwater levels. As shown in Fig. 5.3, additional data can be shown on a hydrograph to indicate the position of measured monthly groundwater levels relative to the long-term average and to historic minima and maxima. Long-term records are invaluable. Climatic effects such as the frequency of wet and dry years can be identified (see Box 8.7) as well as artificial effects, for example the over-exploitation of groundwater resources leading to a gradual decline in groundwater level (see Fig. 5.3b). Groundwater level drawdown data recorded during pumping tests (see Section 5.8.2) can also be corrected for background trends in the regional potentiometric surface by reference to a hydrograph record unaffected by the pumping test.

Large fluctuations in water levels in wells and boreholes in confined aquifers can be caused by changes in atmospheric pressure. With increasing barometric pressure, water levels are noticed to decrease. This phenomenon, which is also seen as a transient effect of external loading from passing trains, construction blasting and earthquakes, relates to a change in the stress field applied to the aquifer (Jacob 1940). Using the principle of effective stress (see Fig. 2.23) and considering the situation shown in Fig. 5.4a, the stress equilibrium at position X at the top of a confined aquifer is given by:

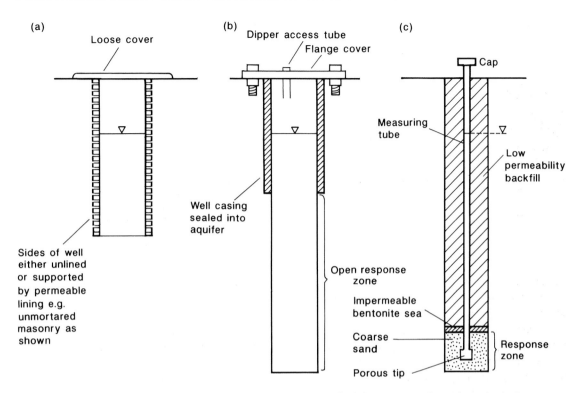

Fig. 5.2 General designs of (a) well, (b) observation borehole and (c) piezometer for the measurement of groundwater level. After Brassington (1998).

$$\sigma_T + P_A = \sigma_e + P_w \qquad \text{eq. 5.1}$$

where P_A is atmospheric pressure, σ_T is the total stress created by the weight of the overlying aquitard, σ_e is the effective stress acting on the aquifer material and P_w is the fluid pressure in the aquifer. The fluid pressure creates a pressure head, ψ, that is measured in the well penetrating the aquifer. At position Y in the well, the balance of pressures is:

$$P_A + \gamma\psi = P_w \qquad \text{eq. 5.2}$$

where γ is the specific weight of water. If, as shown in Fig. 5.4b, the atmospheric pressure is increased by an amount dP_A, the change in the stress field at position X is given by:

$$dP_A = d\sigma_e + dP_w \qquad \text{eq. 5.3}$$

Now, it can be seen that the change in dP_A is greater than the change in dP_w such that at position Y in the well, the new balance of pressures is:

$$P_A + dP_A + \gamma\psi' = P_w + dP_w \qquad \text{eq. 5.4}$$

which, on substitution of equation 5.2 in equation 5.4, gives:

$$dP_A - dP_w = \gamma(\psi - \psi') \qquad \text{eq. 5.5}$$

Since $dP_A - dP_w$ is greater than zero, then $\psi - \psi'$ is also greater than zero, proving that an increase in atmospheric pressure leads to a decrease in water level (Fig. 5.5). In a horizontal, confined aquifer the change in pressure head, $d\psi = \psi - \psi'$ in equation 5.5, is equivalent to the change in hydraulic head, dh, and so provides a definition of barometric efficiency, B, expressed as:

$$B = \frac{\gamma dh}{dP_A} \qquad \text{eq. 5.6}$$

The barometric efficiency of confined aquifers is usually in the range 0.20–0.75 (Todd 1980). Jacob (1940) further developed expressions relating barometric

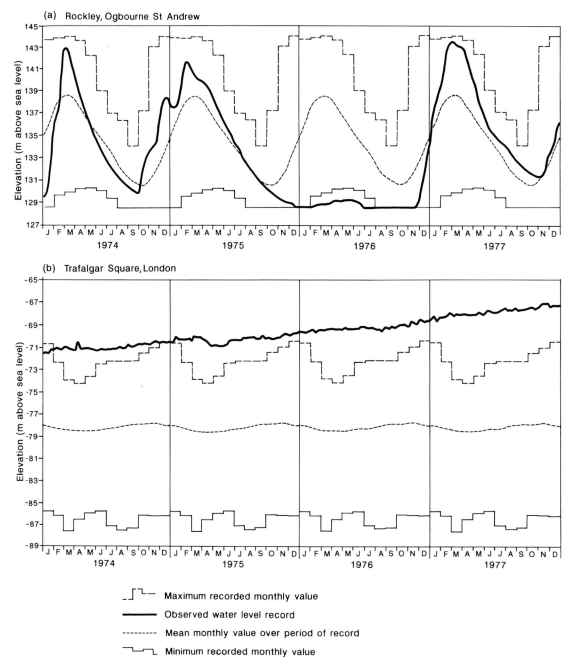

(a) Rockley, Ogbourne St Andrew

(b) Trafalgar Square, London

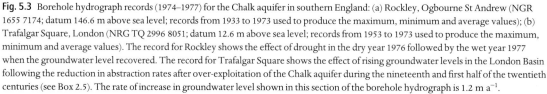

— Maximum recorded monthly value

— Observed water level record

-------- Mean monthly value over period of record

— Minimum recorded monthly value

Fig. 5.3 Borehole hydrograph records (1974–1977) for the Chalk aquifer in southern England: (a) Rockley, Ogbourne St Andrew (NGR 1655 7174; datum 146.6 m above sea level; records from 1933 to 1973 used to produce the maximum, minimum and average values); (b) Trafalgar Square, London (NRG TQ 2996 8051; datum 12.6 m above sea level; records from 1953 to 1973 used to produce the maximum, minimum and average values). The record for Rockley shows the effect of drought in the dry year 1976 followed by the wet year 1977 when the groundwater level recovered. The record for Trafalgar Square shows the effect of rising groundwater levels in the London Basin following the reduction in abstraction rates after over-exploitation of the Chalk aquifer during the nineteenth and first half of the twentieth centuries (see Box 2.5). The rate of increase in groundwater level shown in this section of the borehole hydrograph is 1.2 m a^{-1}.

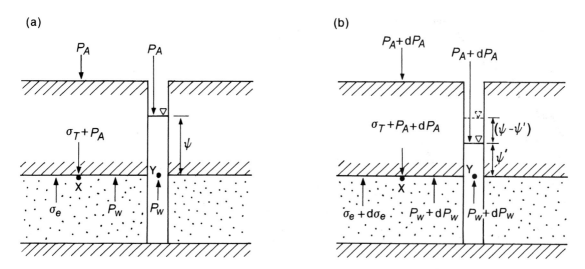

Fig. 5.4 Diagram showing the change in aquifer stress conditions caused by an increase in atmospheric pressure from P_A in (a) to $P_A + dP_A$ in (b) resulting in a decrease in groundwater level ($\psi - \psi'$) in a confined aquifer.

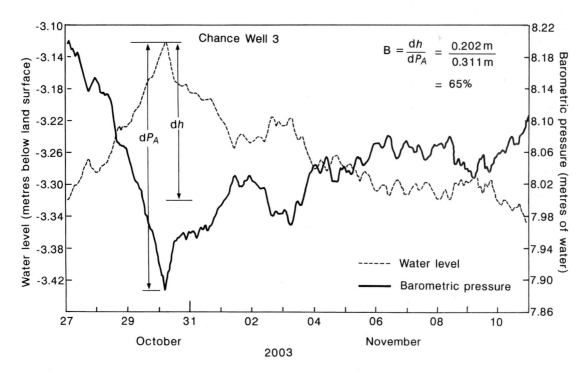

Fig. 5.5 Relationship between groundwater level fluctuation and atmospheric pressure in Chance Well 3 (CW-3), Long Valley Caldera, California (well depth 267 m; altitude 2155.90 m above sea level; latitude 37°38′49″, longitude 118°51′30″). At this site, water level and barometric pressure are monitored using a pressure transducer and barometer, respectively. These data are recorded on a data logger every 30 minutes and the data transmitted by satellite to the United States National Water Information System database.

efficiency of a confined aquifer to aquifer and water properties, including the storage coefficient. In unconfined aquifers, atmospheric pressure changes are transmitted directly to the water table, both in the aquifer and in the well, such that the water level in an observation well does not change. However, air bubbles trapped in pores below the water table are affected by pressure changes and can cause fluctuations similar to but smaller than observed in confined aquifers.

5.2.3 Construction of groundwater level contour maps

To be able to construct a map of the groundwater level and therefore depict the potentiometric surface and determine the direction of groundwater flow, a minimum of three observation points is required as shown in Fig. 5.6. The procedure is first to relate the field groundwater levels to a common datum (map datum or sea level for convenience) and then plot their position on a scale plan. Next, lines are drawn between three groundwater level measurements and divided into a number of short, equal lengths in proportion to the difference in elevation at each end of the line (in the example shown, each division on line AB and BC is 0.2 m, while on line AC each division is 0.1 m). The next step is to join points of equal elevation on each of the lines and then to select a contour interval which is appropriate to the overall variation in water levels in the mapped area (here 0.5 m). The same procedure is followed for other pairs of field observation points until one or two key contour lines can be mapped. At this point, the remaining contour lines can be drawn by interpolating between the field values.

Additional information that can be used in completing a potentiometric surface map is knowledge of the general topography of a region, and records of the elevations of springs known to discharge from an aquifer as well as the elevations of gaining streams and rivers (see Fig. 8.8a) that flow over the aquifer outcrop, since these points represent ground surface interception of the water table. For unconfined aquifers bordering the sea it is usual to represent the coastline as a groundwater contour with an elevation equal to sea level (0 m). Similar assumptions can be

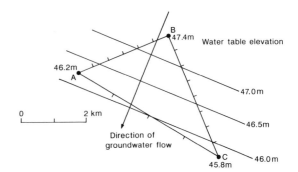

Fig. 5.6 Graphical construction method for determining the direction of groundwater flow from three groundwater level measurements.

made in respect of large surface water bodies at inland locations (Brassington 1998).

The direction of groundwater flow in an isotropic aquifer can be drawn at right angles to the contour lines on the potentiometric surface in the direction of decreasing hydraulic head. This assumes that the aquifer is an isotropic material (see Section 2.4). In anisotropic material, for example fissured or fractured aquifers, the flow lines will be at an angle to the potentiometric contour lines (see Box 2.3). An example of a completed potentiometric surface map for the Chalk aquifer in the London Basin is shown in Box 2.4. Construction of potentiometric surface maps at times of low and high groundwater levels can be of assistance in calculating changes in the volume of water stored in an aquifer and in assessing the local effects of groundwater recharge and abstraction (Brassington 1998).

5.3 Precipitation and evapotranspiration

Near-surface hydrological processes such as precipitation, evapotranspiration and infiltration have a profound influence on streamflow generation and groundwater recharge. Precipitation, namely in the form of rainfall, provides the raw input of water to a catchment but its availability for supporting river flows, replenishing aquifer storage and supporting water supplies depends on catchment conditions such as soil type, geology, climate and land use that affect catchment runoff properties.

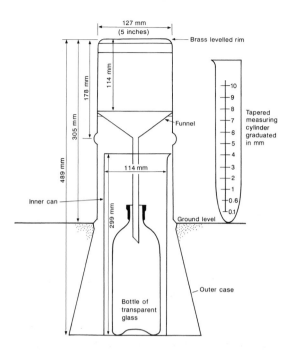

Fig. 5.7 The copper 5-inch standard rain gauge used by the UK Meteorological Office. The rain gauge consists of a 5-inch (127-mm) diameter funnel with a sharp rim, the spout of the funnel being inserted into a glass collecting jar. The jar is in an inner copper can and the two are contained in the main body of the gauge, the lower part of which is sunk into the ground with the rim 12 inches (305 mm) above the surrounding short grass or gravel, this height being chosen so that no rain splashes from the surroundings into the funnel. The funnel has a narrow spout in order to reduce evaporation loss. Normally, the gauge is sited such that its distance from any obstructions (trees, houses, etc.) is at least four times the height of the obstruction.

5.3.1 Precipitation measurement

Precipitation falls mainly as rain but may also occur as hail, sleet, snow, fog or dew. The design of the standard rain gauge used in the United Kingdom is shown in Fig. 5.7 with a cylinder diameter of 5 inches (127 mm). In the United States and Canada, standard rain gauges have diameters of 8 and 9 inches (203 and 229 mm), respectively. Standard rain gauges are read daily, for example at 0900 h in the United Kingdom. In exposed locations, the rain catch of the gauge is affected by high winds and it is generally accepted that more accurate results will be obtained from a rain gauge set with its rim at ground level. Although more expensive, with a ground-level installation it is necessary to house the gauge in a pit and to surround it with an antisplash grid. The measurement of snowfall is also possible with a standard rain gauge, although subject to error due to turbulence around the rim of the gauge. The snow is caught and melted, and the equivalent amount of water recorded.

Recording gauges (or autographic gauges) are able to automatically measure or weigh precipitation and are useful at remote, rarely visited sites. An example is the tipping-bucket rain gauge in which the number of times a small bucket of known volume fills and tips is recorded. Each tip activates a reed switch which sends an electrical pulse to a logger. The tilting siphon autographic gauge has a chart pen that floats up as rainfall fills a chamber, which tips when full, thus returning the pen to the bottom of the chart. Recording gauges are more expensive and more prone to error (for example, very low rainfall amounts are not recorded) but have the advantage of measuring rainfall intensity as well as rainfall total.

The number of rain gauges required to give a reliable estimate of catchment rainfall increases where rainfall gradients are marked. A minimum density of 1 gauge per 25 km² is recommended, considering that large thunderstorm systems may only cover an area of about 20 km². In hilly terrain, where orographic effects may cause large and consistent rainfall variations over short distances, higher rain gauge densities are necessary in the first years of measurement (Table 5.1). In tropical areas there is large spatial variation in daily rainfall, but only a small gradient in annual totals. In such areas, the rain gauge densities in Table 5.1 will be excessive and higher priority should be given to obtaining homogeneous records of long duration at a few reliable sites.

To be able to assess a representative value of rainfall over large areas, it is necessary to employ a method of averaging the individual gauge measurements. The simplest method is to take the arithmetic

Table 5.1 Density of rain gauges required in a hill area.

Catchment area (km²)	Number of gauges
4	6
20	10
80	20
160	30

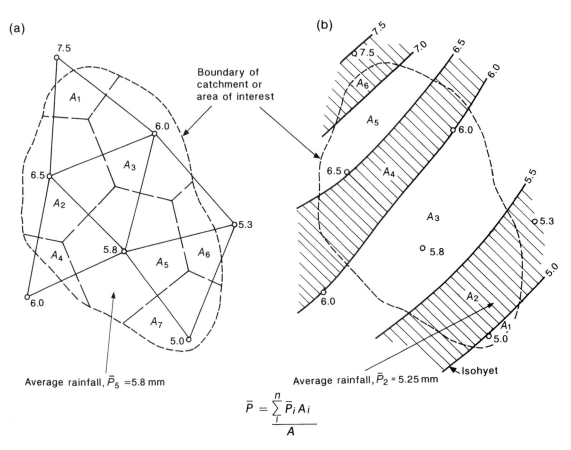

$$\bar{P} = \dfrac{\sum\limits_{i}^{n} \bar{P}_i A_i}{A}$$

Fig. 5.8 Calculation of area-weighted average rainfall amount using the methods of (a) Thiessen polygons and (b) isohyets.

mean of the amounts recorded for all rain gauges in an area. If the distribution of points is uniform and rainfall gradients are small, then this method gives acceptable results. Another method is to use Thiessen polygons to define the zone of influence of each rain gauge by drawing lines between adjacent gauges, bisecting the lines with perpendiculars, and then joining the perpendiculars to form a boundary enclosing the gauge (Fig. 5.8a). It is assumed that the area inside the boundary has received an average rainfall amount equal to the enclosed gauge. A variation of this technique is to draw the perpendiculars by bisecting the height difference between adjacent gauges, although this altitude-corrected approach does not produce a greatly different result. As shown in Fig. 5.8b, a further method is to draw contours of equal rainfall

depth, or isohyets, with the areas between successive isohyets measured and assigned an average rainfall amount.

A map showing rainfall isohyets for the United Kingdom is given in Fig. 5.9 and illustrates the strong west–east gradient in mean annual rainfall as a result of the prevailing, rain-bearing westerly air flow and orographic and rain shadow effects. A yearly compendium of rainfall, river flows, groundwater levels and river water quality data was formerly published by the Natural Environment Research Council in the series entitled *Hydrological Data UK Yearbook*. Yearbooks were published for 1981–1995 but are now superseded by the National Water Archive which is accessed via the Internet at www.ceh.ac.uk/data/NWA.htm.

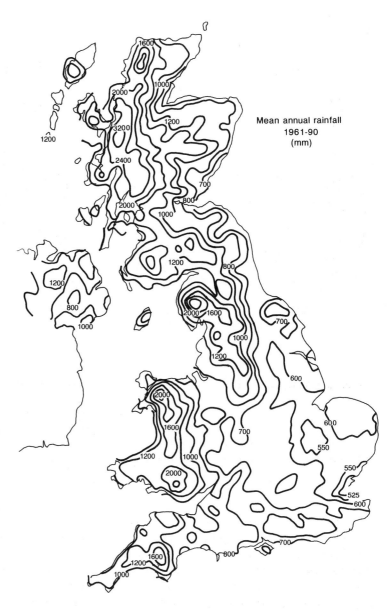

Fig. 5.9 Mean annual rainfall over the United Kingdom, 1961–1990, compiled by the UK Meteorological Office. The mean annual rainfall varies enormously over the British Isles from in excess of 3000 mm in parts of the western highlands of Scotland to about 500 mm in parts of East Anglia and the Thames Estuary. However, there is much less difference in the number of rain days (defined as days when 0.2 mm or more of rain falls). On average, the drier areas have between 150 and 200 rain days annually, while the wetter areas have just over 200 rain days. In most areas, the month of December has the highest number of rain days.

5.3.2 Evapotranspiration measurement and estimation

In the hydrological cycle, evaporation of water occurs from free water surfaces such as lakes, reservoirs and rivers with the rate dependent on factors such as water temperature and the temperature and humidity of the layer of air above the water surface. Wind speed is also a factor in determining the rate at which vapour is carried away from the water surface. Evaporation can be measured directly using an evaporation tank or pan. Tanks are buried in the ground, whereas pans are set above ground on a small plinth. The British standard tank has dimensions of 1.83 m × 1.83 m and is 0.61 m deep. The United States National Weather Service (NWS) Class A pan, with a diameter of 1.22 m and a depth of 0.25 m, is made of unpainted, galvanized metal and is generally accepted as an international standard. Evaporation tanks tend to give a more accurate measurement of true evaporation, unlike pans that are affected by higher losses of vapour caused by heating of their exposed walls and shallow water. Pan coefficients (correction factors) can be applied to correct measurements to 'true evaporation'. Pan coefficients for the British standard tank and US NWS Class A pan range between 0.93–1.07 and 0.60–0.80, respectively (WMO 1994).

For vegetated surfaces, water is lost by evaporation from bare soil and also by transpiration through the leaf stomata of plants. The term evapotranspiration is used to describe the combination of these effects and is a significant process in terms of catchment water balances, often being the principal loss of water from a catchment.

The concept of potential evapotranspiration (PE) is defined as the amount of water that would be removed from a vegetated surface if sufficient water were available in the soil to meet the water demand. The PE may be met in areas where the soil is saturated, for example after a rainfall event or in an area of high water table, say in a groundwater discharge area; otherwise, the actual evapotranspiration rate will be less than PE. A direct measurement of evapotranspiration can be made using a lysimeter, a large container holding a monolith of soil and plants that is set outdoors. Evapotranspiration is estimated for the lysimeter from the balance of precipitation and irrigation inputs, change in soil moisture content and loss of water as soil drainage. Alternatively, field estimates of soil moisture content can be combined with precipitation, river discharge and groundwater monitoring data to calculate catchment-scale evapotranspiration losses.

PE is dependent on the evaporative capacity of the atmosphere and can be calculated theoretically using meteorological data. The most commonly used methods for calculating PE are those of Blaney and Criddle (1962) and Thornthwaite (1948), which are based on empirical correlations between evapotranspiration and climatic factors, and Penman (1948) and Penman–Monteith (Monteith 1965, 1985) which are energy-budget approaches requiring further meteorological data.

The Penman formula for the estimation of evaporation from meteorological data is based on two requirements which must be met if continuous evaporation is to occur. The first is that there must be a supply of energy (radiation) to provide latent heat of vaporization and, second, there must be an aerodynamic mechanism (wind and humidity) for removing vapour, once produced. The Penman formula for evaporation from open water, E_o, in mm day^{-1}, is given as:

$$E_o = \frac{\dfrac{\Delta}{\gamma}H + f(u)(e_{sat} - e_{act})}{\dfrac{\Delta}{\gamma} + 1} \qquad \text{eq. 5.7}$$

where H = net radiation balance in mm of water equivalent, Δ = rate of change of saturated vapour pressure with temperature, γ = psychrometric or hygrometric constant (different values depending on the temperature units and the method of ventilation (aspiration) of the wet and dry bulb thermometers), $f(u)$ = aerodynamic coefficient (function of wind speed), e_{sat} = saturated vapour pressure of air at temperature, t, e_{act} = actual vapour pressure of air at temperature, t.

Penman's formula for evaporation (eq. 5.7) has been adapted to calculate potential evapotranspiration by the application of empirically derived factors. Penman introduced the empirical formula:

$$PE = fE_o \qquad \text{eq. 5.8}$$

where f is a seasonal correction factor that includes the effects of differing solar insolation intensity, day length, plant stomatal response and geometry. For example, the evaporation rate from a freshly wetted bare soil is about 90% of that from an open water surface exposed to the same weather conditions. For a grassed surface in temperate latitudes the value of PE is, on average, about 75% of the open water evaporation rate. A more process-based approach to calculating PE, following Penman's method and extended by experimental work, is given by the Penman–Monteith formula (eq. 5.9) that incorporates canopy stomatal and aerodynamic resistance effects, to calculate evapotranspiration rate in mm day^{-1} as follows:

$$ET = \frac{\Delta H + \dfrac{\rho c_p (e_{sat} + e_{act})}{r_{sfc}}}{\lambda \left\{ \Delta + \dfrac{\gamma (r_{sfc} + r_{aero})}{r_{aero}} \right\}} \qquad \text{eq. 5.9}$$

where, further to the symbols applied in equation 5.7, ρ = density of water, λ = latent heat of vaporization, c_p = specific heat capacity of water, r_{sfc} = surface resistance, r_{aero} = aerodynamic roughness.

The Penman–Monteith formula is used as the basis for the national computerized system, MORECS, the United Kingdom Meteorological Office Rainfall and Evaporation Calculation System (Thompson et al. 1981). MORECS provides an areal-based estimate of evapotranspiration which supplements the Penman approach with simulation of soil water flux and a consideration of local vegetation cover. The model is based on a two-layer soil and provides estimates of areal precipitation (P), PE, actual evapotranspiration (AE), soil moisture deficit (SMD) and hydrologically effective rainfall (P − AE − SMD) for 40 × 40 km grid squares on a weekly basis.

5.4 Soil water and infiltration

Understanding soil water distribution, storage and movement is important in hydrology in predicting when flooding will occur and also in irrigation scheduling. In hydrogeology, understanding infiltration of water in the unsaturated zone is a necessary prerequisite to quantifying groundwater recharge to the water table. The branch of hydrology dealing with soil water and infiltration is studied in detail by soil physicists and suggested further reading in this topic is provided at the end of this chapter.

5.4.1 Soil moisture content and soil water potential

Water contained in the soil zone is held as a thin film of water adsorbed to soil grains and also as capillary water occupying the smaller pore spaces (Fig. 5.10). The main forces responsible for holding water in the soil are those of capillarity, adsorption and osmosis. Adsorption is mainly due to electrostatic forces in

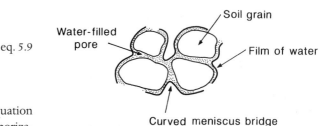

Fig. 5.10 Sketch of the occurrence of water within unsaturated material showing both soil grains coated with a film of adsorbed water and soil pores filled with capillary water.

which the polar water molecules are attracted to the charged surfaces of soil particles. Osmosis is often ignored but acts to retain water in the soil as a result of osmotic pressure due to solutes in the soil water. This occurs particularly where there is a difference in solute concentration across a permeable membrane such as the surface of a plant root, making water less available to plants, especially in saline soils, and is of importance when considering irrigation water quality (Section 6.2.2).

Capillary forces result from surface tension at the interface between the soil air and soil water. Molecules in the liquid are attracted more to each other than to the water vapour molecules in the air, resulting in a tendency for the liquid surface to contract. This effect creates a greater fluid pressure on the concave (air) side of the interface than the convex (water) side such that a negative pressure head, indicated by $-\psi$ (in centimetres or metres head of water), develops relative to atmospheric pressure. The smaller the neck of the pore space, the smaller the radius of curvature and the more negative the pressure head. In soil physics, the negative pressure is often termed the suction head or tension head and describes the suction required to obtain water from unsaturated material such as soils and rocks of the unsaturated zone. With increasing moisture content, the larger pore spaces become saturated and the radius of curvature of the menisci increases creating a more positive pressure head (i.e. the suction head is reduced). Close to the water table, the pore space is fully saturated but the pressure head is still negative as a result of water being drawn up above the water table by the capillary effect (see Fig. 2.14). Hence, the measurement of soil moisture content and pressure

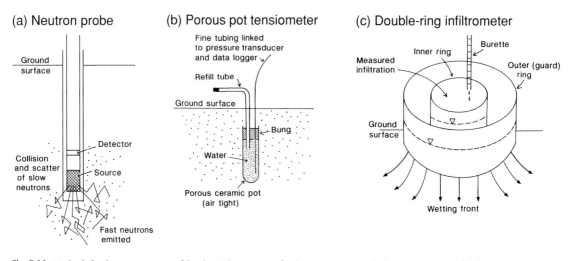

Fig. 5.11 Methods for the measurement of the physical properties of soils: (a) a neutron probe for measurement of soil moisture content; (b) a porous pot tensiometer for measurement of pressure head (suction head or tension head); (c) a double-ring infiltrometer for measurement of infiltration rate.

head and the understanding of their interrelationship are important in the understanding of soil water movement.

The measurement of soil moisture content, θ, can be undertaken in the laboratory by gravimetric determination or in the field with a neutron probe. The gravimetric method requires a known volume, V_t, of soil to be removed and the total mass, m_t, found. The sample is then oven-dried at 105°C until the final mass of the dried sample, m_s, is constant. The difference in mass, $m_t - m_s$, before and after drying is equal to the mass of water originally contained in the sample, m_w. The water content of the soil by mass is then equal to:

$$\theta_m = \frac{m_w}{m_s} \qquad \text{eq. 5.10}$$

By calculating the bulk density of the soil sample, ρ_b, from m_s/V_t, it is then possible to express the water content by volume, as:

$$\theta = \frac{V_w}{V_t} = \theta_m \frac{\rho_b}{\rho_w} \qquad \text{eq. 5.11}$$

where ρ_w is the density of water.

A disadvantage of the gravimetric determination is that the sample is destroyed providing only a one-off measurement. An alternative method is to use a neutron probe to give a direct measurement of soil moisture content. As shown in Fig. 5.11a, a radioactive source (for example Am-Be) is lowered into an augered hole and the impedance of emitted, fast neutrons, caused by impact with hydrogen nuclei contained in the soil water, is determined from the scatter of slow neurons measured by a detector. By calibrating the density of scattered slow neutrons with measurements of soil moisture content made by gravimetric determination, it is then possible to make repeated measurements of soil moisture content at various depths in the access tube. The neutron probe is calibrated for each soil type investigated, with each field site having its own relationship of probe reading versus soil moisture content, and used to make regular measurements of vertical profiles of water content.

The field measurement of soil water pressure (the amount of suction) is made with a tensiometer. One design is shown in Fig. 5.11b and consists of a ceramic pot filled with deionized water. The pressure within the pot equalizes with the fluid pressure in the surrounding soil and is measured via a hydraulic link to a manometer or pressure transducer and data logger. Porous pots operate best in relatively wet conditions up to suctions of about 0.85 bar. In drier conditions there is a danger of air entering the pot and affecting pressure transmission. Porous pots also work under positive fluid pressure, when a vertical bank of

(a)

(b)

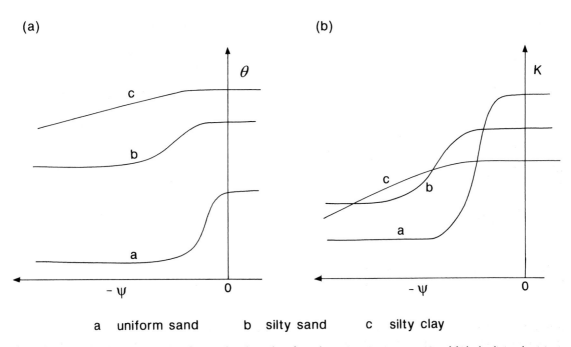

a   uniform sand   b   silty sand   c   silty clay

Fig. 5.12 Single-value characteristic curves showing the relationship of (a) volumetric moisture content, θ, and (b) hydraulic conductivity, K, against pressure head, ψ, for three hypothetical soils. Values of pressure head less than zero indicate unsaturated soil conditions. After Freeze and Cherry (1979).

tensiometers can be used to monitor the position of the water table surface (Jones 1997). To measure suctions above the limit of a tensiometer (about 0.85 bar), an electrical resistance block is used. A porous block of gypsum, nylon or fibreglass mesh containing two electrodes is situated in the ground in close contact with the soil and hydraulic equilibrium established between the block and soil. The water content of the block is calibrated against the electrical resistance between the two electrodes to provide a means of measurement of the soil moisture content. Disadvantages of the electrical resistance block is that it has a long response time for water to seep through the block and reach hydraulic equilibrium and measurements can be affected by dissolved salts in the soil water. The technique is less reliable at suctions of <4 m but does have the advantage of operating at high suctions.

The relationship between soil moisture content and pressure head is conveniently displayed by the presentation of soil characteristic curves or retention curves. Also, given that the degree of interconnection between saturated pore spaces will affect the hydraulic

conductivity, K, of the soil, characteristic curves can also be constructed to show hydraulic conductivity as a function of pressure head. Single-value characteristic curves for hypothetical uniform sand, silty sand and silty clay soils are shown in Fig. 5.12 and demonstrate the different behaviour of the soil types with decreasing pressure head. When the soil is completely saturated, the maximum volumetric moisture content, θ, is equal to the soil porosity such that equation 5.11 equates to equation 2.1. As the soil dries, the uniform sand with its larger diameter pore space is quickly drained and both the water content and hydraulic conductivity decrease rapidly. On the other hand, the larger clay content of the silty sand and silty clay creates a larger porosity as a result of the smaller grain size distribution. Because of the smaller pore diameter, the water is held in the silty soils for longer until a greater suction creating an increasingly negative pressure head is able to drain the pore water. As a result of the finer soil texture, the silty clay retains both a higher moisture content and hydraulic conductivity compared with the uniform sand under dry soil conditions. The slope of the line representing

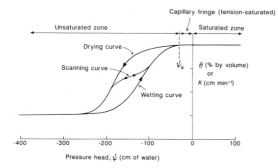

Fig. 5.13 Soil characteristic curves showing the variation of volumetric moisture content, θ, and hydraulic conductivity, K, with pressure head, ψ. The unequal variation of the drying and wetting curves is caused by the effects on soil structure of the drying (soil shrinkage) and wetting (soil swelling) processes. ψ_a is the air-entry or bubbling pressure.

the soil moisture curve, $d\theta/d\psi$, is referred to as the specific moisture capacity, C, and is a measure of the storage behaviour of the unsaturated zone.

The process of drying a soil causes the soil grains to compact and alters the soil structure through shrinkage and air entrapment. On wetting, the soil swells as water is added. Consequently, the resulting characteristic curves for the drying and wetting phases are different and a hysteretic effect is observed as shown in Fig. 5.13. Further important causes of hysteresis are the 'ink-bottle' effect and the 'contact angle' effect. As explained by Ward and Robinson (2000), the 'ink bottle' effect results from the fact that a larger suction is necessary to enable air to enter the narrow pore neck, and hence drain the pore, than is necessary during wetting. The 'contact angle' effect results from the fact that the contact angle of fluid interfaces on the soil particles tends to be greater when the interface is advancing during wetting than when it is receding during drying, such that a given water content tends to be associated with a greater suction in drying than in wetting.

At the start of the drying process, the soil is tension-saturated, a condition analogous to a capillary fringe above a water table, until the air-entry or bubbling pressure is reached, at which point the soil begins to drain. For soils that are only partially dried, then wetted, and vice versa, the soil characteristic curves follow the scanning curves shown in Fig. 5.13.

From the above explanation of characteristic curves, it is clear that moisture content, specific moisture

capacity and hydraulic conductivity in soils and unsaturated rocks and sediments are themselves a function of pressure head. Expressed in mathematical notation, $\theta = \theta(\psi)$, $C = C(\psi)$ and $K = K(\psi)$. It also follows that $K = K(\theta)$. As a result, it is recognized that flow in the unsaturated zone will vary according to the value of hydraulic conductivity at a given soil moisture content. In general, and by considering one-dimensional flow, the quantity of discharge, Q, across a sectional area, A, is found from Darcy's law applied to the unsaturated zone:

$$Q = -AK(\psi)\frac{dh}{dx} \qquad \text{eq. 5.12}$$

where h is the fluid potential or soil water potential. As discussed in Section 2.8, fluid potential is the work done in moving a unit mass of fluid from the standard state to a point in a flow system. Ignoring osmotic potential, the total soil water potential at a given point comprises the sum of the gravitational (or elevation) potential, ψ_g, and pressure potential, ψ_p, as follows:

$$\Phi = \psi_g + \psi_p \qquad \text{eq. 5.13}$$

By applying gravitational acceleration, equation 5.13 is identical to equation 2.22, emphasizing that the gradient of potential energy for subsurface water is continuous throughout the full depth of the unsaturated and saturated zones. In studies of the unsaturated zone it is common to use the ground surface as the datum level for soil water potential values. As shown in Table 5.2, gravitational potential declines uniformly with depth below the ground surface and is negative when referred to a ground surface value of zero.

5.4.2 Calculation of drainage and evaporation losses

From the above explanation of soil water potential, it follows that water will move from a point where the total potential energy is high to one where it is lower. Hence, by plotting a profile of soil water potential it is possible to identify the direction of water movement in the unsaturated zone. As shown in Fig. 5.14b, there

Table 5.2 Values of soil water potential ($\Phi = \psi_g + \psi_p$) relative to a ground surface datum and volumetric moisture content (θ) for a representative clay soil measured on two dates in August and used to plot Fig. 5.14. Calculated evaporation and drainage losses during this 7-day period are given in Table 5.3.

Gravitational potential, ψ_g (cm)	Pressure potential, ψ_p (cm of water)		Soil water potential, Φ (cm of water)		Volumetric moisture content, θ	
	1 August	8 August	1 August	8 August	1 August	8 August
−5	−2493	−3399	−2498	−3404	0.356	0.349
−15	−1124	−1834	−1139	−1849	0.375	0.363
−25	−521	−1011	−546	−1036	0.394	0.378
−35	−197	−618	−232	−653	0.418	0.390
−45	−122	−608	−167	−653	0.428	0.390
−55	−219	−598	−274	−653	0.413	0.390
−70	−388	−645	−458	−715	0.399	0.387
−90	−694	−946	−784	−1036	0.385	0.378
−110	−833	−1029	−943	−1139	0.380	0.375

is one level in the unsaturated zone where there is no potential gradient and, therefore, no vertical soil water movement. At this level, known as the zero flux plane (ZFP), the soil profile is divided into a zone with an upward flux of water above the ZFP and a zone of downward flux below this level. Such a divergent ZFP initially develops at the soil surface, as a result of evaporation exceeding rainfall, and moves downwards into the soil during warm weather as the profile dries out, stabilizing at a depth, typically between 1 and 6 m below ground level, depending on climate and soil conditions and the depth of the water table (Wellings & Bell 1982). If the dry period is followed by a wet period, then a convergent ZFP develops at the ground surface, moving down the soil profile until it reaches the original, divergent ZFP, at which point both ZFPs disappear and downward drainage of the soil water can then take place throughout the soil profile (Ward & Robinson 2000).

By combining measurements of soil moisture content and soil water potential, and adopting a soil water balance approach, it is possible to use these data to quantify both the amounts of deep drainage downwards to the water table and also upward flux to the ground surface due to evapotranspiration. This approach is demonstrated in Table 5.3 using the data contained in Table 5.2 and plotted in Fig. 5.14. In this example, the profiles of volumetric moisture content

show that between the two measurements dates (1 and 8 August) the soil has become drier (Fig. 5.14a) and that a ZFP, identified by the profiles of soil water potential, has descended to a depth of 45 cm below ground level (Fig. 5.14b). By dividing the profiles of volumetric moisture content into convenient depth intervals, d, an approximate amount of water lost during the 7-day period across a single depth interval is $d \times \Delta\theta$. By starting at the level of the ZFP, the total amount of water draining or evaporating from the soil can be calculated by accumulating the individual amounts of water lost from each separate depth interval. In the example calculation shown in Table 5.3, by the end of the 7-day period the total evaporation and drainage losses are approximately 8.22 mm and 8.88 mm, respectively, expressed as a depth of water for a unit area of soil surface.

5.4.3 Infiltration theory and measurement

Water in the soil zone is generally replenished by precipitation or surface runoff at the ground surface. The process by which water enters the soil is known as infiltration and can be defined as the entry into the soil of water made available at the ground surface, together with the associated flow away from the ground surface within the unsaturated zone (Freeze & Cherry 1979). The infiltration rate of a soil will depend on the

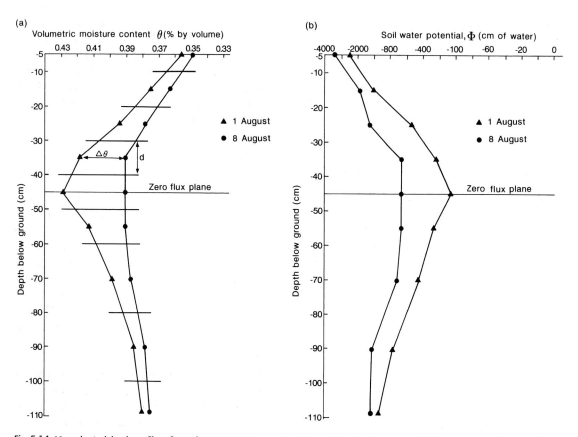

Fig. 5.14 Hypothetical depth profiles of (a) volumetric moisture content, θ, and (b) soil water potential, Φ, for a representative clay soil measured on two dates in August. Notice the zero flux plane (ZFP) developed at −45 cm relative to ground surface.

initial soil moisture condition and the physical properties of the soil. For sandy, open-textured soils infiltration will be rapid whereas for fine-textured, low-permeability clay soil infiltration is limited. Soil structure is also important with the type of vegetated surface, type of cultivation, whether compacted by machinery or poached by cattle, as well as location with respect to proximity to a stream or on a hillside, all affecting the soil's ability to accept infiltrating water.

The infiltration process is represented graphically by the Horton curve of Fig. 5.15 and shows that infiltration rate, f_t, declines rapidly at the start of a rainfall event as the soil pores fill with water, eventually reaching a constant value once the soil has become completely saturated at which point the soil is said to have reached its infiltration capacity, f_c. This constant infiltration rate is equivalent to the saturated hydraulic conductivity of the soil at field capacity (the

volumetric moisture content of the soil after the saturated soil has drained to an equilibrium under the influence of gravity). As also shown in Fig. 5.15, if the rainfall event continues at an intensity greater than the ability of the soil to accept infiltration, then surface ponding of rainfall will occur, potentially leading to the generation of overland flow and the possibility of sheet or gully erosion.

The field measurement of infiltration rate is made using an infiltrometer ring. In the example of a double infiltrometer (Fig. 5.11c), at the start of the experiment water is added to the inner ring, either from a graduated burette, volumetric flask or constant head device, followed by further, measured additions of water to restore the water to a constant level at regular time intervals. The infiltration rate for each time interval is then calculated and measurements continued until a constant rate is achieved. The outer

Table 5.3 Evaporation and drainage losses during a 7-day period for a representative clay soil calculated using the soil water data given in Table 5.2 and with reference to the zero flux plane identified in Fig. 5.14.

Depth range (cm)	d (mm)	$\Delta\theta$	$\Delta\theta \times d$ (mm)	Flux OUT = Flux IN + [$\Delta\theta \times d$] (mm)
Evaporation losses*				
40–45	50	0.036	1.80	1.80 = 0 + 1.80
30–40	100	0.028	2.85	4.65 = 1.80 + 2.85
20–30	100	0.016	1.67	6.32 = 4.65 + 1.67
10–20	100	0.012	1.19	7.51 = 6.32 + 1.19
0–10	100	0.007	0.71	8.22 = 7.51 + 0.71
Drainage losses†				
45–50	50	0.035	1.75	1.75 = 0 + 1.75
50–60	100	0.023	2.37	4.12 = 1.75 + 2.37
60–80	200	0.012	2.38	6.50 = 4.12 + 2.38
80–100	200	0.007	1.44	7.94 = 6.50 + 1.44
100–120	200	0.005	0.94	8.88 = 7.94 + 0.94

* Total evaporation losses = 8.22 mm.
† Total drainage losses = 8.88 mm.

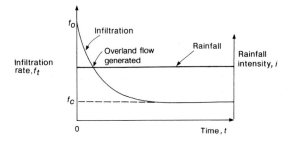

Fig. 5.15 Diagram showing a decrease in infiltration rate, f_t, with time during a rainfall event with intensity, i. As the rain continues the soil becomes saturated and the soil reaches its infiltration capacity, f_c, equivalent to the saturated hydraulic conductivity. At the point where the rainfall intensity equals the infiltration rate, ponding of water occurs at the ground surface and overland flow is generated.

ring is also flooded with water, with the water draining from this ring intended to prevent lateral seepage from beneath the inner ring which, if it occurred, would lead to an overestimation of infiltration rate. Other complicating factors, such as soil heterogeneity, soil swelling and shrinkage and soil aggregation, which limit the interpretation of field measurements based on simple soil physical theory, are discussed by Youngs (1991).

The classical description and analysis of the infiltration process and the uptake of water in dry soils is contained in work by Philip (1969) and Rubin (1966), respectively, with the analytical solution to one-dimensional, vertical infiltration obtained by differentiation of the Philip infiltration equation to give, for large values of time, t:

$$f_t = \frac{1}{2}St^{-1/2} + K_s \qquad \text{eq. 5.14}$$

where f_t is infiltration rate, S is sorptivity and K_s is saturated hydraulic conductivity. The first term on the right-hand side describes the temporal pattern of the absorption of water resulting from the sudden application of water at the surface of a homogeneous soil at time $t = 0$. The second term describes the steady conduction of flow under a potential gradient between the flooded surface and the drier soil below. For further discussion of the theory of infiltration, the reader is referred to Smith (2002).

The classical explanation of infiltration envisages a front of infiltrating water moving down through the soil profile leading to a gradual increase in soil moisture content and pressure head. An inverted water table is conceived to propagate downwards leading to a rise of the true water table in response to the moisture infiltrating from above. At this point, recharge of the groundwater below the water table occurs. The chance of infiltration becoming recharge is greater if a number of conditions are met, namely: low-intensity rainfall of long duration; a shallow water table; wet antecedent moisture conditions; and for soils whose characteristic curves exhibit a high hydraulic conductivity, K, and low specific moisture capacity, C, the ability to maintain a high moisture content over a wide range of values of pressure head, ψ. Although the calculation of the amount of recharge from infiltration measurements is possible, other methods that are less time-intensive and more applicable at a catchment scale are available (Section 5.2.2).

In reality, the infiltration process is affected by soil heterogeneities. Compared with the classical description of a homogeneous infiltration front moving downwards through the soil profile by a piston-flow-type displacement, in reality macropores creating high permeability pathways, such as drying cracks, rootlet channels and worms burrows (Fig. 5.16a), can

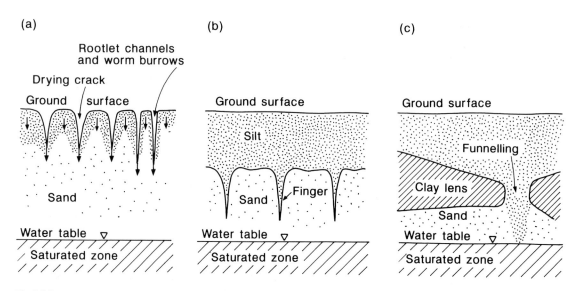

Fig. 5.16 Diagram showing the distribution of infiltrating soil moisture content (densely stippled area) and nature of preferential flow of water in the vadose zone above a water table caused by: (a) macropores; (b) instability due to a permeability contrast creating fingering of flow; (c) a stratified soil profile producing funnelling.

short-circuit the flow of water both vertically and horizontally leading to preferential flow (Beven & German 1982). Equally, a layered heterogeneity with a lower permeability layer overlying a higher permeability layer can cause the flow to concentrate in downward-reaching fingers (Fig. 5.16b) as a result of flow instability at the boundary between the two layers (Hillel & Baker 1988). A further form of preferential flow is reported by Kung (1990a, 1990b) who observed funnelling in the vadose zone of the Central Sand Area of Wisconsin. Funnelled flow occurs where coarse sand layers or densely packed fine layers present in an interbedded soil profile behave like the wall of a funnel which can concentrate an initially unsaturated flow into irregularly spaced columns (Fig. 5.16c). Instead of penetrating through the funnel walls, water will flow laterally on top of these layers, eventually becoming a more concentrated columnar flow after passing the lower edges of these layers.

Apart from affecting the infiltration mechanism in the unsaturated zone, soil heterogeneities such as macropores, fingering and funnelling are also significant in transporting dissolved solutes or contaminants. Hence, soil heterogeneity can increase the risk of groundwater pollution, but the resulting uneven dis-

tribution of contaminant mass in the soil profile makes the location of such groundwater pollution difficult to predict.

5.5 Recharge estimation

Groundwater recharge is the amount of surface water which reaches the permanent water table either by direct contact in the riparian zone or by downward percolation through the unsaturated zone. Groundwater recharge is the quantity which, in the long term, is available for both abstraction and supporting the baseflow component of rivers (Rushton & Ward 1979). The calculation of effective rainfall, the amount of rainfall remaining after evapotranspiration, and the partitioning of this hydrological excess water between surface water and groundwater is an important consideration in a catchment water balance (Section 8.2.1). Methods for estimating direct recharge to the water table and indirect recharge, the latter via fractures and fissures in hard rock or limestone terrain, as localized infiltration below water-filled surface depressions or as lateral runoff to an aquifer at the edge of a confining layer, are discussed in detail by Lerner et al. (1990).

The available methods for calculating direct recharge, three of which are discussed in the following sections, can be classified as: direct measurement using lysimeters over areas up to 100 m² (for further details, see Kitching et al. 1980; Kitchen & Shearer 1982); Darcian approaches to calculate flow in the unsaturated zone above the water table (eq. 5.12); borehole and stream hydrograph analysis; empirical methods that simplify recharge as a function of rainfall amount; soil water budget methods either at a field (Section 5.4.2) or catchment scale; and application of environmental or applied tracers to follow the saturated movement of water in the unsaturated zone (see Fig. 4.10a).

5.5.1 Borehole hydrograph method

The borehole hydrograph method, in conjunction with stream hydrograph separation, provides a convenient means of calculating the partitioning of effective rainfall between surface water runoff and groundwater discharge during a recharge season. Fluctuations in borehole hydrographs represent changes in aquifer storage and, as shown in Fig. 5.17, multiplication of the amplitude of water level change, Δh, by the aquifer storage coefficient provides a value for the net recharge. The total recharge is equal to the addition of net recharge and groundwater outflows (baseflow, found by hydrograph separation (Section 5.7.1) and spring flow). The method is useful in the preparation of a preliminary catchment water balance or in support of regional groundwater flow modelling, but is limited by the need for a good distribution of observation boreholes in the catchment of interest.

5.5.2 Soil water budget method

The conventional method of estimating recharge using a soil water budgeting approach is based on the studies of Penman and Grindley (Penman 1948, 1949; Grindley 1967, 1969). The method is conceptually simple. Water is held in a soil moisture store, precipitation adds to the store and evapotranspiration depletes it. When full, the conceptual quantity of soil moisture deficit (SMD), a measure of the amount of water required to return the soil to field capacity,

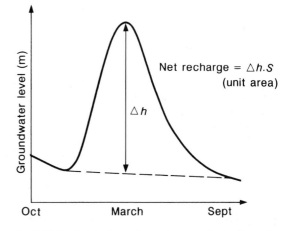

Fig. 5.17 Borehole hydrograph record representing the change in aquifer storage during a single recharge period. The net recharge is equal to the product of the amplitude of groundwater level rise (Δh) and the aquifer storage coefficient, S.

is zero and surplus precipitation (the hydrological excess, HXS) is routed to surface water and groundwater as recharge. The most difficult aspect is to calculate actual evapotranspiration (AE) and, in general, a quantity known as potential evapotranspiration (PE) is first defined as the maximum rate of evapotranspiration under prevailing meteorological conditions over short-rooted vegetation with a limitless water supply. A budgeting procedure is used to convert PE to AE with the degree to which potential and actual evapotranspiration rates diverge being controlled by a root constant (RC), a function of soil and vegetation characteristics and a measure of readily available water within the root range. Representative values of RC expressed as an equivalent depth of water are given in Table 5.4.

The extent to which PE and AE diverge is a matter of debate, with various models having been proposed to represent the reduction in plant transpiration with decreasing soil moisture content. An example of a drying curve is shown in Fig. 5.18, illustrating the decline in AE as plant wilting occurs until finally die-off is reached. The complexity of the wilting process is simply represented as a single step function in Fig. 5.18 with the AE equal to the full PE until a SMD equal to RC + 0.33RC is reached, at which point the AE decreases drastically to one-tenth of the PE rate. No further decline in AE is shown until die-off occurs. By adopting a daily, weekly or monthly budgeting

Table 5.4 Monthly root constant (RC) values for the Penman–Grindley method. Values in millimetres. Based on Grindley (1969).

Month	Crop type													
	1	2	3	4	5	6	7	8	9	10	11	12	13	14
Jan. and Feb.	25	25	25	25	25	25	25	25	25	25	56	76	13	203
Mar.	56	56	56	25	25	56	25	25	25	25	56	76	13	203
Apr.	76	76	76	76	56	56	56	25	25	25	56	76	13	203
May	97	97	97	56	56	56	56	56	25	25	56	76	13	203
Jun. and Jul.	140	140	140	76	76	25	56	56	56	25	56	76	13	203
Aug.	140	140	25	97	97	25	25	56	56	25	56	76	13	203
Sept.	140	25	25	97	25	25	25	25	56	25	56	76	13	203
Oct.	25	25	25	97	25	25	25	25	56	25	56	76	13	203
Nov. and Dec.	25	25	25	25	25	25	25	25	25	25	56	76	13	203

Notes: Valid for England and Wales (temperate, maritime climate). Crop types are: 1, cereals, Sept. harvest; 2, cereals, Aug. harvest; 3, cereals, July harvest; 4, potatoes, Sept. harvest; 5, potatoes, May harvest; 6, vegetables, May harvest; 7, vegetables, July harvest; 8, vegetables, Aug. harvest; 9, vegetables, Oct. harvest; 10, bare fallow; 11, temporary grass; 12, permanent grass; 13, rough grazing; 14, woodland; 15, riparian (not shown) since RC effectively infinite.

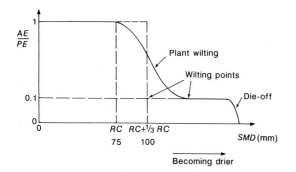

Fig. 5.18 Drying curve for a vegetated soil of short-rooted grass with a root constant (RC) of 75 mm showing the decline in the ratio of actual to potential evapotranspiration (AE/PE) for increasing values of soil moisture deficit (SMD).

approach, it is possible to account for AE by following the development of SMD during the drying phase. An example of the method is shown in Table 5.5 with the calculation started on 1 April when it is assumed that the SMD = 0 and the soil is at field capacity. The effect of introducing the step function shown in Fig. 5.18 is to reduce the rate at which the SMD develops with a cap effectively reached for values above a SMD = 100 mm. No further reduction in SMD is expected at the point of die-off. At the end of the accounting procedure, and in the absence of surface runoff, the cumulative hydrological excess that is predicted to occur for those times when the

SMD = 0 provides an estimate of groundwater recharge for the period of interest (291 mm in the example shown in Table 5.5).

According to the conventional soil water budgeting method, recharge cannot occur when a SMD exists. A critique by Rushton and Ward (1979) showed that recharge amount calculated using the Penman–Grindley method often under-predicts the amount calculated by other methods based on lysimeters, tracers and hydrograph analysis. In particular, the Penman–Grindley method appears to underestimate summer and early autumn recharge. From a sensitivity analysis of recharge calculations for an area of Chalk aquifer in Lincolnshire, England, Rushton and Ward (1979) found that the time-step of the accounting procedure, the estimate of PE (Section 5.3.2), choice of root constant, the functional relationship between AE and PE and the date of harvesting produced an error in calculated recharge by up to 15%. To account for an underestimate in their recharge calculation compared with known outflows and borehole hydrograph records, Rushton and Ward (1979) permitted a direct component of recharge, conceptualized as a bypass flow component via Chalk fissures, equivalent to 15% of actual precipitation in excess of 5 mm plus 15% of the effective precipitation, with the remainder of the recharge calculated with the conventional Penman–Grindley method. However, this method must not be seen as giving the

Table 5.5 Example of the Penman–Grindley soil water budget method to calculate values of actual evapotranspiration (AE), soil moisture deficit (SMD) and recharge (hydrological excess, HXS) using precipitation (P) and potential evapotranspiration (PE) data. The calculation assumes that the permeable soil is covered by a short-rooted grass with a root constant (RC) of 75 mm and that the soil is at field capacity on 1 April (SMD = 0).

Month	P (mm)	PE (mm)	P – PE (mm)	AE (mm)	ΔSMD (mm)	SMD (mm)	HXS (mm)
Apr.	20	48	−28	48	28	28	0
May	12	56	−44	56	44	72	0
Jun.	24	72	−48	$(24 + [100 − 72] + 0.1[20]) = 54$	$(28 + 2) = 30$	102	0
Jul.	9	68	−59	$(9 + 0.1[59]) = 15$	6	108	0
Aug.	31	42	−11	$(31 + 0.1[11]) = 32$	1	109	0
Sep.	60	28	32	28	−32	77	0
Oct.	75	20	55	20	−55	22	0
Nov.	106	10	96	10	−22	0	74
Dec.	94	5	89	5	0	0	89
Jan.	69	5	64	5	0	0	64
Feb.	40	18	22	18	0	0	22
Mar.	72	30	42	30	0	0	42
Apr.	18	50	−32	50	32	32	0
Total recharge							291

correct daily recharge. On a monthly basis, and as input to a regional groundwater flow model, the monthly distribution of recharge calculated this way is regarded as acceptable.

5.5.3 Chloride budget method

Soil water budgeting methods as described above were developed for temperate climates and therefore have less validity in semi-arid and arid zones where these methods normally underestimate recharge, often giving zero values. An alternative, geochemical method is to use a conservative tracer species such as chloride to estimate the amount of recharge and, in favourable circumstances, the recharge history. An example of the application of the chloride budget method is given in Box 5.1.

5.6 Stream gauging techniques

The recording of streamflow data is fundamental to water resources studies and management, flood studies and water quality management. In hydrogeology, the importance of river flow data extends to groundwater resources, with aquifer recharge deduced from

the balance of a number of measurements including baseflow, which is assessed directly from streamflow gauging. Flow data are also necessary for the derivation and application of operating rules for surface reservoirs, groundwater resources and river regulation (Section 8.2.3). Historic flow data are used in setting the minimum residual flow in a river in order to support the aquatic ecology or provide sufficient dilution to achieve water quality standards. A number of simple and advanced techniques are employed to measure or estimate river flows (discharge), with the main techniques described in the following sections. General guidelines for the selection of methods of discharge measurement (velocity–area, slope–area, dilution, ultrasonic, electromagnetic, weirs and flumes) are included in BSI (1998).

5.6.1 Velocity–area methods

Surface floats

This velocity–area method is particularly useful when conditions, for example during a flood, make it dangerous to deploy other discharge measurement procedures. The method requires the choice of a length of river reach sufficient to allow accurate timing of a float released in the middle of the channel and far

The water balance of the coastal Quaternary sand aquifer in the West African sahel region of Senegal is sensitive to short- and long-term climatic change, with groundwater resources in many areas dependent on recharge during former wet periods (Edmunds & Gaye 1994). Hence, it is important to be able to determine the recharge amount in this region in order to quantify the available groundwater resource. For this purpose, the chloride budget method was employed and included collection of well waters and porewaters from the unsaturated zone recovered from hand-augered material down to depths of 35 m (Fig. 1). From analysis of the well water and extracted porewater, and assuming both

negligible surface runoff in the sandy terrain and a chloride source from atmospheric sources only, the direct recharge, R_d, is estimated from:

$$R_d = P \frac{C_p}{C_s}$$

eq. 1

where P is the mean annual precipitation amount, C_p is the mean chloride concentration in rainfall and C_s is the mean chloride concentration in the well water or interstitial porewater. The ratio C_p/C_s represents the evaporative concentration of chloride in

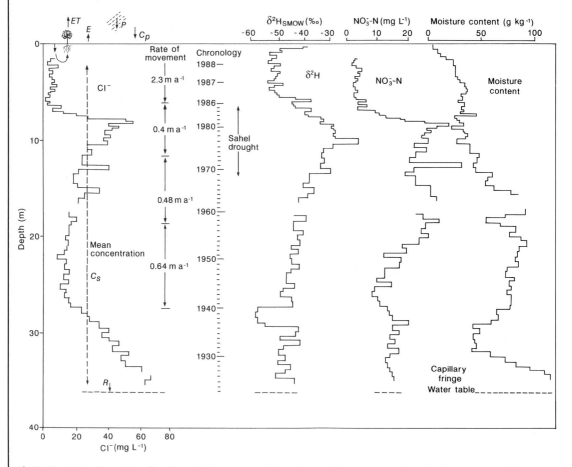

Fig. 1 Chemical and isotopic profiles of water in the unsaturated zone of a coastal Quaternary sand aquifer from a site west of Louga in Senegal, West Africa, interpreted by the chloride budget method to provide a 60-year recharge chronology. The symbols C_p, C_s, E, ET, P and R represent mean chloride concentration in rainfall, mean chloride concentration in well water or interstitial porewater, evaporation, evapotranspiration, precipitation and recharge, respectively. After Edmunds (1991).

BOX
5 . 1

Continued

porewater or shallow groundwater and so measures the effect of evapotranspirative loss of precipitation. For example, using representative data for Senegal, where $P = 443.2$ mm, $C_p = 1.1$ mg L^{-1} and $C_s = 11.2$ mg L^{-1}, then:

$$R_d = 443.2 \frac{1.14}{11.23} = 45 \text{ mm a}^{-1} \qquad \text{eq. 2}$$

Seven unsaturated zone profiles from a small, 1 km^2 area to the west of Louga in Senegal gave a mean C_s value of 82 mg L^{-1} corresponding to a long-term recharge rate of 13 mm a^{-1}. When considered on a regional scale, this long-term recharge represents a sufficient volume to supply the present domestic water needs of the traditional villages, even during a drought when the recharge rate may be halved (Edmunds 1991).

The recharge history can also be calculated if the moisture content profile of the unsaturated zone is known. The rate of downward water movement, $\bar{v}$, can be calculated from:

$$\bar{v} = \frac{i}{\theta} \qquad \text{eq. 3}$$

where i is the recharge or infiltration rate and θ is the moisture content. For the recharge rate calculated above (eq. 2), and for a moisture content of 0.02 measured in 1988, a calculated downward water movement is $0.045/0.02 = 2.25$ m a^{-1}. Thus, for this year, the soil water is estimated to infiltrate a distance of 2.25 m. By inspecting the chloride value in each sampling interval, it is then possible to calculate, from dividing the sampling interval by the rate of downward water movement, the residence time of the water in the interval. Hence, it is possible to build up a recharge chronology for the unsaturated zone profile.

The oscillations in chloride concentration shown in Fig. 1 indicate that the recharge rate has not been constant during the constructed, 60-year recharge chronology for Senegal. The period of Sahel Drought (1968–1986) is clearly visible as a zone of higher chloride concentrations. The drought is also emphasized by the deuterium stable isotope data (δ^2H in ‰) which show that water from this period is enriched in the heavier isotope as a result of greater evaporation. The water quality data for nitrate show values often in excess of 10 mg L^{-1} as N, although these high values are unrelated to surface pollution and instead arise from natural fixation of nitrogen by plants and micro-organisms (see Fig. 6.25) with subsequent concentration by evaporation.

---

enough upstream to attain ambient velocity before entering the reach. By measuring the distance of the reach and the time taken for the float to travel the length of the reach, the water velocity can be calculated by dividing the length by the time. The procedure is repeated a number of times to obtain the average maximum surface velocity, converted to mean velocity using coefficients (Table 5.6). By measuring the flow area upstream and downstream of the reach and taking the average value, the mean flow area for the reach is obtained. The river discharge is then found by multiplying the mean velocity by the mean flow area.

Current metering

Current metering of stream discharge is another velocity–area method and commonly employs one of two types of current meter (cup type or propeller type) to obtain point measurements of velocity (Fig. 5.19). At each measurement point the meter is allowed to run for about 60 seconds and the number of revolutions made by the cup or propeller is obtained from a counter. The velocities can subsequently be

Table 5.6 Coefficients by which the maximum surface velocity of a river should be multiplied to give the mean velocity in the measuring reach.

Average depth in reach (m)	Coefficient
0.3	0.66
0.6	0.68
0.9	0.70
1.2	0.72
1.5	0.74
1.8	0.76
2.7	0.77
3.7	0.78
4.6	0.79
≥6.1	0.80

calculated given a calibration equation linking velocity and count rate. Another type of measuring device is the electromagnetic current meter which measures the voltage resulting from the motion of a conductor (water flow velocity) through a magnetic field. The magnetic field is produced by a coil in the sensor and the voltage is detected by electrodes on the surface of the sensor. The sensor has no moving parts and the

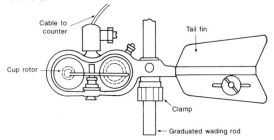

(a) Cup type current meter

Cable to counter

Tail fin

Cup rotor

Clamp

Graduated wading rod

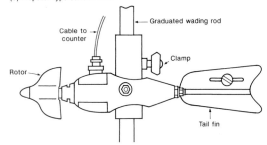

(b) Propeller type current meter

Graduated wading rod

Cable to counter

Clamp

Rotor

Tail fin

Fig. 5.19 Design of two types of current meter for measuring stream discharge. The cup type (a) has an impeller consisting of six small cups which rotate on a horizontal wheel. The propeller type (b) has a rotor as an impeller. In both types, the rate of rotation of the impeller is recorded by an electrically operated counter and is converted to a velocity using a calibration equation. After Brassington (1998).

meter has the advantage of giving a direct reading of velocity. Cleaning with clean water and mild soap is recommended to remove dirt and non-conductive grease and oil from the sensor's electrodes and surface. For a comparison of the performance of different current meters used for stream gauging, the reader is referred to Fulford et al. (1993).

Field methods for obtaining values of stream discharge across a section are described in detail by Rantz et al. (1982). The main objective is the systematic measurement of point velocities across the river channel. For shallow rivers, wading techniques can be employed, while for deeper sections the meter is suspended from a cableway, boat or bridge. The procedure for stream gauging using wading rods is as follows:

1 Choose a straight, uniform channel so that the flow is parallel to the banks.

2 Set up a tag line and measuring tape across the channel perpendicular to the line of the bank and secure.

3 Measure water depth, d, and current meter count rate at a depth of $0.4\,d$ from the bed of the river, if $d \leq 0.75$ m, or $0.2\,d$ and $0.8\,d$ if $d \geq 0.75$ m, at 20 equal intervals across the section. Provided that the velocity profile is logarithmic, the point velocity at $0.4\,d$ or $(0.2\,d + 0.8\,d)/2$ represents the mean velocity for the vertical. When taking velocity readings it is necessary to stand 0.5 m downstream and to one side of the meter, to hold the wading rod vertically, and to ensure that the meter is aligned perpendicular to the section.

4 Convert count rate values to velocities using a calibration equation.

5 Assuming that the average velocity at a vertical is representative of the area bounded by the mid-points between adjacent verticals, calculate the discharge (velocity × area) for each segment. The discharges for each segment are then summed to obtain the total discharge of the section (Fig. 5.20).

5.6.2 Dilution gauging

This technique relies on the dilution of a tracer to measure stream discharge and is a useful technique in smaller, upland streams where flows and mixing of the tracer solution are rapid and the shallow and irregular bed form exclude current metering. The tracer is usually a fluorescent dye such as fluorescein or rhodamine WT (Table 5.11) that can be measured at trace concentrations (μg L^{-1}) using a fluorometer. The fact that low concentrations can be measured is an advantage in that only a few grams of harmless dye solution are required to obtain a measurement of tracer concentration above the background fluorescence. Salt can also be used but is less environmentally satisfactory since a much larger mass is required to detect chloride concentrations above background concentrations. Steam discharge measured by dilution gauging can be estimated to within 2% of current metering results provided that the tracer is fully mixed.

There are two principal dilution gauging techniques: steady-state (constant rate) and slug injection. With the steady-state method, a tracer solution of known concentration is run into the stream at a constant rate using a constant flow device such as a Mariotte bottle. By conservation of tracer mass:

Current meter measurement point

Sectional area for measurement vertical

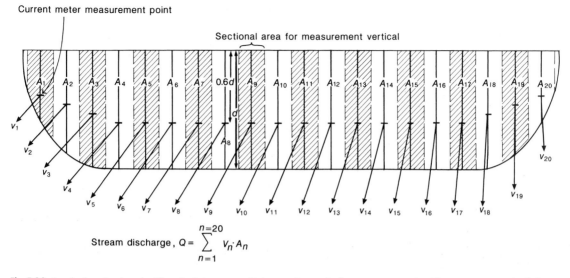

$$\text{Stream discharge, } Q = \sum_{n=1}^{n=20} V_n \cdot A_n$$

Fig. 5.20 Standard sectional method for calculating stream discharge using results from current metering. The cross-section is divided into 20 sections (or about 5% of the total width and containing less than 10% of the total discharge) and the current meter set in the middle of each section (mid-section method) or at either side and the velocities averaged (mean section method).

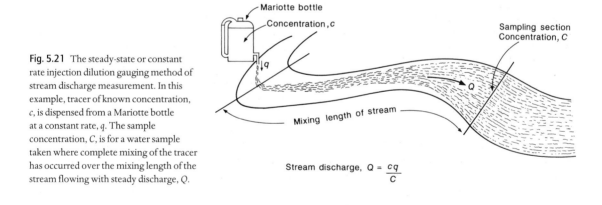

Fig. 5.21 The steady-state or constant rate injection dilution gauging method of stream discharge measurement. In this example, tracer of known concentration, c, is dispensed from a Mariotte bottle at a constant rate, q. The sample concentration, C, is for a water sample taken where complete mixing of the tracer has occurred over the mixing length of the stream flowing with steady discharge, Q.

Mariotte bottle

Concentration, c

Sampling section Concentration, C

Mixing length of stream

$$\text{Stream discharge, } Q = \frac{cq}{C}$$

$$cq = CQ \qquad\qquad \text{eq. 5.15}$$

where c is the initial tracer concentration, q is the rate of injection, C is the concentration in the stream at some downstream point and Q is the unknown discharge. Thus, the discharge can be determined from a single sample taken at a point far enough downstream for full mixing to have occurred over the stream cross-section (Fig. 5.21). The degree of mixing must be checked first by taking several samples across the section. The method assumes steady discharge

that does not vary along the length of the reach and also no losses of tracer, for example in dead zones along the banks of the river.

In the slug injection method, a known volume of tracer solution, v, of known concentration, c, is injected instantaneously into the main flow of the stream. The concentration is measured in samples taken at frequent intervals from a point far enough downstream for full mixing to have occurred. Sampling should begin before the tracer arrives at the sampling point and end after the cloud of tracer has passed. As

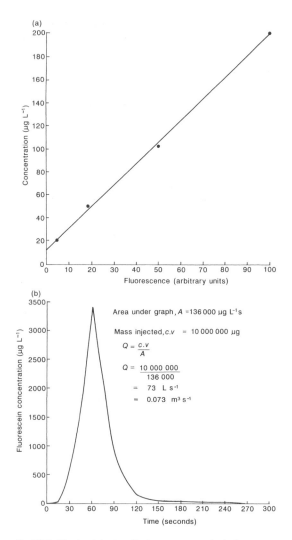

The stream discharge is given by:

$$Q = \frac{cv}{A}$$ eq. 5.17

As with the steady-state method, the slug injection method is based on the assumptions of complete mixing in the stream cross-section, no tracer losses and steady, uniform discharge.

5.6.3 Ultrasonic, electromagnetic and integrating float methods

These three methods rely on the velocity–area measurement approach, but permit automated monitoring. Ultrasonic gauging uses pulses of high-frequency ultrasound which are transmitted from both banks at an angle of 45° to the flow, one upstream and one downstream. The difference in time taken for the sound waves to travel in either direction and be received by transducers is proportional to the average velocity of flow across the stream. Sampling can be at one or more depths but measurements can be affected by suspended sediment and other matter. The technique does not obstruct navigation, with the measurement section usually smoothed and lined to create a stable, rectangular cross-section.

Electromagnetic gauging depends on an electric cable buried in the stream bed. An applied electric current creates an electromotive force (emf) in the above-flowing water which is proportional to the average velocity in the cross-section as measured by bankside probes. The technique is expensive and requires a mains electricity supply.

The integrated float technique, or bubble line method, uses bubbles of compressed air released at regular intervals from a pipe laid across the stream bed. Photographic monitoring reveals the amount of displacement of the bubbles by the flowing water, with the vertical pattern of displacement proportional to the stream velocity profile.

5.6.4 Slope–area method

It is possible to estimate the average velocity of flow through a channel using a friction equation provided

Fig. 5.22 The slug injection dilution gauging method of stream discharge measurement. The results shown were obtained using a slug injection of 10 g of the fluorescent dye fluorescein. The laboratory calibration curve used to obtain fluorescein concentrations in the steam water samples is shown in (a). The stream discharge is calculated by finding the area, A, under the concentration–time curve shown in (b) and equating the area to the mass of injected fluorescein (equal to the product of concentration, c, and volume, v, of injected fluorescein).

shown in Fig. 5.22, results are plotted as a graph of concentration versus time and the area beneath the curve found, as follows:

$$\text{Area, } A = \int_{0}^{\infty} C \cdot dt$$ eq. 5.16

Table 5.7 Table of typical values of Manning's n for application in the estimation of stream discharge. After Wilson (1990).

Type of channel	n
Smooth timber	0.011
Cement-asbestos pipes, welded steel	0.012
Concrete-lined (high-quality formwork)	0.013
Brickwork well laid and flush jointed	0.014
Concrete and cast iron pipes	0.015
Rolled earth: brickwork in poor condition	0.018
Rough-dressed-stone paved, without sharp bends	0.021
Natural stream channel, flowing smoothly in clean conditions	0.030
Standard natural stream or river in stable condition	0.035
River with shallows and meanders and noticeable aquatic growth	0.045
River or stream with rocks and stones, shallow and weedy	0.060
Slow-flowing meandering river with pools, slight rapids, very weedy and overgrown	0.100

that the roughness coefficient for the stretch of channel can be determined. This estimation method is particularly useful for reconnaissance surveys and for estimating flood flows after the peak discharge has subsided. The procedure is as follows:

1 Estimate the roughness coefficient (Manning's n) for the channel from a set of photographs of similar channels with known values of n or from a table of values (Table 5.7).

2 Measure the slope, S, of the water surface over a distance of approximately 200 m.

3 Survey the cross-section of the channel at a representative site to obtain the hydraulic radius, R, equal to the cross-sectional area of flow divided by the wetted perimeter.

4 Calculate the average velocity, v, in units of m s^{-1} using the Manning formula:

$$v = \frac{R^{2/3}S^{1/2}}{n} \qquad \text{eq. 5.18}$$

5 Calculate the stream discharge from $Q = vA$.

Another formula for application in estimating peak discharge is given by the Darcy–Weisbach equation for pipe flow:

$$v = \sqrt{\frac{8RgS}{f}} \qquad \text{eq. 5.19}$$

where f is the Darcy–Weisbach friction factor, g is gravitational acceleration and v, R and S are as defined for equation 5.18. The friction factor, or flow resistance, is dependent on the flow geometry, the roughness height of the stream bed and the cross-sectional variation in roughness heights.

5.6.5 Weirs and flumes

A gauging station is a site on a river which has been selected, equipped and operated to provide the basic data from which systematic records of water level and stream discharge may be derived. Essentially, a gauging station consists of an artificial river cross-section (a weir) where a continuous record of stage (water level upstream of the weir crest) can be obtained and where a relation between the stage and discharge, known as the rating curve, can be determined.

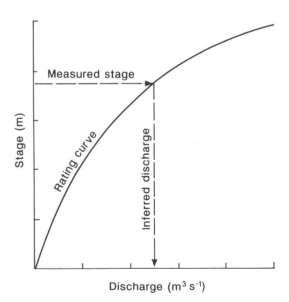

Fig. 5.23 A rating curve to convert measurements of depth of flow (stage or water level) to stream discharge. A rating curve can be established for a gauging station with a fixed gauging structure (a weir or flume) or in a straight, uniform stream section (the rated section) that does not contain a gauging structure by taking a series of discharge measurements at different levels of flow. In the latter case, the time taken to make the discharge measurements may require correction for channel storage and water surface slope effects, especially when measuring higher discharges during flood events, in order to obtain a steady-state rating curve.

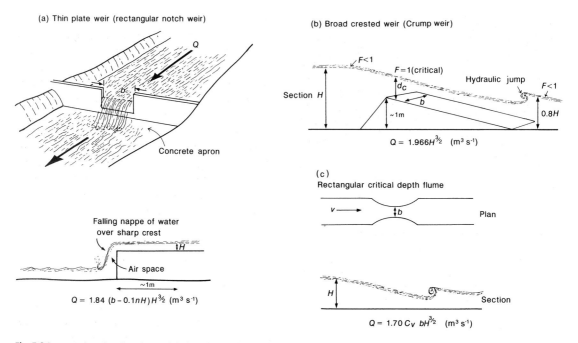

Fig. 5.24 Examples of (a) thin plate and (b) broad crested weirs, and (c) a critical depth flume for the measurement of stream discharge. In the formula given for the rectangular notch weir shown in (a), n is the number of side constrictions ($= 2$ for a rectangular notch; $= 0$ if notch across whole channel), b is the width of the notch and H is upstream head of water above the crest. In (b) the critical depth, d_c, is shown where the Froude number, F, is equal to 1 in the equation $F = v/(gd_c)^{1/2}$ where g is gravitational acceleration and v is average velocity. In (c) the term C_v is the approach velocity coefficient. The introduction of a shape factor is also possible for trapezoidal or U-shaped throats.

A rating curve allows the discharge to be found from the stage alone (Fig. 5.23) with measurements made by a water level recorder positioned over a stilling well linked to the stream in which water turbulence is reduced. The successful operation of a gauging station therefore involves the production of a reliable, accurate and continuous record of stage level.

Fixed gauging structures such as weirs and flumes (Fig. 5.24) are designed so that stream discharge is made to behave according to well-known hydraulic laws of the general form:

$$Q = KbH^a \qquad \text{eq. 5.20}$$

where H is the measured depth, or head, of water, K and a are coefficients reflecting the design of the structure, and b is the width of flow over the weir crest or in the throat (the constricted section) of a flume.

Many specialized weirs, such as V-notch, rectangular notch, compound and Crump weirs, provide accurate discharge data by observations of water level upstream of the weir. The same applies to flumes, where a stream is channelled through a geometrically, often trapezoidal-shaped regular channel section. Flumes are designed so that the point of transition from subcritical to critical flow, when a standing wave is formed accompanied by an increase in velocity and a lowering in water level, occurs at a fixed location at the upstream end of the throat of the flume. Flumes are self-flushing and can be used in streams that carry a high sediment load, unlike a weir that can become silted up. Generally, weirs and flumes are restricted in application to streams and small rivers (Fig. 5.25) since, for large flows and wide rivers, such structures become expensive to construct.

5.7 Hydrograph analysis

Rain falling on a catchment is classically considered to partition between overland flow, interflow and baseflow. These three components of total runoff are shown schematically in Fig. 5.26a and combine to

Fig. 5.25 Marham gauging station on the River Nar, Norfolk, England (NGR TF 723 119; catchment area 153.3 km²; mean flow 1.15 m³ s⁻¹). The gauging structure is a critical depth flume, 7.16 m wide. The stilling well is positioned behind the metal fence on the downstream, left wall of the flume. Prior to April 1982, the flume (7.47 m wide) contained a low flow notch at the centre. Weed growth can be a problem during summer if not cut regularly.

generate the storm hydrograph shown in Fig. 5.26b. Overland flow is rarely observed on natural, vegetated surfaces but may occur where soils are compacted by vehicle movement or are completely saturated, for example at the bottom of a slope next to a stream

channel. Interflow is water moving laterally within the soil zone in the direction of the topographic slope and is potentially accelerated by flow through field drains. Together, overland flow and interflow represent the quickflow or surface runoff from a stream catchment. Baseflow is the component of total runoff contributed by groundwater discharge as springs or seepages and supports surface flows during dry periods when there is little or no rainfall. From the perspective of groundwater resources investigations, techniques of baseflow separation from the quickflow component are useful in contributing to an assessment of groundwater recharge (see Section 5.5.1).

5.7.1 Quickflow and baseflow separation

As shown by the storm hydrograph in Fig. 5.26b, and following passage of the flood peak, surface runoff declines along the recession limb until its contribution to total runoff may eventually disappear. During the storm, infiltration and percolation of water continue, resulting in an elevated groundwater table which enhances the rate of baseflow. With time, as the aquifer drains following the cessation of infiltration, the baseflow component also declines along

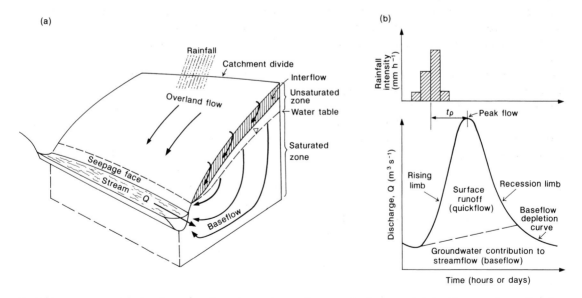

Fig. 5.26 Components of total catchment runoff contributing to streamflow. In (a) the directions of overland flow, interflow and baseflow are shown, and in (b) the flood hydrograph from a rainfall event is shown. At the start of rainfall there is an initial period of interception and infiltration after which runoff reaches the stream and continues until a peak value occurs at time t_p.

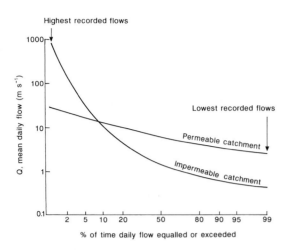

Fig. 5.27 Flow duration curves for permeable and impermeable catchments. Notice the higher peak flows generated in the impermeable catchment and the higher low flows in the permeable catchment. In permeable catchments, rainfall infiltration reduces the effects of flooding and long-term groundwater discharge as baseflow acts to lessen the impacts of droughts.

its depletion curve. In catchments with a permeable geology, for example a limestone or sandstone, the baseflow component is a large fraction of the total surface runoff, but in clay-dominated or hard-rock terrains the storm hydrograph shows a flashy response with a lower percentage of baseflow. These variations can be seen clearly when the record of streamflow is plotted as a flow duration curve (Fig. 5.27).

In practice, hydrograph separation can be conveniently achieved by plotting the streamflow data on semilogarithmic graph paper, with the baseflow depletion curves identified as a series of straight lines, such as shown in Fig. 5.28. The variability encountered in the recession behaviour of individual segments represents different stages in the groundwater discharge and presents a problem in deriving a characteristic recession. One method, as demonstrated in Fig. 5.29, is to derive a master depletion curve and apply this to individual hydrograph peaks in order to separate surface runoff and baseflow discharge.

In general, the section of the hydrograph representing baseflow recession follows an exponential curve and the quantity at any time may be represented by:

$$Q_t = Q_o e^{-at} \qquad \text{eq. 5.21}$$

where Q_o is discharge at the start of baseflow recession, Q_t is discharge at later time t and a is an aquifer coefficient. Once the aquifer coefficient is found using equation 5.21 for two known stream discharge values, the volume of water discharged from an aquifer in support of baseflow during a recession period, t, can either be found by re-plotting the straight line baseflow depletion curve on linear graph paper and then finding the area under the curve, or by integrating equation 5.21 to give:

$$\text{Volume of baseflow} = \frac{Q_o}{a}[1 - e^{-at}] \qquad \text{eq. 5.22}$$

From the analysis of long-term records of baseflow, when variations in aquifer storage are assumed to be

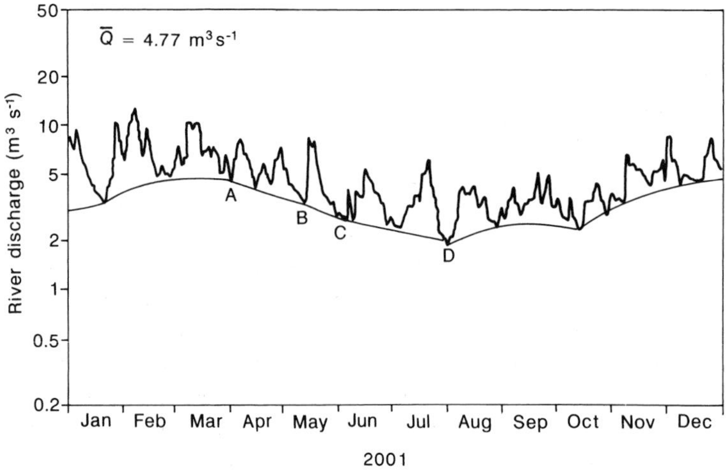

Fig. 5.28 Hydrograph of daily gauged flows during 2001 for Costessey Mill on the River Wensum, Norfolk, England (NGR TG 177 128; catchment area 570.9 km²). The period of baseflow recession is identified by the straight lines AB, BC and CD.

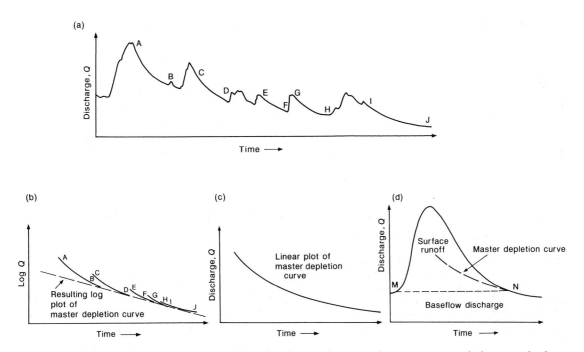

Fig. 5.29 Derivation of a master depletion curve by hydrograph analysis. Firstly, examine the continuous stream discharge record and identify the individual sections of streamflow recession (a). Secondly, plot the individual sections of baseflow recession on semilogarithmic graph paper (b). This is most easily achieved by moving tracing paper over the plots, keeping the axes parallel, until each log Q plot in successively increasing magnitude fits into the growing curve, extending it fractionally upwards. The tangential curve then established to the highest possible discharge is next converted back to linear graph paper and called the master depletion curve (c). The master depletion curve can now be applied to any particular storm hydrograph in which the depletion curves are matched at their lower ends and the point of divergence, N, marked to represent the point at which surface runoff has effectively finished (d). The line MN shown in (d) represents the baseline of the hydrograph of surface runoff, the area below which can then be analysed to find the volume of baseflow discharge for a particular gauging station. After Wilson (1990).

zero, it is possible to establish the volume of groundwater discharge (G_R in the water balance equation, eq. 8.1) and, in the absence of groundwater abstractions, equate this volume to the quantity of groundwater recharge (see Section 5.5.1). Further discussion of baseflow recession analysis and different ways of characterizing the baseflow recession rate is provided by Tallaksen (1995).

Hydrograph separation is also possible by hydrochemical means by adopting a two-component mixing model, or mass balance equation, where the pre-event water (baseflow) and event (quickflow) chemical compositions can be easily distinguished as follows:

$$C_T Q_T = C_P Q_P + C_E Q_E \qquad \text{eq. 5.23}$$

rearranging equation 5.23 and recognizing that $Q_T = Q_P + Q_E$, then:

$$\frac{Q_P}{Q_T} = \frac{C_T - C_E}{C_P - C_E} \qquad \text{eq. 5.24}$$

where Q is discharge, C is tracer concentration and P, E and T represent the pre-event component, event component and total (peak) discharges.

To illustrate this method, the study by Durand et al. (1993) used dissolved silica and stable isotopes of oxygen ($\delta^{18}O$) to separate storm hydrographs in small, granitic mountainous catchments in south-east France. In one event that occurred in early autumn, the following data were obtained for a total stream discharge, Q_T, of 0.6 m^3 s^{-1}:

$\delta^{18}O$ of pre-event streamwater $= -7‰$
$\delta^{18}O$ of event water $= -9‰$
$\delta^{18}O$ of total discharge $= -8‰$

Substituting in equation 5.24 gives:

$$\frac{Q_P}{Q_T} = \frac{-8 - (-9)}{-7 - (-9)} = \frac{1}{2} \qquad \text{eq. 5.25}$$

and it becomes apparent that $Q_P = Q_T/2$ is equal to 0.3 m^3 s^{-1} and $Q_E = (Q_T - Q_P)$ is also equal to 0.3 m^3 s^{-1}. In this example, the baseflow component calculated by hydrochemical means was a higher percentage of the total discharge than was interpreted by the graphical hydrograph separation method.

The hydrochemical separation technique was readily applied in the above example in that the pre-event and event stable isotope compositions were easily distinguishable as a result of evaporative enrichment of the heavier isotope (^{18}O) in groundwater stored in the peaty catchment soils. In general, the technique is best applied to small catchments of the order of 10 km^2. In larger catchments, variation in catchment geology may obscure the chemical signatures of individual components of baseflow and storm runoff.

5.8 Field estimation of aquifer properties

5.8.1 Piezometer tests

Piezometer tests are small in scale and relatively cheap and easy to execute and provide useful site information, but are limited to providing values of hydraulic conductivity representative of only a small volume of ground in the immediate vicinity of the piezometer.

It is possible to determine the hydraulic conductivity of an aquifer by tests carried out in a single piezometer. Tests are carried out by causing a sudden change in the water level in a piezometer through the rapid introduction (slug test) or removal (bail test) of a known volume of water or, to create the same effect, by the sudden introduction or removal of a solid cylinder of known volume. Either way, the recovery of the water level with time subsequent to the sudden disturbance is monitored and the results interpreted.

For point piezometers that are open for a short interval at their base (Fig. 5.30a), the interpretation of the water level versus time data commonly employs the Hvorslev (1951) method. Hvorslev (1951) found that the return of the water level to the original, static level occurs at an exponential rate, with the time taken dependent on the hydraulic conductivity of the porous material. Also, the recovery rate depends on the piezometer design; piezometers with a large area available for water to enter the response zone recover more rapidly than wells with a small open area. Now, if the height to which the water level rises above the static water level immediately at the start of a slug test is h_o and the height of the water level above the static water level is h after time, t, then a semilogarithmic plot of the ratio h/h_o versus time should yield a straight line (Fig. 5.30b). In effect, using the ratio h/h_o normalizes the recovery between zero and one. If the length of the piezometer, L, is more than eight times the radius of the well screen, R, then the hydraulic conductivity, K, can be found from:

$$K = \frac{r^2 \log_e (L/R)}{2LT_o} \qquad \text{eq. 5.26}$$

where r is the radius of the well casing and T_o is the time lag or time taken for the water level to rise or fall to 37% of the initial change (Fig. 5.30b).

The Hvorslev method as presented here assumes a homogeneous, isotropic and infinite material and can be applied to unconfined conditions for most piezometer designs where the length is typically greater than the radius of the well screen. Hvorslev (1951) also presented formulae for anisotropic material and for a wide variety of piezometer geometries and aquifer conditions. For slug tests performed in fully or partially penetrating open boreholes or screened wells, the reader is referred to the method of Bouwer and Rice (1976) for unconfined aquifers and Bouwer (1989) for confined aquifers. The approach is similar to the Hvorslev method but involves using a set of curves to determine the radius of influence of the test.

5.8.2 Pumping tests

Pumping tests are generally of larger scale and duration compared with piezometer tests and are

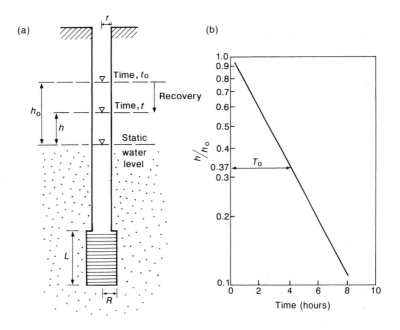

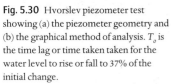

Fig. 5.30 Hvorslev piezometer test showing (a) the piezometer geometry and (b) the graphical method of analysis. T_o is the time lag or time taken taken for the water level to rise or fall to 37% of the initial change.

therefore more expensive, but can provide measurements of aquifer transmissivity and storativity (Section 2.11.1) that are representative of a large volume of the aquifer. In addition to measuring aquifer properties, pumping tests of wells and boreholes are also carried out to measure the variation of well performance with the discharge rate. Long-term pumping tests are invaluable in identifying boundary conditions, effectively describing the units of the aquifer providing water to the borehole being pumped. Pumping tests also provide a good opportunity to obtain information on water quality and its variation in time and perhaps with discharge rate.

When water is pumped from a well, the groundwater level in the well is lowered, creating a localized hydraulic gradient which causes water to flow to the well from the surrounding aquifer. The head in the aquifer is reduced and the effect spreads outwards from the well forming a cone of depression. The shape and growth of the cone of depression of the potentiometric surface depends on the pumping rate and on the hydraulic properties of the aquifer. Hence, by recording the changes in the position of the potentiometric surface in observation wells located around the pumping well it is possible to monitor the growth

of the cone of depression and so determine the aquifer properties.

Different types of pumping test are undertaken with the most common being the step drawdown (variable discharge) and constant discharge tests. Step drawdown tests measure the well efficiency and the well performance. Constant discharge tests measure well performance and aquifer characteristics and help to identify the nature of the aquifer and its boundaries.

In a step drawdown test, the drawdown of water level below the pre-test level, s, in the pumped well is measured while the discharge rate, Q, is increased in steps. Observation boreholes are not required and analysis of the data provides a measure of the variation in specific capacity (Q/s) of the well with discharge rate, information that is invaluable in choosing the pump size and pump setting for the well in long-term production. Further discussion of the interpretation of step drawdown data is provided by Clark (1977) and Karami and Younger (2002).

The usual procedure for a constant discharge test is for water to be pumped at a constant rate from one well (the production well or pumped well) and the resulting change in the potentiometric surface to be monitored in one or more observation wells in close proximity to the pumped well (Fig. 5.31a). The constant discharge test programme has three parts: pre-test observations; pumping test; and observations during potentiometric recovery after the pumping has stopped. Prior to the start of the test, the initial water levels relative to a local datum must be measured and monitored for effects external to the pumping test, for example tidal fluctuations and barometric variations (Section 5.2), and the details of the site hydrogeology recorded, for example well depths and diameters, strata penetrated and the location of nearby streams that could act as recharge boundaries. From the start of the test, the pumping rate is monitored and paired values of drawdown and time are measured in the pumped and observation wells at specific time intervals that increase as the test progresses (Fig. 5.31b). Initially, the cone of depression expands rapidly but the expansion slows logarithmically with time as the volume of aquifer contributing to the pumped well increases. Hence, the measurement time interval can also increase approximately logarithmically as the cone of depression grows. Eventually, a state of quasi- or actual equilibrium conditions may be reached, when the rate of recharge to the borehole catchment balances the rate of abstraction.

The last stage of the constant discharge test is the recovery phase after the pump has been switched off. On cessation of pumping, groundwater levels will recover to a static water level following a drawdown versus time curve that is approximately the converse of the drawdown curve (Fig. 5.31b). The groundwater levels should be measured from the time the pump is switched off with a similar, logarithmically increasing time interval as in the pumping test. In theory, the length of the recovery test is the same as the pumping test but, in practice, the recovery is monitored until the water level is within about 10 cm of the original static water level (Clark 1988).

For further details on the requirements of a pumping test, the reader is referred to the code of practice produced by the British Standards Institution (BSI 1992) and the procedures given by Walton (1987), Clark (1988) and Brassington (1998). One important consideration is the disposal of the discharged water. In tests of shallow aquifers, this water may infiltrate back into the aquifer and interfere with the test

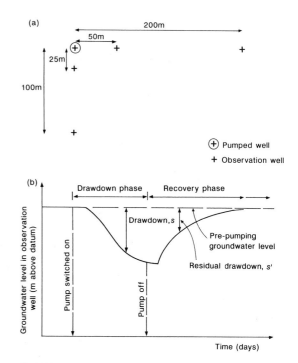

Fig. 5.31 (a) Suggested cruciform borehole array and (b) diagrammatic representation of the drawdown response in an observation well for a constant discharge test followed by a recovery test. By siting observation boreholes along two radii at right angles to each other, this enables the aquifer characteristics to be measured and an indication of aquifer geometry and anisotropy to be obtained (Toynton 1983). As a minimum, there should be at least one observation borehole in order to obtain reliable pumping test data. The recommended minimum period of a constant discharge test where observation wells are more than 100 m from the pumped well, or in an unconfined aquifer where response time is slow, is 3 days. A 5-day test may be advised in situations where derogation effects on neighbouring wells need to be measured. Constant discharge tests involving more than one pumping well (a group test) are used to test wide areas of aquifer and require extended periods of pumping (Clark 1988).

results. To avoid this problem, the pumped water should be abstracted to a point beyond the range of influence of the test. The following sections provide a description of the common methods of pumping test analysis that are applied to confined aquifers and, where conditions allow, unconfined aquifers. For a fuller treatment of the various solutions and techniques for more complex aquifer conditions, the reader is directed to the specialist handbook of Kruseman and de Ridder (1990).

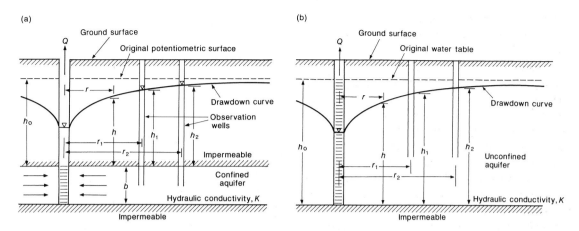

Fig. 5.32 Nomenclature and set-up of radial flow to (a) a well penetrating an extensive confined aquifer and (b) a well penetrating an unconfined aquifer.

Thiem equilibrium method

Depending on whether equilibrium or transient conditions apply, the field data collected during a constant discharge test can be compared with theoretical equations (the Thiem and Theis equations, respectively) to determine the transmissivity of an aquifer. The Theis equation can also be applied to the transient data to determine aquifer storativity, which cannot be determined from equilibrium data.

For equilibrium conditions, and assuming radial, horizontal flow in a homogeneous and isotropic aquifer that is infinite in extent, the discharge, Q, for a well completely penetrating a confined aquifer can be expressed from a consideration of continuity as:

$$Q = Aq = 2\pi rbK\frac{dh}{dr}\qquad\text{eq. 5.27}$$

where A is the cross-sectional area of flow ($2\pi rb$), q is the specific discharge (darcy velocity) found from Darcy's law (eq. 2.9), r is radial distance to the point of head measurement, b is aquifer thickness and K is the hydraulic conductivity. Rearranging and integrating equation 5.27 for the boundary conditions at the well, $h = h_w$ and $r = r_w$, and at any given value of r and h (Fig. 5.32a), then:

$$Q = 2\pi Kb\frac{(h - h_w)}{\log_e(r/r_w)}\qquad\text{eq. 5.28}$$

which shows that drawdown increases logarithmically with distance from a well. Equation 5.28 is known as the equilibrium, or Thiem, equation and enables the hydraulic conductivity or the transmissivity of a confined aquifer to be determined from a well being pumped at equilibrium, or steady-state, conditions. Application of the Thiem equation requires the measurement of equilibrium groundwater heads (h_1 and h_2) at two observation wells at different distances (r_1 and r_2) from a well pumped at a constant rate. The transmissivity is then found from:

$$T = Kb = \frac{Q}{2\pi(h_2 - h_1)}\log_e\frac{r_2}{r_1}\qquad\text{eq. 5.29}$$

A similar equation for steady radial flow to a well in an unconfined aquifer can also be found for the set-up shown in Fig. 5.32b. For a well that fully penetrates the aquifer, and from a consideration of continuity, the well discharge, Q, is:

$$Q = 2\pi rKh\frac{dh}{dr}\qquad\text{eq. 5.30}$$

which, upon integrating and converting to heads and radii at two observation wells and rearranged to solve for hydraulic conductivity, K, yields:

$$K = \frac{Q}{\pi(h_2^2 - h_1^2)} \log_e \frac{r_2}{r_1}$$ eq. 5.31

This equation provides a reasonable estimate of K but fails to describe accurately the drawdown curve near to the well where the large vertical flow components contradict the Dupuit assumptions (see Box 2.9). In practice, the drawdowns caused by pumping should be small (<5%) in relation to the saturated thickness of the unconfined aquifer before equation 5.31 is applied.

As an example of the application of the Thiem equation to find aquifer transmissivity, consider a well in a confined aquifer that is pumped at a rate of 2500 m^3 day^{-1} with the groundwater heads measured at two observation boreholes, A and B, at distances of 250 and 500 m, respectively, from the well. Once equilibrium conditions are established, the groundwater head measured at observation well A is 40.00 m and at observation well B is 43.95 m, both with reference to the horizontal top of the aquifer. Using this information, the aquifer transmissivity can be found from equation 5.29 as follows:

$$T = \frac{2500}{2\pi(43.95 - 40.00)} \log_e \frac{500}{250} = 70 \ m^2 \ day^{-1}$$

eq. 5.32

Theis non-equilibrium method

Application of the Thiem equation is limited in that it does not provide a value of the aquifer storage coefficient, S, it requires two observation wells in order to calculate transmissivity, T, and it generally requires a long period of pumping until steady-state conditions are achieved. These problems are overcome when the transient or non-equilibrium data are considered. In a major contribution to hydrogeology, Theis (1935) provided a solution to the following partial differential equation that describes unsteady, saturated, radial flow in a confined aquifer with transmissivity, T, and storage coefficient, S:

$$\frac{\partial^2 h}{\partial r^2} + \frac{1}{r}\frac{\partial h}{\partial r} = \frac{S}{T}\frac{\partial h}{\partial t}$$ eq. 5.33

By making an analogy with the theory of heat flow, and for the boundary conditions $h = h_o$ for $t = 0$ and

$h \to h_o$ as $r \to \infty$ for $t \geq 0$ where h_o is the constant initial piezometric surface (Fig. 5.32a), Theis derived an analytical solution to equation 5.33, known as the non-equilibrium or Theis equation, written in terms of drawdown, s, as:

$$s = \frac{Q}{4\pi T} \int_u^\infty \frac{e^{-u} du}{u}$$ eq. 5.34

where

$$u = \frac{r^2 S}{4Tt}$$ eq. 5.35

For the specific definition of u given by equation 5.35, the exponential integral in equation 5.34 is known as the well function, $W(u)$, such that equation 5.34 becomes:

$$s = \frac{Q}{4\pi T} W(u)$$ eq. 5.36

A table of values relating $W(u)$ and u is provided in Appendix 6 and the graphical relationship of $W(u)$ versus $1/u$, known as the Theis curve, is given in Fig. 5.33.

The assumptions required by the Theis solution are:

1 The aquifer is homogeneous, isotropic, of uniform thickness and of infinite areal extent.

2 The piezometric surface is horizontal prior to the start of pumping.

3 The well is pumped at a constant discharge rate.

4 The pumped well penetrates the entire aquifer, and flow is everywhere horizontal within the aquifer to the well.

5 The well diameter is infinitesimal so that storage within the well can be neglected.

6 Water removed from storage is discharged instantaneously with decline of groundwater head.

These assumptions are rarely met in practice but the condition that the well is pumped at a constant rate should be checked during the field pumping test in order to limit calculation errors.

The Theis equations (eqs 5.35 and 5.36) can be used to predict the drawdown in hydraulic head in a confined aquifer at any distance, r, from a well at any time, t, after the start of pumping at a known rate, Q.

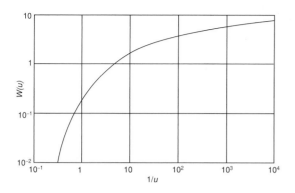

Fig. 5.33 The non-equilibrium type curve (Theis curve) for a fully confined aquifer.

For example, if a confined aquifer with a transmissivity of 500 m² day⁻¹ and a storage coefficient of 6.4 × 10⁻⁴ is pumped at a constant rate of 2500 m³ day⁻¹, the drawdown after 10 days at an observation well located at a distance of 250 m can be calculated as follows. First, a value of u is found from equation 5.35:

$$u = \frac{250^2 \cdot 6.4 \times 10^{-4}}{4.500 \cdot 10} = 2.0 \times 10^{-3} \qquad \text{eq. 5.37}$$

and using the table in Appendix 6, the respective value of $W(u)$ is found to be 5.64. Substituting this value of $W(u)$ in equation 5.36 gives the value of drawdown:

$$s = \frac{2500}{4.500 \cdot 10}5.64 = 0.705 \text{ m} \qquad \text{eq. 5.38}$$

Conversely, the Theis equation enables determination of the aquifer transmissivity and storage coefficient by analysis of pumping test data. The Theis non-equilibrium method of analysis is based on a curve matching technique. An example of the interpretation of pumping test from a constant discharge test is given in Box 5.2.

Cooper–Jacob straight-line method

A modification of the Theis method of analysis was developed by Cooper and Jacob (1946) who noted that for small values of u ($u < 0.01$) at large values of time, t, the sum of the series beyond the term $\log_e u$ in the expansion of the well function, $W(u)$ (eq. A6.1 in Appendix 6) becomes negligible , so that drawdown, s, can be approximated as:

$$s = \frac{Q}{4\pi T}(-0.5772 - \log_e u) \qquad \text{eq. 5.39}$$

By substituting equation 5.35 for u in equation 5.39 and noting that $\log_e u = 2.3\log_{10}u$, gives:

$$s = \frac{2.3Q}{4\pi T} \log_{10} \frac{2.25Tt}{r^2S} \qquad \text{eq. 5.40}$$

Since Q, r, T and S have constant values, a plot of drawdown, s, against the logarithm of time, t, should give a straight line. Furthermore, for two values of drawdown, s_1 and s_2, then:

$$s_2 - s_1 = \frac{2.3Q}{4\pi T}\left[\log_{10} \frac{2.25Tt_2}{r^2S} - \log_{10} \frac{2.25Tt_1}{r^2S}\right]$$

$$\text{eq. 5.41}$$

Therefore:

$$s_2 - s_1 = \frac{2.3Q}{4\pi T}\log_{10} \frac{t_2}{t_1} \qquad \text{eq. 5.42}$$

and if $t_2 = 10t_1$, then:

$$s_2 - s_1 = \frac{2.3Q}{4\pi T} \qquad \text{eq. 5.43}$$

Hence, from a semilogarithmic plot of drawdown against time, the difference in drawdown over one log cycle of time on the straight-line portion of the curve will yield a value of transmissivity, T, using equation 5.43. To find a value for the storage coefficient, S, it is necessary to identify the intercept of the straight line plot with the time axis at $s = 0$, whereupon:

$$s = \frac{2.3Q}{4\pi T} \log_{10} \frac{2.25Tt_0}{r^2S} = 0 \qquad \text{eq. 5.44}$$

Therefore:

$$S = \frac{2.25Tt_0}{r^2} \qquad \text{eq. 5.45}$$

Interpretation of a constant discharge pumping test and recovery test

BOX 5.2

As an example of the interpretation of pumping test data, Table 1 gives the results of a constant discharge pumping test for a confined Chalk borehole site at Woolhampton in the south of England. A geological log for the abstraction borehole and a flow log, obtained using an impeller device, are shown in Fig. 1. The abstraction borehole was pumped at a rate of 6×10^3 m^3 day^{-1} and values of drawdown were recorded with time in an observation borehole located at a distance of 376 m. To obtain representative values of transmissivity and storativity, Fig. 2 shows the recorded values of drawdown, s, and time, t, plotted on log-log paper of the same scale as the type curve (see Fig. 5.33). This field curve is then superimposed over the type curve, keeping the axes parallel, and adjusting its position until the best match between field and types curves is achieved (Fig. 2). By selecting any point on the overlap as a

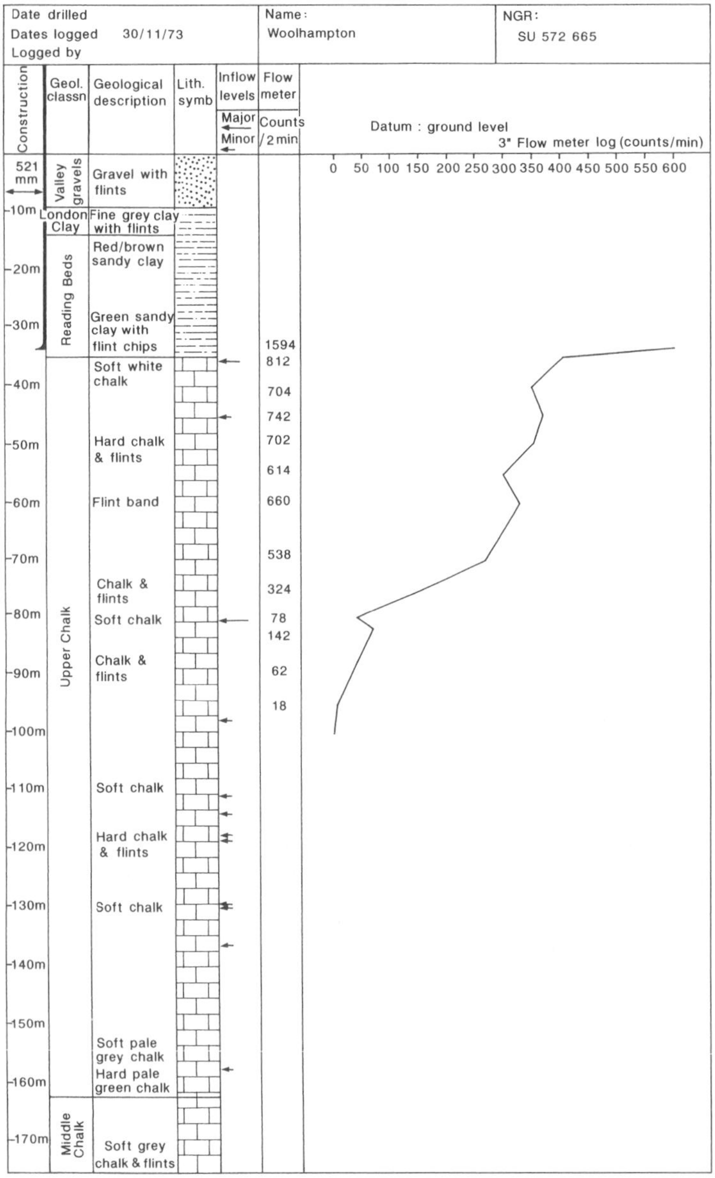

Fig. 1 Geological log and impeller flow log for the Woolhampton river regulation borehole showing the confining beds of Tertiary strata overlying the Cretaceous Chalk aquifer. Notice the increase in flow in the upper section of the Chalk associated with fissured inflow horizons.

BOX
5 . 2

Continued

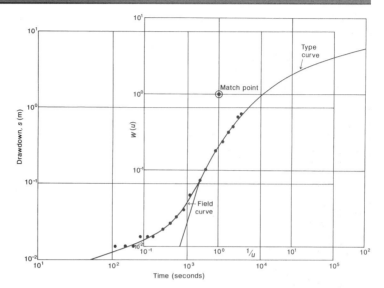

Fig. 2 Diagram showing the match point found from the overlay of a type curve (Theis curve, Fig. 5.33) to the field curve for the constant discharge pumping test data obtained for Woolhampton river regulation borehole (Table 1).

match point, four values are obtained which define $W(u)$, $1/u$, s and t. For Woolhampton, the four match point values are: $W(u) = 1$; $1/u = 1$; $s = 1.5$ m; $t = 2600$ s. Hence, rearranging equation 5.36, and being careful to work in a consistent set of units (here in metres and seconds), a value for the Chalk aquifer transmissivity is found from:

$$T = \frac{Q}{s4\pi} W(u) = \frac{6.0 \times 10^3}{1.5 \times 4 \times \pi \times 86{,}400}$$

$$= 0.003684 \text{ m}^2 \text{ s}^{-1} \text{ or } 318 \text{ m}^2 \text{ day}^{-1} \qquad \text{eq. 1}$$

and using equation 5.35, a value for the confined Chalk storage coefficient is found from:

$$S = \frac{u4Tt}{r^2} = \frac{1 \times 4 \times 0.003684 \times 2600}{376^2} = 2.7 \times 10^{-4} \qquad \text{eq. 2}$$

As an example of the recovery test method, Table 2 lists the residual drawdown recorded following the Chalk borehole constant discharge pumping test. A plot of residual drawdown, s', against the logarithm of t/t' is shown in Fig. 3 and the value of $\Delta s'$ for one log cycle of t/t' is found to equal 2.3. Substitution of this value in equation 5.51 gives, with attention to units in metres and seconds:

$$T = \frac{2.3 \times 6.0 \times 10^3}{4\pi \times 2.3 \times 86{,}400}$$

$$= 0.005526 \text{ m s}^{-1} = 477 \text{ m}^2 \text{ day}^{-1} \qquad \text{eq. 3}$$

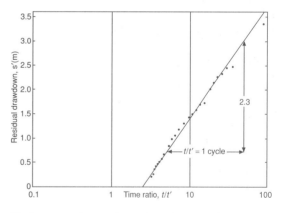

Fig. 3 Diagram showing the semilogarithmic plot of residual drawdown, s', versus the ratio t/t' for the recovery test data obtained for Woolhampton river regulation borehole (Table 2). Time t is measured since the start of the drawdown phase and time t' since the start of the recovery phase.

The calculated value of transmissivity from the recovery test is larger than the value obtained above from the constant discharge test (318 m^2 day^{-1}) and is due to error introduced in the recovery test method in not having satisfactorily met the condition of a large recovery time, t', in order to satisfy a small value of u' as required by the approximation to $W(u)$ (eq. 5.39).

Table 1 Record of the constant discharge pumping test
conducted at the Woolhampton Chalk borehole site, Berkshire.

Pumping test: Woolhampton River Regulation Borehole 56/106
Date: 19 April 1995
Observation borehole: 56/117 Depth: 74 m Distance: 376 m
Location: Woolhampton, Berkshire NGR: SU 574 662
Reference level: 57.223 mOD (flange level)
Initial water level: 56.673 mOD Final water level: 55.953 mOD

Time since start (s)	Depth to water level (m)	Drawdown (m)	Discharge (ML day^{-1})
0	0.55	0	0
30	0.56	0.01	6.07
60	0.56	0.01	6.02
90	0.56	0.01	5.94
120	0.565	0.015	5.96
150	0.565	0.015	6.00
180	0.565	0.015	5.94
240	0.57	0.02	6.02
300	0.57	0.02	5.97
360	0.57	0.02	6.01
480	0.575	0.025	6.00
600	0.58	0.03	5.94
720	0.585	0.035	5.90
900	0.595	0.045	6.02
1200	0.62	0.07	5.95
1500	0.65	0.10	5.95
1800	0.705	0.155	5.93
2400	0.815	0.265	5.89
3000	0.9	0.35	5.83
3600	1.02	0.47	5.91
4200	1.12	0.57	5.91
4800	1.20	0.65	5.91
5400	1.27	0.72	5.92

Table 2 Record of the recovery test conducted at the
Woolhampton Chalk borehole site, Berkshire.

Pumping test: Woolhampton River Regulation Borehole 56/106
Date: 6 April 1999 NGR: SU 572 665
Reference level: 57.21 mOD (top of dip tube)

Time since start of test, t (s)	Time since pump stopped, t' (s)	t/t'	Residual drawdown, s' (m)
5460	60	91.0	3.275
5550	150	37.0	2.475
5580	180	31.0	2.455
5610	210	26.7	2.325
5640	240	23.5	2.235
5670	270	21.0	2.145
5700	300	19.0	2.025
5760	360	16.0	1.725
5820	420	13.9	1.715
5880	480	12.3	1.595
5940	540	11.0	1.500
6000	600	10.0	1.415
6120	720	8.5	1.305
6240	840	7.4	1.195
6360	960	6.6	1.065
6480	1080	6.0	0.985
6600	1200	5.5	0.845
6720	1320	5.1	0.755
6840	1440	4.8	0.675
6960	1560	4.5	0.595
7080	1680	4.2	0.525
7200	1800	4.0	0.475
7320	1920	3.8	0.425
7440	2040	3.6	0.375
7560	2160	3.5	0.275
7800	2400	3.3	0.205

To demonstrate the Cooper–Jacob method of analysis, Table 5.8 provides data for a pumping test in which a well in a confined aquifer is pumped at a rate of 0.01 m^3 s^{-1} and the drawdown in the potentiometric surface is recorded at an observation well situated 30 m away. The drawdown data are plotted in Fig. 5.34 and the difference in drawdown over one log cycle is found to be 1.75 m. Substitution of

this value in equation 5.43, taking care to work in a consistent set of units (here metres and days), gives:

$$T = \frac{2.3 \times 0.01 \times 86,400}{1.75 \times 4\pi} = 90 \text{ m}^2 \text{ day}^{-1} \qquad \text{eq. 5.46}$$

The storage coefficient is found from equation 5.45 with $t_o = 0.0022$ days where $s = 0$ m:

Table 5.8 Record of drawdown in an observation well situated 30 m from a well in a confined aquifer pumping at a rate of 0.01 m³ s⁻¹.

Time since start of pumping (days)	Drawdown (m)
0.010	1.24
0.025	1.89
0.050	2.40
0.075	2.69
0.10	2.92
0.25	3.61
0.50	4.14
0.75	4.45
1.0	4.67
2.5	5.37

$$S = \frac{2.25 \times 90 \times 0.0022}{30^2} = 5 \times 10^{-4} \qquad \text{eq. 5.47}$$

Recovery test method

At the end of a pumping test, when the pump is switched off, the water levels in the abstraction and observation wells begin to recover. As water levels recover, the residual drawdown, s', decreases (Fig. 5.31b). On average, the rate of recharge, Q, to the well during the recovery period is assumed to be equal to the mean pumping rate. Unlike the drawdown phase when the pumping rate is likely

to vary (as seen in Box 5.2, Table 1), an advantage of monitoring the recovery phase is that the rate of recharge can be assumed to be constant and therefore satisfying one of the above Theis solution assumptions.

The Theis method requires that pumping is continuous. Therefore, for the method to be applied to the recovery phase of a pumping test, a hypothetical situation must be conceptualized. If a well is pumped for a known period of time and then switched off, the following drawdown will be the same as if pumping had continued and a hypothetical recharge well with the same discharge was superimposed on the pumping well at the time the pump is switched off. From the principle of superimposition of drawdown (see next section), the residual drawdown, s', can be given as:

$$s' = \frac{Q}{4\pi T}[W(u) - W(u')] \qquad \text{eq. 5.48}$$

where, for time t, measured since the start of pumping and time t', since the start of the recovery phase:

$$u = \frac{r^2 S}{4Tt} \text{ and } u' = \frac{r^2 S}{4Tt'} \qquad \text{eq. 5.49}$$

For small values of u' and large values of t', the well functions can be approximated by the first two terms of equation A6.1 so that equation 5.48 becomes:

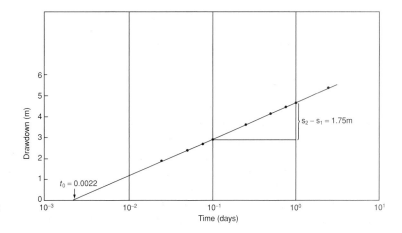

Fig. 5.34 Diagram showing the Cooper–Jacob semilogarithmic plot of drawdown versus time for the data given in Table 5.8.

$$s' = \frac{2.3Q}{4\pi T} \log_{10} \frac{t}{t'} \qquad \text{eq. 5.50}$$

Hence, a plot of residual drawdown, s', versus the logarithm of t/t' should provide a straight line. The gradient of the line equals $2.3Q/4\pi T$ so that $\Delta s'$, the change in residual drawdown over one log cycle of t/t', enables a value of transmissivity to be found from:

$$T = \frac{2.3Q}{4\pi \Delta s'} \qquad \text{eq. 5.51}$$

It is not possible for a value of storage coefficient, S, to be determined by this recovery test method although, unlike the Theis and Cooper–Jacob methods, a reliable estimate of transmissivity can be obtained from measurements in either the pumping well or observation well. An example of the interpretation of recovery test data following a constant discharge pumping test is presented in Box 5.2.

Principle of superposition of drawdown

As discussed in the previous section, the recovery test method relies on the principle of superposition of drawdown. As shown in Fig. 5.35, the drawdown at any point in the area of influence caused by the discharge of several wells is equal to the sum of the drawdowns caused by each well individually, thus:

$$s_t = s_1 + s_2 + s_3 + \ldots + s_n \qquad \text{eq. 5.52}$$

where s_t is the drawdown at a given point and $s_1, s_2, s_3 \ldots s_n$ are the drawdowns at this point caused by the discharges of wells 1, 2, 3 ... n, respectively. Solutions to find the total drawdown can be found using the equilibrium (Thiem) or non-equilibrium (Theis) equations of well drawdown analysis and are of practical use in designing the layout of a well-field to minimize interference between well drawdowns or in designing an array of wells for the purpose of dewatering a ground excavation site.

For example, if two wells are 100 m apart in a confined aquifer ($T = 110$ m^2 day^{-1}) and one well is pumped at a steady rate of 500 m^3 day^{-1} for a long period after which the drawdown in the other well is

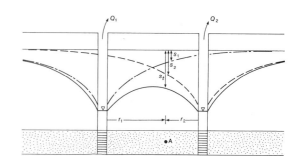

Fig. 5.35 Cross-section through a confined aquifer showing the cones of depression for two wells pumping at rates Q_1 and Q_2. From the principle of superposition of drawdown, at position A between the two pumping wells, the total drawdown, s_t, is given by the sum of the individual drawdowns s_1 and s_2 associated with Q_1 and Q_2, respectively.

0.5 m, what will be the steady-state drawdown at a point mid-way between the two wells if both wells are pumped at a rate of 500 m^3 day^{-1}? First, considering one well pumping on its own, the drawdown at the mid-way position ($r_1 = 50$ m) can be found from the Thiem equation (eq. 5.28). In this case, the head at the mid-way position (h_1) is equal to $h_o - s_1$ where h_o is the original piezometric surface prior to pumping and s_1 is the drawdown due to pumping. Therefore, $s_1 = h_o - h_1$. Similarly, $s_2 = h_o - h_2$ at $r_2 = 100$ m and equation 5.28 becomes, expressed in terms of drawdown:

$$s_1 - s_2 = \frac{Q}{2\pi T} \log_e \frac{r_2}{r_1} \qquad \text{eq. 5.53}$$

In this example, with one well pumping on its own, s_1 is the unknown, $s_2 = 0.5$ m, $r_1 = 50$ m, $r_2 = 100$ m, $Q = 500$ m^3 day^{-1} and $T = 110$ m^2 day^{-1} giving:

$$s_1 = 0.5 + \frac{500}{2\pi 110} \log_e \frac{100}{50} = 1.0 \text{ m} \qquad \text{eq. 5.54}$$

Now, considering both wells pumping simultaneously, and since the discharge rates for both are the same, then from the principle of superposition of drawdown, it can be determined that the total drawdown at the point mid-way between the two wells will be the sum of their individual effects, in other words 2.0 m.

Leaky, unconfined and bounded aquifer systems

The above solution methods for the non-equilibrium equation of radial flow apply to ideal, confined aquifers but for leaky, unconfined and bounded aquifers variations of the curve matching technique must be applied. A summary of aquifer responses is provided here but for a further treatment with worked examples of solution methods, including the case of partially penetrating wells, the reader is referred to Kruseman and de Ridder (1990). The determination of aquifer parameters from large-diameter dug well pumping tests is presented by Herbert and Kitching (1981). Once familiar with the various techniques for the analysis of pumping test data in different hydrogeological situations, it is then possible to use computer programs for the ease of estimating aquifer properties.

In the case of a leaky, or semiconfined, aquifer, when water is pumped from the aquifer, water is also drawn from the saturated portion of the overlying aquitard. By lowering the piezometric head in the aquifer by pumping, a hydraulic gradient is created across the aquitard that enables groundwater to flow vertically downwards. From a consideration of Darcy's law (eq. 2.9) and the sketch in Fig. 5.36b, the amount of downward flow is inversely proportional to the thickness of the aquitard (b') and directly proportional to both the hydraulic conductivity of the aquitard (K') and the difference between the water table in the upper aquifer unit and the potentiometric head in the lower aquifer unit. Compared with an ideal, confined aquifer (Fig. 5.36a), the effect of a leaky aquifer condition on the drawdown response measured at an observation well is to slow the rate of drawdown until a true steady-state situation is reached where the amount of water pumped is exactly balanced by the amount of recharge through the aquitard, assuming the water table remains constant (Fig. 5.36b). Methods of solution for the situation of steady-state and non-equilibrium conditions in a leaky aquifer with or without storage in the aquitard layer are provided by Hantush (1956) and Walton (1960).

Methods of pumping test analysis for confined aquifers can be applied to unconfined aquifers providing that the basic assumptions of the Theis solution are mostly satisfied. In general, if the drawdown is small in relation to the saturated aquifer thickness, then good approximations are possible. Where drawdowns are larger, the assumption that water released from storage is discharged instantaneously with a decline in head is frequently not met. As shown in Fig. 5.36c, the drawdown response of unconfined aquifers typically resembles an S-curve with three distinct sections. At early time, following switching on the pump, water is released from storage due to compression of the aquifer matrix and expansion of the water in an analogous way to a confined aquifer (see Section 2.11.2). A Theis type-curve matched to this early data would give a value for storage coefficient comparable to a confined aquifer. As pumping continues and the water table is lowered, gravity drainage of water from the unsaturated zone in the developing cone of depression contributes delayed yield at a variable rate.

The pattern of drawdown in an unconfined aquifer depends on the vertical and horizontal hydraulic conductivity and the thickness of the aquifer. Once delayed yield begins, the drawdown curve appears to flatten (Fig. 5.36c) compared with the ideal, confined aquifer response. The drawdown is less than expected and resembles the response of a leaky aquifer. At later time, the contribution of delayed yield declines and groundwater flow in the aquifer is mainly radial producing a response that can be matched to a Theis type-curve. Values of storage coefficient calculated for this third segment of the curve provide a value for the specific yield, S_y, of the aquifer (see Section 2.11.3). Graphical methods for interpreting pumping test data in unconfined aquifers which account for the differing aquifer responses are provided by Boulton (1963) and Neuman (1975).

When a well is pumped close to an aquifer boundary, for example an influent river or impermeable geological fault, the assumption that the aquifer is of infinite areal extent is no longer true and the drawdown response is of the type shown in Fig. 5.36d. As shown in Fig. 5.37a, where the boundary is a constant head, for example a surface water body such as the sea, a river or lake, the drawdown around the pumping well is less than expected compared with the ideal, confined aquifer of infinite extent, eventually reaching a steady-state condition with the amount of water pumped balanced by the water recharging from the constant head boundary. Where the

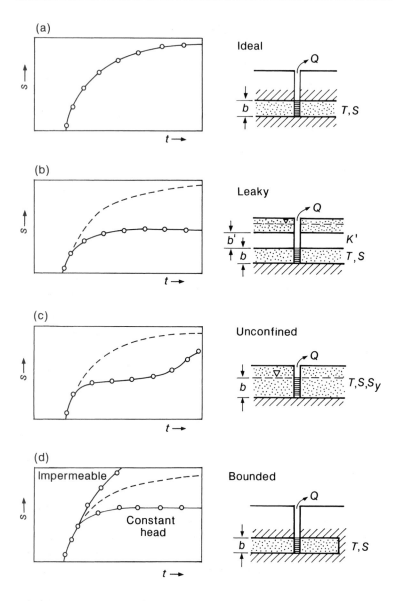

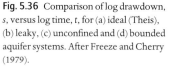

Fig. 5.36 Comparison of log drawdown, s, versus log time, t, for (a) ideal (Theis), (b) leaky, (c) unconfined and (d) bounded aquifer systems. After Freeze and Cherry (1979).

boundary is an impermeable, no-flow boundary then a greater drawdown than expected is observed than would be the case if the aquifer were infinite in extent (Fig. 5.37b).

Analysis of pumping test data affected by boundaries requires application of the principle of superposition of drawdown (see previous section). By introducing imaginary, or image wells with the same discharge or recharge rate as the real well, an aquifer of finite extent can be interpreted in terms of an infinite aquifer so that the solution methods described in the previous sections can be applied. For a well close to a constant head boundary, the image well is a recharging well placed at an equal distance from the boundary as the real well but reflected on the opposite side of the boundary (Fig. 5.37a). For an impermeable boundary, the image well is a discharging well, again placed at an equal distance from the boundary as the real well but on the opposite side of the boundary (Fig. 5.37b). Further explanation of

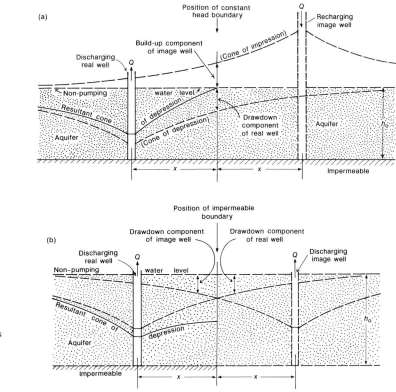

Fig. 5.37 Application of image well theory in the case of pumping wells affected by (a) constant head and (b) impermeable boundaries. The sections show the equivalent hydraulic system required to meet the Theis solution assumption of an aquifer of infinite areal extent. After Ferris et al. (1962).

image well theory is given by Todd (1980) and includes the case of a wedge-shaped aquifer, such as a valley bounded by two converging impermeable boundaries. For complex regional aquifer situations or for the analysis of multiple well systems, then a numerical modelling approach to the solution of the steady-state or non-equilibrium groundwater flow equations is usually required (Section 5.9).

5.8.3 Tracer tests

The principal applications for tracer tests are in the determination of groundwater flowpaths and residence times, in the measurement of aquifer properties and in the mapping and characterizing of karst conduit networks. Much experience has been gained in karstic aquifers to demonstrate connectivity and measure travel times. In intergranular aquifers, tracer tests are used less frequently because of the slower groundwater velocities and the greater potential for dilution. To be of use, an ideal tracer should be non-toxic and easily measured at very large dilutions. The tracer should either be absent or present in very low concentrations in the groundwater system to be studied. An ideal tracer should also follow the same pathway as the substance to be investigated, whether particulates or solutes, and should not react chemically with the groundwater or be adsorbed on to the aquifer rock.

There is no tracer available that meets all of these criteria but there is a wide range of substances and properties of water that can be used as tracers, including: temperature; suspended particles (for example, spores, fluorescent microspheres); solutes (sodium chloride, halogen ions); dyes (fluorescent dyes, optical brighteners); gases (noble gases, sulphur hexafluoride); microbes (bacteriophage); and environmental tracers such as chlorofluorocarbons (CFCs), radiocarbon, tritium, and the stable isotopes of hydrogen, oxygen, nitrogen, sulphur and carbon (Ward et al. 1998). The choice of tracer type will

Table 5.9 Summary of tracer types and their properties. After Hobson (1993).

	Tracer type								
	Lycopodium spores	Colloid	Phage	Inorganic salts	Fluorescent tracers	Fluorocarbons	Organic anions	Radioisotopes	'Natural'
Conservative?	N	N	N	N/C	N/C	N	C	C	N/C
Quantifiable?	N (Yes if dyed)	Y	S	Y	Y	Y	Y	Y	Y
Sampling?	P	S	S	I/S	I/S/P	S	S	I/S	I/S/P
Detectability	?	H	H	L	M	H	M	H	M
Toxicity – actual	A	A	A	A/C	A	A	A	C	A/C
Relative cost (tracer and analysis)	E	E	E	C	C	E	E	E	C
Ease of use	D	D	E	E	E	D	E	D	E
Type of medium for which suitable	K	K/F/G	K/F/G	K/F/G	K/F/G	F/G	K/F/G	K/F/G	K/F/G
Types of test for which suitable*	3	1,6	4	1,2,5,6	1,2,3,4,5,6	6	1,2,4	1,2,5,6	3,4

Conservative?: C, conservative; N, non-conservative. Quantifiable? N, no; Y, yes; S, semiquantifiable. Sampling?: P, passive detector; S, sample of water required; I, in situ measurement possible. Detectability: ?, uncertain; L, low ($\times 10^5$ dilution); M, medium ($\times 10^8$ dilution); H, high ($\times 10^{10}$ dilution). Toxicity: A, acceptable; C, possible concern. Relative cost: E, expensive; C, cheap. Ease of use: D, difficult; E, easy. Type of medium: K, karst; F, fractured; G, granular.

* Types of tracer test for which suitable – see Table 5.10 for key.

depend on its suitability in terms of detectability, toxicity, relative cost and ease of use (Table 5.9) and the choice of tracer test method will be determined by the hydrogeological properties to be measured (Table 5.10).

The chlorofluorocarbons CFC-11 (trichlorofluoromethane), CFC-12 (dichlorofluoromethane) and CFC-113 (trichlorotrifluoroethane) have received attention recently as groundwater tracers with useful reviews of their source, distribution in groundwaters and applications in tracing and age dating modern groundwaters provided by Plummer and Busenberg (2000) and Höhener et al. (2003). CFCs are synthetic, halogenated, volatile organic compounds that were manufactured from 1930 for use as aerosol propellants and refrigerants until banned by the Montreal Protocol in 1996. In general, CFC dating is most likely to be successful in rural settings, with shallow water tables, where the groundwater is aerobic, and not impacted by local contaminant sources such as septic tanks or industrial applications (Plummer & Busenberg 2000). As discussed in Section 4.4.2, age dating techniques have limitations, with greater confidence in deriving apparent age obtained when multiple dating techniques are applied to the same sample, for example CFCs and ^{3}H/^{3}He dating (Section 4.4.5).

Some of the most commonly used tracers in groundwater studies are the fluorescent dyes fluorescein (uranine) and rhodamine WT, the optical brightner photine CU and bacteriophage. The detection and assay of fluorescent dyes is made by illuminating the test solution at an appropriate narrow band of wavelengths (the excitation wavelength) and measuring the amount of fluorescent light emitted at a corresponding longer wavelength band (the emission wavelength). Representative excitation and emission spectra are given in Table 5.11 for eight fluorescent dyes. Measurements are made using a fluorometer and standards of known concentration used to derive a calibration curve relating fluorescence in arbitrary units to concentration, typically measured in µg L^{-1} (Fig. 5.22a).

Smart and Laidlaw (1977) and Atkinson and Smart (1981) evaluated the suitability of different fluorescent dyes for water tracing and note their extensive application in Britain, particularly in the investigation of karstic limestones (Box 5.3). From the topological work of Brown and Ford (1971) and Atkinson et al.

(1973), tracer tests can be used in mapping and characterizing karst conduit networks. By comparing discharges and masses of tracer at the entrance to and exit from a karst system, it is possible to classify the conduit network into one of five topological types (Fig. 5.38).

The scale of investigation of tracer tests can range from laboratory experiments to the site scale (tens or hundreds of metres) and to the regional scale (kilometres) in karst aquifers (Box 5.3). When compared with the scale of influence of other field methods for determining aquifer properties, Niemann and Rovey (2000) noted, in an area of glacial outwash deposits near Des Moines, Iowa, that the hydraulic conductivity values of outwash determined from pumping tests by curve matching techniques (see Section 5.8.2) can be an order of magnitude larger than values found from a tracer test using the conservative solute chloride. This discrepancy may be caused by the different scale and dimensionality of the two test methods (the cones of depression for the pumping tests were about 30–130 m while the tracer tests related to a zone of influence of less than 30 m) with dispersion of the tracer within the glacial outwash preventing the conservative solute from flowing exclusively within smaller, high permeability paths which have a strong influence on the groundwater flow and hydraulic conductivity measured by pumping tests. Hence, the results from Des Moines suggest that the velocity of a conservative solute plume in an intergranular aquifer may be overestimated if values of hydraulic conductivity derived from pumping tests were used in calculations.

At field and regional scales, tracer tests can be carried out with or without wells or boreholes and can be performed under natural hydraulic gradient conditions (see Box 6.3) or under forced gradient (pumping) conditions (for examples see Niemann & Rovey 2000). The advantages and limitations of this range of tests and detailed protocols for conducting two of the most useful tests, the single borehole dilution method and the convergent radial flow tracer test, are discussed by Ward et al. (1998). Both these tests can provide measurements of aquifer properties.

Relative to pumping tests, the advantages of the single borehole dilution method conducted under a natural hydraulic gradient are the low cost of materials and equipment and the simplicity of the

Table 5.10 Hydrogeological properties which may be measured using tracer tests. After Ward et al. (1998).

Property to be determined	Suitable test methods
Measurement of flow paths:	
Connection between two or more points	3, 4, 6
Direction of flow	3, 4
Measurement of velocities:	
Average linear water velocity	3, 4, 5, 6
Specific discharge/darcy velocity	2
Contaminant migration velocity	3, 4, 6
Measurement of aquifer properties:	
Hydraulic conductivity	2
Effective porosity	5
Heterogeneity	4
Fracture characterization	4
Matrix diffusion	1, 6, 4
Measurement of solute/contaminant transport properties:	
Dispersion	3, 4, 5, 6
Sorption	1, 4
Dilution	3, 4, 6
Measurement of recharge/groundwater catchments	3, 4
Measurement of groundwater age	3, 4

Tracer test methods: 1, laboratory tests; 2, single borehole dilution; 3, natural gradient tests (without boreholes); 4, natural gradient tests (multi-well); 5, drift (injection) and pump back; 6, forced gradient (multi-well).

method in determining values for specific discharge, q, and aquifer hydraulic conductivity, K. The method requires injection of tracer into the whole water column in the borehole, or a packered interval, so that a well-mixed column of tracer of uniform initial concentration is obtained. The subsequent dilution of tracer is then monitored by means of an in situ detector or by careful depth sampling of the borehole to minimize disturbance of the concentration profile. The rate of change of concentration at any level, z, in the borehole is given by:

$$\frac{\partial C(z,t)}{\partial t} = \frac{\partial (Cu(z))}{\partial z} - \frac{CQ_o(z)}{\pi r^2} \qquad \text{eq. 5.55}$$

where $C(z,t)$ is the concentration at time t, $u(z)$ is the vertical velocity of water in the borehole, $Q_o(z)$ is the volume of water leaving the borehole per unit depth per unit time and r is the borehole radius. The first term on the right-hand of equation 5.55 is the change in concentration due to vertical flow in the borehole and the second term is the effect of tracer solution leaving the borehole. Solution of equation 5.55 depends upon the form of the functions $u(z)$ and $Q_o(z)$. If it is assumed that there is no vertical flow in the borehole, then $u(z)$ is equal to zero and equation 5.55 becomes:

Table 5.11 Excitation and emission maxima of tracer dyes and filter combinations for their analysis. After Smart and Laidlaw (1977).

Dye	Maximum excitation (nm)	Maximum emission (nm)	Primary filter	Mercury line (nm)	Secondary filter
Blue fluorescent dyes:					
Amino G acid	355(310)	445	7–37*	365	98†
Photine CU	345	435(455)			
Green fluorescent dyes:					
Fluorescein	490	520	98†	436	55†
Lissamine FF	420	515			
Pyranine	455(405)	515			
Orange fluorescent dyes:					
Rhodamine B	555	580	2 × 1-60* + 61†	546	4–97* + 3–66*
Rhodamine WT	555	580			
Sulpho rhodamine B	565	590			

Note: Figures in parentheses refer to secondary maxima. For all spectra, pH is 7.0.
* Corning filter.
† Kodak Wratten filter.

B O X
5 . 3

Dye-tracer test in the Chepstow Block Carboniferous limestone aquifer, south-west England

As part of an investigation into the risk of landfill leachate contamination of the Chepstow Block Carboniferous limestone aquifer in south-west England, a dye-tracer test was conducted to study the groundwater flow system (Clark 1984). The study area is shown in Fig. 1a with groundwater discharge from the Chepstow Block almost entirely focused on the Great Spring located in the Severn rail tunnel. The karstic nature of the limestone suggests that groundwater flow is through solutionally widened fissures or joints, with the joints parallel to the main north-westerly oriented faults that trend towards the Great Spring. The spring is an important water resource for the area, having a mean flow of about 650 L s^{-1} (56×10^3 m^3 day^{-1}) (Fig. 1c).

A dye-tracer test using fluorescein was designed to determine groundwater flowpaths and residence times and confirm the theory of conduit flow. A mass of 30 kg of fluorescein was injected into a sinkhole in the Cas Troggy Brook located to the north-west of the

Great Spring in November 1982 and its recovery monitored at the Great Spring (Fig. 1b). The recovery curve shows a minimum travel time of 42 days and a peak travel time of 130 days over 7 km giving a groundwater velocity of 54 m day^{-1}. The curve can be simulated using an advection-dispersion model with a longitudinal dispersivity (Section 6.3.1) of 720 m. This high dispersivity, even allowing for the scale of the test, suggests that groundwater flow is through a complex fissure system rather than as pure conduit flow (Clark 1984).

The total recovery of the fluorescein tracer was about 23% suggesting losses by adsorption and degradation of some 77%. The maximum concentration of dye-tracer recovered was 0.92 µg L^{-1} giving a dilution of the dye input of 6×10^7 and leading to the conclusion that the impacts of waste disposal activities on the groundwater quality of the Great Spring will be considerably lessened by dilution in the limestone aquifer.

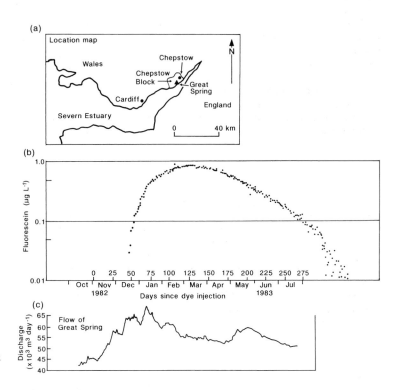

Fig. 1 (a) Location map of the Chepstow Block and Great Spring, (b) recovery of fluorescein dye-tracer and (c) variation in discharge at the Great Spring during the tracer test. After Clark (1984).

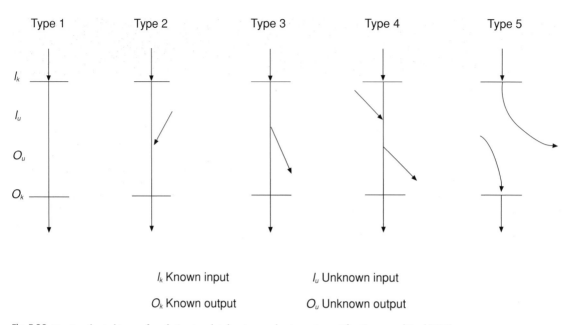

I_k Known input I_u Unknown input

O_k Known output O_u Unknown output

Fig. 5.38 Five topological types of conduit network in karst groundwater systems. After Brown and Ford (1971).

$$\frac{\partial C(z,t)}{\partial t} = -\frac{CQ_o(z)}{\pi r^2} \qquad \text{eq. 5.56}$$

which, for a fixed value of z, yields upon integration:

$$C = C_o \exp\left[-\frac{Q_o t}{\pi r^2}\right] \qquad \text{eq. 5.57}$$

If it is assumed that the water flowing through the borehole is drawn from a width of aquifer equal to twice the borehole diameter, then, for a one-dimensional flow system with parallel streamlines, $Q_o = 2qd$ and equation 5.57 becomes (Lewis et al. 1966):

$$C = C_o \exp\left[-\frac{8qt}{\pi d}\right] \qquad \text{eq. 5.58}$$

where C_o is concentration at time $t = 0$, d is the borehole diameter and q is the horizontal specific discharge or darcy velocity (Section 2.3). Hence, the dilution of tracer in the borehole should be exponential, with a time constant related to the specific discharge. By plotting concentration versus time on semilogarithmic paper, the specific discharge can be calculated by substituting any two values of tracer concentration and the corresponding time interval

into equation 5.58. Calculation of the hydraulic conductivity can then be made if the hydraulic gradient is known using equation 2.9. An example of the application of the single borehole dilution method is given in Box 5.4.

In a convergent radial flow tracer test, tracer is added to a soakaway, well or piezometer and the breakthrough monitored at a pumping well. A divergent radial flow tracer test is also possible where tracer is added to an injection well and the forced plume of tracer observed in an array of surrounding observation points. In comparison, a shortcoming of the convergent radial flow test is that the converging flow field counteracts spreading due to dispersion. An example of the convergent radial flow test is given in Box 5.5.

5.8.4 Geophysical methods

Downhole (borehole) and surface geophysical techniques are now routinely used in hydrogeological investigations and take advantage of modern techniques and instrumentation that have benefited from advances in electronics and digital technology. Although the two types of geophysics can be considered

B O X
5 . 4

Single borehole dilution tracer test conducted at a Chalk aquifer site

As part of an investigation into the flow behaviour of a Chalk aquifer at a site in southern England that is subjected to artificial recharge of treated sewage effluent, 8.5 g of the optical brightner amino G acid were injected into a Chalk observation borehole in a single borehole dilution tracer test. Using a depth sampler to obtain groundwater samples, the decrease in concentration of amino G acid was measured daily at intervals of 1 m in the open section of the borehole (Hiscock 1982). Results for the dilution of amino G acid during the 6 days of the test are shown in Fig. 1a for the Chalk horizon at 30 m below ground level (m bgl). By choosing $C_o = 3600$ $\mu g\,L^{-1}$ at $t = 0$ and $C = 15\,\mu g\,L^{-1}$ at $t = 3.6$ days, the specific discharge, q, at this level is calculated, using equation 5.58 for a borehole diameter of 150 mm, as:

$$15 = 3600 \exp\left[-\frac{8q3.6}{\pi 0.15}\right] \qquad \text{eq. 1}$$

giving $q = 0.09$ m day^{-1}. The calculated specific discharge values for all the sampling levels are shown in Fig. 1b, with the resulting profile indicating the presence of solutionally widened fissures in the Chalk to explain the higher values recorded between 18 and 21 m bgl and at 26 and 30 m bgl. For a regional hydraulic gradient of 0.005 (see Fig. 2.21a), the mean Chalk hydraulic conductivity found using the dye-tracer results is 18 m day^{-1}, ranging from a minimum value of 12 m day^{-1} to a maximum value of 37 m day^{-1}.

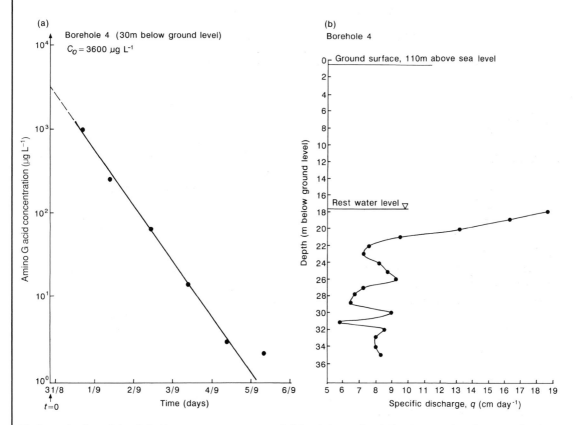

Fig. 1 Results of a single borehole dilution tracer test in an open Chalk borehole at Ludgershall (NGR SU 272 497) showing (a) the logarithmic decrease in amino G acid dye-tracer concentration with time at a depth of 30 m below ground level and (b) the profile of specific discharge values, q, obtained from the tracer test.

In an investigation of the risk of road drainage entering a soakaway and contaminating a well pumping from a sand and gravel aquifer, Bateman et al. (2001) conducted a convergent radial flow test to establish a connection between the soakaway and well and also to determine a value for the aquifer longitudinal dispersivity (Section 6.3.1). The well has an open base with a single horizontal collecting pipe extending 18.3 m to the north. To test a connection between the road soakaway and the pumping well, a distance of 30 m, 60 g of fluorescein were injected into the soakaway and the recovery monitored at the well. Prior to injection, 500 L of water were trickled into the soakaway to wet the unsaturated zone above the shallow water table. The tracer was flushed into the aquifer following injection by trickling a further 2000 L of water into the soakaway.

The fluorescein tracer recovery (Fig. 1) clearly shows that there is a connection between the road soakaway and the well, with fluorescein breakthrough beginning at 6–7 days after injection, with the peak arriving at approximately 16 days. A recovered mass of 37.7 g represents 63% of the injected fluorescein, with some of the loss explained by the fact that pumping of the well ceased while recovery of the fluorescein continued.

Analysis of the aquifer dispersivity from the tracer breakthrough data is possible using the numerical method of Sauty (1980) developed for application in either diverging or converging groundwater flow problems. Sauty (1980) produced a set of type curves with tracer results presented in terms of dimensionless concentration, C_R, versus dimensionless time, t_R, for various Peclet numbers, P, with:

$$C_R = \frac{C}{C_{max}} \qquad \text{eq. 1}$$

$$t_R = \frac{t}{t_c} \qquad \text{eq. 2}$$

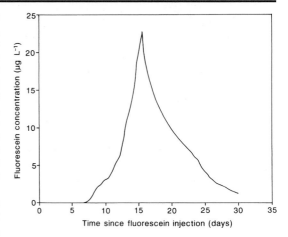

Fig. 1 Recovery of fluorescein dye-tracer at Quay Lane well, Southwold (NGR TM 486 774) recorded during a convergent radial flow test in a sand and gravel aquifer. After Bateman et al. (2001).

$$P = \frac{r}{\alpha} \qquad \text{eq. 3}$$

where C_{max} is equal to the peak tracer concentration, t_c is the time to peak concentration, r is radial distance and α is aquifer longitudinal dispersivity. By superimposing the field data on the Sauty-type curves (Fig. 2), and notwithstanding the presence of the horizontal collecting pipe that disturbs the assumption of symmetrical flow around the well, the best fit line to the data suggests a Peclet number of between 30 and 100, giving an aquifer dispersivity value of between 1.0 and 3.3 m for a radial distance, r, of 30 m.

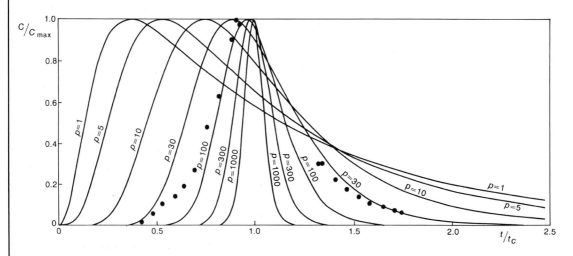

Fig. 2 Fluorescein concentration data from the radial flow test at Quay Lane well, Southwold (Fig. 5.44) plotted on Sauty-type curves (Sauty 1980). After Bateman et al. (2001).

complementary, surface geophysics is generally employed at an early stage in a hydrogeological investigation, before boreholes are drilled, while logging techniques are employed later to obtain detailed information on aquifer and fluid properties (see Fig. 6.28) (Barker 1986). There is a large literature relating to the application of geophysical methods in hydrogeology and the objective here is to direct the reader to textbooks and articles that expound the various downhole and surface techniques. Downhole geophysical techniques are discussed by Beesley (1986), BSI (1988), Chapellier (1992) and Sharma (1997) and surface geophysical techniques by Griffiths and King (1981), Barker (1986) and Kearey and Brooks (1991).

Various hydrogeological applications are demonstrated in the following papers:
- The application of geophysical borehole measurements in crystalline rocks using acoustic televiewer and caliper measurements, electrical resistance, thermal techniques and vertical seismic profiling (Wilhelm et al. 1994a).
- A gravity survey and resulting Bouguer anomaly map of the subsurface position of a buried channel in the Chalk aquifer of East Anglia (Barker & Harker 1984).
- An investigation of saline intrusion using borehole logging, seismic reflection profiling, vertical electrical resistivity soundings (VES) and electromagnetic induction surveying in a coastal sand and gravel aquifer (Holman & Hiscock 1998; Holman et al. 1999).
- An evaluation of lithological, stratigraphical and structural controls on the distribution of aquifers using VES, transient electromagnetic (TEM), tensorial audio-magnetotelluric (AMT) and nuclear magnetic resonance (NMR) depth sounding and inversion measurements (Meju et al. 1999, 2002).
- The detection of leaks from environmental barriers using electrical current imaging (Binley et al. 1997).
- The application of cross-borehole transmission radar and electrical resistance tomography to characterize groundwater flow and solute transport in the unsaturated (vadose) zone (Binley et al. 2002a, 2002b).

5.9 Groundwater modelling

Numerical modelling of groundwater flow can be undertaken at the start or end of a hydrogeological investigation: at the start for conceptualizing the main controls on groundwater flow in the model area and in indicating the type and length of field data that will be required to construct a model; and at the end for predicting future aquifer response under different groundwater conditions (Rushton 1986, 2003). With a well-constructed model, the ability to predict groundwater flow patterns, for example the effects of different groundwater abstraction patterns on sensitive aquatic systems (Box 5.6), or the shape of wellhead capture zones for protecting groundwater quality (see Fig. 7.8), or future aquifer response to changing recharge amounts under climate change (see Section 8.5), makes groundwater modelling an indispensable tool for managing local and regional groundwater resources.

In the process of constructing a groundwater model, the primary aim is to represent adequately the different features of groundwater flow through the aquifer within the model area or domain. In this respect, the important features to consider in governing the response of an aquifer to a change in hydrogeological conditions include: aquifer inflows (recharge, leakage and cross-formational flows); aquifer outflows (abstractions, spring flows and river baseflows); aquifer properties (hydraulic conductivity and storage coefficient); and aquifer boundaries (constant or fixed head, constant flow or variable head, and no-flow boundaries).

Given the complexity of regional groundwater flow problems, the equations of groundwater flow (eqs 2.40, 2.46) cannot be solved by analytical methods (Section 2.12) and, instead, approximate numerical techniques are used. These techniques require that the space and time co-ordinates are divided into some form of discrete mesh and time interval. Common approaches to defining the space co-ordinates in the model domain are the finite-difference and finite-element approximations. The finite-difference approach is based on a rectilinear mesh whereas the finite-element approach is more flexible in allowing a spatial discretization that can fit the geometry of the flow problem. For each cell in the mesh and, for transient simulations, at each time step, the unknown heads are represented by a set of simultaneous equations that can be solved iteratively by specifying initial head conditions. Model runs are performed by a computer program that employs

BOX
5.6

Groundwater modelling of the Monturaqui-Negrillar-Tilopozo aquifer, Chile

The Monturaqui-Negrillar-Tilopozo (MNT) aquifer is located at the foot of the Andes in the extremely arid environment of northern Chile and occupies a north–south oriented graben approximately 60 km long, with surface elevations decreasing from around 3200 m in the southern (Monturaqui) area to 2300 m in the north (Tilopozo wetland) area where the aquifer discharges at the southeastern margin of the Salar de Atacama (Fig. 1a). The MNT basin is hosted by Palaeozoic rocks and infilled with mainly Tertiary alluvial and volcanic sediments to depths of over 400 m. The Tertiary sediment infill to the basin forms the MNT aquifer with transmissivity values ranging between 500 and 4500 $m^2 day^{-1}$, confined storage coefficient values of around 1×10^{-3} and an estimated specific yield of about 0.1. The aquifer is exploited to meet the demands of the important mining industry for potable and ore-processing water. The Tilopozo wetland is the only groundwater-dependent natural feature in the MNT basin (Anderson et al. 2002).

As part of a sustainability assessment of the region, it has been suggested that a decline in water level of approximately 0.25 m, equivalent to a reduction in aquifer outflow of approximately 6%, should not be exceeded in order to protect the flora and fauna of the Tilopozo wetland. The MNT aquifer and the established wellfields (Fig. 1a) have complex flow dynamics in relation to the capture of discharge from the wetland. Given these complexities, a three-dimensional, spatially distributed, time-variant numerical flow model using the MODFLOW code (McDonald & Harbaugh 1988) was developed to investigate a groundwater abstraction strategy that would satisfy the sustainability criteria (Anderson et al. 2002).

The groundwater flow model extended from the southern limit of the Monturaqui Basin north to the Tilopozo wetland, and used a 1000-metre-resolution finite-difference grid with two layers. The model domain and boundary conditions are shown in Fig. 1a.

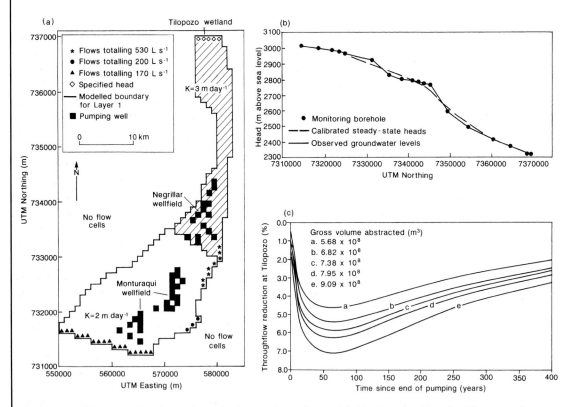

Fig. 1 (a) Model domain and boundary conditions for the groundwater flow model of the Monturaqui-Negrillar-Tilopozo aquifer, Chile. (b) Observed groundwater heads versus modelled pre-abstraction heads of the calibrated steady-state model. (c) Predictive modelling results for different wellfield abstraction volumes leading to simulated groundwater through-flow reductions at the Tilopozo wetland. After Anderson et al. (2002).

BOX
5.6

Continued

Constant inflows on the southern and eastern boundaries were used to represent system recharge, with a range of possible total inflows simulated. The model was refined within the bounds established in the conceptual model so that it reproduced as closely as possible the observed steady-state (pre-abstraction) head distribution in the aquifer (Fig. 1b). Following model calibration, a time-variant model for a period of up to 500 years was developed for predictive modelling of impacts. Groundwater abstractions with a range of durations from both the Monturaqui and Negrillar wellfields were represented in the model with the model simulations used to predict outflow changes with time across the northern boundary representing the Tilopozo wetland. Significant uncertainty in the value of specific yield (varied between 0.05 and 0.2) and recharge rate (between 450 and 1800 L s^{-1}) was addressed at all stages of the modelling through sensitivity analyses.

Analysis of the model output (Fig. 1c) showed that the impacts of groundwater abstraction will reach the wetland between 20 and 40 years after the start of pumping from either wellfield, with maximum impacts likely to occur 75–300 years after abstraction ceases. The location of the Monturaqui wellfield upgradient of approximately 60% of the aquifer recharge was found to significantly reduce its relative impacts on groundwater flows at Tilopozo. From an interpretation of the results of sensitivity analyses, a 'worst case' prediction model was developed which maximized the magnitude of predicted impacts, and model runs executed until a pumping duration of between 10 and 20 years and a gross abstracted volume of 7.38×10^{8} m^3 (Fig. 1c) were found that limited through-flow reduction at the Tilopozo wetland to 6% (Anderson et al. 2002). The resulting model now forms the basis for the sustainable groundwater development of the MNT aquifer which, fortunately, is supported by the extremely large volume of groundwater storage in the aquifer (approximately 10^{10} m^3).

matrix methods for solving the large number of unknowns. Successful completion of a model run is obtained when convergence of the head solution is reached, usually determined by a model error criterion set at the beginning of the run.

In all groundwater modelling investigations, for example the case study presented in Box 5.6, the process of deriving a model for predictive purposes involves the following common steps: conceptualization of the flow mechanisms in the model area based on existing knowledge; acquisition of available field data on groundwater heads, aquifer properties and river flows; discretization of the model domain and construction of the model input file; calibration of the model by comparing simulated steady-state (equilibrium) and transient (time-variant) heads and flows (flow vectors and water balance) against field-measured values; sensitivity analysis of aquifer property values and recharge and boundary conditions; validation of the model against an independent set of data (for example, groundwater head and river flow data not used during model calibration, or hydrochemical data such as salinity); and, lastly, prediction of aquifer response under changed groundwater conditions.

For further discussion of finite-difference and finite-element numerical modelling techniques, the reader is referred to the texts by Wang and Anderson (1982), Spitz and Moreno (1996) and Rushton (2003). A popular finite-difference model for application in two- and three-dimensional groundwater flow problems is the United States Geological Survey's code MODFLOW (McDonald & Harbaugh 1988), with demonstrations of this model presented by Anderson and Woessner (1992) and Chiang and Kinzelbach (2001).

Other groundwater modelling approaches include more specialist applications such as solute transport modelling using the method of characteristics (Konikow & Bredehoeft 1978) or the 'random walk' method (Prickett et al. 1981) for solving the advection-dispersion equation (see eq. 6.7). In the method of characteristics, advective transport is simulated by particles distributed in a geometrically uniform pattern over the entire model area with each particle assigned an initial concentration associated with the concentration in the cell containing the particle. Dispersive transport is simulated by a finite-difference calculation on the rectangular grid after which the particle concentrations are updated according to the changes in the grid concentrations, and advective transport in the next time step is calculated. In a similar way, the random walk method combines a flow submodel, usually based on the finite-difference method, with the use of random variables that are Gaussian distributed and applied to particles introduced into the flow field to simulate dispersion. Each of the particles that is advected and dispersed by groundwater flow is assigned a mass which represents a fraction of the total mass of the chemical constituent involved. At the end of the simulation, the total mass of particles within an overlaid grid cell is divided by the product of the cell volume and porosity to give the average cell concentration.

A popular solute transport model developed by Zheng (1990), and available from the United States Environment Protection Agency, is the code MT3D for simulating reactive mass transport including equilibrium-controlled linear or non-linear sorption (see eq. 6.13) and first-order irreversible decay or biodegradation (see eq. 4.9). The model uses a mixed Eulerian–Lagrangian approach to solve the three-dimensional advection-dispersion-reaction equation with three basic options based on the method of characteristics.

In techniques for simulating saline intrusion, or situations where density variations are significant, for example the underground disposal of brine wastes, then the United States Geological Survey's density-coupled model SUTRA (Voss 1984) is appropriate. Given that hydraulic conductivity and head are functions of density (see eqs 2.7 and 2.22), the model provides a numerical solution to the governing groundwater flow equation written in terms of the pressure potential, ψ, and intrinsic permeability. Hence, models simulating density-dependent flow require initial pressure and density distributions in order to find a solution.

5.10 FURTHER READING

Anderson, M.P. & Woessner, W.W. (1992) *Applied Groundwater Modeling: simulation of flow and advective transport*. Academic Press, San Diego.

Brandon, T.W. (ed.) (1986) *Groundwater: occurrence, development and protection*. Institution of Water Engineers and Scientists, London.

Brassington, R. (1998). *Field Hydrogeology*, 2nd edn. John Wiley, Chichester.

Chapellier, D. (1992) *Well Logging in Hydrogeology*. A.A. Balkema, Rotterdam.

Chiang, W.-H. & Kinzelbach, W. (2001) *3D-Groundwater Modeling with PMWIN: a simulation system for modeling groundwater flow and pollution*. Springer-Verlag, Berlin.

Clark, L. (1988) *The Field Guide to Water Wells and Boreholes*. Geological Society of London Professional Handbook Series, Open University Press, Milton Keynes.

Cook, P.G. & Herczeg, A.L. (2000) *Environmental Tracers in Subsurface Hydrology*. Kluwer Academic, Boston.

Driscoll, F.G. (1986) *Groundwater and Wells*, 2nd edn. Johnson Filtration Systems, St Paul, Minnesota.

Griffiths, D.H. & King, R.F. (1981) *Applied Geophysics for Geologists and Engineers: the elements of geophysical prospecting*, 2nd edn. Pergamon Press, Oxford.

Hillel, D. (1982) *Introduction to Soil Physics*. Academic Press, Orlando, Florida.

Jones, J.A.A. (1997) *Global Hydrology: processes, resources and environmental management*. Addison Wesley Longman, Harlow, Essex.

Jury, W.A., Gardner, W.R. & Gardner, W.H. (1991) *Soil Physics*, 5th edn. John Wiley, New York.

Kearey, P. & Brooks, M. (1991) *An Introduction to Geophysical Exploration*, 2nd edn. Blackwell Science, Oxford.

Kendall, C. & McDonnell, J.J. (eds) (1998) *Isotope Tracers in Catchment Hydrology*. Elsevier, Amsterdam.

Kruseman, G.P. & de Ridder, N.A. (1990) *Analysis and Evaluation of Pumping Test Data*, 2nd edn. Pudoc Scientific Publishers, Wageningen, The Netherlands.

Lerner, D.N., Issar, A.S. & Simmers, I. (1990) *Groundwater Recharge: a guide to understanding and estimating natural recharge*, vol. 8, *International Contributions to Hydrogeology*. Verlag Heinz Heise, Hannover.

Rushton, K.R. (2003) *Groundwater Hydrology: conceptual and computational models*. John Wiley, Chichester.

Schwartz, F.W. & Zhang, H. (2003) *Fundamentals of Ground Water*. John Wiley, New York.

Sharma, P.V. (1997) *Environmental and Engineering Geophysics*. Cambridge University Press, Cambridge.

Shaw, E.M. (1994) *Hydrology in Practice*, 3rd edn. Nelson Thornes, Cheltenham.

Smith, R.E. (2002) Infiltration theory for hydrologic applications. *Water Resources Monograph*. American Geophysical Union, Washington, DC **15**.

Spitz, K. & Moreno, J. (1996) *A Practical Guide to Groundwater and Solute Transport Modeling*. John Wiley, New York.

Todd, D.K. (1980) *Groundwater Hydrology*, 2nd edn. John Wiley, New York.

Twort, A.C., Ratnayaka, D.D. & Brandt, M.J. (2000) *Water Supply*, 5th edn. Butterworth-Heinemann, Oxford.

Wang, H.F. & Anderson, M.P. (1982) *Introduction to Groundwater Modelling: finite difference and finite element methods*. Academic Press, San Diego, California.

Ward, R.C. & Robinson, M. (2000) *Principles of Hydrology*, 4th edn. McGraw-Hill, Maidenhead, Berkshire.

Wilson, E.M. (1990) *Engineering Hydrology*, 4th edn. Macmillan, London.

Groundwater quality and contaminant hydrogeology

6

6.1 Introduction

The occurrence of groundwater contamination is a legacy of past and present land-use practices and poor controls on waste disposal. Many raw materials and chemicals have had a long history of usage before becoming recognized as hazardous. During this time, handling and waste disposal practices have frequently been inadequate. Hence, it must be considered that any industrial site where hazardous materials have been used is now a potential source of contaminated land. In the United States alone, the National Academy of Science (1994) reported that there are an estimated 300,000 to 400,000 hazardous waste sites in the United States and that, over the next three decades, 750 billion US dollars could be spent on groundwater remediation at these sites. In the United Kingdom, with its long industrial history, there are estimated to be as many as 100,000 contaminated land sites covering between 50,000 and 200,000 ha, equivalent to an area larger than Greater London. Added to this picture of industrial contamination, the drive towards self-sufficiency in agricultural production and the increasing urbanization of the world's growing population directly threaten the quality of groundwater through the over-application of agri-chemicals and the often uncontrolled disposal of human and landfill wastes.

As a definition, contaminated groundwater is groundwater that has been polluted by human activities to the extent that it has higher concentrations of dissolved or suspended constituents than the maximum admissible concentrations formulated by national or international standards for drinking, industrial or agricultural purposes. The main contam-inants of groundwater include chemicals such as heavy metals, organic solvents, mineral oils, pesticides and fertilizers, and microbiological contaminants such as faecal bacteria and viruses. Table 6.1 is a compilation of the sources and potential characteristics of ground-water contaminants.

This chapter is first concerned with the quality of water intended for drinking and irrigation purposes as determined by international water quality standards. Given the importance of water hardness to consumers and its apparent health benefits, a section provides background information on this water quality parameter. Following this introduction to water quality, the principles of groundwater contaminant transport are then discussed in relation to non-reactive and reactive solutes and their behaviour in homogeneous and heterogeneous aquifer material. The latter part of the chapter provides an overview of major polluting activities, including industrial, mining, agricultural and municipal sources of contaminants, and concludes with a discussion of the causes and effects of saline intrusion in coastal regions.

6.2 Water quality standards

The chemical composition of natural groundwaters is discussed in Section 3.3. In addition, groundwaters may contain synthetic organic compounds and microbiological organisms, for example organic solvents and pathogenic bacteria, introduced from sources of surface contamination. To limit the possible harmful effects of natural and introduced components of groundwater, various measures have been developed to protect water users. Water quality criteria include

Table 6.1 Potential sources of groundwater pollution arising from domestic, industrial and agricultural activities. Adapted from Jackson (1980).

Contaminant source	Contaminant characteristics
Septic tanks	Suspended solids 100–300 mg L^{-1} BOD 50–400 mg L^{-1} Ammonia 20–40 mg L^{-1} Chloride 100–200 mg L^{-1} High faecal coliforms and streptococci Trace organisms, greases
Storm water drains	Suspended solids ~1000 mg L^{-1} Hydrocarbons from roads, service areas Chlorides or urea from de-icing Compounds from accidental spillages Bacterial contamination
Industry	
Food and drink manufacturing	High BOD. High suspended solids. Colloidal and dissolved organic substances. Odours
Textile and clothing	High suspended solids and BOD. Alkaline effluent
Tanneries	High BOD, total solids, hardness, chlorides, sulphides, chromium
Chemicals	
Acids	Low pH
Detergents	High BOD
Pesticides	High TOC, toxic benzene derivatives, low pH
Synthetic resins and fibres	High BOD
Petroleum and petrochemical	
Refining	High BOD, chloride, phenols, sulphur compounds
Process	High BOD, suspended solids, chloride, variable pH
Plating and metal finishing	Low pH. High content of toxic metals
Engineering works	High suspended solids, hydrocarbons, trace heavy metals. Variable BOD, pH
Power generation	Pulverized fuel ash: sulphate, and may contain germanium and selenium. Fly ash and flue gas scrubber sludges: low pH, disseminated heavy metals
Deep well injection	Concentrated liquid wastes, often toxic brines. Acid and alkaline wastes. Organic wastes
Leakage from storage tanks and pipelines	Aqueous solutions, hydrocarbons, petrochemicals, sewage
Agriculture	
Arable crops	Nitrate, ammonia, sulphate, chloride and phosphates from fertilizers. Bacterial contamination from organic fertilizers. Organochlorine compounds from pesticides
Livestock	Suspended solids, BOD, nitrogen. High faecal coliforms and streptococci
Silage	High suspended solids, BOD 1–6 × 10^4 mg L^{-1} Carbohydrates, phenols
Mining	
Coal mine drainage	High TDS (total dissolved solids), suspended solids. Iron. Low pH. Possibly high chloride
Metals	High suspended solids. Possibly low pH. High sulphates. Dissolved and particulate metals
Household wastes	High sulphate, chloride, ammonia, BOD, TOC and suspended solids from fresh wastes. Bacterial contamination. On decomposition: initially TOC of mainly volatile fatty acids (acetic, butyric, propionic acids), subsequently changing to high molecular weight organics (humic substances, carbohydrates)

BOD, biological oxygen demand; TOC, total organic carbon; pH, $-\log_{10}(H^+)$.

the scientific information with which decisions on water quality can be based, for example toxicity data, information relating to the available water treatment technology and environmental degradation rates. With this information, and taking into account political, legal and socioeconomic issues, policy makers are able to set water quality objectives for the attainment of good quality.

For groundwater, the setting of chemical quality objectives may not always be the best approach since it gives the impression of an allowed level of pollution. An alternative approach is to state that groundwater should not be polluted at all. In the EU, this precautionary approach to protecting the chemical status of groundwater is adopted and comprises a prohibition on direct discharges to groundwater and, to cover indirect discharges, a requirement to monitor groundwater bodies in order to detect changes in chemical composition and reverse any upward trend in pollution. Under the Directive on the Protection of Groundwater Against Pollution Caused by Certain Dangerous Substances (80/68/EEC; Council of European Communities 1980), the most toxic substances are listed under Lists I and II (Appendix 9). List I substances, including organophosphorus compounds, mercury and cadmium, should be prevented from being discharged into groundwater. List II substances, including metals, fluoride and nitrate, should have discharges of these substances into groundwater minimized. Only a few specific directives have been established at European level for particular issues, including the Directive on Diffuse Pollution by Nitrates (91/676/EEC; Council of European Communities 1991). Taken together, it is envisaged that the above measures, which are subsumed under the EU Water Framework Directive (see Section 1.8), should prevent and control groundwater pollution and achieve good groundwater chemical status for the future.

Further to water quality objectives, water quality standards present the detailed rules that govern how the objectives should be met. To be workable, the standards must be relatively simple so that routine monitoring can detect water quality failures. A standard may allow some variability, or derogation, in terms of a given concentration being met for a certain percentage of samples but with no single sample allowed to exceed a maximum allowable concentration. A further consideration is where, in the cycle of water abstraction, treatment and supply, to apply the standard. Standards applied at the tap control the standard of water used for human consumption and are termed drinking water quality standards. A number of large organizations, including the EU, United States EPA and WHO, have published drinking water quality standards and these are summarized in Appendix 9.

The hardness of groundwater can become a water quality issue especially where it affects industrial and domestic uses where the water is heated. There has also been a long debate as to the relative health benefits of drinking hard water. In fact, no health-based guideline value is proposed for hardness since it is considered that the available data on the inverse relationship between the hardness of drinking water and cardiovascular disease (CVD) are inadequate to permit the conclusion that the association is causal (World Health Organization 2002). However, a concentration of 500 mg L^{-1} is at the upper limit of aesthetic acceptability. Further discussion follows in the next section and Box 6.1.

6.2.1 Water hardness

Water hardness is the traditional measure of the capacity of water to react with soap and describes the ability of water to bind soap to form lather, a chemical reaction detrimental to the washing process. Hardness has little significance in terms of hydrochemical studies, but it is an important parameter for water users. Today, the technical significance of water hardness is more concerned with the corrosive effects on water pipes that carry soft water.

Despite the wide usage of the term, the property of hardness is difficult to define exactly. Water hardness is not caused by a single substance but by a variety of dissolved polyvalent metallic ions, predominantly Ca and Mg, although other ions, for example Al, Ba, Fe, Mn, Sr and Zn, also contribute. The source of the metallic ions are typically sedimentary rocks, the most common being limestone ($CaCO_3$) and dolomite ($CaMg(CO_3)_2$). In igneous rock, magnesium is typically a constituent of the dark-coloured ferromagnesian minerals, including olivine, pyroxenes, amphiboles and dark-coloured micas, and slow weathering of these silicate minerals produces water hardness.

Hardness is normally expressed as the total concentration of Ca^{2+} and Mg^{2+} ions in water in units of mg L^{-1} as equivalent $CaCO_3$. For this purpose, hardness can be determined by substituting the concentration of Ca^{2+} and Mg^{2+}, expressed in mg L^{-1}, in the following equation:

$$\text{Total hardness} = 2.5(Ca^{2+}) + 4.1(Mg^{2+}) \qquad \text{eq. 6.1}$$

Each concentration is multiplied by the ratio of the formula weight of $CaCO_3$ to the atomic weight of the

The 'hard-water story'

BOX 6.1

The history and debate surrounding whether hard water protects against cardiovascular disease (CVD) is often referred to as the 'hard-water story' and started with a Japanese agricultural chemist. Kobayashi (1957) had for many years studied the nature of agricultural irrigation water and found a close relation between the chemical composition of river water and the death rate from 'apoplexy' (cerebrovascular disease). The death rate from apoplexy in Japan was extraordinarily high compared to other countries, and the biggest cause of death in Japan. Kobayashi (1957) found that it was especially the ratio of sulphur to carbonate ($SO_4/CaCO_3$) in drinking water that was related to the death rate from apoplexy and suggested that inorganic acid might induce, or $CaCO_3$ prevent, apoplexy.

Since Kobayashi (1957), and in different parts of the world, many studies have been completed on the relation between Ca and Mg in local drinking water, and CVD mortality. These studies are generally based upon death registers and water quality data at regional or municipality levels. Even with all these studies, the results are not conclusive as to the role of Ca and Mg in drinking water for CVD. However, most of these studies are 'ecological', meaning that the exposure to water constituents is determined at group levels with a high risk of misclassification. Often, very large groups, for example all inhabitants in large cities or areas, are assigned the same value of water Ca and Mg, despite the presence of several waterworks or private wells. In addition, the disease diagnoses studied are sometimes unspecific, with wide definitions that include both cardiac and cerebrovascular diseases. In some studies, it is also unclear whether the range of Ca and Mg in drinking water is large enough to allow for appropriate analyses.

One of the most comprehensive studies of the geographic variations in cardiovascular mortality was the British Regional Heart Study. The first phase of this study (Pocock et al. 1980) applied multiple regression analysis to the geographical variations in CVD for men and women aged 35–74 in 253 urban areas in England, Wales and Scotland for the period 1969–1973. The investigation showed that the relationship to water hardness was non-linear, being much greater in the range from very soft to medium-hard water than from medium to very hard water. The geometric mean for the standardized mortality ratio (SMR) for CVD for towns grouped according to water hardness both with and without adjustments (by analysis of covariance) for the effects of four climatic and socioeconomic variables (percentage of days with rain, mean daily maximum temperature, percentage of manual workers and car ownership) is shown in Fig. 1. The adjusted SMR decreased steadily in moving from a hardness of 10 to 170 mg L^{-1} but changed little between 170 and 290 mg L^{-1} or greater. After adjustment, CVD in areas with very soft water, around 25 mg L^{-1}, was estimated to be 10–15% higher than in areas with medium-hard water, around 170 mg L^{-1}, while any further increase in hardness beyond 170 mg L^{-1} did not additionally lower CVD mortality. Hence, it appeared that the maximum effect on CVD lay principally between the very

soft and medium-hard waters. Importantly, adjusting for climatic and socioeconomic differences considerably reduced the apparent magnitude of the effect of water hardness (Pocock et al. 1980).

A problem with correlation studies such as the British Regional Heart Study, as argued by Jones and Moon (1987), is the failure of much of the research to consider the causal mechanism that links independent variables to the disease outcome. Also, many of the calibrated models presented in the literature are socially blind in including only those variables pertaining to the physical environment, often a large number of water quality elements. Even in those better analyses that have included social variables, as in the case of the British Regional Heart Study, the relatively strong correlation found for Ca in England and Wales may be a result of Ca acting as a very good surrogate for social variables. The soft-water areas of the north and west of the country equate to the areas of early industrialization, and today these areas house a disproportionate percentage of the socially disadvantaged (Jones & Moon 1987). Therefore, it is important that further studies undertake the challenge of quantitatively analysing the separate effects of social variables from those of water hardness.

Fewer studies have been carried out in developing countries, but Dissanayake et al. (1982), for example, found a negative correlation between water hardness and various forms of CVD and leukaemia in Sri Lanka.

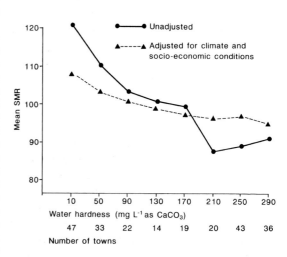

Fig. 1 Geometric means of the standardized mortality ratio (SMR) (for all men and women aged 35–74 with cardiovascular disease) for towns in England, Wales and Scotland grouped according to water hardness (in concentration units of mg L^{-1} as $CaCO_3$). After Pocock et al. (1980).

ion; hence the factors 2.5 and 4.1 are included in the hardness relation (Freeze & Cherry 1979).

Where reported, carbonate hardness includes that part of the total hardness equivalent to the HCO_3^- and CO_3^{2-} content (or alkalinity). If the total hardness exceeds the alkalinity, the excess is termed the non-carbonate hardness and is a measure of the calcium and magnesium sulphates. In older publications, the terms 'temporary' and 'permanent' are used in place of 'carbonate' and 'non-carbonate'. Temporary hardness reflects the fact that the ions responsible may be precipitated by boiling, such that:

$$Ca^{2+} + 2HCO_3^- \rightarrow CaCO_3\downarrow + H_2O + CO_2\uparrow \quad \text{eq. 6.2}$$

'scale'

In Europe, water hardness is often expressed in terms of degrees of hardness. One French degree is equivalent to 10 mg L^{-1} as $CaCO_3$, one German degree to 17.8 mg L^{-1} as $CaCO_3$ and one English or Clark degree to 14.3 mg L^{-1} as $CaCO_3$. One German degree of hardness (dH) is equal to 1 mg of calcium oxide (CaO) or 0.72 mg of magnesium oxide (MgO) per 100 mL of water.

A number of attempts have been made to classify water hardness. Water with hardness values greater than 150 mg L^{-1} as equivalent $CaCO_3$ is designated as being very hard. Soft water has values of less than 60 mg L^{-1}. Groundwaters in contact with limestone or gypsum ($CaSO_4.2H_2O$) rocks can commonly attain levels of 200–300 mg L^{-1}. In water from gypsiferous formations, 1000 mg L^{-1} or more of hardness may be present (Hem 1985).

Hardness in water used for domestic purposes does not become particularly troublesome until a level of 100 mg L^{-1} is exceeded. Depending on pH and alkalinity, hardness of about 200 mg L^{-1} can result in scale deposition, particularly on heating, and increased soap consumption. Soft waters with a hardness of less than about 100 mg L^{-1} have a low buffering capacity and may be more corrosive to water pipes resulting in the presence of heavy metals, such as Cd, Cu, Pb and Zn, in drinking water, depending also on the pH and dissolved oxygen content of the water.

In developing countries reliant on groundwater supplies developed in crystalline bedrock aquifers,

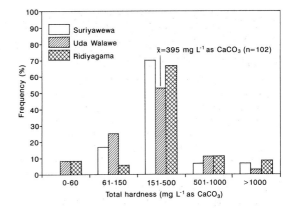

Fig. 6.1 Histogram of total hardness values recorded for groundwaters sampled during the wet season (January and February 2001) from dug wells and tube wells in three subcatchments of the Uda Walawe basin of Sri Lanka. Data courtesy of L. Rajasooriyar.

water hardness is often an important consideration, particularly as silicate weathering not only produces Ca^{2+}, Mg^{2+} and HCO_3^- but can also release elements such as Al, As and F that are hazardous to human health. Studies by Dissanayake (1991) and Rajasooriyar (2003) have highlighted the problem of high fluoride concentrations and associated dental fluorosis in areas of hard water in Sri Lanka.

Rajasooriyar (2003) measured total hardness in dug wells and tube wells in the Uda Walawe basin of southern Sri Lanka in the wide range 7–3579 mg L^{-1} as $CaCO_3$ with an average of 395 mg L^{-1} (Fig. 6.1). Compared with the government water quality limit of 600 mg L^{-1}, of the 102 samples collected during the wet season in 2001, 12% of the samples were in excess of the limit and are considered too hard to drink (values above 100–150 mg L^{-1} are locally considered too hard as a water supply). Soft waters are found in areas with a dense irrigation network supplied by rain-fed surface reservoirs. Irrigation canal waters in the Suriyawewa and Uda Walawe subcatchments were measured in the dry season to have a total hardness in the range 40–90 mg L^{-1} as $CaCO_3$ and it is leakage of this water source that leads to the softening of shallow groundwater. Most groundwaters (63% of samples analysed) in the fractured aquifer are very hard with carbonate hardness in the range 151–500 mg L^{-1} contributed by the weathering of ferromagnesian minerals, anorthite, calcite and dolomite.

Table 6.2 Salinity hazard of irrigation water with basic guidelines for water use relative to dissolved salt content.

Salinity hazard	Dissolved salt content (mg L^{-1})	Electrical conductivity (μS cm^{-1})
Water for which no detrimental effects will usually be noticed	500	750
Water that may have detrimental effects on sensitive crops*	500–1000	750–1500
Water that may have adverse effects on many crops and requiring careful management practices	1000–2000	1500–3000
Water that can be used for salt-tolerant plants† on permeable soils with careful management practices and only occasionally for more sensitive crops	2000–5000	3000–7500

* Field beans, string beans, peppers, lettuce, onions, carrots, fruit trees.
† Sugarbeets, wheat, barley.

The products of this weathering lead to high concentrations of dissolved Ca^{2+}, Mg^{2+} and HCO_3^- in groundwaters. Exceptionally high values of hardness (>1000 mg L^{-1}) typically occur in non-irrigated areas with additional non-carbonate hardness contributed by pyrite oxidation buffered by weathering of Ca-minerals and, in the case of the coastal Ridiyagama coastal catchment, by salt water inputs of Ca^{2+} and SO_4^{2-}.

6.2.2 Irrigation water quality

Crop irrigation is the most extensive use of groundwater in the world and so it is important to consider plant requirements with respect to water quality. The most damaging effects of poor quality irrigation water are excessive accumulation of soluble salts (the salinity hazard) and a high percentage sodium content (the sodium hazard). The salinity hazard increases the osmotic pressure of the soil water and restricts the plant roots from absorbing water, even if the field appears to have sufficient moisture. The result is a physiological drought condition. The salinity hazard is generally determined by measuring the electrical conductivity of the water in μS cm^{-1} and then assessed against the type of criteria given in Table 6.2.

The sodium hazard relates to the accumulation of excessive sodium which causes the physical structure of the soil to breakdown. The replacement by sodium of calcium and magnesium adsorbed on clays results in the dispersion of soil particles. The soil becomes hard and compact when dry and increasingly impervious to water such that the plant roots do not get enough water, even though water may be standing on the surface. The sodium hazard of irrigation water is estimated by the sodium adsorption ratio (SAR) which relates the proportion of Na^+ to Ca^{2+} and Mg^{2+} in the water as follows:

$$SAR = \frac{Na^+}{\sqrt{\dfrac{Ca^{2+} + Mg^{2+}}{2}}} \qquad \text{eq. 6.3}$$

with the ionic concentrations expressed in meq L^{-1}. Generally, irrigation water with a SAR greater than 9 should not be used on crops, even if the total salt content is relatively low. Higher values of SAR may be tolerated if the soil contains an appreciable amount of gypsum ($CaSO_4 \cdot 2H_2O$) or if gypsum can be added to the soil and so provide a source of soluble calcium to decrease the SAR.

The two aspects of salinity and sodium content can be combined and an irrigation water quality classification obtained. An example classification is shown in Fig. 6.2 based on research by the United States Department of Agriculture.

Other water quality considerations for irrigation water include the bicarbonate and carbonate concentrations in water which effectively increase the sodium hazard by precipitating calcium and magnesium carbonates during soil drying, hence increasing the SAR, and the presence of toxic elements, particularly boron and chloride. Excessive levels of boron and chloride are common in groundwater and concentrations of B greater than 1 mg L^{-1} and of chloride above 70 mg L^{-1} can lead to injury in sensitive plants.

Additional factors to consider in deciding the usefulness of water for a specific irrigation purpose include soil texture and structure, drainage conditions,

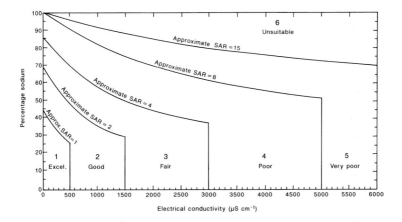

Fig. 6.2 Diagram for classifying irrigation water based on percentage sodium content (sodium hazard) and electrical conductivity (salinity hazard). The sodium adsorption index (SAR) is also shown. The six classes are described as follows: Class 1 (Excellent), suitable for use on all crop types; Class 2 (Good), suitable for use on most crops under most conditions but limiting conditions can develop on poorly draining clayey soils; Class 3 (Fair), suitable for most crops if care is taken to prevent accumulation of soluble salts, including sodium, in the soil; Class 4 (Poor), suitable only in situations having very well-drained soils for production of salt-tolerant crops; Class 5 (Very poor), restricted to irrigation of sandy, well-drained soils in areas receiving at least 750 mm of rainfall; Class 6 (Unsuitable), not recommended for crop irrigation. After Oklahoma State University Extension Facts F-2401 *Classification of Irrigation Water Quality*.

gypsum and lime content of the soil and the irrigation method and management. Further information can be found in UNESCO/FAO (1973).

6.3 Transport of contaminants in groundwater

The type of soil, sediment or rock in which a pollution event has occurred and the physicochemical properties of individual or mixtures of contaminants influence the spread and attenuation of groundwater contaminants. The fundamental physical processes controlling the transport of non-reactive contaminants are advection and hydrodynamic dispersion which create a spreading pollution plume and cause a dilution in the pollutant concentration. For reactive contaminant species, attenuation of the pollutant transport occurs by various processes including chemical precipitation, sorption, microbially mediated redox reactions and radioactive decay. For the class of contaminants known as non-aqueous phase liquids (NAPLs) both immiscible and dissolved phases of the contaminant need to be considered. To explain these processes, it is convenient to divide the following sections into general contaminant classes, namely non-reactive and reactive dissolved contaminants and non-aqueous phase liquids.

6.3.1 Transport of non-reactive dissolved contaminants

Non-reactive contaminants, such as saline wastes containing chloride, are principally affected by the major processes of advection and hydrodynamic dispersion. Advection is the component of solute movement attributed to transport by the flowing groundwater. The advective velocity of the contaminant is the average linear velocity of the groundwater and can be calculated from a consideration of Darcy's law (see eq. 2.10). Hydrodynamic dispersion of contaminants in porous material occurs as a result of mechanical mixing and molecular diffusion as illustrated in Fig. 6.3. The significance of the dispersive processes is to decrease the contaminant concentration with distance from the source. As shown in Fig. 6.4, a continuous pollution source will produce an elongate plume, whereas a single point source will produce a slug that grows with time while becoming less concentrated as a result of dispersion as the plume moves in the direction of groundwater flow.

Molecular diffusion of contaminants is not normally of practical consideration where advection and mechanical dispersion are dominant. This is typically the case for shallow groundwater environments but in situations such as the very long term deep disposal

(a)

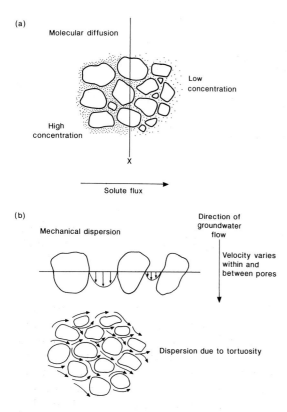

(b)

Fig. 6.3 Diagrammatic representation of (a) molecular diffusion and (b) mechanical dispersion which combine to transport solute within a porous material by the process of hydrodynamic dispersion. Notice that mechanical dispersion results from the variation of velocity within and between saturated pore space and from the tortuosity of the flowpaths through the assemblage of solid particles. Molecular diffusion can occur in the absence of groundwater flow, since solute transport is driven by the influence of a concentration gradient, while mechanical dispersion occurs when the contaminant is advected by the groundwater.

$$\frac{\partial C}{\partial t} = -D^\star \frac{\partial^2 C}{\partial x^2} \qquad \text{eq. 6.4}$$

where $D^\star$ is the molecular diffusion coefficient for the solute in a porous material. The analytical solution for an instantaneous step-change in solute concentration, C, for an infinite aquifer space is given by:

$$\frac{C}{C_0} = \text{erfc}\left(\frac{x}{2\sqrt{D^\star t}}\right) \qquad \text{eq. 6.5}$$

where erfc is the complementary error function (see Appendix 8 for tabulated values of erf (error function) and erfc), C_0 is the initial concentration at $x = 0$ at time $t = 0$, and C is the concentration measured at position x at time t. A graphical solution to equation 6.5 is shown in Fig. 6.5 for values of diffusion coefficient equal to 10^{-10} and 10^{-11} m^2 s^{-1}. Even after 10,000 years, the diffusive breakthrough of contaminant with a relative concentration of 0.01 (or 1% of the initial concentration) has only reached about 25 m from the pollution source.

Of greater importance in terms of contaminant transport in the shallow subsurface, mechanical dispersion of a dissolved solute in a groundwater flow field is represented by:

Mechanical dispersion $= \alpha \bar{v}$

where α is the dispersivity of the porous material and $\bar{v}$ is the advective velocity (average linear velocity) of the groundwater. Dispersivity is a natural physical characteristic of porous material and determines the degree of contaminant spreading. Dispersivity is greatest in the longitudinal direction of groundwater flow but much smaller, typically one-thirtieth to one-fifth of the longitudinal dispersivity, in the direction perpendicular, or transverse, to the flow. Dispersivity is found to be scale dependent. At the microscale, for example in controlled laboratory experiments using sand-filled columns, longitudinal dispersivity is measured to between 0.1 and 10 mm, and is mainly caused by pore-scale effects. By contrast, tracer experiments at the macro, field scale (see Boxes 5.3 and 5.5) give higher values of dispersivity, generally of a few metres and normally less than

of waste in stable geological environments of low hydraulic conductivity and low hydraulic gradient, diffusion is significant. The safe disposal of low- to medium-level nuclear wastes in rock repositories is dependent on the engineered containment of the waste and, should the containment fail over periods of thousands of years, the absence of an advective transport route back to the biosphere.

Diffusion represents the net movement of solute under a concentration gradient (Fig. 6.3) and can be described using Fickian theory. In one dimension, Fick's second law describes the time-varying change in solute concentration for a change in solute flux as:

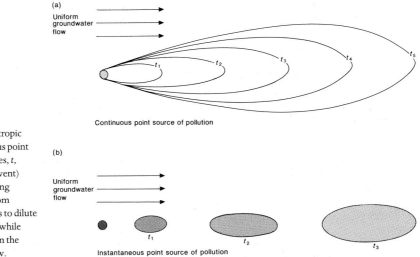

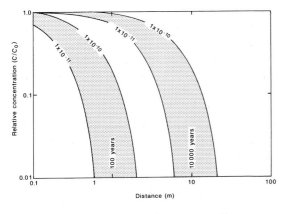

Fig. 6.4 Dispersion within an isotropic porous material of (a) a continuous point source of pollution at various times, t, and (b) an instantaneous (single event) point source of pollution. Spreading of the pollution plumes results from hydrodynamic dispersion and acts to dilute the contaminant concentrations, while advection transports the plumes in the field of uniform groundwater flow.

Fig. 6.5 Positions of a contaminant front transported by one-dimensional molecular diffusion at times of 100 years and 10,000 years for a source concentration, $C = C_0$, at time, $t > 0$, and a diffusion coefficient, D^*, of 10^{-10} and 10^{-11} m^2 s^{-1}. The curves of relative concentration are calculated using equation 6.5. After Freeze and Cherry (1979).

100 m, as a result of the physical heterogeneities in the aquifer encountered during transport.

Both laboratory and field experiments show that contaminant mass spreading by dispersion in a porous material conforms to a normal (Gaussian) distribution, with the position of the mean of the concentration distribution representing transport at the advective velocity of the water. The degree of contaminant dispersion about the mean is proportional to the variance (σ^2) of the concentration distribution.

The variance tensor can be resolved into three principal components that are approximately aligned with the longitudinal, transverse horizontal and transverse vertical directions. Assuming a constant groundwater velocity, dispersivity can be calculated as one half of the gradient of the linear spatial trend in variance.

To illustrate the dispersive transport of groundwater solutes, Hess et al. (2002) conducted a tracer test in the unconfined sand and gravel aquifer of Cape Cod, Massachusetts using the conservative tracer bromide (Br$^-$). As shown in Fig. 6.6a, and with increasing time of transport, physical dispersion of the injected mass of tracer causes the solute plume to become elongate. The tracer plume continued to lengthen as it travelled down-gradient through the aquifer and should conform to a linear increase in the longitudinal variance with distance travelled. The synoptic results of the tracer test showed that the longitudinal Br$^-$ variance increased at a slow rate early in the test but increased at a larger rate after about 70 m of transport. A linear trend fit to the later results (69–109 m of transport) produced a longitudinal dispersivity estimate for Br$^-$ of 2.2 m (Fig. 6.6b). The number of observations ($n = 4$) on which this estimate is based is small and scatter around the trend is apparent such that the dispersive process may not yet have reached a constant asymptotic value of longitudinal dispersivity (Hess et al. 2002). In general, transverse horizontal and vertical dispersivities were much smaller with values of 1.4×10^{-2} m and 5×10^{-4} m, respectively.

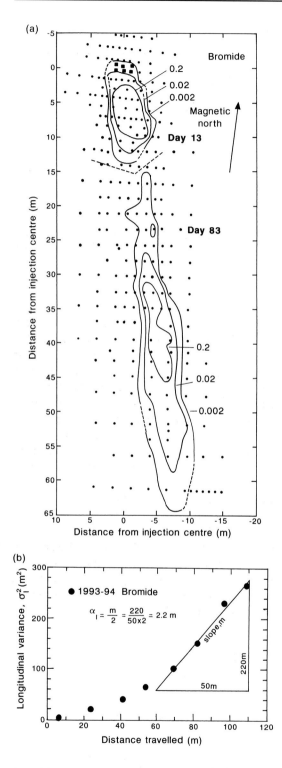

Combining the mathematical description of mechanical dispersion with the molecular diffusion coefficient gives an expression for the hydrodynamic dispersion coefficient, D, as follows:

$$D_l = \alpha_l \bar{v}_l + D^\star$$
$$D_t = \alpha_t \bar{v}_t + D^\star$$

eq. 6.6

where the subscripts $_l$ and $_t$ indicate the longitudinal and transverse directions, respectively.

The relative effects of mechanical dispersion and molecular diffusion can be demonstrated from the results of a controlled column experiment. The breakthrough curve for a continuous supply of tracer fed into a column packed with granular material is shown in Fig. 6.7. At low tracer velocity, molecular diffusion is the important contributor to hydrodynamic dispersion, although with little effect in spreading the tracer front. At high velocity, mechanical dispersion dominates and the breakthrough curve adopts a characteristic S-shape with some of the tracer moving ahead of the advancing front and some lagging behind, as controlled by the tortuosity of the flowpaths. The midpoint of the breakthrough curve occurs for a relative concentration, C/C_0, equal to one-half. This point of half-concentration represents the advective behaviour of the solute transport (shown by the vertical dashed line in Fig. 6.7) as if the tracer were moving by a plug-flow-type mechanism.

One-dimensional solute transport equation

Following from the above description of solute transport processes, the one-dimensional form of the solute transport equation describing the time-varying change in concentration of non-reactive dissolved contaminants in saturated, homogeneous, isotropic material under steady-state, uniform flow conditions

Fig. 6.6 (*left*) Results of a tracer experiment in the Cape Cod sand and gravel aquifer showing: (a) the distributions of relative concentrations (C/C_0) of Br$^-$ observed 13 and 83 days after injection (dashed where inferred); (b) calculated longitudinal variances of the Br$^-$ tracer plume for each synoptic sampling round. Also shown in (a) are locations of multilevel samplers available for groundwater sampling (solid circles) and tracer injection (solid squares). After Hess et al. (2002).

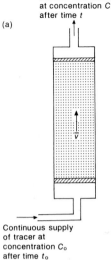

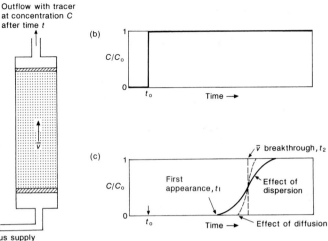

Outflow with tracer
at concentration C
after time t

Fig. 6.7 Results of a controlled laboratory column experiment showing the effect of longitudinal dispersion of a continuous inflow of tracer in a porous material. (a) Experimental set-up. (b) Step-function type tracer input relation. (c) Relative concentration of tracer in outflow from column. At low tracer velocity, molecular diffusion dominates the hydrodynamic dispersion and the breakthrough curve would appear as the dashed curve. At higher velocity, mechanical dispersion dominates and the solid curve would typically result. The vertical dashed line indicates the time of tracer breakthrough influenced by advective transport without dispersion. After Freeze and Cherry (1979).

Continuous supply
of tracer at
concentration C_0
after time t_0

and undergoing advection and hydrodynamic dispersion in the longitudinal direction, is given as:

$$D_l \frac{\partial^2 C}{\partial l^2} - \bar{v}_l \frac{\partial C}{\partial l} = \frac{\partial C}{\partial t} \qquad \text{eq. 6.7}$$

An analytical solution to the advection-dispersion equation (eq. 6.7) was provided by Ogata and Banks (1961) and is written as:

$$\frac{C}{C_0} = \frac{1}{2}\left[\text{erfc}\left(\frac{l - \bar{v}_l t}{2\sqrt{D_l t}}\right) \right.$$

$$\left. + \exp\left(\frac{\bar{v}_l l}{D_l}\right) \text{erfc}\left(\frac{l + \bar{v}_l t}{2\sqrt{D_l t}}\right) \right] \qquad \text{eq. 6.8}$$

for a step-function concentration input with the following boundary conditions:

$$C(l, 0) = 0 \quad l \geq 0$$

$$C(0, t) = C_0 \quad t \geq 0$$

$$C(\infty, t) = 0 \quad t \geq 0$$

For conditions in which the dispersivity of the porous material is large or when the longitudinal distance, l, or time, t, is large, the second term on the right-

hand side of equation 6.8 is negligible. Additionally, if molecular diffusion is assumed small compared to mechanical dispersion, then the denominator $\sqrt{D_l t}$ can also be written as $\sqrt{\alpha_l \bar{v}_l t}$. The expression $\sqrt{\alpha_l \bar{v}_l t}$ has the dimensions of length and may be regarded as the longitudinal spreading length or a measure of the spread of contaminant mass around the advective front, represented by the half-concentration, $C/C_0 = 0.5$.

The Ogata–Banks equation can be used to compute the shape of breakthrough curves and concentration profiles. For example, a non-reactive contaminant species is injected into a sand-filled column, 0.4 m in length, in which the water flow velocity is 1×10^{-4} m s^{-1}. If a relative concentration of 0.31 is recorded at a time of 35 minutes, calculate the longitudinal dispersivity, α_l, of the sand. First, taking the simplified version of equation 6.8 that ignores the second term on the right-hand side, and expressing the denominator as the longitudinal spreading length, then:

$$\frac{C}{C_0} = \frac{1}{2}\left[\text{erfc}\left(\frac{l - \bar{v}_l t}{2\sqrt{\alpha_l \bar{v}_l t}}\right) \right] \qquad \text{eq. 6.9}$$

and substituting the known values gives:

$$0.31 = \frac{1}{2}\text{erfc}\left[\frac{0.4 - 1 \times 10^{-4} \times 35 \times 60}{2\sqrt{\alpha_l \times 1 \times 10^{-4} \times 35 \times 60}}\right] \qquad \text{eq. 6.10}$$

which reduces to:

$$0.62 = \text{erfc}\left[\frac{0.19}{2\sqrt{\alpha_l 0.21}}\right] \qquad \text{eq. 6.11}$$

From Appendix 8, a value of $\beta = 0.35$ produces a $\text{erfc}[\beta] = 0.62$, hence:

$$0.35 = \left[\frac{0.19}{2\sqrt{\alpha_l 0.21}}\right] \qquad \text{eq. 6.12}$$

Rearranging and solving for the longitudinal dispersivity results in a calculated value for $\alpha_l = 0.35$ m or 35 cm.

6.3.2 Transport of reactive dissolved contaminants

Reactive substances behave similarly to conservative solute species, but can also undergo a change in concentration resulting from chemical reactions that take place either in the aqueous phase or as a result of adsorption of the solute to the solid matrix of soil, sediment or rock. The chemical and biochemical reactions that can alter contaminant concentrations in groundwater are acid–base reactions, solution–precipitation reactions, oxidation–reduction reactions, ion pairing or complexation, microbiological processes and radioactive decay. Discussion of a number of these processes is included in Chapter 3 in relation to natural groundwater chemistry and can be equally applied to the fate of dissolved contaminants.

One important type of process affecting the transport of reactive dissolved contaminants is sorption. Sorption processes such as adsorption and the partitioning of contaminants between aqueous and solid phases attenuate, or retard, dissolved solutes in groundwater and are of special relevance to the transport of organic contaminants, particularly hydrophobic compounds.

Attenuation due to sorption can be described by the retardation equation, as follows:

$$R_d = 1 + K_d \frac{(1-n)\rho_s}{\theta} \qquad \text{eq. 6.13}$$

where R_d is known as the retardation factor, K_d is the partition or distribution coefficient with units of mL g^{-1} (the reciprocal of density), ρ_s is the solid mass density of the sorbing material and θ is the moisture content of the unsaturated or saturated porous material. Below the water table, θ equates to the porosity and recognizing that $(1-n)\rho_s$ equates to the bulk density, ρ_b, equation 6.13 can be written as:

$$R_d = 1 + K_d \frac{\rho_b}{n} \qquad \text{eq. 6.14}$$

Bouwer (1991) presented a simple derivation of the retardation equation and also demonstrated applications describing preferential contaminant movement through macropores (see Section 5.4.3) and the shape of the concentration breakthrough curve for macro-dispersion caused by layered heterogeneity (Box 6.2). Limitations of the equation are that it assumes ideal, instantaneous sorption and equilibrium between the chemical sorbed and that remaining in solution.

The dimensionless retardation factor, R_d, is a measure of the attenuated transport of a reactive contaminant species compared to the advective behaviour of groundwater. As such, the retardation factor can be expressed in three ways as follows:

$$R_d = \frac{\bar{v}_w}{\bar{v}_c} = \frac{l_w}{l_c} = \frac{t_c}{t_w} \qquad \text{eq. 6.15}$$

where the subscripts $_w$ and $_c$ indicate the water and dissolved contaminant species, respectively, $\bar{v}$ is the average linear velocity, l is the distance travelled by the water or the central mass of a contaminant plume and t is the arrival time of the water or the midpoint of a contaminant breakthrough curve.

With knowledge of the rate of movement of a non-reactive tracer such as chloride representing the unattenuated flow of water by advection and dispersion, the time axis of a contaminant breakthrough curve can be transformed to a dimensionless time, t/t_{tracer}, where t_{tracer} is the breakthrough time of the tracer. Alternatively, the time axis can be written as the number of pore water flushes (V/V_p, the ratio of feed volume, V, to pore volume, V_p). In doing so, the retardation factor can be read directly from the dimensionless breakthrough time of contaminant at $C/C_0 = 0.5$. Examples of laboratory column breakthrough curves

The effect of macrodispersion in a layered aquifer system on the transport of a reactive contaminant can be demonstrated by combining Darcy's law (eq. 2.5) and the retardation equation (eq. 6.13) to derive the shape of the relative concentration breakthrough curve. For the example of layered heterogeneity shown in Fig. 1, and for a well positioned at 1000 m from the pollution source and pumping at 1.75 m^3 day^{-1}, then assuming saturated flow under a hydraulic gradient of 0.01 in each layer, the relative breakthrough concentration in each layer can be calculated as shown in Table 1. Notice that the greater sorption, as indicated by the larger R_d values, and longest travel times occur in the lower permeability layers. The first breakthrough at 7.5×10^3 days is in the layer with a hydraulic conductivity of 20 m day^{-1}. At the well, the chemically

laden flow in this layer contributes 23% (0.4/1.75) of the actual flow through the aquifer. Thus, the relative concentration of the chemical in the well at this time, which is found from the concentration in the well water, C, as a fraction of the original source concentration, C_0, also equals 0.23. After 25×10^3 days, the flow of contaminated water at the well increases to 1.4 m^3 day^{-1} (0.4 + 1), or 80% of flow from the aquifer. The calculation is continued until 303×10^3 days when the relative concentration C/C_0 of the well water equals 1.

The concentration breakthrough curve for this layered aquifer system is shown in Fig. 2 and resembles a parabola with pronounced tailing, no longer showing the typical symmetrical sigmoidal shape for dispersion in homogeneous porous material (Fig. 6.7). Also, the point $C/C_0 = 0.5$, which is theoretically reached after one pore volume has passed through a homogeneous material with retardation of the chemical, now occurs after about 23×10^3 days. In this time, $1.75 \times 23 \times 10^3 = 40.25 \times 10^3$ m^3 of water are pumped from the well, representing $(40.25 \times 10^3)/(5.6 \times 10^3)$ or 7.2 pore volume flushes of the layered system, where one pore volume is equal to 5.6×10^3 m^3.

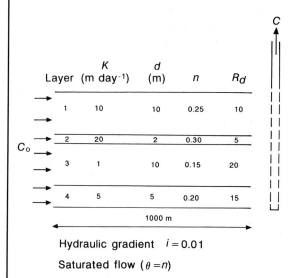

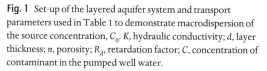

Fig. 1 Set-up of the layered aquifer system and transport parameters used in Table 1 to demonstrate macrodispersion of the source concentration, C_0. K, hydraulic conductivity; d, layer thickness; n, porosity; R_d, retardation factor; C, concentration of contaminant in the pumped well water.

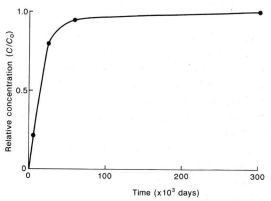

Fig. 2 Concentration breakthrough curve demonstrating macrodispersion caused by layered heterogeneity for the set-up shown in Fig. 1 and calculations given in Table 1.

Table 1 Calculation of relative breakthrough concentrations, C/C_0, for macrodispersion caused by layered heterogeneity in an aquifer of unit width. The layered aquifer system and associated transport parameters are shown in Fig. 1 and the resulting breakthrough curve in Fig. 2. After Bouwer (1991).

Layer (in order of breakthrough)	Darcy's law: $v_w = iK/n$ (m day^{-1})	Retardation equation $v_c = v_w/R_t$ (m day^{-1})	$t = 1000/v_c$ ($\times 10^3$ days)	$Q = -diK$ (m^3 day^{-1})	C/C_0
2	0.6667	0.1333	7.5	0.40	(0.40/1.75) = 0.23
1	0.4000	0.0400	25.0	1.00	(1:40/1.75) = 0.80
4	0.2500	0.0167	59.9	0.25	(1.65/1.75) = 0.94
3	0.0667	0.0033	303.0	0.10	(1.75/1.75) = 1.00
Well				1.75	

for two aromatic amine compounds are shown in Fig. 6.8. Aromatic amines occur as constituents of industrial waste waters, for example from dye production, and also as degradation products (metabolites) of pesticides.

When inspecting contaminant plumes from field data, it is also useful to choose a non-reactive tracer such as chloride to represent the advective transport of groundwater for comparison with the contaminant behaviour. Examples of retardation factors calculated from both spatial and temporal field data for organic solutes in a sand aquifer are given in Box 6.3.

If there is no difference in behaviour, then the retardation factor is equal to 1 and K_d is 0, in other words there is no reaction between the contaminant and the soil, sediment or aquifer material. For values of R_d greater than 1, attenuation of the contaminant species relative to the groundwater is indicated.

The partition coefficient, K_d, describes the process of contaminant sorption from the aqueous phase to the solid phase. Qualitatively, K_d is equal to the mass of chemical sorbed to the solid phase (per unit mass of solid) per concentration of chemical in the aqueous phase. Values of K_d are found empirically from labor-

---

BOX 6.3

Controlled field experiments to investigate transport of organic solutes

Hydrophobic sorption of organic solutes was examined in detail as part of two large-scale field experiments to investigate the natural gradient transport of organic solutes in groundwater. The experiments were conducted in an unconfined sand aquifer at the Canadian Forces Base Borden, Ontario, and are described in detail in papers by Mackay et al. (1986), Roberts et al. (1986), Rivett et al. (2001) and Rivett & Allen-King (2003). The aquifer is about 9 m thick and is underlain by a thick, silty clay aquitard. The water table has an average horizontal gradient of about 0.005 that may vary seasonally by as much as a factor of 2 (Rivett et al. 2001). The aquifer is composed of clean, well-sorted, fine- to medium-grained sand of glacio-lacustrine origin. Although the aquifer is fairly homogeneous, undisturbed cores reveal distinct bedding features. The bedding is primarily horizontal and parallel, although some cross-bedding and convolute bedding occur. Clay size fractions in the sand are very low and the organic carbon content (0.02%), specific surface area (0.8 m^2 g^{-1}) and cation exchange capacity (0.52 meq (100 g)$^{-1}$) of the aquifer solids are all low. The bulk density of the sand was estimated at 1.81 g cm^{-3} and the average porosity at 0.33. The hydraulic conductivity varies by approximately an order of magnitude with depth as a consequence of layering of the sand material. The geometric means of the hydraulic conductivity for cores sampled at two locations were similar at about 10^{-4} m s^{-1} (Mackay et al. 1986; Sudicky 1986).

The first experiment, the Stanford-Waterloo natural gradient tracer experiment (Mackay et al. 1986; Roberts et al. 1986), was designed with nine injection wells, each slotted and screened within the saturated zone (Fig. 1). Five halogenated organic solutes were chosen representing a range of expected mobilities as measured by their octanol-water partition coefficient (K$_{OW}$) values (Table 1).

At the start of the experiment, approximately 12 m^3 of solution were injected over a 15-hour period in order not to disturb the natural hydraulic gradient. The injected volume was chosen to be large relative to the scales of heterogeneity of the aquifer, as well as to ensure that dispersion during transport for several years would not too rapidly reduce the solution concentrations to values close to

background levels. The monitoring network system consisted of a dense network of 340 multilevel sampling devices. The horizontal spacing of the multilevel wells varied from 1.0 to 4.0 m, while the vertical spacing of the sampling points varied from 0.2 to 0.3 m, again chosen to be consistent with the estimated scales of hydraulic conductivity.

The closely spaced array of sampling points gave an unparalleled opportunity to study the morphology of the developing solute plumes and their attenuation by sorption and biodegradation by comparison with chloride introduced as a non-reactive tracer. Equal concentration contour plots of vertically averaged solute concentration for the chloride ion at 647 days after injection and carbon tetrachloride (CTET) and tetrachloroethene (PCE) at 633 days after

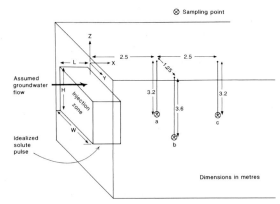

Fig. 1 Configuration of the injected pulse and the time-series sampling points in the natural gradient tracer experiment. The zone initially permeated by injected water is conceptualized as a rectangular prism with dimensions of length, L, 3.2 m, height, H, 1.6 m, and width, W, 6 m. After Roberts et al. (1986).

Table 1 Injected organic solutes used in the natural gradient tracer experiment and their associated sorption properties. After Mackay et al. (1986).

Solute	Injected concentration (mg L^{-1})	Injected mass (g)	Octanol-water partition coefficient K_{ow}
Chloride (tracer)	892	10,700	–
Bromoform (BROM)	0.032	0.38	200
Carbon tetrachloride (CTET)	0.031	0.37	500
Tetrachloroethene (PCE)	0.030	0.36	400
1,2-Dichlorobenzene (DCB)	0.332	4.0	2500
Hexachloroethane (HCE)	0.020	0.23	4000

injection are shown in Fig. 2. Initially, the plumes were nearly rectangular in plan view. The solute plumes moved at an angle to the field co-ordinate system and, with time, became progressively more ellipsoidal due to hydrodynamic dispersion. The chloride plume appeared to move at an approximately constant velocity, yet a distinct bimodality developed during the first 85 days. The centre of the chloride plume exhibited a constant advective velocity of 0.09 m day^{-1} while the organic solutes showed decreasing velocities with time. Significant spreading in the longitudinal direction, and its accompanying dilution, were observed for both the inorganic and organic solute plumes. Relatively little horizontal transverse spreading was evident. As can be seen in Fig. 2, the relative mobility of the CTET was significantly less than that of chloride, providing qualitative evidence of retardation due to sorption. The retardation of the other organic solutes was even greater, as observed for PCE in Fig. 2, generally in accord with their hydrophobicity.

Retardation factors were estimated by two methods. First, by comparing average travel times estimated from concentration breakthrough responses for the organic solutes with that of chloride, based on time-series sampling at the discrete points shown in Fig. 1, and second, by comparing the velocities of the organic solutes with that of the chloride tracer based on analyses taken from the three-dimensional sampling array at a particular time, based on snapshot or synoptic sampling.

A comparison of retardation estimates from temporal and spatial data is given in Table 2 and retardation factors estimated from the synoptic sampling data are shown in Fig. 3. Retardation factors for the organic solutes relative to chloride ranged from 1.5 to 9.0, being generally greater for the more strongly hydrophobic compounds (Table 1). Interestingly, the retardation factors increased over time for the organic solutes: for example CTET increased by 40%, PCE by 120% and 1,2-dichlorobenzene (DCB) by 130% in the period 16–650 days. One possibility to explain the increase in retardation behaviour is a gradual, temporal increase in the partition coefficient (K_d) as a result of slow approach to sorption equilibrium, for example owing to a diffusion rate limitation imposed by stratification or aggregation of the aquifer solids at the particle scale. However, it must be recognized that the temporal behaviour of the retardation factors may also be influenced by the spatial

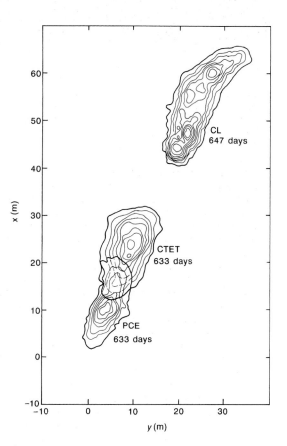

Fig. 2 Movement and dispersive spreading of the carbon tetrachloride (CTET, 633 days), tetrachloroethene (PCE, 633 days) and chloride (CL, 647 days) plumes during the natural gradient tracer experiment. The contour interval for chloride is 5 mg L^{-1} beginning with an outer contour of 10 mg L^{-1}. Contour intervals depicted for CTET and PCE are 0.1 µg L^{-1} beginning with an outer contour of 0.1 µg L^{-1}. After Roberts et al. (1986).

BOX
6.3

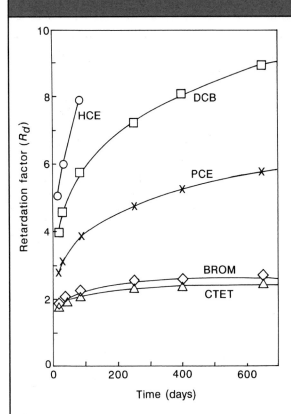

Fig. 3 Comparison of retardation factors estimated from synoptic sampling of organic solute movement during the natural gradient tracer experiment. After Roberts et al. (1986).

variability of the porous material at the field scale and, as shown in the second experiment described below, non-linear and competitive sorption effects (Rivett & Allen-King 2003).

The second experiment, the emplaced source natural gradient tracer experiment (Rivett et al. 2001; Rivett & Allen-King 2003), involved the controlled emplacement below the water table of a block-shaped source of sand with dimensions of length 0.5 m, height 1.0 m and width 1.5 m, and containing the chlorinated solvents perchloroethene (PCE), trichloroethene (TCE) and trichloromethane (TCM) together with gypsum. Gypsum was added to provide a continuous source of conservative inorganic tracer as dissolved sulphate in the aerobic groundwater of the Borden aquifer. Unlike the first experiment conducted about 150 m away, and which involved a finite pulse of dissolved organic solutes at low concentrations, the emplaced source experiment was intended to provide a simplified, yet realistic analogue of actual solvent contaminated sites. Such sites commonly contain residual zones of dense non-aqueous phase liquid (DNAPL) that continuously generate dissolved phase organic solute plumes over long time periods.

The gradual dissolution of the residual, multicomponent chlorinated solvent source under natural aquifer conditions caused organic solute plumes to develop continuously down-gradient. Source dissolution and three-dimensional plume development were again monitored via a dense array of 173 multilevel sampling wells over a 475-day tracer test period. As shown in Fig. 4, organic solute plumes with concentrations spanning 1–700,000 µg L^{-1} were identified. The calculated mean groundwater pore velocity until 322 days was 0.085 m day^{-1} inferring a travel distance due to advection alone of 27 m. The dissolved solvent plumes were observed to be narrow (less than 6 m width after 322 days) due to weak transverse dispersion processes and much more elongate (the TCM plume migrated 50 m in 322 days) due to advection and

Table 2 Comparison of retardation estimates for organic solutes from temporal and spatial data from the natural gradient tracer experiment. After Roberts et al. (1986).

Organic solute	Temporal data			Spatial data	
	Retardation factor*		Time range* (days)	Instantaneous retardation factor†	Ratio of travel distances‡
	Mean	Range			
CTET	1.73	1.6–1.8	48–119	2.0–2.1	1.8–1.9
BROM	1.70	1.5–1.8	46–122	2.1–2.3	1.9–2.0
PCE	3.30	2.7–3.9	83–217	3.8–4.7	3.0–3.7
DCB	2.73	1.8–3.7	55–245	5.2–7.2	4.0–5.6

* Retardation factor and average travel time from time-series sampling data.
† Retardation factor from Fig. 3 evaluated over the range of times given in column 4 of this table.
‡ Ratio of travel distances (chloride : organic) evaluated to conform to the time interval of column 4 in this table.

BOX
6 . 3

Continued

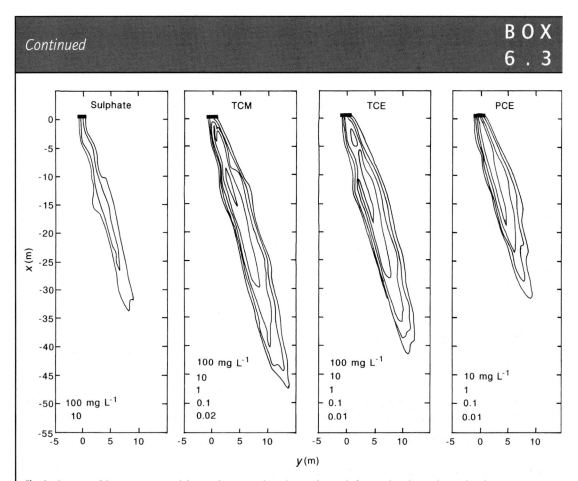

Fig. 4 Plan view of the conservative sulphate and organic solute plumes observed after 322 days during the emplaced source tracer experiment and based on maximum concentrations measured in each multilevel sampling well. Contour values are shown at the base of the plots, with the minimum value corresponding to the outer plume contour. After Rivett et al. (2001).

significant longitudinal dispersion. TCM was observed to be the most mobile dissolved organic solute, closely followed by TCE. PCE was comparatively more retarded. Background sulphate concentrations in the aquifer restricted sulphate plume detection to a range of two orders of magnitude, such that the lower detection limit for the organic solute plumes caused these plumes to be detected beyond the conservative sulphate tracer plume.

Hence, in the emplaced source tracer experiment, all the organic solute plumes appeared to have relatively high mobility, including the most retarded solute, PCE. This higher mobility was confirmed by the estimated retardation factors; TCM was essentially conservative ($R_d = \sim 1.0$) and TCE almost so ($R_d = \sim 1.1$), much lower than values estimated for the organic solutes injected in the Stanford-Waterloo tracer experiment. The sorptive retardation of PCE of about 1.4–1.8 (Rivett & Allen-King 2003) was observed to be greater than TCE, consistent with its greater hydrophobicity. Interestingly, PCE, the most retarded solute in the emplaced source

experiment, exhibited a retardation factor up to three times lower than observed in the previous Stanford-Waterloo natural gradient tracer experiment. Further laboratory and modelling studies by Rivett and Allen-King (2003) indicated that sorption of PCE was non-linear and competitive with reduced sorption behaviour observed in the presence of co-solvents such as TCE.

Significantly, plume attenuation due to abiotic chemical reactions or biodegradation processes was not apparent in the emplaced source tracer experiment. The absence of biodegradation was primarily attributed to a lack of additional, easily biodegradable carbon within the emplaced source plumes, and the prevailing aerobic conditions which did not favour the sulphate-reducing or methanogenic conditions required for dechlorination. Although attenuation due to biodegradation may be a significant factor at some sites, the emplaced source experiment demonstrated that dissolved DNAPL plumes can be extremely mobile, persistent and very capable of causing extensive aquifer contamination.

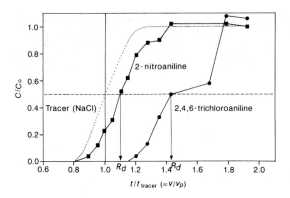

Fig. 6.8 Column breakthrough curves for 2-nitroaniline (C_0 = 6 µg L^{-1}) and 2,4,6-trichloroaniline (C_0 = 6 µg L^{-1}). t_{tracer} is the breakthrough time of the tracer and V/V_p is the ratio of feed volume, V, to pore volume, V_p. The retardation factor, R_d, can be read directly from the dimensionless breakthrough time (t/t_{tracer}) of contaminant at C/C_0 = 0.5. After Worch et al. (2002).

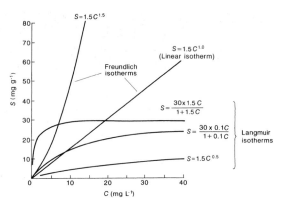

Fig. 6.9 Examples of Freundlich and Langmuir isotherms. S is the mass of chemical sorbed per unit mass of solid and C is the dissolved chemical concentration. After Domenico and Schwartz (1998).

atory batch tests conducted at a constant temperature to derive sorption 'isotherms'. As shown in Fig. 6.9, two common relationships describing the sorption of dissolved contaminants are the Freundlich isotherm described by the equation:

$$S = K_d C^n \qquad \text{eq. 6.16}$$

and the Langmuir isotherm described by:

$$S = \frac{Q°K_d C}{1 + K_d C} \qquad \text{eq. 6.17}$$

where K_d is the partition coefficient reflecting the degree of sorption, S is the mass of chemical sorbed per unit mass of solid, C is the dissolved chemical concentration, $Q°$ is the maximum sorptive capacity of the solid surface and n is a constant usually between 0.7 and 1.2. A Freundlich isotherm with $n = 1$ is a special case known as the linear isotherm (Fig. 6.9). The gradient of the straight line defining the linear isotherm provides a value of K_d that is the appropriate value for inclusion in the retardation equation (eq. 6.13).

In reality, K_d is not a constant but changes as a function of the mineralogy, grain size and surface area of the solid surface, the experimental conditions of the batch experiments, for example temperature, pressure, pH and Eh conditions, and undetected chemical

processes such as mineral precipitation. Not surprisingly, it is difficult to control all the relevant variables in order to give reproducible results and it is therefore unrealistic to represent all the processes affecting the sorptive behaviour of contaminants in porous material by a simple one-parameter model defined by K_d. Even so, and as explained under 'Hydrophobic sorption of non-polar organic compounds' in Section 6.3.3, the attenuation process can be successfully modelled using values of K_d to derive R_d.

The retardation equation can also be used to study cation exchange reactions (see Section 3.8), an important consideration in the attenuation of heavy metals (Box 6.4). By defining the partition coefficient as a function of the properties of the exchanger and the solution as found from laboratory experimentation, the partition coefficient can be written as:

$$K_d = \frac{K_s \text{CEC}}{\tau} \qquad \text{eq. 6.18}$$

where K_s is the selectivity coefficient, CEC is the cation exchange capacity (meq per mass) and τ is the total competing cation concentration in solution (meq per mass). Selectivity coefficients are found from mass-action equations. For example, the cation exchange reaction involving exchangeable ions A and B can be written as:

$$aA_{aq} + bB_{ad} = aA_{ad} + bB_{aq} \qquad \text{eq. 6.19}$$

BOX
6 . 4

Groundwater contamination by heavy metals in Nassau County, New York

The heavy metals of concern in drinking water supplies include Ni, Zn, Pb, Cu, Hg, Cd and Cr. In reducing and acidic waters, heavy metals remain mobile in groundwater; but in soils and aquifers that have a pH buffering capacity, and under oxidizing conditions, heavy metals are readily adsorbed or exchanged by clays, oxides and other minerals. Sources of heavy metals include, in general, the metal processing industries, particularly electroplating works with their concentrated acidic electrolytes, and other metal surface treatment processes.

Infiltration of metal plating wastes through disposal basins in Nassau County on Long Island, New York, since the early 1940s has formed a plume of contaminated groundwater (Fig. 1). The plume contains elevated Cr and Cd concentrations. The area is within an undulating glacial outwash plain, and there are two major hydrogeological units: the Upper glacial aquifer of Late Pleistocene age, and the Magothy aquifer of Late Cretaceous age, which supplies all local municipal water supplies.

The Upper glacial aquifer is between 24 and 43 m thick, with a water table from 0 to 8 m below ground level. The aquifer comprises medium to coarse sand and lenses of fine sand and gravel.

Stratification of the deposits means that the vertical permeability is up to five to ten times less than the horizontal permeability.

As part of the historical investigations in the South Farmingdale-Massapequa area, a number of test wells were installed in 1962 driven to depths ranging from 2 to 23 m below ground level. Water samples were collected at depth intervals of 1.5 m by hand pump during installation. The results of this investigation are shown in Fig. 1, and define a pollution plume that is about 1300 m long, up to 300 m wide, and as much as 21 m thick. The upper surface of the plume is generally less than 3 m below the water table. The plume is thickest along its longitudinal axis, the principal path of flow from the basins, and is thinnest along its east and west boundaries. The plume appears to be entirely within the Upper glacial aquifer.

Differences in chemical quality of water within the plume may reflect the varying types of contamination introduced in the past. In general, groundwater in the southern part of the plume reflects conditions prior to 1948 when extraction of Cr from the plating wastes, before disposal to the basins, commenced. Since the start of Cr treatment, the maximum observed concentrations in the plume have decreased from about 40 mg L^{-1} in 1949 to about 10 mg L^{-1}

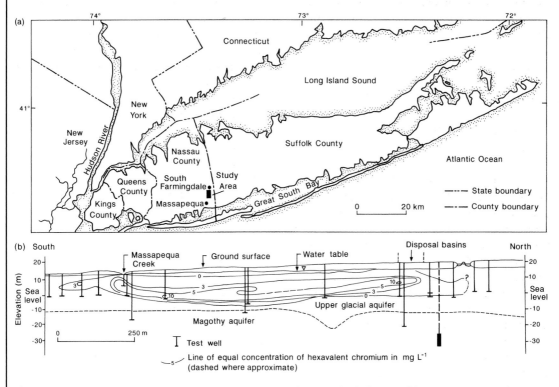

Fig. 1 Groundwater contamination by metal plating wastes, Long Island, New York. The location of the investigation area is shown in (a) and a cross-section of the distribution of the hexavalent chromium plume in the Upper glacial aquifer in 1962, South Farmingdale-Massapequa area, Nassau County, is illustrated in (b). After Ku (1980).

in 1962. The WHO guide value for Cr is 0.05 mg L^{-1}. Concentrations of Cd have apparently decreased in some places and increased in others, and peak concentrations do not coincide with those of Cr. These differences are probably due partly to changes in the chemical character of the treated effluent over the years and partly to the influence of hydrogeological factors such as aquifer permeability and sorption characteristics. A test site near the disposal basins recorded as much as 10 mg L^{-1} of Cd in 1964. The WHO guide value for Cd is 0.003 mg L^{-1}.

At the time of operation, the pattern of movement of the plating waste was vertically downwards from the disposal basins, through the unsaturated zone, and into the saturated zone of the Upper glacial aquifer. From here, most of the groundwater contamination

moved horizontally southwards, with an average velocity of about 0.5 m day^{-1}, and discharged to the Massapequa Creek.

Analysis of cores of aquifer material along the axis of the plume showed that the median concentrations of Cr and Cd per kg of aquifer material were, respectively, 7.5 and 1.1 mg, and the maximum concentrations were, respectively, 19 and 2.3 mg. Adsorption occurs on hydrous iron oxide coatings on the aquifer sands. The ability of the aquifer material to adsorb heavy metals complicates the prediction of the movement and concentration of the plume. Furthermore, metals may continue to leach from the aquifer material into the groundwater long after cessation of plating waste discharges, and so necessitating continued monitoring of the site.

where a and b are the number of moles and $_{ad}$ and $_{aq}$ indicate the adsorbed and aqueous phases, respectively. The equilibrium coefficient for this reaction is:

$$K_{A-B} = K_s = \frac{[A_{ad}]^a [B_{aq}]^b}{[A_{aq}]^a [B_{ad}]^b} \qquad \text{eq. 6.20}$$

Thus, for a problem involving binary exchange, the retardation equation is now:

$$\frac{\bar{v}_w}{\bar{v}_c} = 1 + \frac{\rho_b K_s \text{CEC}}{\theta \tau} \qquad \text{eq. 6.21}$$

For some exchange systems involving electrolytes and clays, K_s is found from experimentation to be constant over large ranges of concentrations of the adsorbed cations and ionic strength. For the Mg^{2+}-Ca^{2+} exchange pair, K$_{\text{sMg-Ca}}$ is typically in the range 0.6–0.9, meaning that Ca^{2+} is adsorbed preferentially to Mg^{2+} (Freeze & Cherry 1979).

Clay minerals, metal oxides and organic material exhibit preferential exchange sites for ion occupation and attempts have been made to establish a selectivity sequence, particularly with respect to heavy metals, where equivalent amounts of cations are arranged according to their relative affinity for an exchange site. In general, the greater the charge on a cation, the greater the affinity for an exchange site. Account must also be taken of the pH of contaminant

leachate and the competitive effect between the heavy metals present.

Yong and Phadungchewit (1993) demonstrated that a change in soil solution pH results in a corresponding change in the dominant retention mechanism of heavy metals in soils. At high pH values, precipitation mechanisms, for example precipitation of hydroxides and carbonates, dominate. As pH decreases, precipitation becomes less important and cation exchange becomes dominant. It was also shown that the selectivity order governing the retention of heavy metals in soils depends on the soil solution pH. At pH values above 4–5, when precipitation prevails, the selectivity order was found to be Pb > Cu > Zn > Cd, as demonstrated for illite, montmorillonite and natural clays soils. At lower soil solution pH, the selectivity order was Pb > Cd > Zn > Cu, as shown in the case of kaolinite and montmorillonite.

6.3.3 Transport of non-aqueous phase liquids

The transport of non-aqueous phase liquids (NAPLs) concerns the contamination of groundwater by organic compounds and includes dense non-aqueous phase liquids (DNAPLs) with a density greater than water, for example chlorinated hydrocarbon solvents, and light non-aqueous phase liquids (LNAPLs) with a density less than water, for example hydrocarbons (refined mineral oils). NAPLs have relatively

 (a) Water-wet sand   (b) Oil-wet sand

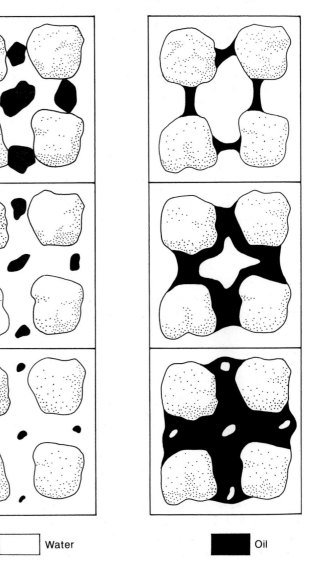

Fig. 6.10 Comparison of fluid wetting states for a porous sand containing water and oil. In (a) water is the wetting fluid and in (b) oil is the wetting fluid. After Fetter (1999).

Water   Oil

low solubility in water and partition preferentially towards organic material contained in soils, sediments and rocks and in doing so demonstrate hydrophobic sorption behaviour (see next section).

The presence of individual phases of water, NAPL and air (in the vadose zone) leads to the condition of multiphase flow where each phase is competing for the available pore space. In the presence of two fluids, wettability is defined as the tendency for a given fluid to be attracted to a surface (solid or liquid)

in preference to another fluid. Thus, if fluid A has a higher attraction to a given surface than fluid B, then fluid A is the 'wetting fluid' with respect to fluid B. For the purpose of considering groundwater pollution by organic compounds within porous material, then water can always be considered the wetting fluid with respect to NAPLs or air.

If a porous material is water-wet and a compound such as oil is introduced, the water will continue to occupy the capillary space in preference to the oil phase (Fig. 6.10). During simultaneous flow of two immiscible fluids, and as shown in Fig. 6.10, part of the available pore space will be filled with water and the remainder with oil such that the cross-sectional area of the pore space available for each fluid is less than the total pore space. This situation leads to the concept of relative permeability and is defined as the ratio of the permeability for the fluid at a given saturation to the total permeability of the porous material. A relative permeability exists for both the wetting and non-wetting phases (Fig. 6.11).

Chlorinated solvents such as trichloroethene (TCE), tetrachloroethene (PCE) and 1,1,1-trichloroethane (TCA) are DNAPLs that are volatile and of low viscosity, and consequently are more mobile than water in a porous material (Table 6.3). On infiltrating through the unsaturated zone, solvents leave behind a residual contamination which partitions into a vapour phase that subsequently migrates upwards and laterally by diffusion. The remaining contaminant mass migrates downwards under its own weight and through the water table until halted by the base of the aquifer, or by some other intermediate impermeable barrier (Fig. 6.12a). At the point of reaching an aquitard layer, the pore openings are so small that the weight of DNAPL cannot overcome the pore water pressure. A small, residual amount of solvent, or residual DNAPL saturation, is left in the pore spaces through which

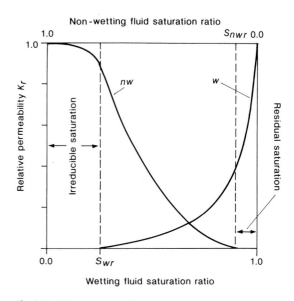

Fig. 6.11 Relative permeability curves for a two-phase system of wetting (w) and non-wetting (nw) liquids. After Fetter (1999).

the solvent body has passed (Fig. 6.13). In fractured material, and as long as the weight of DNAPL exceeds the displacement pressure of water contained in a fracture, the DNAPL can potentially migrate to significant depths (Fig. 6.14a).

Refined mineral oils such as petrol, aviation fuel, diesel and heating oils are LNAPLs that behave in a

Table 6.3 Physical and chemical properties of five common chlorinated solvents. Values from Vershueren (1983), Devitt et al. (1987) and Schwille (1988).

Chlorinated solvent	Chemical formula	Molecular weight	Density (g cm^{-3})	Kinematic viscosity (mm^2 s^{-1})	Solubility (mg L^{-1})	Vapour pressure (mm at 20°C) (kPa)	K_{oc}* (cm^3 g^{-1})	Henry's law constant (kPa m^3 mol^{-1} at 25°C)
Trichloroethene (TCE)	CCl$_2$=CHCl	131.5	1.46	0.4	1100 at 25°C	60 8.0	150	1.2
1,1,1-trichloroethane (TCA)	CCl$_3$CH$_3$	133.4	1.35	0.6	4400 at 20°C	100 13.3	113	2.8
Tetrachloroethene (perchloroethene) (PCE)	CCl$_2$CCl$_2$	165.8	1.63	0.5	150 at 25°C	14 1.0	364	2.3
Tetrachloromethane (carbon tetrachloride) (CTET)	CCl$_4$	153.8	1.59	0.6	800 at 20°C	90 12.1	110	2.4
Trichloromethane (chloroform) (TCM)	CHCl$_3$	119.4	1.49	0.4	8000 at 20°C	160 32.8	29	0.4

* Partition coefficient between organic carbon and water.

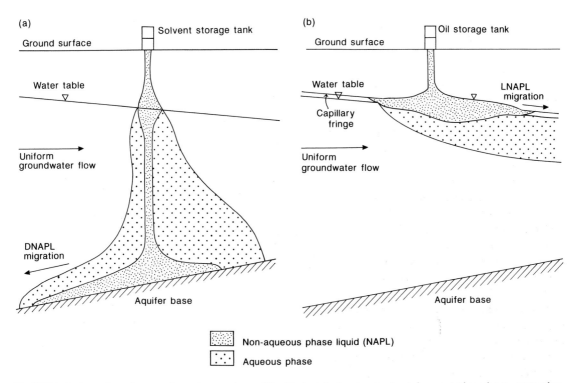

Fig. 6.12 Behaviour of organic contaminants in groundwater. The chlorinated solvent contaminant shown in (a) has a density greater than water (a dense non-aqueous phase liquid or DNAPL) and so sinks to the base of the aquifer. Here, transport of the DNAPL is controlled by the slope of the base of the aquifer, while the dissolved aqueous phase moves in the direction of groundwater flow. The hydrocarbon contaminant shown in (b) has a density less than water (a light non-aqueous phase liquid or LNAPL) and so floats on the water table. In this case, transport of the LNAPL is controlled by the slope of the water table, while the dissolved aqueous phase moves in the direction of groundwater flow.

similar manner to chlorinated solvents except, as shown in the Fig. 6.12b, by reason of their density, they float on the water table. The aromatic BTEX compounds, benzene, toluene, ethylbenzene and xylene, are released in significant amounts by petroleum and can be transported by groundwater in the aqueous phase. When spilled at the land surface, oil will migrate vertically in the vadose zone under the influence of gravity and capillary forces, in an analogous manner to water, until it reaches the top of the capillary fringe. Much of the LNAPL will be left trapped in the vadose zone, but on reaching the capillary fringe the LNAPL will accumulate and an 'oil table' will develop. As the weight of LNAPL increases, the capillary fringe will become thinner until mobile or 'free' product accumulates. Eventually, the capillary fringe may disappear completely and the

oil table will rest directly on the water table. In the case of a thick zone of mobile LNAPL, the water table may be depressed by the weight of the LNAPL (Fig. 6.12b). The mobile LNAPL can migrate in the vadose zone, following the slope of the water table, while the dissolved components can disperse with the advecting groundwater. The residual LNAPL phase in the vadose zone can partition into the vapour phase as well as the water phase, with the degree of partitioning dependent on the relative volatility of the hydrocarbon and its solubility in water. In fractured rocks, LNAPL will typically resist migration below the water table but where there is sufficient weight, LNAPL can penetrate below the water table to a limited extent when the pressure exerted by the LNAPL exceeds the displacement pressure of the water in the fractures (Fig. 6.14b).

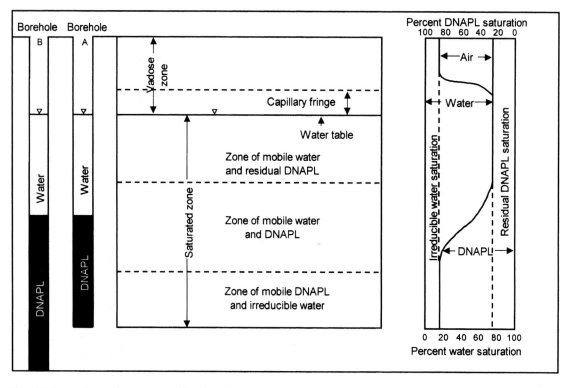

Fig. 6.13 Zones of DNAPL and water, and the relationship of mobile DNAPL and non-mobile (residual) DNAPL to the degree of DNAPL saturation. The relationship of mobile DNAPL thickness to the thickness of DNAPL measured in two monitoring boreholes (A and B) is also shown, with monitoring borehole B giving a false measure of the DNAPL thickness since it contains DNAPL filled below the level of the aquitard top. After Fetter (1999).

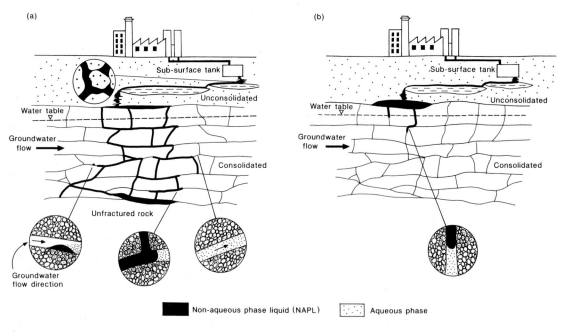

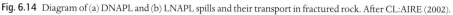

Fig. 6.14 Diagram of (a) DNAPL and (b) LNAPL spills and their transport in fractured rock. After CL:AIRE (2002).

Table 6.4 Empirical correlations between $\log_{10}K_{OC}$ and $\log_{10}K_{OW}$ for non-class-specific and class-specific organic compounds undergoing hydrophobic sorption.

Organic compound	Study
Non class-specific:	
(a) $\log_{10}K_{OC} = 0.544\log_{10}K_{OW} + 1.377$	Kenaga & Goring (1980)
(b) $\log_{10}K_{OC} = 0.679\log_{10}K_{OW} + 0.663$	Gerstl (1990)
(c) $\log_{10}K_{OC} = 0.909\log_{10}K_{OW} + 0.088$	Hassett et al. (1983)
(d) $\log_{10}K_{OC} = 0.903\log_{10}K_{OW} + 0.094$	Baker et al. (1997)
Class-specific:	
(e) Chloro and methyl benzenes	
$\log_{10}K_{OC} = 0.72\log_{10}K_{OW} + 0.49$	Schwarzenbach & Westall (1981)
(f) Benzene, PAHs	
$\log_{10}K_{OC} = 1.00\log_{10}K_{OW} - 0.21$	Karickhoff et al. (1979)
(g) Polychlorinated biphenyls	
$\log_{10}K_{OC} = 1.07\log_{10}K_{OW} - 0.98$	Girvin & Scott (1997)
(h) Aromatic amines	
$\log_{10}K_{OC} = 0.42\log_{10}K_{OW} + 1.49$	Worch et al. (2002)

Hydrophobic sorption of non-polar organic compounds

Non-polar organic molecules, for example low molecular weight volatile organic compounds (VOCs), polycyclic aromatic hydrocarbons (PAHs), polychlorinated benzenes and biphenyls (PCBs), and non-polar pesticides and herbicides, have a low solubility in water, itself a polar molecule (Section 3.2). These immiscible organic compounds tend to partition preferentially into non-polar environments, for example on to small quantities of solid organic carbon such as humic substances and kerogen present as discrete solids or as films on individual grains of soil, sediment and rock. The organic carbon content of sediments varies depending on lithology and can range from a few per cent in the case or organic-rich alluvial deposits to less than one-tenth of a per cent for clean sands and gravels. At low concentrations, the sorption of non-polar compounds on to organic material can very often be modelled with a linear isotherm (eq. 6.16 with $n = 1$).

In order to provide a rapid assessment of the sorption behaviour of solid organic carbon and to minimize experimental work, it is useful to find empirical correlations between K_{OC}, the organic carbon-water partition coefficient, and the properties of known substances. In studies of sorption processes, it is also useful to correlate K_{OC} with the octanol-water partition coefficient, K_{OW}, a measure of hydrophobicity. Such an approach is possible given that the partitioning of an organic compound between water and organic carbon is not dissimilar to that between water and octanol. An extensive compilation of K_{OC} and K_{OW} values, as well as other physicochemical properties of organic compounds, is provided by Mackay et al. (1997). A number of empirical correlations are given in the literature with a selection shown in Table 6.4. The partition coefficient, K_d, for application in the retardation equation (eq. 6.13) can now be normalized to the weight fraction organic carbon content of the sediment, f_{OC}, assuming that adsorption of hydrophobic substances occurs preferentially on to organic matter, as follows:

$$K_{OC} = \frac{K_d}{f_{OC}} \qquad \text{eq. 6.22}$$

As shown in Table 6.4, a number of studies have proposed non-class-specific correlations between K_{OC} and K_{OW} which should be applicable to many types of solid organic carbon. However, as noted by Worch et al. (2002), the parameters of the non-class-specific compounds differ significantly and it is not clear which correlation is most reliable. Hence, these correlations which were obtained for different sediments and thus different organic matter composition should only be used as first approximations to the sorption behaviour of specific compounds. For more exact K_{OC} estimations, experimental determinations of the $\log K_{OW}$–$\log K_{OC}$ correlation, or column experiments to determine the retardation coefficient and K_d, for the substance class of interest are required. Currently, however, class-specific correlations exist only for a limited number of substance classes.

As an example calculation of the application of the hydrophobic sorption model, consider a sand aquifer, with a solid density of 2.65 g cm^{-3}, an organic carbon content of 1.5×10^{-4} kg kg^{-1} ($f_{OC} = 0.00015$) and porosity of 0.35, that is contaminated with the aromatic amine 2,4,6-trichloroaniline. The $\log_{10}K_{OW}$ value for this compound found from laboratory experimentation is equal to 3.7 (Worch et al. 2002) and this enables an estimate of the retardation factor as follows.

From Table 6.4, the adsorption behaviour of aromatic amines is described by the equation:

$$\log_{10}K_{OC} = 0.42 \log_{10}K_{OW} + 1.49 \qquad \text{eq. 6.23}$$

By substituting $\log_{10}K_{OW} = 3.7$ in equation 6.23, the resulting value for $\log_{10}K_{OC}$ is:

$$\log_{10}K_{OC} = (0.42 \times 3.7) + 1.49 = 3.04$$

and:

$$K_{OC} = 1107$$

From equation 6.22, the value of the partition coefficient K_d is:

$$K_d = 1107 \times 0.00015 = 0.17 \text{ mL g}^{-1}$$

and using equation 6.13, the retardation factor, R_d is:

$$R_d = 1 + 0.17 \frac{(1 - 0.35)2.65}{0.35} = 1.84$$

With a retardation factor of less than 2, the retardation of 2,4,6-trichloroaniline by adsorption in the sand aquifer is low and so it is likely that this type of aromatic compound, which is poorly biodegradable, will persist as a groundwater contaminant.

The hydrophobic sorption model is recommended where the solubility of the organic compound is less than about 10^{-3} M. More soluble compounds, such as methanol, show much less affinity for organic carbon and preferentially partition to the aqueous phase. Also, if the fraction of organic carbon, f_{OC}, is small, it is possible that organic compounds will show a tendency to sorb to a small but significant extent on to inorganic surfaces, particularly where a clay fraction presents a large surface area to the contaminant. A further limitation is the possible effect of the presence of mixtures of organic compounds where co-solvents alter the solubility characteristics of individual compounds and compete in sorption-desorption reactions.

Overall in the above approach, it is assumed that the higher the hydrophobicity of a contaminant, the greater the adsorption behaviour. In advancing models to describe the sorption of hydrophobic organic chemicals to heterogeneous carbonaceous matter in soils, sediments and rocks, Allen-King et al. (2002) considered more specifically the relative importance of partitioning and adsorption of hydrophobic contaminants in the environment. Partitioning is likely to control contaminant solid-aqueous behaviour when the solid phase contains a substantial proportion of 'soft' humic substances, such as in modern soils and sediments high in organic carbon content, and/or concentrations of any adsorbing solutes are sufficiently high to effectively saturate all the adsorbent sites present. Under these circumstances, K_d values estimated from K_{OW}, K_{OC} and total f_{OC} values (eq. 6.22) are likely to produce a reasonable estimate of K_d, usually accurate to within a factor of 3. Such estimates can be successfully applied in circumstances such as the attenuation of hydrophobic pesticides in agricultural soils and for chlorinated solvents near a source containing non-aqueous phase liquid. Deeper in the subsurface environment, other types of organic matter with high adsorption capacities are present, for example, thermally altered 'hard' carbonaceous material (pieces of coal, soot, char or kerogen), and under these circumstances it is likely that adsorption will contribute significantly to the total sorption over a broad concentration range of contaminants. Therefore, hydrophobic sorption of organic compounds in soils, sediments and rocks occurs as a combination of phase partitioning and surface adsorption, with the former typically more linearly dependent on aqueous concentration (Allen-King et al. 2002).

In the process of remediating a contaminated aquifer, it should be noted that desorption involving the partitioning of an organic compound back into the aqueous phase yields a different isotherm to the sorption process. This so-called sorption 'hysteresis' is often attributed to slow processes of diffusion within sediment particles and soil aggregates. Usually, desorption exhibits a greater affinity for partitioning to the solid phase than is the case during sorption.

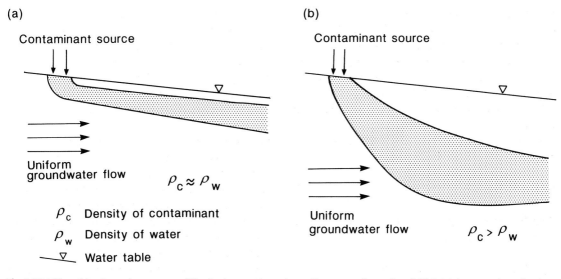

Fig. 6.15 Effect of density on the transport of dissolved contaminants in a uniform groundwater flow field. In (a) the contaminant density (ρ_c) is slightly greater than the density of the groundwater (ρ_w). In (b) the contaminant density is greater than the groundwater density.

6.3.4 Effects of density and heterogeneity

The processes of advection, dispersion and retardation all influence the pattern of contaminant distribution away from a pollution source. Other considerations include the effects of contaminant density and, moreover, the influence of aquifer heterogeneity, both of which make the overall monitoring and prediction of the extent of groundwater pollution very difficult. A density contrast between the contaminant and groundwater will affect the migration of a pollution plume, with a contaminant denser than water tending to sink steeply downwards into the groundwater flow field as shown in Fig. 6.15.

The description of contaminant transport given in Section 6.3.1 assumes a homogeneous porous material with steady, uniform groundwater flow. This is a simplification of real situations in nature where heterogeneities within the aquifer lithology create a pattern of solute movement considerably different to that predicted by the theory for homogeneous material. If a pollution source contains multiple solutes and occurs within a heterogeneous aquifer containing beds, lenses and fractures of differing hydraulic conductivity, then there will be a number of contaminant fronts and pathways such that the morphology of the resulting plume will be very complex indeed (Fig. 6.16).

In fractured material, aquifer properties are spatially variable and are often controlled by the orientation and frequency of fractures. As shown in Fig. 6.17, when contamination occurs in fractures, there is a gradient of contaminant concentration between the mobile groundwater in the fracture and the static water in the adjacent rock matrix. Under this condition, part of the contaminant mass will migrate by molecular diffusion from the fracture into the porewater contained in the rock matrix, so effectively removing it from the flowing groundwater. Such dual-porosity aquifers are notoriously difficult to remediate since the contaminant stored in the matrix can gradually diffuse back into the moving groundwater in the fracture, long after the source of contamination has been removed.

6.4 Sources of groundwater contamination

The sources of groundwater contamination are, as shown in Table 6.1, as varied as the range of polluting activities. The purpose of this section is to introduce the principal sources and classes of groundwater contaminants with respect to urban and industrial contaminants, municipal and septic wastes, agricultural contaminants, and saline intrusion in coastal aquifers.

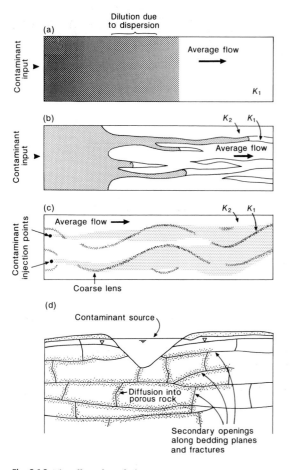

Fig. 6.16 The effect of aquifer heterogeneity on contaminant zones influenced by hydrodynamic dispersion. In (a) dilution occurs in the direction of advancing contaminant in a homogeneous intergranular material. In (b) the presence of greater hydraulic conductivity beds and lenses causes fingering of the contaminant transport ($K_1 > K_2$). In (c) contaminant spreading is created by the presence of irregular lenses of lesser hydraulic conductivity ($K_1 > K_2$). In (d) contaminant migration is dispersed throughout the network of secondary openings developed in a fractured limestone with molecular diffusion into the porous rock matrix. After Freeze and Cherry (1979).

6.4.1 Urban and industrial contaminants

Urban expansion and industrial activity, in some cases in the United Kingdom since the industrial revolution in the 1700s, is accompanied by continual disposal and spillage of potentially polluting wastes. The types of wastes are diverse, ranging from inorganic contaminants associated with mining and foundry wastes to organic compounds produced by the petrochemical and pharmaceutical industries. Accompanying urbanization is the need to dispose of domestic municipal and septic wastes leading to the risk of contamination from toxic materials and sewage. Other sources of pollution in the urban environment include salt and urea used in de-icing roads, paths and airport runways (Howard & Beck 1993), highway runoff potentially directed to soakaways (Price et al. 1989), the application of fertilizers and pesticides in parks and gardens, and the presence of chlorinated compounds such as trihalomethanes caused by leakage of chlorinated mains water that has reacted with organic carbon either in the distribution system or in the subsurface. Atmospheric emissions of sulphur dioxide and nitrogen oxides from urban areas contribute to wet and dry deposition of sulphur and nitrogen in adjacent regions that can impact soils, vegetation and freshwaters as a result of acid deposition and eutrophication (NEGTAP 2001).

The regulated control of waste disposal in urban areas is now practised in many developed countries but because of the slow transmission time of contaminants in the unsaturated zone, the legacy of historical, uncontrolled disposal of wastes may present a potential for groundwater pollution. A rise in groundwater levels, caused by a reduction in groundwater abstraction in postindustrial urban centres, may lead to remobilization of this pollution. In the Birmingham Triassic sandstone aquifer in the English Midlands, a region of metal manufacturing and processing and mechanical engineering, samples from shallow piezometers, tunnels and basements show that groundwater concentrations at shallow depths are often heterogeneous in distribution and much higher than in groundwater pumped from greater depth (Ford & Tellam 1994).

Generalizations as to the likely contaminants to be found in urban areas are not always possible except that the dominant inorganic contaminants are likely to be chloride and nitrate associated with a long history of supply and a wide range of multiple point sources. Other than these two, contamination is normally correlated to land use. In the Birmingham Triassic sandstone aquifer, the highest major ion concentrations (Fig. 6.18) and levels of boron and total heavy metal concentrations are associated with metal working sites (Ford & Tellam 1994). In other areas,

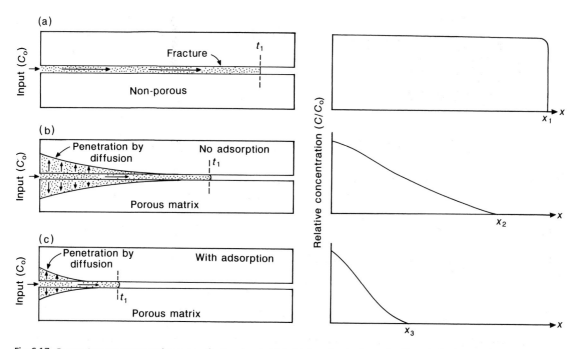

Fig. 6.17 Contaminant transport within porous fractured material. In (a) the solute is advected, without hydrodynamic dispersion, with the groundwater flowing through a fracture where the matrix porosity is insignificant. In (b) the solute transport is retarded by the instantaneous molecular diffusion of the solute into the uncontaminated porous matrix. Further attenuation occurs in (c) where adsorption of a reactive solute occurs, accentuated by the greater surface area of contact resulting from migration of the solute into the porous matrix. The position of the leading edge of the contaminant front within the fracture is shown for time t_1 in each case. After Freeze and Cherry (1979).

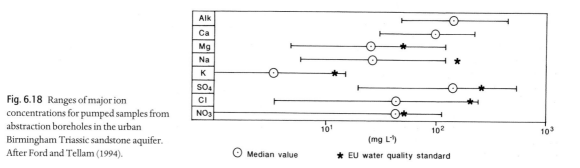

Fig. 6.18 Ranges of major ion concentrations for pumped samples from abstraction boreholes in the urban Birmingham Triassic sandstone aquifer. After Ford and Tellam (1994).

the discharge of acidic mine water from disused mine workings and spoil heaps is a potential cause of heavy metal pollution (Box 6.5).

Organic contamination of urban aquifers mainly concerns industrial solvents (DNAPLs) and petroleum hydrocarbons (LNAPLs) and encompasses the chemical classes and sources listed in Table 6.5. Contaminants escape to the subsurface environment

mainly as a result of careless handling, accidental spillages, misuse, poor disposal practices and inadequately designed, poorly maintained or badly operated equipment. These contaminants are of concern in groundwater because of their toxicity and persistence, particularly with respect to solvents.

Chlorinated solvents were developed in the early years of the last century as a safe, non-flammable

BOX 6.5

Mine water pollution

Mine water pollution is a widespread problem in present and former mining districts of the world with numerous cases of severe water pollution having been reported from base metal mines, gold mines and coal mines. In the United Kingdom, concern centres on aquatic pollution from the major coalfields and most of the base metal ore fields, such as those of Cornwall, upland Wales, northern England and Scotland. A general conceptual model for sources of mine water pollution, transport pathways for soluble contaminants and potentially sensitive receiving waters at risk of contamination is shown in Fig. 1. Mining activities contribute significantly to the solute load of receiving surface waters and aquifers. To illustrate, the contribution of sulphide mineral weathering associated with mine sites to the sulphate ion load is estimated at 12% of the global fluvial sulphate flux to the world's oceans (Nordstrom & Southam 1997). Associated with this weathering flux are dissolved metal ions (for example Fe, Zn and Al), sulphate and, in the case of pyrite, acidity. When discharging into the wider environment, acid mine drainage can coat stream beds with orange precipitates of iron hydroxides and oxyhydroxides ('ochre') as well as white aluminium hydroxide deposits (Gandy & Younger 2003).

The oxidative weathering and dissolution of contaminant source minerals such as pyrite and sphalerite associated with abandoned deep or opencast coal and metal mines and surface spoil heaps is described by the following two equations:

$$FeS_2(s) + 7/2O_2(aq) + H_2O \rightarrow Fe^{2+} + 2SO_4^{2-} + 2H^+$$
(pyrite weathering)   eq. 1

$$ZnS(s) + 2O_2(aq) \rightarrow Zn^{2+} + SO_4^{2-} \text{ (sphalerite weathering)}$$ eq. 2

The study of long-term changes in the quality of polluted mine water discharges from abandoned underground coal workings in the Midland Valley of Scotland shows that mine water pollution is most severe in the first few decades after a discharge begins (the 'first flush'), and that the largest systems settle down to a lower level of pollution, particularly in terms of iron concentration, within 40 years (Wood et al. 1999). As shown in Fig. 2, long-term iron concentrations of less than 30 mg L^{-1} are typical, with many less than 10 mg L^{-1}. In the Scottish coalfield, low pH values do not generally persist due to the rapid buffering of localized acidic waters by carbonates.

The Durham coalfield of north-east England (Fig. 3) was one of the first coalfields in the world to be commercially exploited and has left a legacy of acid mine discharge as the mines have closed. The worked Coal Measures comprise Carboniferous strata of fluvio-lacustrine and fluvio-deltaic facies. High-sulphur coals, which might be expected to be prolific generators of acid mine water, are associated with shale bands of marine origin (Younger 1995). In the more easterly districts of the coalfield, the Coal Measures are unconformably overlain by Permian strata including the Magnesian limestone, an important public supply aquifer. Mining ceased in the exposed coalfield to the west of the Permian scarp in the early 1970s and the deep mines beneath Permian cover on the coast closed in 1993. Following closure, pumping water from the coastal mines ceased and the water table has begun to rebound. In the far west of the coalfield, uncontrolled discharges of acid mine drainage occurs. The water quality of the five most significant discharges is given in Table 1.

Remediation technologies for the treatment of acidic mine waters can be divided into active techniques, such as alkali dosing,

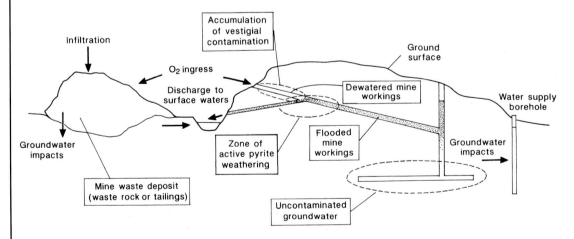

Fig. 1 Schematic diagram of contaminant sources, transport pathways and potential receiving waters in mining environments. A distinction is made between juvenile contamination arising from active weathering of sulphide minerals above the water table where oxygen ingress occurs, and vestigial contamination that accumulates as secondary mineral precipitates of metal ions and sulphate that arise from sulphide mineral weathering in dewatered void spaces within mine environments. After Banwart et al. (2002).

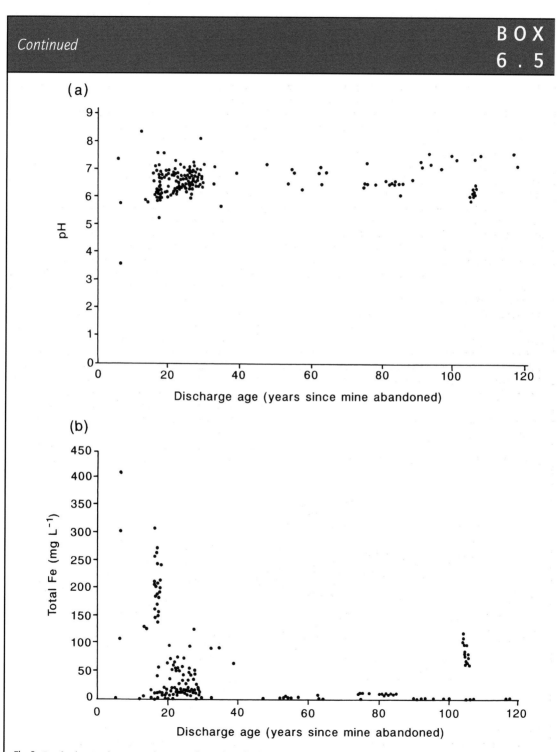

Fig. 2 Graphs showing long-term changes in the quality of polluted mine water discharges from abandoned underground coal workings in the Midland Valley of Scotland in terms of the variation of (a) pH and (b) total iron concentration. After Wood et al. (1999).

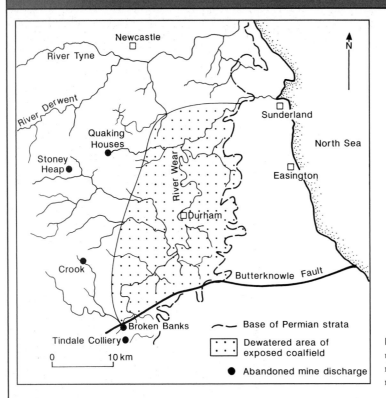

Fig. 3 Map showing the location of major uncontrolled discharges from abandoned mine workings in the Durham coalfield, north-east England. After Younger (1995).

Table 1 Hydrochemistry of the Durham coalfield mine drainage discharges at first emergence at the locations shown in Fig. 3. After Younger (1995).

Hydrochemistry	Site and grid reference				
	Broken Banks (NZ 197 295)	Crook (NZ 185 356)	Quaking Houses (NZ 178 509)	Stoney Heap (NZ 147 515)	Tindale Colliery (Brusselton) (NZ 197 269)
Flow rate (m^3 s^{-1}) on 15 Apr. 1994	0.14	0.002	0.007	0.0256	0.01
Calcium (mg L^{-1})	100.7	185	255	83.6	262
Magnesium (mg L^{-1})	61.23	93	103	49.7	107
Sodium (mg L^{-1})	26.9	21.5	463.6	27.9	80
Potassium (mg L^{-1})	10.7	6.8	57.0	6.7	13
Iron (total) (mg L^{-1})	1.8	79.8	18.0	26.3	1.8
Manganese (mg L^{-1})	1.0	6.9	4.8	1.2	1.7
Aluminium (mg L^{-1})	0.26	4.2	12.9	0.16	0.04
Zinc (mg L^{-1})	0.023	0.045	0.040	0.022	0.0184
Copper (mg L^{-1})	0.11	0.23	0.23	0.09	0.01
Alkalinity (mg L^{-1} as CaCO$_3$)	364.0	0.0	0.0	188	357
Sulphate (mg L^{-1})	137.0	810	1358	325	890
Chloride (mg L^{-1})	60	65	1012	102	75
pH	6.5	4.8	4.1	6.3	6.4
Temperature (°C)	10.9	11.8	11.2	10.3	12.0
Eh (mV)	39	264	327	36	−50
Conductivity (μS cm^{-1})	1177	1563	3560	1134	2360

aeration, flocculation and settlement, and passive techniques, such as constructed wetlands, inorganic media passive systems (Fig. 4) and subsurface flow, bacterial sulphate reduction systems. Unless the metal loadings are particularly high, in which case active treatment can be a cost-effective solution in the long term, passive treatment is increasingly the preferred option (Gandy & Younger 2003). In practice, mine water remediation should allow for active treatment of discharges for the first decade or two, followed by long-term passive treatment after asymptotic pollutant concentrations are attained (Wood et al. 1999).

Fig. 4 (*right*) Limestone filter installation sited upstream of Cwm Rheidol, west Wales, to precipitate metals contained in ochreous acid mine drainage resulting from the flushing of ferrous sulphate and sulphuric acid (the products of oxidation of pyrite and marcasite during dry-working of mines) and issuing from the abandoned mine adits and spoil heaps in the Ystumtuen area. The acidified water dissolves heavy metals such as Pb, Zn, Cd and Al which enter the river system, the effect of which has been detected 16 km downstream of the mined area with a pH as low as 2.6 (Fuge et al. 1991).

alternative to petroleum-based decreasing solvents in the metal processing industry. Until about 1970, trichloroethene (TCE) and tetrachloroethene (PCE) were predominantly used, the latter also in dry cleaning applications. TCE and PCE degrade extremely slowly, and some of the degradation products may be more toxic, soluble and mobile than the parent compounds. For example, tetrachloroethene can be progressively de-halogenated, first to trichloroethene, then to dichloroethene, and finally to carcinogenic vinyl chloride. From the mid-1970s, concern was expressed about the potentially carcinogenic effects of TCE, PCE and carbon tetrachloride (CTC) at trace level concentrations in drinking water, and the WHO set guide values of $70\,\mu g\,L^{-1}$ for TCE, $40\,\mu g\,L^{-1}$ for PCE and $2\,\mu g\,L^{-1}$ for CTC (Appendix 9). Since the

1960s, both TCE and PCE have begun to be replaced by the less toxic 1,1,1-trichloroethane (TCA) and 1,1,2-trichlorotrifluoroethane (Freon 113).

Hydrocarbons include petrol, aviation fuel, diesel and heating oils. As a group, their physical characteristics are variable, particularly that of viscosity; but all have a density less than water, and a heterogeneous composition dominated by pure hydrocarbons. In the context of groundwater, regulation is aimed primarily at taste and odour control (guideline values are given in Appendix 9). Sources of contamination include oil storage depots, cross-country oil pipelines, service stations, tanker transport and airfields.

In a further survey of the Birmingham Triassic sandstone aquifer in the English Midlands (Rivett et al. 1990), almost half of the 59 supply boreholes

Table 6.5 Sources of organic contaminants found in urban groundwaters. After Lloyd et al. (1991).

Chemical class	Sources	Examples
Aliphatic and aromatic hydrocarbons (including benzenes, phenols and petroleum hydrocarbons)	Petrochemical industry wastes Heavy/fine chemicals industry wastes Industrial solvent wastes Plastics, resins, synthetic fibres, rubbers and paints production Coke oven and coal gasification plant effluents Urban runoff Disposal of oil and lubricating wastes	Benzene Toluene Iso-octane Hexadecane Phenol
Polynuclear aromatic hydrocarbons	Urban runoff Petrochemical industry wastes Various high temperature pyrolytic processes Bitumen production Electrolytic aluminium smelting Coal-tar coated distribution pipes	Anthracene Pyrene
Halogenated aliphatic and aromatic hydrocarbons	Disinfection of water and waste water Heavy/fine chemicals industry waste Industrial solvent wastes and dry cleaning wastes Plastics, resins, synthetic fibres, rubbers and paints production Heat-transfer agents Aerosol propellants Fumigants	Trichloroethylene Trichloroethane Para-dichlorobenzene
Polychlorinated biphenyls	Capacitor and transformer manufacture Disposal of hydraulic fluids and lubricants Waste carbonless copy paper recycling Heat transfer fluids Investment casting industries PCB production	Pentachlorobiphenyls
Phthalate esters	Plastics, resins, synthetic fibres, rubbers and paints production Heavy/fine chemicals industry wastes Synthetic polymer distribution pipes	

sampled that contained chlorinated solvents were located on the sites of metal manufacturing and processing and mechanical engineering industries. The results are shown in Table 6.6 and Fig. 6.19 and indicate that chlorinated solvents are widespread, in particular TCE which is detected in 78% of boreholes. TCE is frequently observed at high levels with 40% of boreholes contaminated above 30 µg L^{-1} to a maximum of 5500 µg L^{-1} (Rivett et al. 1990). Occasional high values are also observed for TCA and PCE, the latter associated with dry cleaning laundry sites. Contamination of the Birmingham aquifer by organic chemicals other than chlorinated solvents is low in the supply boreholes and, where present, is often associated with degraded lubricating oils.

Several factors were shown by Rivett et al. (1990) to explain the distribution of organic contaminants including the historic use of solvents, point source inputs close to sampling boreholes, thickness of the unsaturated zone, the presence or absence of confining deposits above the aquifer and the depth of groundwater sampling. Higher concentrations of solvents were found at deeper sampling levels, by virtue of their DNAPL behaviour, while the opposite was the case for hydrocarbons (LNAPLs).

Urban districts can extend over wide areas of aquifer outcrop below which is a large volume of potentially available water; a contrasting situation with adjacent rural areas where groundwater sources are increasingly fully developed (Lerner 2002). However, research

Table 6.6 Summary of chlorinated solvents detected in groundwater in the Birmingham Triassic sandstone aquifer. Solvent abbreviations are given in Table 6.3. After Rivett et al. (1990).

Industrial solvents	TCE	TCA	PCE	TCM	CTET
% boreholes with solvent detected	78	46	44	53	37
Proportion of boreholes exceeding (%)					
1 µg L^{-1}	62	22	9	17	2
10 µg L^{-1}	43	13	4	0	0
100 µg L^{-1}	30	5	2	0	0
Maximum concentration detected (µg L^{-1})	5500	780	460	5	1

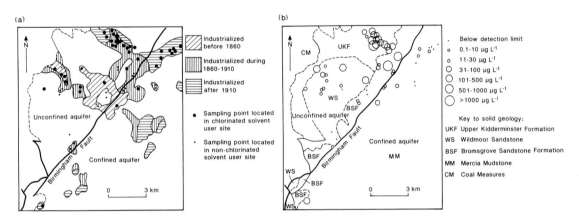

Fig. 6.19 Organic solvent contamination of the urban Birmingham Triassic sandstone aquifer showing: (a) distribution of groundwater sampling points and industrialized areas; (b) distribution of maximum trichloroethene (TCE) concentrations. After Rivett et al. (1990).

in the Nottingham Triassic sandstone aquifer in the English Midlands has shown that although the inorganic quality of the groundwater may be acceptable for water supply, high nitrate concentration and microbial contamination relating to untreated sewage inputs from broken or leaking sewers are causes for concern (Barrett et al. 1999; Cronin et al. 2003; Powell et al. 2003; Fukada et al. 2004). Also, research in Berlin has shown that polar pharmaceutical compounds such as clofibric acid, diclofenac, ibuprofen, propyphenazone, primidone and carbamazepine are detectable at individual trace (µg L^{-1}) levels in surface waters and groundwaters with a source in municipal sewage treatment plants (Heberer 2002). Waste discharge to surface water courses infiltrating to groundwater was highlighted as a transport pathway for these soluble polar organic compounds, enabling migration over large distances in impacted aquifers.

6.4.2 Municipal landfill, domestic and cemetery wastes

The landfilling of municipal wastes is common practice, with older sites up until the late 1980s formerly operated under a 'dilute and disperse' principle where leachate generated in the waste was allowed to migrate away from the site and disperse in groundwater below the water table. This practice is no longer allowed in Europe, with engineered solutions required to contain leachate in the landfill site. A typical solution is to line the site with an artificial liner with a low design permeability of no higher than 10^{-9} m s^{-1} and to install a leachate collection system for subsequent treatment and discharge of the leachate (Department of the Environment 1995). After completion of filling, the landfill site may be restored to agricultural land by sealing the site with a low permeability clay cap that prevents water infiltration into the buried waste.

Table 6.7 Composition of landfill leachate (after compilation of Christensen et al. 2001). Data for October 1985 and September 1986 describe the leachate composition pumped from containment cell 4 at Compton Bassett landfill site, west of England (after Robinson 1989). All results in mg L^{-1}, except pH and electrical conductivity (μS cm^{-1}).

Parameter	Range	October 1985 (acetogenic or acid phase)	September 1986 (methanogenic phase)
pH	4.5–9	6.5	7.4
Electrical conductivity	2500–35,000	15,000	10,400
Total solids	2000–60,000	550	230
Organic matter			
Total organic carbon (TOC)	30–29,000		
Biological oxygen demand (BOD)	20–57,000	15,750	580
Chemical oxygen demand (COD)	140–152,000	20,700	2000
BOD/COD ratio	0.02–0.80	0.76	0.29
Organic nitrogen	14–2500	–	–
Inorganic macrocomponents			
Total phosphorus	0.1–23	–	–
Chloride	150–4500	1710	2020
Sulphate	8–7750	–	–
Bicarbonate	610–7320	–	–
Sodium	70–7700	–	–
Potassium	50–3700	–	–
Ammonium-N	50–2200	930	840
Calcium	10–7200	1410	143
Magnesium	30–15,000	–	–
Iron	3–5500	787	24
Manganese	0.03–1400	–	–
Silica	4–70	–	–
Inorganic trace elements			
Arsenic	0.01–1	–	–
Cadmium	0.0001–0.4	–	–
Chromium	0.02–1.5	–	–
Cobolt	0.005–1.5	–	–
Copper	0.005–10	–	–
Lead	0.001–5	–	–
Mercury	0.00005–0.16	–	–
Nickel	0.015–13	–	–
Zinc	0.03–1000	8.4	2.4

In the absence of engineered controls, the extent to which leachate from a landfill site contaminates groundwater is dependent on the hydraulic, geochemical and microbiological properties of the hydrogeological system (Nicholson et al. 1983). For a comprehensive review of the biogeochemistry of landfill leachate plumes, the reader is referred to Christensen et al. (2001).

Considering the example of a landfill receiving a mixture of municipal, commercial and mixed industrial wastes, landfill leachate may be characterized as a water-based solution containing four groups of

pollutants: dissolved organic matter; inorganic macrocomponents; heavy metals; and xenobiotic organic compounds (Christensen et al. 2001). Ranges of values for the first three of these groups are given in Table 6.7. Xenobiotic organic compounds (aromatic hydrocarbons, halogenated hydrocarbons, phenols and pesticides) originating in household or industrial chemicals are usually present in relatively low concentrations in the leachate (<1 mg L^{-1} for individual compounds).

Landfill leachate contamination of groundwater has a source in the internal, biogeochemical decomposition processes that take place during the breakdown of putrescible materials contained in domestic wastes. Three phases of decomposition are recognized. In the first phase, aerobic decomposition rapidly uses the available oxygen in the wastes. The reaction is common with septic systems (eq. 7 in Table 6.8) with the process usually lasting for up to 1 month, in which time significant quantities of CO_2 and some H_2 are produced. In the second phase, anaerobic and facultative organisms (acetogenic bacteria) hydrolyse and ferment cellulose and other putrescible materials, producing simpler, soluble compounds such as volatile fatty acids. This phase (represented by eq. 3 in Table 6.8) can last for several years producing an acidic leachate (pH of 5 or 6) high in biological oxygen demand (BOD) of greater than $10,000\ mg\ L^{-1}$, and NH_4^+ in the range $500-1000\ mg\ L^{-1}$. The aggressive leachate assists in the dissolution of other waste components, such that the leachate can contain high levels of Fe, Mn, Zn, Ca and Mg. Gas production consists mainly of CO_2 with lesser quantities of CH_4 and H_2. The third phase experiences slower-growing methanogenic bacteria that gradually consume simple organic compounds, producing a mixture of CO_2 and CH_4 gases which is released as landfill gas that can be recovered and used as an energy source. This phase of methanogenesis (eqs 6a and 6b in Table 6.8) will continue for many years, if not decades, until the landfill wastes are largely decomposed and atmospheric oxygen can once more diffuse into the landfill. Leachates produced during this last phase are characterized by relatively low BOD values. However, NH_4^+ continues to be released and will be present at high values in the leachate. Inorganic substances such as Fe, Na, K, sulphate and chloride may continue to dissolve and leach from the landfill for many years.

These various phases of landfill decomposition are represented in the analyses of leachate given in Table 6.7. The landfill cell at Compton Bassett was filled over a 2- to 3-year period from late 1983 during which time waste was filled to a depth of 15–20 m, compacted, capped with clay and restored to grassland in 1987. Table 6.7 demonstrates rapid changes in leachate composition from a strongly acetogenic to methanogenic state during a 12-month period from October 1985. The change is accompanied by a rise in pH from acid to slightly alkaline conditions, but more obviously by steep declines in BOD and chemical oxygen demand (COD) concentrations as

organic compounds such as volatile fatty acids are being degraded at rates faster than they are produced. The less aggressive nature of this leachate is associated with substantial reductions in concentrations of Fe and Ca being solubilized from the wastes. Concentrations of conservative determinands such as chloride remain stable or even increase, demonstrating that dilution is of little importance in the results obtained. High concentrations of NH_4^+ are also maintained, demonstrating the continuing biological activity and rapid acetogenic processes occurring to decompose the solid wastes (Robinson 1989).

In the event of leachate contamination of the unsaturated zone below a landfill site, attenuation of the leachate by physical, chemical and biological processes is possible. However, the extent of penetration of organic and other components of landfill leachate in an aquifer will depend on its buffering capacity leading to the development of favourable conditions for microbial degradation. This buffering effect is shown in Figs 6.20 and 6.21 for two contrasting landfill sites located above sandstone and Chalk aquifers, respectively. In the Nottinghamshire Triassic sandstone aquifer at the Burntstump municipal landfill site, sequential profiling of the unsaturated zone shows a downward migration of a conservative chloride front over a 9-year period (Fig. 6.20). A similar movement is observed for the organic compounds as indicated by the TOC (total organic carbon) values, mainly attributable to total volatile acids. A zone of reduced pH values migrated downwards consistently with the TOC, with the lower edge of the low pH zone corresponding closely with the leading edge of the TOC front. The profiles for TOC and pH are interpreted to result from the combination of the low buffering capacity of the sandstone due to its limited carbonate content, the high concentrations of organic acids, and the dissolution of CO_2 generated within the wastes, all of which depress the pH of the interstitial water at the leachate front. Hence, at this site, there is significant penetration of the organic and other components of leachate due to the low buffering capacity and the persistence of conditions unfavourable to microbial degradation at low pH (Williams et al. 1991).

By contrast, the landfill site at Ingham, Suffolk, in eastern England and situated on the Cretaceous Chalk aquifer gave the sequential profiles shown in Fig. 6.21. Reinstatement of the site to agricultural soil

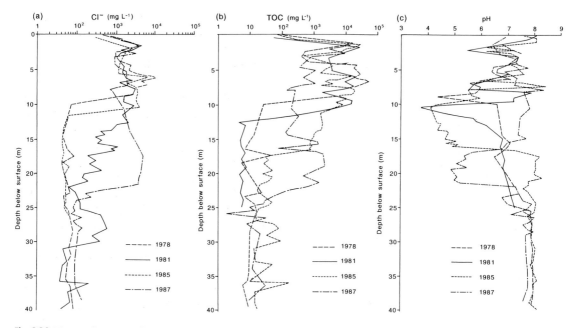

Fig. 6.20 Temporal variation of concentration profiles of (a) chloride, (b) total organic carbon (TOC) and (c) pH in the unsaturated zone of the Triassic sandstone aquifer at the Burntstump landfill site, Nottinghamshire, 1978–1987. After Williams et al. (1991).

was completed in 1977 after which the chloride front effectively stagnated as a result of the effectiveness of the completed cap in limiting infiltration to very low values. During an 8-year period of stagnation until 1986, a persistent decrease in the TOC : Cl ratio occurred in the unsaturated zone indicating the continuous removal of organic carbon. Early surveys of organic compounds at the site showed the presence of phenols (absent post-1984), readily degradable volatile fatty acids (absent post-1977), mineral oils and halogenated solvents; the latter remaining at levels of up to 50 $\mu g\,L^{-1}$ beneath the landfill. The results showed that the high buffering capacity of the Chalk is conducive to microbial metabolism to explain the decrease in TOC : Cl ratios with time and the disappearance of readily degradable organic compounds. However, the organic solvents remained as persistent contaminants (Williams et al. 1991).

Below the water table, the general shape of a landfill leachate plume is determined by the advective-dispersive nature of groundwater flow in the aquifer, the amount of recharge from the leachate mound developed below the landfill and the increased density of the leachate (Fig. 6.15). In most cases, the leachate plumes are relatively small: a few hundred

metres wide, corresponding to the width of the landfill; and restricted to less than 1000 m in length as a result of attenuation processes within the leachate plume, although potentially longer if contamination occurs in fissured or fractured material. According to Christensen et al. (2001), the infiltrating leachate creates a sequence of redox zones in the groundwater, with methanogenic conditions close to the landfill and oxidized conditions at the outer boundary of the plume. The anaerobic zones are driven by microbial utilization of dissolved organic matter in the leachate in combination with reduction of oxidized species in the aquifer, particularly iron oxides, which provides substantial redox buffering by reducing iron oxides and precipitating reduced iron species. Other important attenuating mechanisms include dilution, ion exchange, complexation and precipitation, such that heavy metals are not normally considered a major groundwater pollution problem in leachate plumes. The attenuation mechanism for NH_4^+ is not well understood but probably involves anaerobic oxidation of this persistent pollutant in landfill leachate. Xenobiotic organic compounds in leachate are not extensively attenuated by sorption processes, but there is increasing evidence that many organic com-

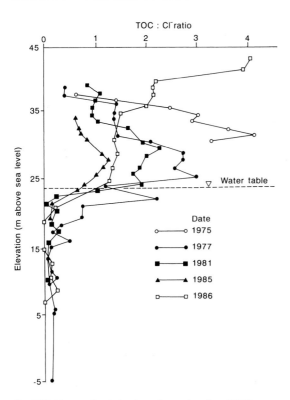

Fig. 6.21 Temporal variation in total organic carbon (TOC) : chloride ratios in the unsaturated zone of the Chalk aquifer at the Ingham landfill site, Suffolk, 1975–1986. After Williams et al. (1991).

pounds are degradable in the strongly anaerobic iron-reducing zone of leachate plumes (Christensen et al. 2001).

To illustrate the sequence of redox reactions in a landfill leachate plume, Fig. 6.22 shows the Villa Farm landfill site near Coventry in the English Midlands. The site became disused in 1980 having received a wide variety of industrial wastes over 30 years, including oil/water mixtures and effluent treatment sludges containing heavy metals, acids, alkalis, organic solvents and paint wastes. These liquid wastes were disposed of directly into lagoons in hydraulic continuity with a shallow lacustrine sand aquifer. The inorganic reactions and organic compounds observed in the groundwater are identical to those reported for domestic waste and co-disposal sites. The results of extensive monitoring at the site are shown schematically in Fig. 6.22, with chloride acting as a conservative tracer to delimit the extent of the groundwater leachate plume (Fig. 6.22a). A geochemical zonation, based on redox reactions, is

observed in the transition from oxidizing conditions in the background, uncontaminated groundwater to the heavily polluted and highly reducing zone near the lagoons at the base of the aquifer (Fig. 6.22b). These zones follow the theoretical sequence of redox reactions predicted from thermodynamic considerations in a closed, organically polluted system (Table 3.11). Heavy metals are attenuated near the lagoon as carbonates and as sulphides in the zone where sulphate is reduced to sulphide.

Measurements of TOC suggested little gross change in the organic carbon content of the pollution plume at Villa Farm, although biotransformations did occur. Aromatic hydrocarbons were broken down with increasing distance from the lagoons as suggested by the presence of benzoic acid derivatives which were not present in the lagoons, but synthesized from the primary disposal of phenol. The highly reducing conditions throughout the plume below the zone of sulphate reduction led to the production of CH_4. However, at the same time, CH_4 appeared to be consumed during oxidation to CO_2 in anaerobic conditions in the overlying zone of sulphate reduction. Where sulphate reduction was limited, such as at the leading edge of the plume, CH_4 was probably able to diffuse upwards without oxidation, and so explaining the CH_4 found in the soil atmosphere at concentrations of up to 55% by volume (Williams et al. 1991).

On-site septic systems for the disposal of domestic waste are common in rural areas without a connected sewerage system. In the United States, it is estimated that on-site systems dispose of approximately one-third of the population's domestic waste water. Since the domestic waste water in septic systems contains many environmental contaminants, septic systems in North America constitute approximately 20 million potential point sources for groundwater contamination (Wilhelm et al. 1994b). A conceptual model of the biogeochemical evolution of domestic waste water in conventional on-site septic systems is given in Fig. 6.23. As described by Wilhelm et al. (1994b), the evolution of waste water is driven by microbially catalysed redox reactions involving organic carbon and nitrogen and occurs in as many as three different redox zones (Table 6.8).

Anaerobic digestion of organic matter and production of CO_2, CH_4 and NH_4^+ predominate in the first zone, which consists mainly of the septic tank. In the second zone, gaseous diffusion through the unsatur-

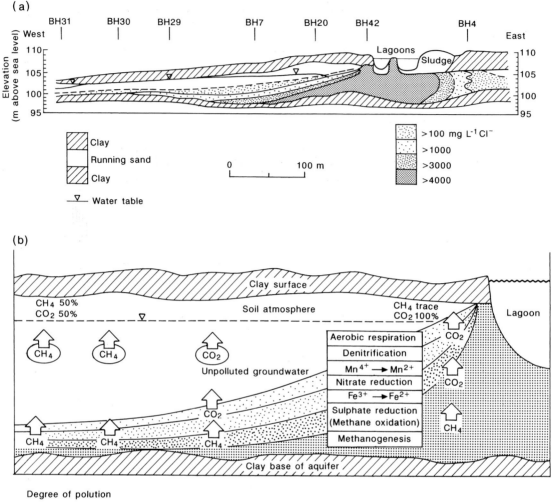

Fig. 6.22 Landfill leachate contamination of groundwater in a lacustrine sand aquifer at Villa Farm, West Midlands, showing (a) the distribution of chloride concentration along a vertical cross-section of the pollution plume as determined by borehole (BH) sampling and (b) a schematic diagram showing the transition from oxidizing conditions in the background, uncontaminated groundwater to the heavily polluted and highly reducing zone near the lagoons at the base of the aquifer. After Williams et al. (1991).

ated sediments of the drain field supplies oxygen for aerobic oxidation of organic carbon and NH_4^+ and a consequent decrease in waste water alkalinity. The nitrate formed by NH_4^+ oxidation in this zone is the primary and generally unavoidable adverse impact of septic systems at most sites (Fig. 6.24). In the third zone, nitrate is reduced to N_2 by the anaerobic process of denitrification, although rarely found below septic systems due to a lack of labile organic carbon in natural settings. Without natural attenuation by denitrification, it is quite likely that in unconfined sand aquifers common in North America, the typical minimum permissible distance between a well and septic tank (25–35 m) will not be sufficient to provide protection against nitrate contamination by dispersive dilution alone (Robertson et al. 1991).

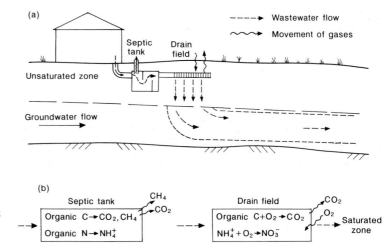

Fig. 6.23 Diagrams showing (a) a schematic cross-section of a conventional septic system, including septic tank, distribution pipe and groundwater plume and (b) the sequence of simplified redox reactions in the two principal zones of a conventional septic system, the septic tank and drain field. After Wilhelm et al. (1994b).

Table 6.8 Hydrochemical and biogeochemical reactions in septic systems. After Wilhelm et al. (1994b).

Reaction	Equation
Anaerobic zone (septic tank and biological mat)	
Organic molecule hydrolysis:	(1)
Proteins + $H_2O \rightarrow$ amino acids	
Carbohydrates + $H_2O \rightarrow$ simple sugars	
Fats + $H_2O \rightarrow$ fatty acids and glycerol	
Ammonium release:	
Urea $[CO(NH_3^+)_2] + H_2O \rightarrow 2NH_4^+ + CO_2$	(2a)
Amino acids + $H_2O \rightarrow NH_4^+$ + organic compounds	(2b)
Fermentation:	
Amino acids, simple sugars $\rightarrow H_2$, acetate (CH_3OO^-), other organic acids	(3)
Anaerobic oxidation:	
Fatty acids + $H_2O \rightarrow H_2$, CH_3OO^-	(4)
Sulphate reduction:	
$SO_4^{2-} + 2CH_2O^* + 2H^+ \rightarrow H_2S + 2CO_2 + 2H_2O$	(5)
Methanogenesis:	
CH_3OO^- (acetate) $+ H^+ \rightarrow CH_4 + CO_2$	(6a)
$CO_2 + 4H_2 \rightarrow CH_4 + 2H_2O$	(6b)
Aerobic zone (unsaturated zone and saturated zone to lesser extent)	
Organic matter oxidation:	
$CH_2O + O_2 \rightarrow CO_2 + H_2O$	(7)
Nitrification:	
$NH_4^+ + 2O_2 \rightarrow NO_3^- + 2H^+ + H_2O$	(8)
Sulphide oxidation:	
H_2S (or organic sulphide) $+ 2O_2 \rightarrow SO_4^{2-} + 2H^+$	(9)
Carbonate buffering:	
$H^+ + HCO_3^- \rightarrow H_2CO_3$	(10a)
$CaCO_3 + H^+ \rightarrow Ca^{2+} + HCO_3^-$	(10b)
$CaCO_3 + CO_2 + H_2O \rightarrow Ca^{2+} + 2HCO_3^-$	(10c)
Second anaerobic zone (saturated or near-saturated conditions)	
Denitrification:†	
$4NO_3^- + 5CH_2O + 4H^+ \rightarrow 2N_2 + 5CO_2 + 7H_2O$	(11)

* Organic matter is simplified as CH_2O throughout. Actual organic matter contains C of various oxidation states and other elements such as N, P and S, and therefore actual reaction products vary.
† Nitrate reduction can also be accomplished via oxidation of reduced sulphur compounds.

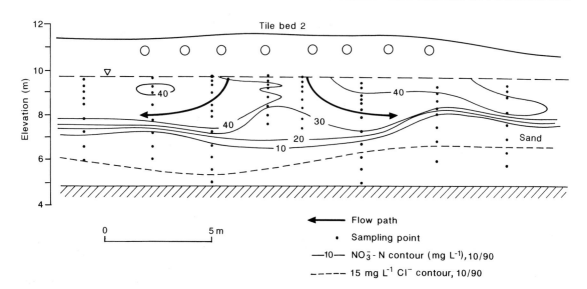

Fig. 6.24 Distribution of nitrate concentration in an unconfined medium sand aquifer below a large septic system located on the north shore of Lake Erie, Ontario. The water-table depth below the infiltration pipes of Tile bed 2 is about 1.2 m and is sufficient to allow for almost complete oxidation of the sewage constituents during migration through the sandy vadose zone. After Aravena and Robertson (1998).

Other waste water constituents are influenced by the major changes in redox and pH conditions that occur in the reaction zones of septic systems. Calcite ($CaCO_3$) is often dissolved in drain fields in order to buffer the acidity released during NH_4^+ oxidation and this results in increased Ca^{2+} concentrations in the effluent. Other cations may also be released from the solid phase during buffering reactions such as mineral dissolution or cation exchange. Wilhelm et al. (1994b) also identified trace metal cations such as Cu, Cr, Pb and Zn in concentrations in the range of 2–300 μg L^{-1} in many domestic waste waters as a result of the changes in redox and pH. Although their specific behaviour in septic systems is less well understood, a large fraction of trace metals is likely to be retained in particulate matter in the septic tank where they form insoluble sulphides.

Domestic waste water contains pathogenic bacteria and viruses and overflow and seepage of waste water is a major cause of disease outbreaks associated with groundwater (Craun 1985; Pedley & Howard 1997). Bacteria are retained in septic systems primarily by straining in the biological mat; the layer of accumulated organic matter found directly beneath the distribution pipes. In general, the mobility of bacteria and viruses is much greater in saturated than unsaturated flow, making unsaturated conditions below septic systems desirable for both oxygen supply and pathogen retention.

In terms of hydrophobic organic contaminants such as halogenated aliphatics and aromatics, partitioning on to the accumulated organic matter in the septic tank and the drain field will act to retain these contaminants.

Until recently overlooked, a potential source of contamination can arise from cemetery operations. Detailed studies within cemeteries in Australia (Knight & Dent 1998; Dent 2002) have principally identified forms of nitrogen but also sodium, magnesium, strontium, chloride, sulphate and forms of phosphorus as characterizing cemetery groundwaters. Cemetery functions are best understood conceptually as a special kind of landfill operation that is strongly influenced by temporal and spatial variability of cemetery practices. Dent (2002) found that the amounts of decomposition products leaving cemeteries are very small and that well-sited and managed cemeteries have a low environmental impact and are a sustainable activity. The most serious pollution situation is for the escape of pathogenic bacteria or viruses into the environment. The potential for such contamination can only be assessed by a comprehensive hydrogeological investigation.

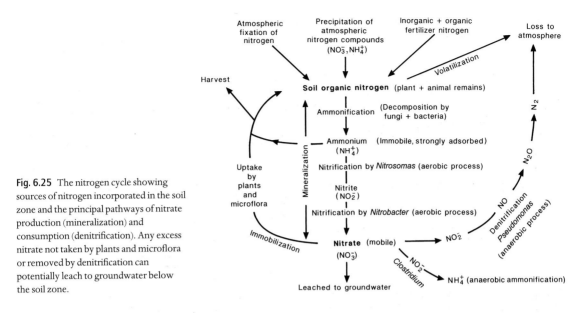

Fig. 6.25 The nitrogen cycle showing sources of nitrogen incorporated in the soil zone and the principal pathways of nitrate production (mineralization) and consumption (denitrification). Any excess nitrate not taken by plants and microflora or removed by denitrification can potentially leach to groundwater below the soil zone.

6.4.3 Agricultural contaminants

Agricultural contaminants include nitrate and pesticides used in intensive farming practices that often affect wide areas of aquifer outcrop. As such, nitrate and pesticides are diffuse pollutants in the environment and can lead to serious consequences for the quality of groundwater resources and surface waters receiving contaminated groundwater. Other sources of contamination arise from livestock and poultry farming through the intensive management of grazing pasture and the operation of concentrated animal feeding operations (COFAs) (Mallin & Cahoon 2003).

Nitrate contamination of groundwater has resulted from the desire for greater self-sufficiency in food supply that has led to the ploughing up of grassland and the application of nitrogen-based fertilizers and organic manure. Ploughing up grassland stimulates the natural process of mineralization and the release of nitrate from the organic-bound nitrogen in the soil zone (Fig. 6.25). The use of fertilizers during the period of crop growth often coincides with the onset of the rainfall season and the use of irrigation water. If fertilizer and manure applications are not applied following good agricultural practice (see Section 7.3.4), soil nitrate leaching losses can occur causing high nitrate concentrations in the unsaturated (Fig. 4.10b) and saturated zones (Box 6.6) of aquifers.

The health effects of high nitrate concentrations in excess of European and WHO water quality standards (50 mg L^{-1} as nitrate) are concerned with methaemoglobinaemia ('blue baby' syndrome) in infants (Walton 1951; Craun et al. 1981) and gastric cancer (National Academy of Sciences 1981; Nomura 1996). Environmental impacts associated with excessive nitrate in the aquatic environment are eutrophication of inland and coastal waters and the consequent loss of biodiversity (Hecky & Kilham 1988; European Environment Agency 2003). Approaches to controlling diffuse contamination of groundwater by nitrate include 'end of pipe' technological solutions such as blending a contaminated source with a low-nitrate water, biological reactor beds and anion exchange resins (Hiscock et al. 1991). To prevent further contamination and achieve environmental standards, as for example set by the EU Nitrates Directive, approaches include reductions in fertilizer and manure applications following good agricultural practice and, more radically, changing land use from arable cropping to low intensity grassland or forestry (Section 7.3.4).

Pesticides refer to the group of synthetic organic chemicals used mainly as fungicides, herbicides and insecticides. Herbicides are used in the largest quantities and generally have much greater water solubility compared with insecticides such that critical concentrations, in excess of water quality guidelines and standards, may be exceeded. The European Union has adopted a maximum admissible concentration of 0.1 μg L^{-1} for any individual pesticide and 0.5 μg L^{-1} for the sum of all individual pesticides,

Nitrate contamination of the Jersey bedrock aquifer

BOX
6.6

The island of Jersey is the largest of the British Channel Islands. The island setting comprises a plateau, formed largely of Precambrian crystalline rocks, with a steep topographic rise along the coastline. The temperate maritime climate encourages early flowers and vegetables, with intensive agricultural production sustained by large fertilizer applications. During winter and early spring, applications of nitrogen fertilizer to early-cropping potatoes and horticultural crops may exacerbate the problem of nutrient leaching, with estimates of leaching losses of up to 100 kg N ha^{-1} expected from Jersey potato crops.

Mains water supply is principally from surface water storage, but there are large areas, particularly in the rural north of the island, that are reliant on well and borehole supplies, typically yielding less than 0.5 L s^{-1}, to meet domestic, agricultural and light industrial demands. The main aquifer and isolated perched aquifers occur within a shallow zone of weathering in the bedrock, up to 25 m in depth below the water table surface, with groundwater flow almost entirely dependent on secondary permeability, imparted by dilated fractures.

The chemical composition of groundwater is controlled by maritime recharge inputs and water–rock interaction, although the effects of anthropogenic pollution, particularly from nitrate, are in places severe. A 1995 survey of groundwater quality at 46 locations across the island produced the regional distribution of nitrate shown in Fig. 1. Elevated nitrate concentrations occurred across much of the island and ranged from undetected to 215 mg L^{-1}, with a mean value of 71 mg L^{-1} (Fig. 2). Of the total, 67% of samples exceeded the European Union water quality standard for nitrate in drinking water of 50 mg L^{-1}. The high nitrate concentrations appear to decrease in a coastal direction, especially in those samples from the central southern area and in the valley areas around St Saviour, probably as a result of localized denitrification (Green et al. 1998). Otherwise, local variations in well and borehole depths and local land-use and agricultural practices, combined with the physical heterogeneity of the aquifer, produced no obvious pattern in the distribution of nitrate. It is concluded that the source of dissolved nitrate in the Jersey bedrock aquifer is primarily a result of the intensive agricultural and horticultural practices and high livestock densities on the island. In some areas, domestic pollution from septic tank discharges is a further potential hazard.

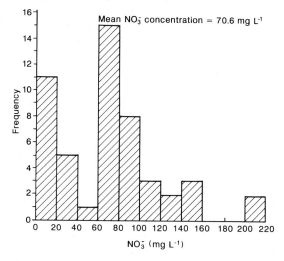

Fig. 2 Histogram of groundwater nitrate concentrations in the Jersey bedrock aquifer sampled in June 1995. After Green et al. (1998).

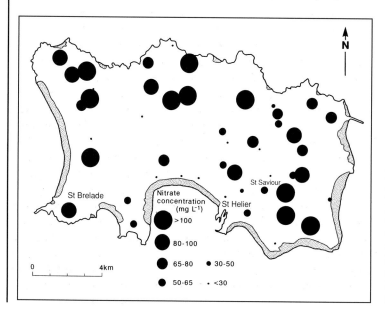

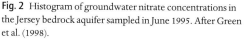

Fig. 1 Regional distribution of groundwater nitrate concentrations in the Jersey bedrock aquifer sampled in June 1995. The absence of a regional pattern is due to differences in land use and inorganic nitrogen fertilizer and cattle manure inputs, variable recharge rates across the island and the physical heterogeneity of the weathered bedrock aquifer that leads to unpredictable borehole yields and groundwater flowpaths. After Green et al. (1998).

and in North America guidelines are applied to individual pesticides (Appendix 9). The factors affecting the leaching of pesticides from soils include the timing of application, the quantity reaching the target area and the physical and chemical properties of the soil. Newer formulations of pesticides are tailored to have short half-lives of less than 1 month in the soil through retention and elimination of compounds by hydrophobic sorption (see 'Hydrophobic sorption of non-polar organic compounds' in Section 6.3.3) and degradation by chemical hydrolysis and bacterial oxidation. Caution is required, however, in that quoted half-lives appropriate to a fertile clay-loam soil may not be representative of permeable sandy soils developed on aquifer outcrops. Below the soil zone, pesticide mobility will again be affected by the availability of sorption sites for attenuation and the viability of micro-organisms for bacterial degradation.

Sorption is promoted by organic carbon, iron oxides and clay minerals and is a significant mechanism in the attenuation of pesticides with depth such that the amount of pesticide leached to groundwater is generally less than the amount lost to surface runoff (Rodvang & Simpkins 2001). Total herbicide losses in subsurface drainage on fine-textured soils are usually less than 0.3%, but occasionally 1.5% of the amount applied.

Contamination of groundwater by pesticides is common in agricultural and urban areas. In a survey of groundwater in 20 of the major hydrological basins in the United States in which 90 pesticide compounds (pesticides and degradates) were analysed, one or more pesticide compounds were detected at 48% of the 2485 sites sampled. The pesticide concentrations encountered were generally low, with the median total concentration being 0.05 $\mu g\,L^{-1}$. Pesticides were commonly detected in shallow groundwater beneath both agricultural (60%) and urban (49%) areas and so highlighting urban areas as a potential source of pesticides (Kolpin et al. 2000).

In Iowa, which has some of the most intensive applications of herbicides in the United States, herbicide compounds were detected in 70% of 106 municipal wells sampled; with degradation products comprising three of the four most frequently detected compounds (Kolpin et al. 1997). The highest herbicide concentrations in groundwater were found in areas of greatest intensity of herbicide use (Table 6.9).

Factors explaining the distribution of herbicides included an inverse relation to well depth and a positive correlation with dissolved oxygen concentration that appear to relate to groundwater age, with younger groundwater likely to contain herbicide compounds. The occurrence of herbicide compounds was substantially different among the major aquifer types across Iowa, being detected in 83% of the alluvial, 82% of the bedrock/karst region, 40% of the glacial till and 25% of the bedrock/non-karst region aquifers. Again, the observed distribution was partially attributed to variations in groundwater age among these aquifer types. A significant, inverse relationship was identified between total herbicide compound concentrations in groundwater and the average soil slope within a 2-km radius of the sampled wells. Steeper soil slopes may increase the likelihood of surface runoff occurring rather than transport to groundwater by infiltration (Kolpin et al. 1997).

In the United Kingdom, isoproturon is the most extensively used pesticide with over 3×10^6 ha treated in 1996 (Thomas et al. 1997). Concentrations of isoproturon greater than the European Union limit have been found in groundwater abstracted from the major Chalk aquifer (Table 6.10). Although concentrations are generally low there is concern that significant quantities of isoproturon may be moving through the unsaturated zone only to contaminate groundwater in the future. Clark and Gomme (1992) recovered unsaturated Chalk cores for pore water analysis and showed that the uron herbicides (isoproturon, chlortoluron and linuron) left the base of the profile at very low concentrations (Table 6.11) and had not penetrated beyond 2 m into the unsaturated zone. If correct, these results would suggest that pollution of Chalk groundwater by the uron herbicides through intergranular flow in the Chalk matrix is unlikely except in areas where the water table is close to the surface. Where uron herbicides are detected in Chalk groundwater, it is possible that pesticide transport has occurred by flow through the fissure system. Support for these results is provided by Besien et al. (2000), who measured recovery rates of isoproturon of 48–61% in laboratory column experiments using Chalk cores eluted with non-sterile groundwater containing an initial mass of 1.5 mg of isoproturon. The column results also illustrated the importance of microbial degradation in removing isoproturon during the 162-day experiment.

Table 6.9 Pesticides and their degradation products in samples collected during the summer of 1995 from 106 municipal wells in Iowa. After Kolpin et al. (1997).

Compound	Per cent detection	Reporting limit ($\mu g\ L^{-1}$)	Maximum concentration ($\mu g\ L^{-1}$)	Maximum contaminant level* ($\mu g\ L^{-1}$)	Health advisory level* ($\mu g\ L^{-1}$)	Use or origin
Alachlor-ESA†	65.1	0.10	14.8	–	–	Alachlor degradation product
Atrazine	40.6	0.05	2.13	3.0	3.0	Herbicide
Deethylatrazine	34.9	0.05	0.59	–	–	Triazine degradation product (atrazine, propazine)
Cyanazine amide	19.8	0.05	0.58	–	–	Cyanazine degradation product
Metolachlor	17.0	0.05	11.3	–	70	Herbicide
Prometon	15.1	0.05	1.0	–	100	Herbicide
Deisopropylatrazine	15.1	0.05	0.44	–	–	Triazine degradation product (atrazine, cyanazine, simazine)
Alachlor	7.5	0.05	0.63	2.0	–	Herbicide
Cyanazine	5.7	0.05	0.30	–	1.0	Herbicide
Acetochlor	0.9	0.05	0.77	–	–	Herbicide
Metribuzin	0.9	0.05	0.27	–	100	Herbicide
Ametryn	0.0	0.05	–	–	60	Herbicide
Prometryn	0.0	0.05	–	–	–	Herbicide
Propachlor	0.0	0.05	–	–	90	Herbicide
Propazine	0.0	0.05	–	–	10	Herbicide
Simazine	0.0	0.05	–	4	4	Herbicide
Terbutryn	0.0	0.05	–	–	–	Herbicide

* US Environmental Protection Agency.
† Alachlor ethanesulfonic acid.

Table 6.10 Summarized analytical results of pesticides in Chalk boreholes in the Granta catchment, Cambridgeshire. After Gomme et al. (1992).

Site	Number of samples	Pesticides detected	Number of detections	Concentration range ($\mu g\ L^{-1}$)
Babraham	3	Atrazine	3	<l.q.–0.07
		Simazine	2	<l.q.
Sawston	3	Atrazine	3	<l.q.–0.13
		Simazine	3	<l.q.–0.07
Linton	3	Atrazine	1	<l.q.
Fleam Dyke	3	n.d.	–	–
45/12	1	Atrazine	1	0.31
		Simazine	1	0.40
45/17	1	n.d.	–	–
54/28	1	n.d.	–	–
54/99	1	n.d.	–	–
54/101	1	Simazine	1	<l.q.
54/112	1	Atrazine	1	<l.q.
54/116	3	Atrazine	2	0.05–0.06
		Chlortoluron	2	0.17–0.35
		Isoproturon	2	0.49–0.61
		Simazine	3	0.09–0.12
54/119	1	Chlortoluron	1	<l.q.
		Isoproturon	1	<l.q.
		Simazine	1	0.05
55/84	1	Simazine	1	<l.q.
64/40	1	n.d.	–	–

n.d., none detected; <l.q., below limit of quantification.

Table 6.11 Analytical results of Chalk core profiling, Granta catchment, Cambridgeshire. After Clark and Gomme (1992).

Depth of sample (m below ground)	Isoproturon concentration (μg kg^{-1})	Chlortoluron concentration (μg kg^{-1})	Linuron concentration (μg kg^{-1})
0.0–0.5	0.11	0.53	–
0.5–1.0	–	–	–
1.0–1.5	<0.03	0.21	–
1.5–2.0	–	3.04	–
2.5–3.0	–	–	n.r.
4.5–5.0	–	–	n.r.
6.5–7.0	–	–	n.r.
9.5–10.0	–	–	n.r.

n.r., not recorded; – below detection limit (0.03 μg L^{-1}).

Livestock farming produces waste containing many pathogenic micro-organisms associated with serious gastrointestinal disease, including bacteria such as *Escherichia coli* and *Streptococcus*, viruses such as enterovirus, and protozoa such as *Cryptosporidium* and *Giardia*. The presence of faecal coliform bacteria indicates that other disease-causing organisms may be present. In Ontario, 17% and 20% of farm wells in coarse- and fine-textured sediments were contaminated with faecal coliform and *E. coli*, respectively (Goss et al. 1998). Coliform bacteria were also present in 17% of domestic water wells in loess and till deposits in eastern Nebraska (Gosselin et al. 1997).

Not all strains of *E. coli* are harmful but some strains, such as O157:H7, are serious pathogens. A stark example illustrating that pathogen occurrence is not only restricted to developing countries, is the case of Walkerton, Ontario, when in May 2000, *E. coli* O157:H7 and *Campylobacter jejuni* contaminated the drinking water supply leading to the death of seven individuals and illness in over 2000 others. *Escherichia coli* bacteria were found to have entered the Walkerton drinking water supply through a well which had been contaminated by cattle manure spread on a nearby farm. Normally, water can be treated using chlorine which acts to kill *E. coli* bacteria, but in the case of the Walkerton outbreak chlorine levels had not been sufficiently maintained. Also, exceptional environmental factors contributed to the outbreak with heavy rainfall in early May that assisted transport of the contaminants to Well 5, located in a shallow fractured aquifer vulnerable to surface-derived contamination. A full judicial inquiry was set up by the Ontario Provincial Government into the circumstances surrounding the outbreak and it also acted to introduce a new drinking water regulation to protect water supplies (Holme 2003).

Cryptosporidiosis is a significant cause of gastroenteritis in the United Kingdom with an estimated 42,000 cases in England and Wales in 1995 (Adak et al. 2002). Cryptosporidiosis is caused by the protozoan pathogen *Cryptosporidium parvum* which is widespread in the environment and is found in the intestinal regions of most humans and animals. It is excreted from infected individuals as an oocyst which can survive for long periods in the environment and is resistant to disinfection by conventional water treatment. Outbreaks of cryptosporidiosis have occurred due to oocyst-contaminated groundwater supplied by wells, mainly in hydrogeological settings characterized by fractured material. These outbreaks have occurred in karst limestone aquifers, for example the Edwards Aquifer in Texas (Bergmire-Sweat et al. 1999) and the Chalk aquifer in the north London Basin (Willocks et al. 1999). *Cryptosporidium* contamination hazard assessment and risk management for British groundwater sources are discussed by Morris and Foster (2000).

6.4.4 Saline water intrusion in coastal aquifers

Intrusion of saltwater into an aquifer occurs where seawater displaces or mixes with fresh groundwater. The intrusion of saltwater is one of the most common pollutants of fresh groundwater (Todd 1980; Custodio 1987) and often results from human activities which reduce groundwater flow towards the sea. In an aquifer where freshwater is flowing towards the sea, the Ghyben–Herzberg relation predicts, for freshwater and seawater densities (ρ_f and ρ_s) of 1000 and 1025 kg m^{-3}, respectively, that the depth below sea level to the saline water interface, z_s, is approximately 40 times the height of the freshwater table above sea level, z_f. This can be shown with reference to Fig. 6.26a and assuming simple hydrostatic conditions in a homogeneous, unconfined coastal aquifer in which:

$$\rho_s g z_s = \rho_f g (z_f + z_s)$$

eq. 6.24

or

$$z_s = \frac{\rho_f}{\rho_s - \rho_f} z_f \qquad \text{eq. 6.25}$$

which for $\rho_f = 1000$ kg m^{-3} and $\rho_s = 1025$ kg m^{-3} gives the Ghyben–Herzberg relation:

$$z_s = 40 z_f \qquad \text{eq. 6.26}$$

The Ghyben–Herzberg relation can also be applied to confined aquifers by substituting the water table by the potentiometric surface.

It can be seen from equation 6.26 that small variations in the freshwater head will have a large effect on the position of the saltwater interface. If the water table in an unconfined aquifer is lowered by 1 m, the saltwater interface will rise 40 m. The freshwater–saltwater equilibrium established requires that the water table (or potentiometric surface) lies above sea level and that it slopes downwards towards the sea. Without these conditions, for example when groundwater abstraction reduces the freshwater table in coastal boreholes below sea level, seawater will advance directly inland causing saline intrusion to occur.

It can be shown that where the groundwater flow is nearly horizontal, the Ghyben–Herzberg relation gives satisfactory results, except near the coastline where vertical flow components are more pronounced leading to errors in the position of the predicted saltwater interface. In most real situations, the Ghyben–Herzberg relation underestimates the depth to the saltwater interface. Where freshwater flow to the sea occurs, a more realistic picture is shown in Fig. 6.26b for steady-state outflow to the sea. The exact position of the interface can be determined for any given water table configuration by graphical flow net construction (Box 2.3), noting the relationships shown in Fig. 6.26b for the intersection of equipotential lines on the freshwater table and at the interface (Freeze & Cherry 1979).

The saltwater interface shown in Figs 6.26a and b is assumed to be a sharp boundary, but in reality a brackish transition zone of finite thickness separates the freshwater and saltwater. This zone develops from dispersion caused by the flow of freshwater and unsteady movement of the interface by external influences such as tides, groundwater recharge and pumping wells. In general, the thickest transition zones are found in highly permeable coastal aquifers subject

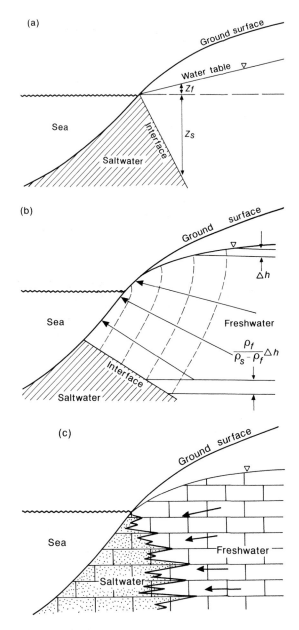

Fig. 6.26 Development of a saline interface in an unconfined coastal aquifer under (a) a hydrostatic condition and (b) a condition of steady-state seaward freshwater flow. In (c) the absence of a simple saline interface is caused by complex flow conditions in a fissured aquifer.

to large abstractions. An important consequence of the development of a transition zone and its seaward flow is the cyclic transport of saline water back to the sea (Fig. 6.27). This saline water component originates from the underlying saline water and so, from

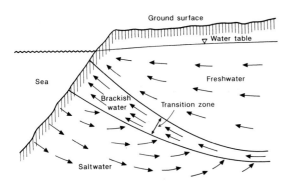

Fig. 6.27 Vertical cross-section showing flow patterns of freshwater and saltwater in an unconfined coastal aquifer illustrating the development of a brackish transition zone and the cyclic flow of saline water to the sea. After Todd (1980).

continuity considerations, there exists a small landward flow in the saltwater wedge (Todd 1980).

Saline intrusion problems are known from around the world with well-documented examples including the Biscayne aquifer in Florida (Klein & Hull 1978), the Quaternary sand aquifers of Belgium and The Netherlands (De Breuck 1991), the Chalk aquifer of South Humberside (Howard & Lloyd 1983) and the Llobregat delta confined aquifer of Spain (Iribar et al. 1997) (Box 6.7).

Methods for controlling saline intrusion are described by Todd (1980) and include: (i) changing the locations of pumping wells, typically by moving them inland; (ii) artificial recharge from a supplemental water source to raise groundwater levels; (iii) an

---

Saltwater intrusion in the Llobregat delta aquifer system, Spain

BOX 6.7

The Lower Llobregat aquifer system is formed by the Lower Valley and deep delta aquifers located a few kilometres south-west of Barcelona (Fig. 1). The Lower Valley aquifer is formed from Quaternary sands and coarse gravels and extends over an area of 100 km². The aquifer formation continues below the present morphological delta towards the coast (Fig. 2). At the sides of the delta, the aquifer materials change to sediments from local creek alluvial fans and beach deposits. The delta aquifer is formed by these deposits and the deep formation shown in Fig. 2. This formation extends seawards with decreasing thickness and outcrops on the sea floor at around 100 m depth and 4–5 km offshore. The deep delta aquifer is confined by wedge-shaped clay, silt and fine sand sediments that act as an aquitard. Above this aquitard, the shallow delta aquifer is formed by sands, gravels and silt.

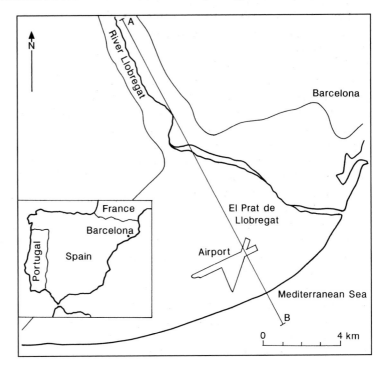

Fig. 1 Location map of the lower valley and delta of the River Llobregat, Spain. After Iribar et al. (1997).

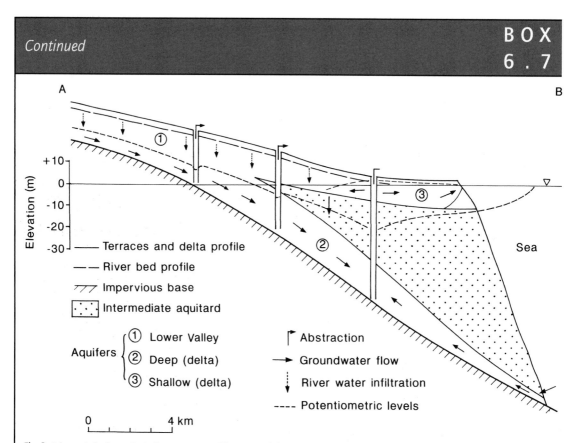

Fig. 2 Schematic hydrogeological cross-section of the Lower Llobregat aquifer system showing the flow pattern in the Lower Valley and delta aquifers. The line of section A–B is shown in Fig. 1. After Iribar et al. (1997).

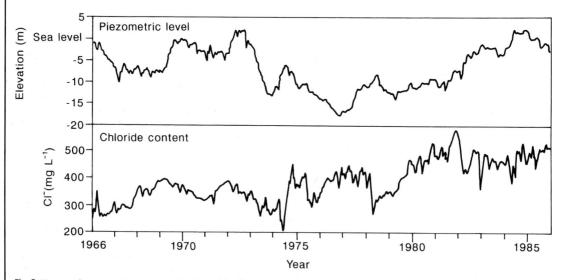

Fig. 3 Temporal variation in piezometric level and chloride content at the position where the Lower Valley aquifer meets the deep delta aquifer. After Iribar et al. (1997).

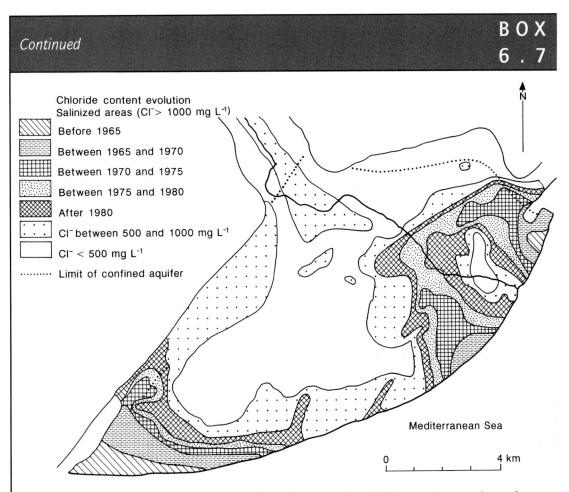

Chloride content evolution
Salinized areas (Cl⁻> 1000 mg L⁻¹)

Before 1965

Between 1965 and 1970

Between 1970 and 1975

Between 1975 and 1980

After 1980

Cl⁻between 500 and 1000 mg L⁻¹

Cl⁻ < 500 mg L⁻¹

········· Limit of confined aquifer

Mediterranean Sea

0   4 km

Fig. 4 Evolution of saline intrusion in the Lower Llobregat deep delta aquifer as indicated by the progressive encroachment of water with a chloride content above 1000 mg L⁻¹. After Iribar et al. (1997).

Since 1950, the Lower Valley and the delta aquifers have been intensively exploited for groundwater as an important supply source and an emergency reserve. However, over-exploitation has caused a depression of the potentiometric surface in the central area (Fig. 2) and the salinization of 30% of the confined aquifer below the delta. Wells near the greatest part of the potentiometric depression draw water both supplied as recharge from the Lower Valley and groundwater from the seaward margin. Monitoring of water levels and chloride concentrations is well documented at the position where the Lower Valley aquifer meets the deep delta aquifer (Fig. 3).

The displacement of the 1000 mg L⁻¹ isochlor (often chosen to delineate a saltwater front) is indicated for the Lower Valley and delta by comparing the time evolution of chloride content in numerous wells and piezometers (Fig. 4). Saline water has penetrated inland from the sea following three preferential paths with the plumes pointing towards the main extraction wells of the delta. These plumes relate to the sedimentological features of the delta. The plume in the central part of the delta intrudes through a high

permeability zone, coinciding with the pre-glaciation palaeovalley of the Llobregat river. At the eastern boundary of the delta, the deep aquifer is covered by a sandy formation, and seawater penetration is only hindered by thin muddy deposits present on the sea bed. At the south-west delta boundary, a former saline water body existed as the remnant of incomplete flushing of marine water by freshwater. Since 1965, abstractions along this south-west boundary have reversed the process of slow flushing and saline water has now penetrated towards the main wells in the area, causing the unconfined aquifer in this coastal zone to become brackish. Two of the plumes merge at the delta centre, leaving a freshwater pocket surrounded by saline water, the surface area of which is decreasing owing to groundwater abstractions within it.

A very wide saltwater–freshwater transition zone with little or no vertical salinity stratification is evident as a result of the high aquifer permeability and dispersivity in the heterogeneous aquifer sediments, the small aquifer thickness of about 5 m compared to the flowpath lengths, and the long displacement of saline water inside a confined area without flushing (Iribar et al. 1997).

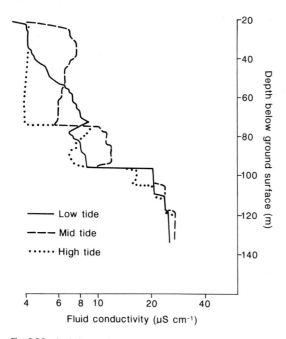

Fig. 6.28 Fluid electrical conductivity logs recorded by downhole geophysical borehole logging at a site on the Brighton sea front, Sussex, illustrating the inland penetration of saline water along discrete horizontal fissures in the Chalk aquifer. After Headworth and Fox (1986).

extraction barrier created by a continuous pumping trough with a line of wells adjacent to the sea; (iv) an injection barrier to maintain a pressure ridge along the coast by a line of recharge wells injected with high quality imported water; and (v) an impermeable subsurface barrier constructed parallel to the coast and through the vertical extent of the aquifer.

An example of the first of the above methods is practised in the Brighton Chalk Block on the Sussex coast of southern England. For fissured aquifers such as the Chalk, there is no simple saline interface and over-abstraction can induce seawater to invade the aquifer along discrete fissure zones, often for discernable distances (Fig. 6.26c). This complex form of intrusion is illustrated by the downhole fluid electrical conductivity logs for a coastal borehole at Brighton (Fig. 6.28). The geophysical borehole logs reveal freshwater moving seawards and saltwater moving inland along discrete horizontal fissures extending to 100 m below sea level. Below a depth of about 130 m the fluid logs indicate the existence of a saline water zone. Analysis of porewater in the Chalk

between fissured zones revealed it to be composed predominantly of freshwater. Analysis of seasonal changes in the fluid log profiles showed that the salinity increases in response to the natural depletion of groundwater storage in the Chalk during the summer and can respond rapidly to changes in abstraction rates from wells located as much as 6 km inland (Headworth & Fox 1986).

Groundwater management to limit saline water intrusion in the Brighton Chalk Block is based on abstracting from pumping stations located around the margins of the aquifer in order to intercept outflows from the aquifer, while at the same time reducing abstractions from inland pumping stations in order to conserve aquifer storage. In addition, aquifer losses to the sea in winter are reduced as far as possible to assist inland storage levels to recover to be able to support increased output in summer. In drought years, following winters with below-average recharge, coastal outflows from the aquifer decline, and inland storage levels, increased as a result of the operating policy, allow greater use to be made of inland pumping stations so as to meet high summer demand for water in the coastal resorts (Headworth & Fox 1986; Miles 1993).

6.5 FURTHER READING

Appelo, C.A.J. & Postma, D. (1994) *Geochemistry, Groundwater and Pollution*. A.A. Balkema, Rotterdam.

Bitton, G. & Gerba, C.P. (eds) (1984) *Groundwater Pollution Microbiology*. John Wiley, New York.

Domenico, P.A. & Schwartz, F.W. (1998) *Physical and Chemical Hydrogeology*, 2nd edn. John Wiley, New York.

Fetter, C.W. (1999) *Contaminant Hydrogeology*, 2nd edn. Prentice-Hall, New Jersey.

Freeze, R.A. & Cherry, J.A. (1979) *Groundwater*. Prentice-Hall, Englewood Cliffs, New Jersey.

Hemond, H.F. & Fechner, E.J. (1994) *Chemical Fate and Transport in the Environment*. Academic Press, San Diego.

Pankow, J.F. & Cherry, J.A. (eds) (1996) *Dense Chlorinated Solvents and other DNAPLs in Groundwater: history, behavior, and remediation*. Waterloo Press, Portland, Oregon.

Todd, D.K. (1980) *Groundwater Hydrology*, 2nd edn. John Wiley, New York.

Williams, P.T. (1998) *Waste Treatment and Disposal*. John Wiley, Chichester.

Younger, P.L., Banwart, S.A. & Hedin, R.S. (2002) *Mine water: hydrology, pollution, remediation*. Kluwer Academic Publishers, Dordrecht.

Groundwater pollution remediation and protection

7

7.1 Introduction

Compared with the consequences of groundwater contamination described in Chapter 6, it is without doubt better to prevent pollution from occurring in the first instance in order to avoid expensive aquifer remediation costs, damage to the environment and the necessity of finding alternative water supplies. In other words, prevention is better than cure. Strategies for preventing groundwater pollution are typically divided between protecting individual groundwater sources and the wider aquifer resource. Differences are also apparent in dealing with point sources of contamination, for example from waste disposal sites, and diffuse sources such as agrochemicals. Groundwater contamination is a global problem and it should also be recognized that strategies developed in the more technologically advanced industrialized countries will need adapting for application in developing countries.

The following sections of this chapter address the above issues and discuss groundwater remediation techniques, risk assessment methods for deciding remedial measures and the location of potentially polluting activities, and methods for protecting groundwater sources and aquifer resources (groundwater bodies). The last section presents examples of how spatial planning, including fundamental changes in land use, can be used to protect groundwater from long-term contamination from diffuse agricultural contaminants.

7.2 Groundwater pollution remediation techniques

The remediation of groundwater can involve an attempt at the total clean-up of a contaminated aquifer or the containment of a groundwater pollution source. In certain circumstances, for example in areas of low risk of human or environmental exposure to contaminants, a further option may be to leave the aquifer to recover through natural attenuation processes.

The successful remediation of contaminated groundwater must address both the source of pollution and remediation of the contaminant plume. Conventional remediation techniques employ pump-and-treat methods, but these have been shown to be less successful, particularly with respect to the clean-up of pools of trapped organic pollutants such as crude oil and chlorinated solvents that act as long-term sources of groundwater contamination. Newer technologies include soil vapour extraction, air sparging and bioremediation for the enhanced removal of organic pollutants. A useful overview is provided by Fetter (1999). Of increasing interest are passive techniques such as permeable reactive barriers that provide an innovative, cost-effective and low-maintenance solution to the clean-up of contaminated land and groundwater.

The following sections provide an introduction to the pump-and-treat and passive techniques of groundwater remediation and also review the case for monitored natural attenuation. The final choice of remediation technique for a given pollution incident will be decided on the basis of a thorough site investigation giving consideration to the type of pollution source, the hydrogeological characteristics and natural attenuation capacity of the affected aquifer, and a cost–benefit analysis to achieve an acceptable reduction in the environmental risks.

7.2.1 Pump-and-treat

The conventional pump-and-treat method of aquifer clean-up is to extract the contaminated groundwater and, following treatment to remove and possibly recover the contaminant source, for the treated water to be either injected into the aquifer or, if a discharge consent is obtained, released to a surface water course. Once the cause of the groundwater pollution has been eliminated, and depending on the shape, extent and concentration distribution of a contaminant plume, the following design criteria must be considered in order to choose the least expensive pumping arrangement for capturing the plume (Javandel & Tsang 1986):

1 What is the optimum number of pumping wells or boreholes required?
2 Where should the wells or boreholes be sited so that no contaminated water can escape between the pumping wells?
3 What is the optimum pumping rate for each well or borehole?
4 What is the optimum water treatment method?
5 How should the treated water be disposed of?

Depending on available resources, a detailed site investigation, including the installation and testing of monitoring wells to provide information on aquifer properties and contaminant distribution, may result in a numerical groundwater model for the site. The model can then be used with particle tracking methods to simulate capture zones (see Section 7.3.2) for one or more pumping wells that encompass the zone of contamination.

At an earlier stage in the investigation, a desk study using the following straightforward method may assist in the initial selection of design criteria. The method is presented by Javandel and Tsang (1986) and is based on the application of complex potential theory to provide an analytical solution to the problem of flow to a fully penetrating well in a homogeneous and isotropic aquifer of uniform thickness. Uniform and steady regional groundwater flow is also assumed. The simplest case is to assume a single pumping well, although the method can be applied to any number of wells, with the solution to a problem obtained using type curves for either single-, double-, three- or four-well capture zones.

In terms of design criteria, the objective of the analytical method is to select the type curve that encompasses the specified concentration contour that delimits the contaminant plume within the capture zone of the well or array of wells. The following procedure, demonstrated for the problem given in Box 7.1, and using the type curves for a single well given in Appendix 10 (Fig. A10.14), explains the method:

1 Prepare a site map using the same scale as the type curves and showing the direction of regional groundwater flow and the contour of the maximum allowable concentration in the aquifer of a given contaminant that defines the contour line of the plume.
2 Superimpose the map on the set of type curves making sure that the direction of regional groundwater flow is aligned with the direction of regional flow shown in the type curves. Now move the contour line of the plume towards the head of the capture zone type curves and read the value of Q/bq from the particular curve which completely encompasses the contour line of the plume.
3 Calculate the value of Q, the well discharge rate, by multiplying Q/bq obtained in the previous step by bq, the product of the aquifer thickness, b, and the magnitude of regional groundwater flow, or specific discharge, q (Section 2.3), to provide a value for the pumping rate for a single well.
4 If the well is able to produce the required discharge rate Q, then a solution has been reached and a single well, with the location copied directly from the position of the well on the type curves to the site map at the matching position, is the optimum design.
5 If the single well is unable to produce at the calculated rate, then the above procedure has to be followed using the type curves for two or more wells (see type curves in Javandel & Tsang 1986) with the optimum distance between two wells given by $Q/\pi bq$.

The above procedures can also be used to explore the position of an injection well for the treated water at

BOX 7.1

Pump-and-treat system design using capture-zone type curves

In this hypothetical example, a pump-and-treat system is under consideration for the restoration of an aquifer at the site of a former sand quarry used for industrial waste disposal (Fig. 1). The site, which is now closed, is unlined and is known to have received drums of the organic solvent TCE for disposal at the site. Some of these drums have ruptured and monitoring wells around the site show a plume of dissolved TCE contaminating the sand aquifer. The aquifer is confined by a clay aquitard, except where this has been removed by quarrying, and preliminary site investigation has revealed that the natural hydraulic gradient in the aquifer is 0.001 from east to west. A short constant-rate pumping test using a pair of monitoring wells has yielded a value of aquifer transmissivity, T, of 10^{-3} m^2 s^{-1} which gives a hydraulic conductivity value of 10^{-4} m s^{-1} for a general saturated aquifer thickness, b, of 10 m in the vicinity of the site. The storage coefficient, S, calculated from the pumping test data is 2×10^{-4}.

Given the above information, find the position and required pumping rate, Q, of a single extraction well to capture of the TCE plume. The target clean-up standard for TCE dissolved in water is 10 µg L^{-1}.

Following the procedure outlined in Section 7.2.1, by superimposing the single-well type curves given in Appendix 10, Fig. A10.14, and moving the contour line of the plume represented by the 10 µg L^{-1} TCE concentration towards the head of the curves, the chosen type curve that encompasses the contaminant plume has a value of $Q/bq = 1200$ m. The result of the overlay of the type curves on to the site map is represented in Fig. 1. In performing this overlay operation, the x-axis of the type curves should be parallel with the direction of regional groundwater flow shown on the site map.

The specific discharge, q, for the regional flow is calculated from Darcy's law (see eq. 2.9) as:

$$q = -K\frac{dh}{dl} = 0.001 \times 10^{-4} = 1 \times 10^{-7} \, m \, s^{-1} \qquad \text{eq. 1}$$

Now, given that $Q/bq = 1200$ m, then with the calculated value of q and for a given aquifer thickness of 10 m, the required discharge rate for the extraction well is found from:

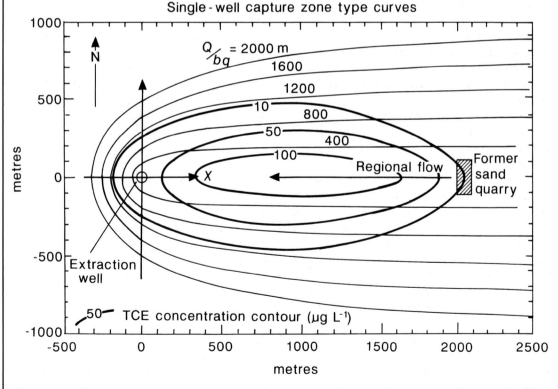

Fig. 1 Single-well capture zone curves overlain on a site map of a sand quarry used for industrial waste disposal. The contour line of the 10 µg L^{-1} concentration of TCE defines the limit of a pollution plume emerging from the waste site and is encompassed by the matching position of the capture zone curve with a value of $Q/bq = 1200$ m.

$Q = 1200 \times bq = 1200 \times 10 \times 1 \times 10^{-7} \text{ m s}^{-1}$

$= 1.2 \times 10^{-3} \text{ m}^3 \text{ s}^{-1} \text{ or } 1.2 \text{ L s}^{-1}$ ⁣ eq. 2

The location of the extraction well is identified by copying directly from the position of the well on the type curves to the site map at the matching point, as indicated in Fig. 1. To check that this pumping rate produces an acceptable water level drawdown at the well, then with a chosen well radius, r_w, of 0.075 m, the following non-equilibrium radial flow equation (see eq 5.40) can be applied for large values of time, t, of say 1 year:

$$s = \frac{2.3Q}{4\pi T} \log_{10} \frac{2.25Tt}{r_w^2 S}$$ ⁣ eq. 3

which, on substitution of the above values gives:

$$s = \frac{2.3 \times 1.2 \times 10^{-3}}{4\pi \times 1 \times 10^{-3}} \log_{10} \frac{2.25 \times 1 \times 10^{-3} \times 3.1536 \times 10^7}{0.075^2 \times 2 \times 10^{-4}}$$

$= 2.37$ m ⁣ eq. 4

This is an acceptable drawdown but, for the required discharge rate, the time required to remove the estimated volume of $2.92 \times 10^6 \text{ m}^3$ of contaminated water from within the 10 µg L^{-1} contour line of the plume (for an aquifer porosity of 0.20) is 77 years, assuming that no water with a concentration below 10 µg L^{-1} is extracted by the well. Therefore, the pump-and-treat system will require a long investment of time, ongoing maintenance and water treatment costs, and substantial energy inputs to maintain this remediation approach.

the upper end of the plume in order to force the contaminated water towards the extraction well, and so shorten the total clean-up time of the aquifer.

Limitations of the above analytical method are the basic assumptions of a homogeneous and isotropic aquifer and a fully penetrating well open over the entire thickness of the aquifer. The method can be applied to unconfined aquifers where the amount of drawdown relative to the total saturated thickness of the aquifer is small, but in heterogeneous aquifers such as fluvial deposits with low permeability clay lenses and high permeability gravel beds, the technique may give erroneous results.

An example of the successful application of the pump-and-treat method for groundwater remediation is described in Box 7.2 for an airport site that experienced a leak of kerosene. This example of organic contamination is common worldwide but it is now recognized that non-aqueous phase liquids (NAPLs), such as oil products and organic solvents, are not treated satisfactorily by the pump-and-treat approach. In a review of the technology, Mackay and Cherry (1989) considered that the rate of contaminant mass removal by extraction wells is exceedingly slow compared with the often large mass of the contaminant source. In such cases, the pump-and-treat option is best considered as a method of hydraulic manipulation of the aquifer to prevent continuation of contaminant migration. The pump-and-treat

method often effectively shrinks the plume towards its source, but for the shrinkage to persist it is necessary for pumping to continue. However, the long-term cost of such pumping becomes expensive and without sufficient detail concerning the mass of NAPL and its distribution at or below the water table in heterogeneous aquifer material, then it becomes difficult to predict reliably the time necessary for permanent clean-up. Hence, groundwater remediation by pump-and-treat may last for a very long time. It is these shortcomings that have provided a strong incentive for the development of alternative remediation technologies, such as passive treatment using permeable reactive barriers.

7.2.2 Permeable reactive barriers

Following recognition that the pump-and-treat approach can prove expensive and in many cases ineffective, research since the late 1980s has focused on alternative, in situ approaches such as permeable reactive barriers (PRBs). In outline, PRBs are constructed by excavating a portion of the aquifer and then replacing the material excavated with a permeable mixture designed to react with the contaminant. Typically, PRBs are installed in trenches, but barriers have also been constructed by jetting reactive materials into the ground, or by generating fractures within

BOX
7.2

Jet fuel clean-up at Heathrow International Airport

This case study concerns hydrocarbon contamination of ground-water adjacent to Technical Block L at Heathrow Airport. Heathrow Airport is built on the Taplow Terrace adjacent to the River Thames floodplain. The geology is formed by 4.5 m of coarse clean gravels overlying low permeability London Clay. The water table is shallow, about 2.5 m from the surface, with groundwater flow southwards beneath the airport towards the River Thames at Shepperton. A leak of jet fuel (kerosene, a light non-aqueous phase liquid, or LNAPL) occurred from a cracked fuel pipe leading to an engine mainten-ance facility, the leak having occurred over a number of years. The leak was discovered when fuel was observed floating on drainage water in a manhole north of Technical Block M (Fig. 1). In response, a large concrete-lined well (Well 1), about 1.5 m in diameter, was installed close to the manhole and revealed about 10 cm of kero-sene floating on the water table. As a first step in remediating the contaminated site, the leak was traced to the cracked pipe and the fracture repaired.

A detailed site investigation, including the installation of 14 monitoring boreholes, showed that the 'pancake' of floating kerosene was about 100 m in diameter and at its thickest point measured a depth of 0.95 m in borehole 5, with further 'free product' measured in boreholes 1, 11 and 13 and in Well 1 (Fig. 1). Odour was reported during the drilling of boreholes 3, 9 and 10, indicative of kerosene. No kerosene was detected in the outlying observation boreholes, including borehole 12.

The basic remediation structures used included large diameter wells lined with perforated concrete rings about 1.5 m in diameter (Fig. 2). Wells 1 and 2, installed close to borehole 1 where a consid-erable thickness of fuel was shown to be floating on the water table, were used to begin recovery of the floating kerosene. The kerosene was removed by floating oil-skimmer pumps. Surface-mounted cen-trifugal pumps, installed with their intakes in the two wells, were also used to lower the water table and encourage the kerosene to move towards the recovery wells. The waste water pumped from the wells was discharged to the drainage system of the airport, which leads to a balancing reservoir before flowing into the River Thames. The balancing reservoir provided settlement and dilution of the remediation waste water.

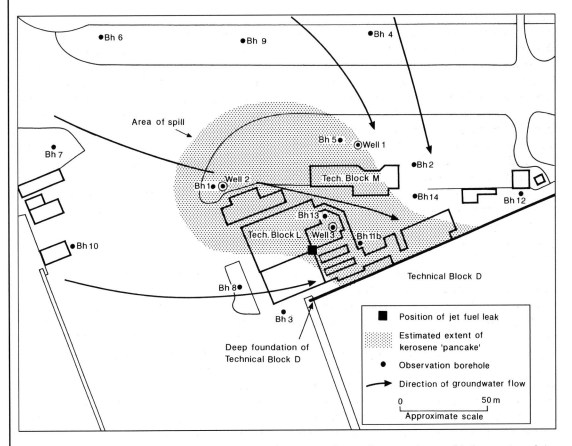

Fig. 1 Site of a jet fuel leak adjacent to Technical Block L at Heathrow Airport showing the estimated extent of the kerosene 'pancake' resting on the gravel aquifer water table. After Clark and Sims (1998).

Groundwater levels were monitored regularly, and within 2 months the cone of depression in the water table produced by the pumping of Wells 1 and 2 encompassed the estimated area of the kerosene 'pancake'. Initially, the recovery rate was such that 19,200 L of kerosene were removed (Fig. 3) and sold to be blended into commercial heating oil. The recovery rate then dropped substantially, yet the kerosene layer in borehole 11 still remained unaltered, suggesting that Wells 1 and 2 were not affecting the southern part of the kerosene 'pancake'. Later, Well 3 was installed. It is believed that Well 3 tapped a 'pool' of kerosene isolated from the effects of Wells 1 and 2 by the foundations of the Technical Blocks. A further 10,100 L of kerosene were removed from Well 3 (Fig. 3), making a total recovery of 29,300 L of kerosene in 4 years at which point the removal of the original kerosene 'pancake' was considered to be complete. Although active remediation by pump-and-treat ended at this time, kerosene recovery using passive collectors, for example absorbent mops, in the three wells continued for about another year. The clean-up project officially ended in 1994 (Clark & Sims 1998).

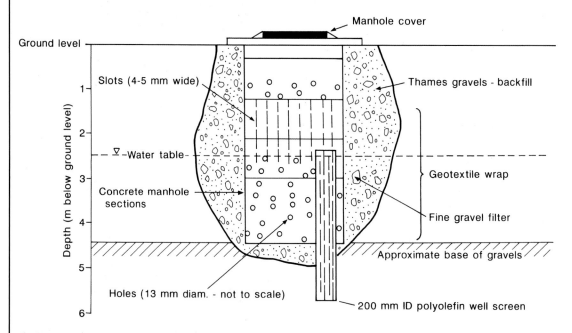

Fig. 2 Construction details of recovery Well 2. The well was installed by excavating a pit by back-hoe as deeply as possible, about 2 m below the water table, lowering the perforated concrete rings into position and then backfilling around the rings using the gravel excavated from the pit. After Clark and Sims (1998).

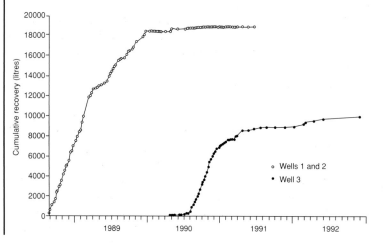

Fig. 3 Cumulative recovery of hydrocarbon from Wells 1, 2 and 3 at Heathrow Airport during clean-up of the gravel aquifer. After Clark and Sims (1998).

an aquifer and filling the fractures with reactive materials (Hocking et al. 2000; Richardson & Nicklow 2002).

The reactive material contained in the barrier is selected to retain the contaminant within the barrier. PRBs containing zero-valent iron (iron filings) have been used to treat hexavalent chromium, uranium and technetium (Blowes et al. 2000) and chlorinated ethenes (PCE and TCE) (O'Hannesin & Gillham 1998). Solid phase organic carbon in the form of municipal compost has been used to remove dissolved constituents associated with acid mine drainage, including sulphate, iron, nickel, cobalt and zinc. Dissolved nutrients, including nitrate and phosphate, have also been removed from domestic septic-system effluent and agricultural drainage in this way (Blowes et al. 2000).

In treating inorganic and organic contaminants, a range of processes has been used such as: manipulation of the redox potential to enhance biological reductive dechlorination and to change the chemical speciation of metals; chemical (abiotic) degradation; precipitation; sorption to promote organic matter partitioning and ion exchange; and biodegradation. Further information and guidance on the use of PRBs for the remediation of contaminated groundwater is given by Carey et al. (2002).

The most common designs of PRBs are the 'funnel and gate' and 'continuous wall' reactive barriers illustrated in Fig. 7.1. Funnel and gate PRBs are described by Starr and Cherry (1994) and consist of low hydraulic conductivity cut-off walls such as sheet piles and slurry walls with gaps that contain in situ reactors for removal of contaminants. Funnel and gate systems can be installed in front of plumes to prevent further plume growth, or immediately downgradient of contaminant source areas to prevent contaminants from creating plumes. Cut-off walls (the funnel) modify the groundwater flow pattern so that groundwater flows primarily through the high conductivity gaps of the gates. Continuous PRBs transect the contaminant plume with an unbroken wall of permeable material which is combined with the reactive material, for example a pea-gravel and reagent-filled trench. The majority of PRBs have been placed at relatively shallow depths, around 10–20 m deep, although a few have been placed to depths of 40 m.

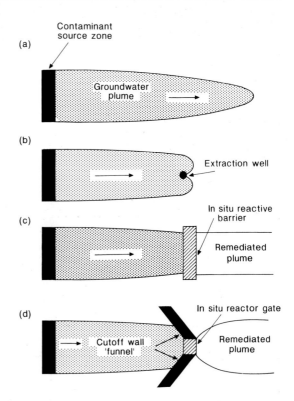

Fig. 7.1 Three options for remediation of contaminated groundwater. (a) Unremediated contaminant plume. (b) Pump-and-treat system. (c) In situ 'continuous wall' reactive barrier. (d) In situ 'funnel and gate' reactive barrier. In the case of the funnel and gate system, a balance must be achieved between maximizing the size of the capture zone for a gate and maximizing the retention time of contaminated groundwater in the gate. In general, capture zone size and retention time are inversely related. After Starr and Cherry (1994).

To be successful, it is likely that PRBs will need to be operated over extended periods, possibly decades, but unlike the pump-and-treat method of remediation, the low operating and maintenance requirements of PRBs make such long-term clean-up a possibility. The treatment process may result in a change in the in situ biological and geochemical environment within and downgradient of the PRB, for example a change from oxidizing to reducing conditions, that may cause secondary reactions including precipitation of mineral phases such as hydroxides and carbonates. The long-term hydraulic and chemical performance of PRBs can be affected by biofouling, chemical precipitation and the production of gases, and long-term studies are required for the full evaluation of PRBs (Box 7.3).

7.2.3 Monitored natural attenuation

The United States Environmental Protection Agency (1997) defined natural attenuation as a variety of physical, chemical or biological processes that, under favourable conditions, act without human intervention to reduce the mass, toxicity, mobility, volume or concentration of contaminants in soil or groundwater. These in situ processes include biodegradation, dispersion, dilution, sorption, volatilization, radioactive decay, and chemical or biological stabilization, transformation or destruction of contaminants (Fig. 7.2).

In situ permeable reactive barrier for remediation of chlorinated solvents

BOX 7.3

A field demonstration of a 'continuous wall' in situ permeable reactive barrier (PRB) was conducted in the Borden sand aquifer, Ontario, and downgradient of the emplaced source of mixed chlorinated solvents (perchloroethene (PCE), trichloroethene (TCE) and trichloromethane (TCM)) described in Box 6.3. At the time of the long-term test of the PRB, the plume from the emplaced source was approximately 1 m thick and 1 m wide near the source, with peak PCE and TCE concentrations of about 50 and 270 mg L^{-1}, respectively. However, most of the TCM had been dissolved from the source, resulting in very low concentrations. As shown in Fig. 1, the PRB was installed 5.5 m downgradient from the emplaced source and positioned below the water table. The reactive material used to construct the wall consisted of 22% by weight of granular iron mix with 78% by weight of coarse sand, and had dimensions of $5.5 \times 1.6 \times 2.2$ m, giving a volume of 19.4 m^3. Permeameter measurements on samples of the iron–sand mixture gave hydraulic conductivity values of 4.37×10^{-4} m s^{-1} (O'Hannesin & Gillham 1998). Metal-enhanced degradation of chlorinated organic compounds is an abiotic redox reaction involving reduction of the organic compound and oxidation of the metal (Johnson et al. 1996). The reaction appears to be pseudo-first-order with respect to the organic concentration and the products of the reaction are chloride, iron (Fe^{2+}) and non-chlorinated, or less chlorinated hydrocarbons (dichloroethene and vinyl chloride). In the case of chlorinated ethenes such as PCE and TCE, dechlorination is complete with ethene and ethane as the final carbon-containing compounds, while for CTET, a fraction of the parent compound persists as dichloromethane.

The results of the field experiment are shown in Fig. 2 and show that for both TCE and PCE there is a substantial decline in concentration of the core of the contaminant plume at the position of the first sampling fence (50 cm into the wall), followed by a gradual decrease with further distance into the wall. As a result, TCE declined from an influent concentration of 268,000 µg L^{-1} to an effluent value measured at the 7.5 m fence (50 cm downgradient of the wall) of 23,350 µg L^{-1}. Similarly, PCE declined from 58,000 µg L^{-1} to 10,970 µg L^{-1}. Thus, based on the maximum observed concentrations at each sampling fence, 91% of the TCE and 81% of the PCE were removed from solution with passage through the reactive material. It is reasonable to expect that, had a higher percentage of iron been used in the iron–sand mixture, or had a more

reactive material been used, then the quality of the effluent leaving the wall could have been further improved.

Potential chlorinated degradation products were also analysed, including chloride, 1,1-dichloroethene (1,1-DCE), *trans* 1,2-dichloroethene (tDCE), *cis* 1,2-dichloroethene (cDCE) and vinyl

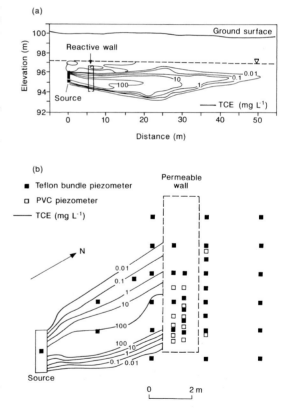

Fig. 1 (a) Cross-section of the emplaced source of chlorinated solvents, reactive wall and TCE plume situated in the Borden aquifer test site, Ontario. (b) Plan view of the permeable wall test site, monitoring network and TCE plume. After O'Hannesin and Gillham (1998).

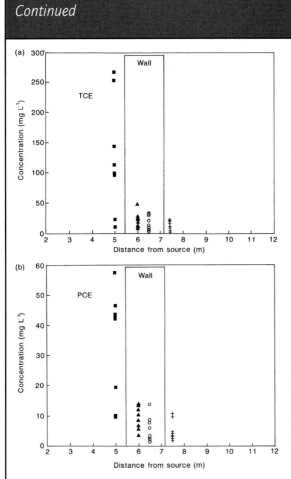

Fig. 2 Longitudinal section through the Borden aquifer test site showing the maximum chlorinated solvent concentrations obtained for 10 sampling sessions over a 5-year period along the flowpath of (a) the TCE and (b) the PCE plumes. Note that the variation in concentration at a particular distance does not reflect variation over time but is a consequence of plume position. O'Hannesin and Gillham (1998).

chloride (VC). Of these, the major product within the iron–sand mixture was cDCE (2110 µg L^{-1}), with substantially lesser amounts of 1,1-DCE (453 µg L^{-1}) and tDCE (146 µg L^{-1}); the sum of which is equivalent to about 1% of the influent TCE. However, measurements of the effluent leaving the wall showed that the DCE isomers were also degraded within the PRB. Concentrations of VC above the limit of the analytical method were not detected.

Changes in water chemistry as a result of abiotic reduction of organic compounds involve the oxidation of zero valent iron (Fe^0) by water-producing Fe^{2+}, an increase in H$^+$ and OH$^-$ (eqs 1 and 2), and a decrease in redox potential and dissolved oxygen. The H$^+$ forms hydrogen gas and the OH$^-$ remaining in solution causes an increase in pH that can cause precipitation of iron hydroxides and carbonate minerals (eqs 4–6).

$$2Fe^0 + O_2 + 2H_2O \rightarrow 2Fe^{2+} + 4OH^- \quad \text{(aerobic conditions)} \quad \text{eq. 1}$$

$$Fe^0 + 2H_2O \rightarrow Fe^{2+} + H_2 + 2OH^- \quad \text{(anaerobic conditions)} \quad \text{eq. 2}$$

$$Fe^{2+} + 2OH^- \rightarrow Fe(OH)_2 \quad \text{(iron hydroxide)} \quad \text{eq. 3}$$

$$HCO_3^- + OH^- \rightarrow CO_3^{2-} + H_2O \quad \text{eq. 4}$$

$$Fe^{2+} + CO_3^{2-} \rightarrow FeCO_3 \quad \text{(siderite)} \quad \text{eq. 5}$$

$$Ca^{2+} + CO_3^{2-} \rightarrow CaCO_3 \quad \text{(calcite)} \quad \text{eq. 6}$$

The field results showed that iron concentrations entering the wall were <0.5 mg L^{-1}, while within the PRB concentrations were generally within the range of 5–10 mg L^{-1} before decreasing to <0.5 mg L^{-1} downgradient of the wall. Dissolved oxygen and Eh values within the treatment zone were nearly always recorded as zero and within the range −200 to −350 mV, respectively. The pH increased from a background value of 8.0 to 8.7 in the PRB as a result of the reduction of water. After 4 years of operation, only trace amounts of iron oxides and calcium and iron carbonates were found within the first few millimetres of the wall at the upgradient interface between the aquifer sand and the wall, such that there was no evidence of a decline in performance of the PRB over the duration of the study.

Overall, O'Hannesin and Gillham (1998) concluded that the results of this long-term field study provide good evidence that in situ use of granular iron can provide a long-term, low-maintenance cost solution for groundwater contamination problems.

As an alternative to more expensive pump-and-treat and engineered solutions to groundwater contamination, reliance on monitored natural attenuation (MNA) appears attractive but opponents claim that natural attenuation conveniently avoids the high costs of installing clean-up systems. The feasibility of MNA as a strategy depends on whether the regulatory aim is to clean up a contaminant plume to drinking water standards or whether a less stringent,

risk-based goal applies, such as preventing a plume from spreading. Since the mid-1990s, the use of MNA as a remedial solution for benzene, toluene, ethylbenzene and xylene (BTEX compounds) has increased dramatically (National Academy of Sciences 2000). Natural attenuation has been proposed for chlorinated solvents, nitroaromatics, heavy metals, radionuclides and other contaminants for which further research and scientific understanding is required before the

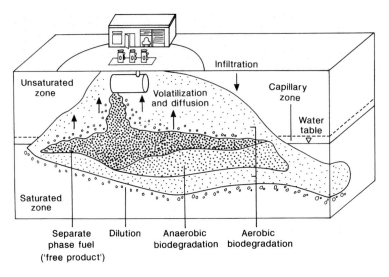

Fig. 7.2 Conceptual diagram illustrating the important natural attenuation processes that affect the fate of petroleum hydrocarbons in aquifers. After Bekins et al. (2001a).

technique can be considered robust (Bekins et al. 2001a).

In considering the case for application of MNA at a contaminated site, a substantial degree of understanding of the subsurface processes must be developed. Thus, the major expense is likely to shift from the design and operation of an active pump-and-treat or passive PRB to detailed investigation and modelling of the site in order to understand the natural groundwater flow and biogeochemical reactions responsible for attenuating the contamination (Box 7.4).

Key to the future success of MNA is further research into the practical issues regarding the performance of natural attenuation over long time periods. This research should include the effects of active remediation efforts on the natural attenuation process, the design of long-term monitoring networks to verify that natural attenuation is working, and proving the natural attenuation capacity of the aquifer over the lifetime of the contaminant source.

7.3 Groundwater pollution protection strategies in industrialized countries

7.3.1 Groundwater vulnerability mapping and aquifer resource protection

As illustrated in Fig. 7.3, the vulnerability of groundwater to surface-derived pollution is a function of the nature of the overlying soil cover, the presence and nature of overlying superficial deposits, the nature of the geological strata forming the aquifer and the depth of the unsaturated zone or thickness of confining deposits. This approach has been used by the Environment Agency in England and Wales to produce a series of 53 regional groundwater vulnerability maps showing vulnerability classes determined from the overlay of soils and hydrogeological information at a scale of 1 : 100,000 (Fig. 7.4). The maps form part of the Environment Agency's strategy for protecting groundwater resources (Robins et al. 1994; Environment Agency 1998) with the intention of encouraging the development of potentially polluting activities in those areas where it will present least concern. As regional maps, the control of diffuse pollution can be readily related to zones of aquifer vulnerability. The overlay operation of soils and hydrogeological information can be conveniently manipulated within a geographical information system (GIS) to provide specific groundwater vulnerability maps, such as the nitrate vulnerability map shown in Fig. 7.5.

In the United States, the Environmental Protection Agency has developed a similar methodology to evaluate groundwater vulnerability designed to permit the systematic evaluation of the groundwater pollution potential at any given location (Aller et al. 1987). The system has two major components: first, the designation of mappable units, termed

Monitored natural attenuation of a crude oil spill, Bemidji, Minnesota

A demonstration site for monitored natural attenuation (MNA) within the United States Geological Survey Toxic Substances Hydrology Program (http://toxics.usgd.gov) is located near Bemidji, Minnesota, where a buried pipeline located in a glacial outwash plain ruptured in 1979 spilling crude oil into the subsurface. The oil is entrapped as a residual non-aqueous phase in the vadose zone and also forms two bodies of oil floating on the water table. The largest oil body was estimated to contain 147,000 L of oil in 1998. As shown in Fig. 1a, the oil forms a long-term, continuous source of hydrocarbon contaminants that dissolve in and are transported with the groundwater. Microbial degradation of the petroleum hydrocarbons in the plume has resulted in the growth of aquifer microbial populations dominated by aerobes, iron-reducers, fermentors and methanogens (Fig. 1b). The biodegradation reactions cause a number of geochemical changes near the dissolved aqueous plume which include decreases in concentrations of oxygen and hydrocarbons and increases in concentrations of dissolved iron, manganese and methane (Fig. 1c).

Modelling of the natural attenuation processes simulates initial aerobic degradation followed by the development of an anoxic zone in which manganese and iron reducers and methanogens begin to grow, consuming solid phase Mn(IV) and Fe(III) and releasing dissolved Mn(II), Fe(II) and methane (Fig. 1c). The modelling predicts that 40% of the hydrocarbon degradation occurs aerobically and 60% anaerobically. Combined with field data and the measurement of microbial populations, the results suggest that the natural attenuation capacity of the glacial outwash sands is being slowly consumed by depletion of the intrinsic, electron-accepting capacity of the aquifer (Bekins et al. 2001b; Cozzarelli et al. 2001).

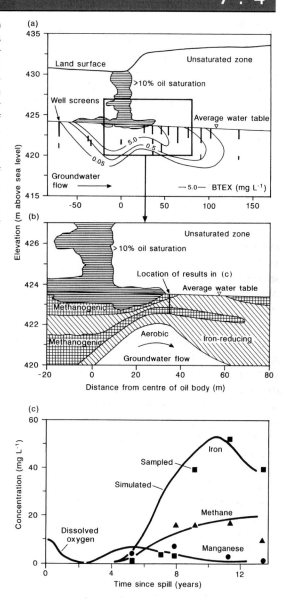

Fig. 1 Illustration of natural attenuation of crude oil contamination by aerobic and anaerobic biodegradation in a glacial outwash aquifer located near Bemidji, Minnesota. (a) The 1995 concentration of BTEX compounds defines the extent of contamination in the aquifer for a vertical cross-section along the plume axis. (b) The cross-section shows the distribution of microbial populations inferred from most probable number data. (c) Modelled and observed concentrations versus time plots for a well positioned at the water table, 36 m downgradient from the contaminant source and illustrating the loss of oxygen and production of reduced electron acceptors (Fe(II), Mn(II) and methane) during the temporal evolution of redox conditions in the aquifer. After Bekins et al. (2001a).

hydrogeologic settings; and second, the superposition of a relative rating system having the acronym DRASTIC. Inherent in each hydrogeologic setting are the physical characteristics that affect groundwater pollution potential. The most important mappable

factors considered to control the groundwater pollution potential are: depth to water (D); net recharge (R); aquifer media (A); soil media (S); topography (slope) (T); impact of the vadose zone (I); and hydraulic conductivity of the aquifer (C). The numerical

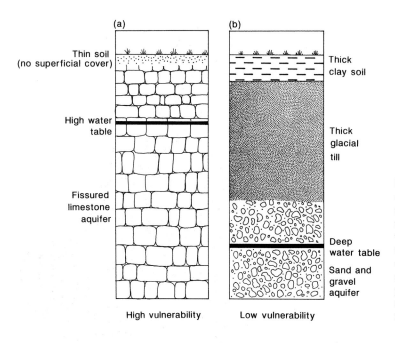

Fig. 7.3 Illustration of two situations of contrasting groundwater vulnerability to surface-derived pollution. In (a) the unconfined, fissured limestone aquifer with a permeable soil cover and high water table (thin unsaturated zone) has a high apparent vulnerability. In (b) the sand and gravel aquifer, overlain by a low permeability soil and glacial till cover, has a low apparent vulnerability. After Environment Agency (1998).

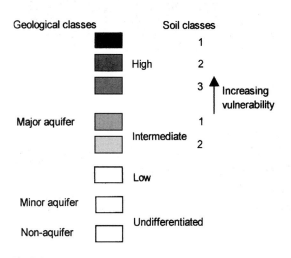

Fig. 7.4 Groundwater vulnerability classification scheme used by the Environment Agency in England and Wales derived from the overlay of information on geological strata and soil type. Major aquifers of regional importance include highly permeable formations usually with significant fracturing. Minor aquifers and non-aquifers are classed as having variable or negligible permeability that generally support local or very small abstractions. Soil classes are divided on the basis of leaching potential (high, intermediate and low) depending on the physicochemical properties of soil types with respect to diffuse source pollutants and liquid discharges. After Environment Agency (1998).

ranking system which is applied to the DRASTIC factors contains three significant parts: weights, ranges and ratings. Weights relate to the relative importance of each of the seven factors on a scale of 1–5, where 5 is the most important. Each factor is divided into ranges (or significant media types) depending on the impact each has on pollution potential. Each range is then assigned a rating (1–10) to differentiate the significance of each range with respect to pollution potential. The factor for each range receives a single value, except the ranges for factors A and I, for which a typical rating and a variable rating have been provided. The following equation is then used to provide each hydrogeologic setting with a relative numerical value:

$$\text{Pollution potential} = D_R D_W + R_R R_W + A_R A_W + S_R S_W$$
$$+ T_R T_W + I_R I_W + C_R C_W$$

$$\text{eq. 7.1}$$

where $_R$ is rating and $_W$ is weight. The greater the DRASTIC score the greater the pollution potential. The scores are applied to their respective hydrogeologic settings and are mapped.

A limitation of the above approaches to mapping apparent and specific groundwater vulnerability

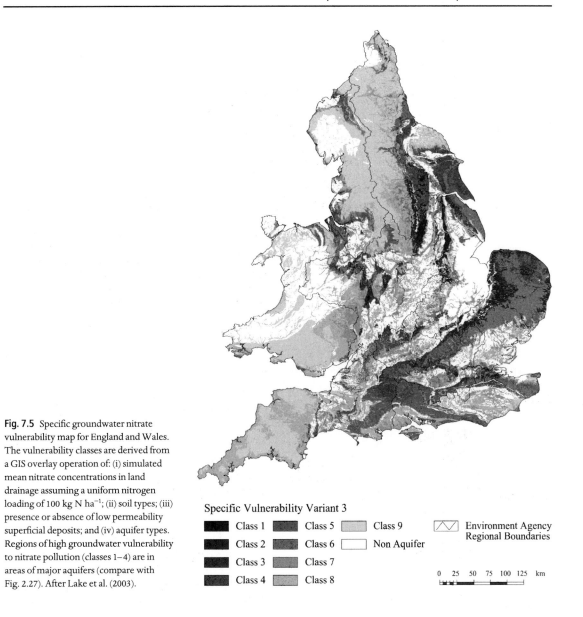

Fig. 7.5 Specific groundwater nitrate vulnerability map for England and Wales. The vulnerability classes are derived from a GIS overlay operation of: (i) simulated mean nitrate concentrations in land drainage assuming a uniform nitrogen loading of 100 kg N ha^{-1}; (ii) soil types; (iii) presence or absence of low permeability superficial deposits; and (iv) aquifer types. Regions of high groundwater vulnerability to nitrate pollution (classes 1–4) are in areas of major aquifers (compare with Fig. 2.27). After Lake et al. (2003).

Specific Vulnerability Variant 3

Class 1	Class 5	Class 9
Class 2	Class 6	Non Aquifer
Class 3	Class 7	
Class 4	Class 8	

Environment Agency Regional Boundaries

0 25 50 75 100 125 km

is that they provide a regional picture that is insufficiently detailed to demonstrate the actual threat to the groundwater resource at a local scale. The true vulnerability can only be established with confidence through supporting, site-specific field investigations. Even so, groundwater vulnerability maps are instrumental in conveying groundwater pollution potential to planners and can help achieve water quality objectives by influencing land-use management.

7.3.2 Source protection zones

In the definition of groundwater source protection zones, the proximity of a hazardous activity to a point of groundwater abstraction (including springs, wells and boreholes) is one of the most important factors in assessing the pollution threat to an existing groundwater source. In principle, the entire recharge area in the vicinity of a groundwater source should be protected, but this is unrealistic on socioeconomic

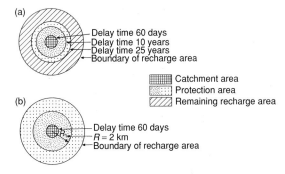

Delay time 60 days
Delay time 10 years
Delay time 25 years
Boundary of recharge area

Catchment area
Protection area
Remaining recharge area

Delay time 60 days
R = 2 km
Boundary of recharge area

Fig. 7.6 Examples of protection zones for groundwater sources in (a) a porous, permeable aquifer and (b) a fissured, karstic aquifer. See Table 7.1 for land-use restrictions applied in each area. After van Waegeningh (1985).

Table 7.1 Land-use restrictions for the source protection zones shown in Fig. 7.6. After van Waegeningh (1985).

Catchment area	Protection area	Remaining recharge area
60 days and ≥30 m	10- and 25-year delay-time or 2 kilometres	
Protection against pathogenic bacteria and viruses and against chemical pollution sources	Protection against hardly degradable chemicals	Soil and groundwater protection rules
Only activities in relation to water supply are admissible	As a rule, the following are not admissible: Transport and storage of dangerous goods Industrial sites Waste disposal sites Building Military activities Intensive agriculture and cattle breeding Quarrying Waste water disposal	

grounds. In this situation, a system of zoning of the recharge area, or protection area, is desirable and this approach has been adopted in Europe and the United States. For example, in the Netherlands, abstraction of drinking water supplies is concentrated in wellfields tapping mainly uniform, horizontally layered aquifers of unconsolidated sands and clays. As illustrated in Fig. 7.6 and Table 7.1, the zoning system includes a first zone based on a delay time of 60 days from any point below the water table in order to protect against pathogenic bacteria and viruses and rapidly degrading chemicals. This zone typically extends some 30–150 m from an individual borehole. For the continuity of water supplies in the event of a severe pollution incident requiring remedial action, and in order to exclude public health risks, a delay time of at least 10 years is needed in the next zone. In many cases, even 10 years is not sufficient to guarantee the continuity of safe water supplies, and a protection zone of 25 years is necessary. The 10- and 25-year protection zones extend to about 800 m and 1200 m from the borehole, respectively, and constitute the source protection area.

In the United States, the Wellhead Protection Program (United States Environmental Protection Agency 1993) aims to delineate the area from which an abstraction well obtains its water and then limit potentially hazardous activities from taking place in this area. The first area, the zone of influence (ZOI), is almost synonymous with the cone of depression, while the second area, the well capture zone or zone of contribution (ZOC), is defined as the region surrounding a pumping well that encompasses all areas or features that supply groundwater to the well (Fig. 7.7). The size and shape of the ZOI and ZOC are dependent on well design, aquifer properties and boundaries, and the position and hydraulic loading of the contaminant source. The ZOC can be further delineated by the zone of contaminant transport (ZOT), generally presented as isochrones (contours of equal travel time) that indicate the time required for a contaminant to reach a pumping well from a source within the ZOC (Fig. 7.7). The time of travel depends on the groundwater flow velocity, the contaminant characteristics and the properties and composition of the aquifer material (Livingstone et al. 1995).

Mapping of wellhead protection area (WHPA) criteria can be performed at different costs and levels of complexity, ranging from arbitrary radii to numerical flow and transport models (Fig. 7.8), including the capacity of the aquifer to assimilate contaminants (Livingstone et al. 1995). The overall objectives of wellhead protection are to produce a remedial action

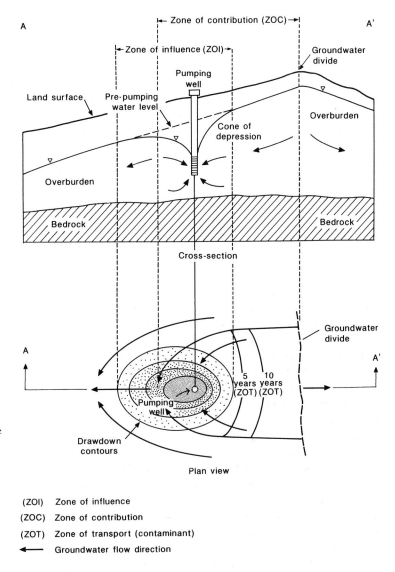

Fig. 7.7 Conceptual model of a wellhead protection area and associated terminology. As shown, the zone of contribution (ZOC) and the zone of influence (ZOI) do not coincide. For the ZOC, groundwater is removed from the pumping well from only a relatively small portion of the downstream area of the well, but it may extend as far as the groundwater divide on the upgradient side of the well. By contrast, the downgradient portion of the groundwater within the ZOI is not drawn towards the pumping well but continues downgradient, while the ZOI does not extend to the upgradient limit of the ZOC. However, experience shows that if the ZOC is small, then the ZOC and ZOI will generally overlap (Livingstone et al. 1995).

zone, an attenuation zone and a management zone. Specific guidance is available in more complex hydrogeologic settings such as in confined aquifers (United States Environmental Protection Agency 1991a) and fractured rocks (United States Environmental Protection Agency 1991b).

In England and Wales, the Environment Agency (1998) has established source protection zones (SPZs) that are applied to public water supplies (there are nearly 2000 major sources) and private water supplies, including bottled water, and commercial food and drink production. As illustrated schematically in Fig. 7.9, the orientation, shape and size of the SPZs are determined by the hydrogeological characteristics and the direction of groundwater flow around each source. Steady-state groundwater flow modelling is used to define three zones (Zones I, II and III) in each SPZ and a set of groundwater protection policy statements set out the acceptability of various polluting activities in each zone, for example landfill operations and the application of liquid effluents to land.

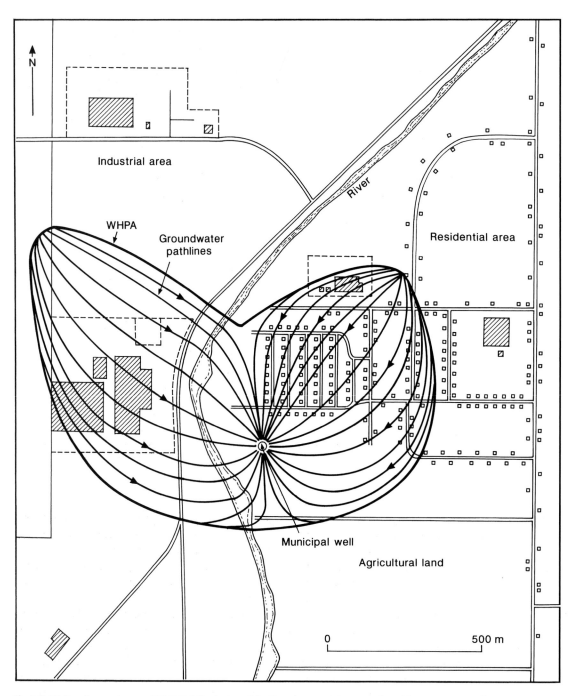

Fig. 7.8 Wellhead protection area (WHPA) defined using a fully three-dimensional numerical model (MODFLOW and MODPATH) to simulate groundwater flow and contaminant migration using particle tracking. In this hypothetical example, the municipal well pumps at a rate of 250 m^3 day^{-1}. The aquifer comprises sand and gravel with a transmissivity of 26 m^2 day^{-1} and is 30 m thick. The well partially penetrates the aquifer with the well screen extending 20–30 m below ground surface. The well is located close to a river that is situated within an area of alluvial deposits with a transmissivity of 0.1 m^2 day^{-1} and a thickness of 10 m. The groundwater model explicitly represents the alluvial unit overlying the sand and gravel aquifer near the river. The results of the simulation define a capture zone configuration that suggests that groundwater protection efforts should be concentrated in the areas to the west and east of the river. After Livingstone et al. (1995).

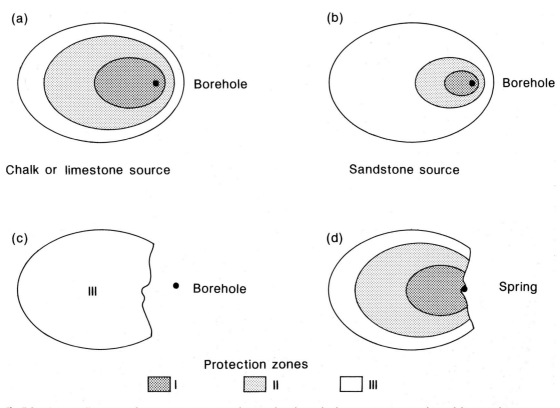

Fig. 7.9 Schematic illustration of source protection zones showing the relationship between Zones I, II and III and the groundwater source in four idealized hydrogeological situations representing: (a) a low effective porosity limestone aquifer; (b) a high effective porosity sandstone aquifer; (c) a confined aquifer; (d) a spring. In reality, the size, shape and relationship of the zones will vary significantly depending on the soil, geology, amount of recharge and volume of water abstracted. See text for an explanation of the definitions of Zones I, II and III. After Environment Agency (1998).

Zone I, or the inner source protection zone, is located immediately adjacent to the groundwater source, and is designed to protect against the impacts of human activity which might have an immediate effect upon the source. The area is defined by a 50-day travel time from any point below the water table to the source and as a minimum 50-metre radius from the source. This rule of thumb is used in other countries and is based on the presumed time taken for biological contaminants to decay in groundwater. The land immediately adjacent to the source and controlled by the operator of the source is included within this zone.

Zone II, or the outer protection zone, is the area around the source defined by a 400-day travel time, and is based on the requirement to provide delay and attenuation of slowly degrading pollutants. In high groundwater storage aquifers, such as sandstones, it is necessary, in order to provide adequate attenuation, to define further the outer protection zone to be the larger of either the 400-day travel time area or the recharge catchment area calculated using 25% of the long-term abstraction rate for the source.

Zone III, or the source catchment, is the remaining catchment area of a groundwater source, and is defined as the area needed to support an abstraction from long-term annual groundwater recharge. For wells and boreholes, the source catchment area is defined by the authorized abstraction rate while, for springs, it is defined by the best known value of average annual total discharge. In practice, the size of Zone III will vary from tens to a few thousands of hectares depending on the volume of groundwater abstraction and the amount of recharge. In areas

Table 7.2 Calculation of the toxicological index, I_{tox}, for three sites located at the pulp and paper mill complex at Sjasstroj, north-west Russia. Analytical results are given in mg L^{-1}. After Schoenheinz et al. (2002).

Sample number	Sample date	Constituent, i	BOD	COD	SO$_4^{2-}$	Cl$^-$	NH$_4^+$	NO$_2^-$	NO$_3^-$	Fe	Al	Phenols	Surfactants	I_{tox}
		LAC$_i$	3	30	500	350	2	3	45	0.3	0.5	0.001	0.1	1
I	09/99			3044	160	15	5.6	0.04	1.0	5	90			301
II	11/99		5.4	61	13	42.7	3.5	0.005	0.45	53	0.2	0.001	0.1	185
	02/00		0.9	44	6.9	83	5	0.002	0.15	2.3	4.6	0.003	0.13	25
III	02/00		0.6	7.7	5.8	4	1.45	0.005	0.0	0.6	4.5	0.002	0	14

I, excess sludge; II, groundwater close to active sludge basin in upper sand aquifer; III, groundwater in lower aquifer; LAC$_i$, Russian limit of the admissible concentration of constituent, I; BOD, biochemical oxygen demand; COD, chemical oxygen demand.

where the aquifer is confined beneath impermeable cover, the source catchment may be some distance from the actual abstraction.

7.3.3 Risk assessment methods

Of increasing relevance to managing aquifers, risk assessment methods are applied in the decision-making process, both with reference to the choice of aquifer remediation technology in cases where pollution has already occurred, for example in areas of contaminated land, and also in the siting of new containment facilities, such as municipal landfills. Influential publications concerning the definition of risk assessment include the United States National Research Council (1983) and the Royal Society (1992). Petts et al. (1997) stated that risk assessment is a process comprising hazard identification, hazard assessment, risk estimation and risk evaluation and, in general, is the study of decisions subject to uncertain consequences.

A basic risk assessment calculation can be performed by the determination of a toxicological index, I_{tox}, for a given site using the following equation:

$$I_{tox} = \sum_{i=1}^{n} c_i / \text{LAC}_i \qquad \text{eq. 7.2}$$

where $i = 1 \ldots n$ represents the contaminant constituent, c_i is the measured concentration of constituent i and LAC is the limit of the admissible concentration of constituent i. An example calculation of I_{tox} values

for a large pulp and paper mill complex in north-west Russia is given in Table 7.2.

In calculating I_{tox}, constituents are chosen arbitrarily, mainly as a function of laboratory and financial capabilities, such that the importance of different compounds in terms of their hazard potential is not evaluated. The results of the risk assessment allow a comparative, quantitative assessment of analytical results for different measurement points but are neither source nor target related. For the calculations shown in Table 7.2, it is clear that all three samples are predicted to be at a high potential risk given the values of I_{tox} in excess of 1 and would therefore suggest that remedial action is necessary. However, shortcomings of the data presented in Table 7.2 are that substances with no toxicological potential, for example chemical oxygen demand (COD) for sample 1, can determine the outcome of the toxicological index calculation, and that concentrations of phenols, surfactants and the biological oxygen demand (BOD) were not always available (Schoenheinz et al. 2002).

More sophisticated approaches to groundwater pollution risk assessment recognize a source–pathway–target paradigm and adopt a cost-effective, tiered approach to risk assessment. In contaminated land studies, risk assessment identifies the pathway term as the route that the contaminant takes from the pollution source to a receiving well or borehole receptor. A refinement is to divide the pathway into an environmental pathway between the source and groundwater receptor and an exposure pathway between the receptor and an ecological or human target. The objective of assessing the effects of the environmental pathway is to determine the concentration of the contaminant at an abstraction site while the exposure pathway assesses the effects of exposure to contaminated water.

To gain the most efficient use of time and financial constraints, tiered approaches to risk assessment are widely applied. Increasing tiers equate to increasing levels of sophistication with respect to site-specific risk assessment. In general, four levels of assessment can be identified:

• Tier 1: Preliminary investigations: qualitative desk study; determination of source–pathway–target chains; limited intrusive investigation and sampling.
• Tier 2: Site characterization: semiquantitative; some intrusive site investigation and sampling; prioritization and screening methods.
• Tier 3: Generic risk assessment: qualitative/semiquantitative; comparison of estimated contaminant concentrations with generic guidelines and standards; assessment of environmental pathway; intrusive site investigations; computer simulations/modelling; stochastic approaches.
• Tier 4: Quantitative risk assessment: use of derived contaminant concentrations for exposure pathway assessment; exposure assessment models; integrated approaches to risk assessment.

This type of tiered approach is the basis of the American Society for Testing and Materials (ASTM 2001) standard guide to RBCA (risk-based corrective action) for site investigation and remediation at petroleum-contaminated sites, although the process can be applied to any contaminant and release scenario. The framework has a site characterization stage followed by a three-tier approach where the human health and environmental risks are equally accounted for. As each tier is completed there is an evaluation to determine if more information is required. If so, then the next level of assessment is conducted; if not, then an assessment of the corrective action is required. A tiered approach is also used by the Environment Agency in England and Wales in the hydrogeological risk assessment of landfills (Leeson et al. 2003) as outlined in Fig. 7.10.

A practical example of a Tier 3 risk assessment is provided by Davison et al. (2002), who presented a management tool to identify the best use for urban groundwater pumped from a user-defined location. A probabilistic catchment zone model and land-use model are combined to provide contaminant source data that are then used in a pollution risk model to calculate a probability distribution for the concentration of a contaminant at a selected pumped borehole for comparison with water quality standards.

7.3.4 Spatial planning and groundwater protection

The problem of nitrate and pesticides leaching from regions of intensively managed arable farming and cattle grazing affects wide areas of western Europe and other regions of the world and it has become clear that more stringent controls on land-use activities, integrated into local and regional spatial planning, are required if groundwater quality is to improve. Such an approach has been applied in the United Kingdom and Denmark (Box 7.5) and provides experience for the future formulation of strategies for reducing the impacts of diffuse groundwater contamination.

In England, the Pilot Nitrate Sensitive Area (NSA) scheme was started in 1990 and by the time of the Main NSA scheme in 1998, 80% of the land area comprising 35,000 ha in 32 catchments was included in this voluntary, compensated agri-environment scheme. Two levels of payment were offered to farmers entering into the scheme with higher compensation given under the Premium scheme for arable land conversion to grass with total nitrogen inputs of less than 150 kg ha^{-1}. Under the alternative, Basic scheme, farmers were paid less for sowing winter cover crops, restricting organic manure inputs and their timing, and limiting nitrogen fertilizer inputs to below the recommended optimum. Table 7.3 summarizes the results of long-term monitoring of the effectiveness of the scheme in the first 10 catchments included in the Pilot NSA scheme with respect to measured soil nitrate losses prior to and during the scheme. The results indicate an overall reduction of about 50% in nitrate concentrations and 28% in nitrogen fluxes leaving the root zone compared to baseline values. Corresponding values for the further catchments included in the Main NSA scheme are 34% and 16%, respectively, between 1994 and 1996 and 1998 and 2000 (ADAS 2003). Take-up and conversion of arable land under the Premium scheme

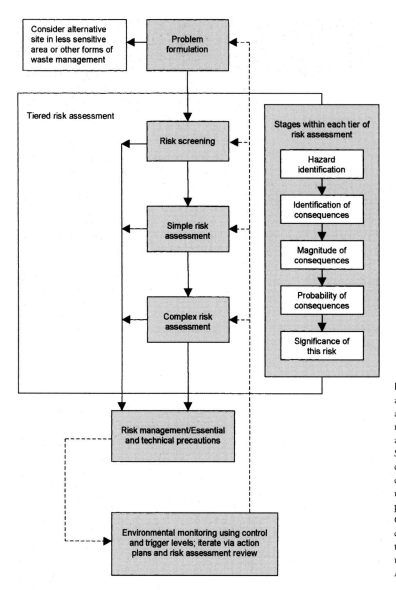

Fig. 7.10 Framework for a tiered approach to hydrogeological risk assessment for application in waste management. Two levels of risk assessment are recognized in the scheme. Simple risk assessment consists of quantitative calculations, typically deterministic analytical solutions using conservative (worst-case) input parameters, assumptions and methods. Complex risk assessment consists of quantitative, stochastic (probabilistic) techniques applied to analytical solutions using site-specific characterization data. After Leeson et al. (2003).

option was limited but had the most effect in reducing nitrate leaching losses by at least 80% and made an important contribution to the total reduction in nitrate losses in a catchment.

In response to the EU Directive on Diffuse Pollution by Nitrates (91/676/EEC; Council of European Communities 1991), a Nitrate Vulnerable Zone (NVZ) scheme has now been implemented in England (Fig. 7.11) with mandatory, uncompensated measures based on 'good agricultural practice' now applied to

about 55% of the land area of England (DEFRA 2002). The NVZ areas have been identified as lands draining into polluted waters, including surface waters and groundwaters which contain or could contain, if preventative action is not taken, nitrate concentrations in excess of 50 mg L^{-1}; and natural freshwater lakes, or other freshwater bodies, estuaries, coastal waters and marine waters which are eutrophic or may become so in the near future if action is not taken (DEFRA 2002).

BOX
7.5

The Drastrup Project, north Jutland, Denmark

A unique example of spatial planning to achieve sustainable land use in a groundwater catchment area with the aim of improving and protecting groundwater quality is provided by the Drastrup Project in north Jutland, Denmark. The motivation for the project was decreasing water quality from the diffuse input of nitrate and pesticides, with concentrations of nitrate up to 125 mg L^{-1} in groundwater. The Drastrup area, covering about 870 ha, is one of two large groundwater catchment areas for the city of Aalborg, contributing 20%, or approximately 1.7×10^6 m^3, of the annual municipal water supply demand (Fig. 1).

To protect the underlying Chalk aquifer and demonstrate the effects of land use change on nitrate leaching, it was decided to apply a municipal planning act to execute voluntary land distribution among farmers in the groundwater catchment area near Frejlev and Drastrup. The farm owners received compensation in the form of either payments or as land outside of the groundwater catchment area. The project also had the objective of establishing a recreational area close to Aalborg for the benefit of its citizens through the creation of a new, 230-ha recreational forest comprising a non-rotation, mixed age forest with some permanent grass

cover with low intensity grazing (Fig. 1). Planting began in the late 1990s and the new forest was inaugurated in September 2001 (Municipality of Aalborg 2001).

The initial preparation of soil in the new forest area by deep ploughing caused an initial flush of nitrate (up to 40 mg L^{-1} as N) but it is considered that as the trees grow, soil nitrate concentrations will begin to decline after 4–6 years. Trees demand a high nitrogen uptake during the first 15 years of growth, and so effectively reduce soil nitrate leaching, but later soil nitrate concentrations may increase as a result of atmospheric inputs of nitrogen to the ageing forest canopy. In areas of grassland, soil monitoring has shown that conversion to grassland quickly reduces nitrate concentrations to effectively zero within the first 2–3 years of conversion (Fig. 2a).

Since the implementation of land-use change in 1994, depth sampling of groundwater may now be showing signs of improvement in water quality, although further years of monitoring are required to identify unambiguously an improvement in water quality from natural background variation in nitrate content (Fig. 2b).

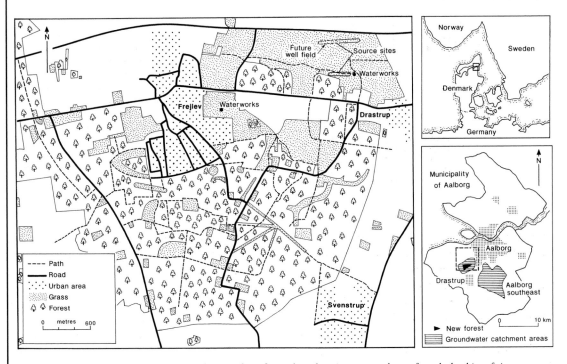

Fig. 1 The Drastrup area, Denmark, showing the area of new forest planted to protect groundwater from the leaching of nitrate. After the Municipality of Aalborg (2001).

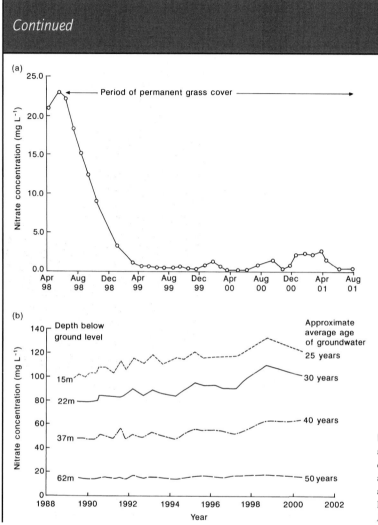

Fig. 2 The reduced nitrate concentrations achieved in (a) soil water beneath an area converted from arable to permanent grass and (b) in depth samples of groundwater in an area of new forest in the Drastrup area, Denmark. After the Municipality of Aalborg (2001).

Table 7.3 Summary of measured soil nitrate losses prior to (1990/91) and during the Pilot Nitrate Sensitive Area (NSA) scheme. Fluxes have been adjusted to mean rainfall conditions. After ADAS (2003).

Crop type	N loss winter 1990/91			Mean N loss in winters of 1992/3, 1993/4, 1994/5		
	kg N ha^{-1} (adjusted)	mg L^{-1} as NO$_3^-$	n	kg N ha^{-1} (adjusted)	mg L^{-1} as NO$_3^-$	n
Potatoes, sugar beet	92 ± 23	228 ± 52	10	38 ± 4	87 ± 9	46
Cereals	40 ± 4	102 ± 11	61	53 ± 3	85 ± 5	142
Grass	44 ± 7	102 ± 16	27	42 ± 8	69 ± 14	79
Premium scheme	–	–	0	6 ± 4	9 ± 6	70
All sites	65 ± 12	163 ± 29	108	47 ± 3	79 ± 5	380

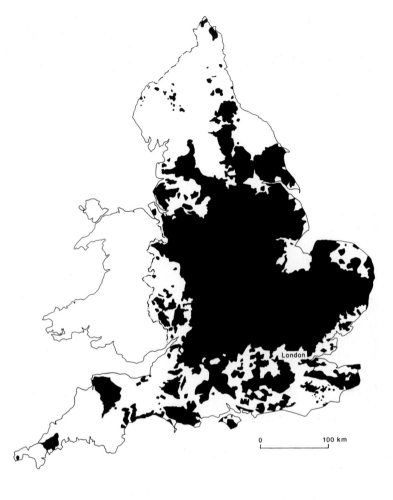

Fig. 7.11 Areas of Nitrate Vulnerable Zones (NVZs) in England. The following rules apply under this mandatory, uncompensated scheme based on 'good agricultural practice': (i) closed periods for inorganic nitrogen (fertilizer) applications during the autumn and winter and for organic nitrogen (manure) applications during the autumn for both arable and grass lands; (ii) nitrogen limits applied for inorganic and organic nitrogen applied to arable and grass lands that do not exceed crop requirements; (iii) spreading controls to restrict fertilizer and manure applications on steep slopes or close to water courses; (iv) slurry storage for manure during the autumn closed period; and (v) record keeping of agricultural practices for at least 5 years. Further, specific details are provided by DEFRA (2002).

It is likely that the impact of NVZs on reducing nitrate leaching will be modest and probably smaller than the 16–28% reduction achieved in nitrate leaching fluxes reported for the earlier NSA scheme. In critical areas, land-use change similar to the Premium NSA scheme will be needed if nitrate concentrations in surface waters and groundwaters are to meet the EU drinking water quality standard of 50 mg L^{-1} (Appendix 9).

7.4 Groundwater protection strategies in developing countries

Groundwater is extensively used for drinking water supplies in developing countries, especially in smaller towns and rural areas where it is often the cheapest and safest source. Waste water disposal is often by means of unsewered, pour-flush pit latrines that provide adequate waste disposal at a much lower cost than main sewerage systems. In cases where thin soils are developed on aquifer outcrops, then there is the risk of direct migration of pathogenic microbes, especially viruses, to underlying groundwater resources. The inevitable result will be the transmission of water-borne diseases. A further problem with human wastes is the organic nitrogen content which can cause widespread and persistent problems of nitrate in water, even where dilution and biological reduction processes occur.

Groundwater pollution problems are exacerbated in less developed areas without significant regional groundwater flow to provide dilution and by the use of inorganic fertilizers and pesticides in an effort to

Jaffna municipal area		Valigamam region		Valigamam region	
Distance (m)	% dug wells	Distance (m)	% dug wells	Lining (m)	% dug wells
<1.5	5.7	<10	13.6	<1.0	7.4
1.6–3.0	8.0	10.1–20.0	48.2	1.1–3.0	41.5
3.1–4.5	5.7	>20.1	38.2	>3.1	38.5
4.6–6.0	6.8			Damaged	12.6
>6.1	73.8				

Table 7.4 Percentage distributions of distances between pit latrines and dug wells and length of lining in dug wells in the Jaffna Peninsula, Sri Lanka. After Rajasooriyar et al. (2002).

secure self sufficiency in food production. Also, the use of irrigation to provide crop moisture requirements poses the risk of leaching of nutrients, especially from thin, coarse-textured soils. Increases in chloride, nitrate and trace elements will result from excessive land application of waste water, sewage effluent and sludge, and animal slurry.

Other pollution sources occur in urban areas where increasing numbers of small-scale industries, such as textiles, metal processing, vehicle maintenance and paper manufacture, are located. The quantities of liquid effluent generated by these industries will generally be discharged to the soil, especially in the absence of specific control measures and the prohibitive cost of waste treatment. Larger industrial plants generating large volumes of process water will commonly have unlined surface impoundments for the handling of liquid effluents.

Unless shallow dug wells have adequate protection from surface water runoff and are sufficiently distant from pit latrines, this type of groundwater source is vulnerable to both water table decline in drought periods, and to contamination. Although simple measures such as boiling can help combat waterborne diseases, it is understandable that the large aid programmes in the last few decades have focused on drilling deeper boreholes and installing simple pumping apparatus. As a result, hand-pumped tube wells are very common across much of Africa and Asia, but even these sources are now associated with problems, as illustrated graphically by the natural occurrence of arsenic in groundwater in Bangladesh and West Bengal in India (Box 7.6).

The natural soil profile can be effective in purifying human wastes, including the elimination of faecal microbes, and also in the adsorption, breakdown and removal of many chemicals. Given the potential for groundwater pollution in developing countries, protection of water supplies requires a broad-based approach that should include a strategy of minimum separations, depending on the hydro-geological situation, between a groundwater supply source and pit latrines for microbiological protection. The water laws and codes of practice of many countries require a minimum spacing between groundwater supply source and waste disposal unit of 15 m. There is, however, considerable pressure to reduce this permitted spacing to as little as 5 m in some developing countries such as Bangladesh and parts of India and Sri Lanka (Table 7.4), often resulting from the lack of space in very densely populated settlements. This example of law governing the location of waste disposal units demonstrates that criteria for groundwater pollution protection is rather arbitrary, based on limited or no technical data.

Other practical recommendations include the delineation of dilution zones of modified land use to alleviate the impact of polluting activities (Foster 1985) and the replacement of unsanitary municipal dumpsites, or tips, by controlled landfills using simple technology at a sustainable and realistic cost appropriate to gross domestic product (GDP). For example, in Tanzania and the Gambia, controlled but unlined landfills at existing quarry sites have been proposed that will operate on a dilute and disperse basis. Risk assessments demonstrate that local aquifers are not at risk, such that some local groundwater contamination is acceptable in return for major improvements in health and hygiene resulting from the removal of the current dumpsites (Griffin & Mather 1998).

BOX
7 . 6

Arsenic pollution of groundwater in southern Bangladesh

The Quaternary alluvial aquifers of Bangladesh provide drinking water for 95% of the population and also most of the water used for irrigation (Rahman & Ravenscroft 2003). Relative to surface water, the groundwater is bacteriologically safe, and its increased exploitation since the late 1970s has probably saved many millions of lives that would otherwise have been lost to water-borne diseases resulting from the use of contaminated surface water sources.

Arsenic was first detected in groundwater in Bangladesh in 1993, when analysis was prompted by increasing reports of contamination and sickness in the adjoining state of West Bengal in India. Prolonged exposure to inorganic arsenic in water causes

a variety of ailments including melanosis (a darkening of the skin), keratosis (a thickening of the skin, mostly on hands and feet), damage to internal organs and, ultimately, cancer of the skin or lungs.

Groundwater studies have demonstrated the wide extent of arsenic occurrence in Bangladesh (Dhar et al., 1997; DPHE 1999) at concentrations greater than the Bangladesh regulatory limit for arsenic in drinking water of 50 μg L^{-1} and the World Health Organization (1994) recommended limit of 10 μg L^{-1}. These regional surveys have shown that aquifers of the Ganges, Meghna and Brahmaputra floodplains are all affected in parts, making this the most extensive occurrence of groundwater pollution in the world (Table 1, Fig. 1). It is estimated that at least 21 million people

(a) (b)

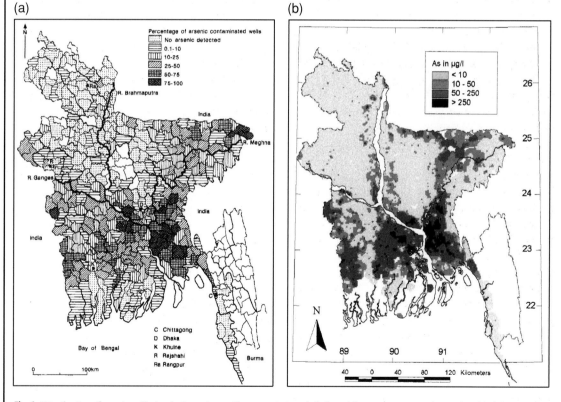

Fig. 1 Distribution of arsenic pollution in the main aquifer system in Bangladesh. (a) The map represents the results of the DPHE (1999) survey showing the percentage of contaminated wells in the subdistricts (upazilas) of Bangladesh that exceed the Bangladesh regulatory limit for arsenic in drinking water of 50 μg L^{-1}. The map may misrepresent the true percentage of wells that are contaminated by arsenic because of sampling bias (many samples were collected at wells where arsenic poisoning was suspected) and measurement inaccuracies (field test kits where used do not reliably indicate values exceeding 50 μg L^{-1} when arsenic concentrations are in the range 50–200 μg L^{-1}). Although the affected areas are unlikely to change, the percentages shown may be revised in the future. (b) The interpolated surface, generated using ARCVIEW SPATIAL ANALYST software on log-transformed data, shows the spatial distribution of groundwater arsenic concentrations in wells less than 150 m deep using data based on the surveys of DPHE (1999, 2000) which are available from http://www.bgs.ac.uk/arsenic/Bangladesh.

Table 1 Frequency distribution of arsenic concentrations in groundwater from a regional survey of the 41 worst affected districts of Bangladesh. After DPHE (1999).

Arsenic concentration class* (μg L^{-1})	Frequency	Percentage frequency	Percentage in or above concentration class
<10	998	49	100
10–50	319	16	51
50–100	209	10	35
100–250	268	13	25
250–500	168	8	11
500–1000	57	3	3
>1000	3	0.1	0.1

* WHO recommended limit = 10 μg L^{-1}; Bangladesh regulatory limit = 50 μg L^{-1}.

are presently drinking water containing more than 50 μg L^{-1} of arsenic, while probably more than double this number are drinking water containing more than 10 μg L^{-1} of arsenic (DPHE 2000; Nickson et al. 2000; Burgess et al. 2002). The number of persons who must be considered 'at risk' of arsenic poisoning is even higher because testing of the 5–10 million tubewells in Bangladesh will take years to complete.

The cause of the elevated arsenic concentrations in the Ganges–Meghna–Brahmaputra deltaic plain is thought to relate to the microbial reduction of iron oxy-hydroxides contained in the fine-grained Holocene sediments and the release of the adsorbed load of arsenic to groundwater. It has been proposed that the reduction is driven by microbial metabolism of buried peat deposits (McArthur et al. 2001; Ravenscroft et al. 2001). The presence of abundant organic matter is expected in deltaic or fluvial areas that supported peat formation during climatic optimums.

Severely polluted aquifers are all of Holocene age, although at a local scale the distribution of arsenic pollution is very patchy. There are cases of grossly polluted boreholes, pumping groundwater with arsenic at concentrations greater than 1000 μg L^{-1} being separated spatially by only a few tens of metres from boreholes pumping groundwater with arsenic at concentrations less than 10 μg L^{-1} (Burgess et al. 2002). However, there are many cases where almost all wells in a village contain more than 50 μg L^{-1}.

Data presented by DPHE (1999) show that the highest percentage of wells that contain arsenic concentrations above the regulatory limits of 10 and 50 μg L^{-1} occur at depths between 28 and 45 m (Table 2). Hand-dug wells are mostly less than 5 m deep and are usually unpolluted by arsenic, but the risk of bacteriological contamination is high. Below 45 m there is a decrease in the percentage of wells that are polluted, but the risk remains significant until well depths exceed 150 m, the maximum depth of river channel incision during the Last Glacial Maximum at 18 ka.

Even so, across much of southern Bangladesh, more than 50% of boreholes in the shallow aquifer have arsenic levels that comply with the 50 μg L^{-1} limit and so continued development of the alluvial aquifers may still be possible, at least in the medium term (Burgess et al. 2002; Ravenscroft et al. 2004).

The problem of arsenic in groundwater is not only confined to Bangladesh and West Bengal. According to Smedley and Kinniburgh (2002), the areas with large-scale problems of arsenic in groundwater tend to be found in two types of environment: inland or closed basins in arid and semi-arid areas; and strongly chemically reducing alluvial aquifers. The hydrogeological situation in these areas is such that the aquifers are poorly flushed and any arsenic released from the sediments following burial tends to accumulate in the groundwater.

Areas containing high-arsenic groundwaters are well known in Argentina, Chile, Mexico, China and Hungary, but the problems in Bangladesh, West Bengal and, additionally, Vietnam are more recent (Smedley & Kinniburgh 2002). In Vietnam, the capital Hanoi is situated at the upper end of the Red River Delta and analysis of raw groundwater pumped from the lower Quaternary alluvial aquifer gave arsenic concentrations of 240–320 μg L^{-1} in three of the city's eight treatment plants and 37–82 μg L^{-1} in another five plants (Berg et al. 2001). In surrounding rural districts, high arsenic concentrations found in tubewells in the upper aquifers (48% above 50 μg L^{-1} and 20% above 150 μg L^{-1}) indicate that several million people consuming untreated groundwater might be at a high risk of chronic arsenic poisoning (Berg et al. 2001).

As in Bangladesh, the source of arsenic in the Red River Delta sediments is believed to be associated with iron oxy-hydroxides that release arsenic to groundwater under chemically reducing conditions. A characteristic feature of arsenic contamination of wells in both Bangladesh and Vietnam is the large degree of spatial variability in arsenic concentrations at a local scale. As a result, it is difficult to know when to take action to provide arsenic-free water sources. For now, it appears safer to analyse each well until further research has been completed into the sources, controls and distribution of arsenic in susceptible areas.

Table 2 Distribution of arsenic concentrations in groundwater by well depth from a regional survey of the 41 worst affected districts of Bangladesh. After DPHE (1999).

Well depth (m)	Number of samples	Number of samples above 50 μg L^{-1}	Percentage of samples above 50 μg L^{-1}
<10	36	12	33
10–30	582	339	58
30–100	1032	334	32
100–200	92	18	20
>200	280	2	0.7
Total	2022	705	35

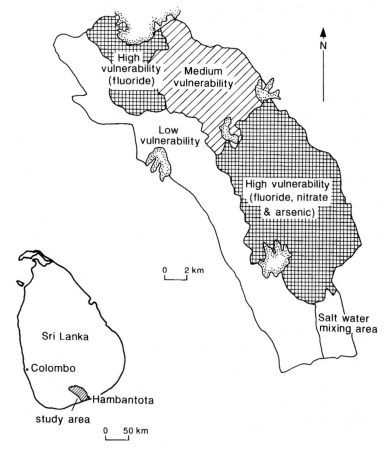

Fig. 7.12 Groundwater vulnerability map for the lower Walawe river basin, Sri Lanka, derived qualitatively with regard to the locally important human health risk parameters of fluoride, arsenic and nitrate in groundwater from dug wells and tube wells. After Rajasooriyar (2003).

The application of groundwater vulnerability mapping using available information and scientific knowledge is a valuable aid in educating people about the potential risks of groundwater pollution and explaining the need for appropriate land-use management. An example groundwater vulnerability map for the Uda Walawe basin in Sri Lanka is shown in Fig. 7.12. The map was developed with regard to the locally important risk parameters of fluoride, arsenic and nitrate in groundwater from dug wells and tube wells (Rajasooriyar 2003). Microbiological factors, although important, were not included in the derivation of the map due to insufficient information. The following factors were combined, using expert judgement, to classify qualitatively areas as low, medium and high groundwater vulnerability:

1 Geological factors: mineralogy (fluoride- and arsenic-bearing minerals); geological structure (divided into the fractured Highland Series and less fractured Eastern Vijayan Complex).

2 Hydrogeological factors: shallow regolith (weathered) aquifers; deep, hard rock (fractured) aquifers with low/moderate and high transmissivity zones.

3 Recharge conditions: areas subject to low rainfall recharge only; areas subject to high rainfall and irrigation recharge.

4 Salt water mixing: areas close to the coast subject to saline intrusion and sea-salt spray; unaffected areas away from the coast.

Interestingly, it was found that nitrate and phosphate do not pose an immediate groundwater pollution risk in areas of banana and paddy cultivation in the Highland Series and Eastern Vijayan Complex regions, even though these crops are subject to irrigation recharge and the application of large amounts of fertilizer input. It appears that denitrification occurs where waterlogged paddy field soils are present to explain the absence of widespread nitrate contamination (Rajasooriyar 2003).

7.5 FURTHER READING

Addiscott, T.M., Whitmore, A.P. & Powlson, D.S. (1991) *Farming, Fertilizers and the Nitrate Problem*. C.A.B International, Wallingford, Oxon.

Ellis, B. (ed.) (1999) Impacts of urban growth on surface water and groundwater quality. In: *Proceedings of IUGG 99, Symposium HS5, University of Birmingham, July 1999*. IAHS Publ. No. **259**.

Environment Agency (1998) *Policy and Practice for the Protection of Groundwater*. Environment Agency, Bristol.

Fetter, C.W. (1999) *Contaminant Hydrogeology*, 2nd edn. Prentice-Hall, New Jersey.

Hemond, H.F. & Fechner, E.J. (1994) *Chemical Fate and Transport in the Environment*. Academic Press, San Diego.

Morris, B.L., Lawrence, A.R.L., Chilton, P.J.C., Adams, B., Calow, R.C. & Klinck, B.A. (2003) *Groundwater and its Susceptibility to Degradation: a global assessment of the problem and options for management*. Early Warning and Assessment Report Series, RS. 03-3. United Nations Environment Programme, Nairobi, Kenya.

National Academy of Sciences (2000) *Natural Attenuation for Groundwater Remediation*. National Academy Press, Washington, DC.

Ward, C.H., Cherry, J.A. & Scalf, M.R. (1997) *Subsurface Restoration*. CRC Press, Boca Raton, Florida.

Wickramanayake, G.B., Gavaskar, A.R. & Chen, A.S.C. (eds) (2000) *Chemical Oxidation and Reactive Barriers: remediation of chlorinated and recalcitrant compounds*. Battelle Press, Columbus, Ohio.

Groundwater resources and environmental management

8

8.1 Introduction

The development of groundwater resources for public, agricultural and industrial uses can create environmental conflicts. Groundwater abstractions capture recharge water that might otherwise flow to springs and rivers and so diminish the freshwater habitats dependent on groundwater discharge. In the current era of integrated river basin management (see Section 1.8), sufficient volume of water is required to maintain freshwater (or saline) ecosystems. In this way, the fraction of available recharge needed for environmental benefits is accounted for, together with the fraction required for human and economic benefits in order to achieve sustainable groundwater development (see Fig. 1.1).

In this chapter, examples of sustainable and non-sustainable groundwater resources development schemes at large and regional scales are discussed together with examples of modern groundwater management techniques including artificial storage and recovery and riverbank filtration schemes. The adverse environmental impacts of groundwater exploitation are illustrated with reference to the sensitivity of river flows and wetlands to groundwater inputs. Lastly, the possible changes in groundwater resources, both quantity and quality, as a result of climate change are reviewed based on current knowledge of predicted scenarios of future climate.

8.2 Groundwater resources schemes

The assessment and development of groundwater resources is central to hydrogeology. Groundwater resources have a number of positive advantages compared to surface reservoir developments that include: (i) a large storage volume that can be developed in stages as demand for water arises; (ii) resilience to drought conditions because of the large storage volume; (iii) relatively low environmental impact of well-field developments; and (iv) no loss of storage volume to evaporation. With current awareness that surface water and groundwater resources should be managed together, it is useful to conceive this approach in the context of a water balance equation that equates demand for water against abstraction requirements and environmental needs.

8.2.1 Water balance equation

The basic raw resource within a catchment is precipitation (P) and, as shown in Fig. 8.1, precipitation is either lost to evaporation and transpiration (usually combined and referred to as evapotranspiration, ET) (see Section 5.3) or routed through the hydrological pathways of overland flow and interflow to give surface water runoff (S_R) and groundwater discharge (G_R) (see Fig. 5.26). The groundwater discharge component is supplied by groundwater recharge and includes natural discharge to springs and rivers (the river baseflow, Q_G) and artificial abstractions (Q_A). Depending on the catchment geology, the groundwater catchment may or may not coincide with the surface catchment area such that additional components of cross-formational groundwater flow (Q_U) may need to be considered. Now, assuming that the surface water and groundwater catchments coincide, the following catchment water balance equation can be written:

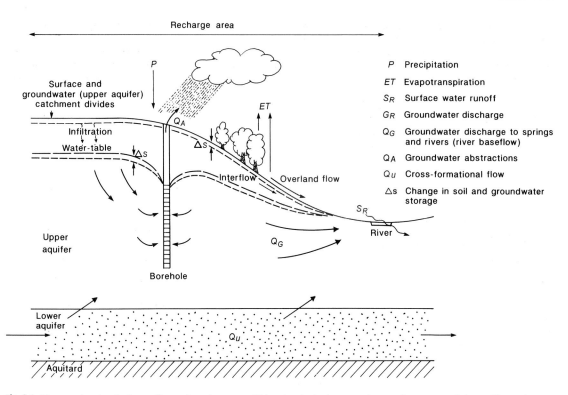

Fig. 8.1 Diagram showing the inputs, flowpaths and outputs within a river basin that comprise a catchment water balance. The total groundwater discharge, G_R, is the sum of $Q_G + Q_A + Q_U$.

$$P = ET + S_R + G_R \pm \Delta S \qquad \text{eq. 8.1}$$

On a short timescale of weeks to months, equation 8.1 is balanced by changes in the water held in soil and groundwater storage (ΔS) as represented by changes in soil moisture content and groundwater levels. Over longer timescales of several years, changes in storage balance out to zero and, expressing $G_R = Q_G + Q_A$, equation 8.1 becomes:

$$P - ET = S_R + Q_G + Q_A \qquad \text{eq. 8.2}$$

In other words, the difference between precipitation and evapotranspiration, or effective precipitation, supports surface runoff, groundwater discharge as river baseflow and borehole abstractions. The total flow in a river is calculated as the sum of S_R and Q_G. Methods for calculating the amount of effective precipitation (hydrological excess) are discussed in Section 5.5. Clearly, if equation 8.2 is to be balanced, any increase in the amount of groundwater abstrac-

tion (Q_A) will be at the expense of river baseflow (Q_G) and so potentially causing environmental impacts. As a general rule, a useful measure of the quantity of water required to maintain acceptable minimum river flows is the Q95 low flow statistic. This means that during the baseflow recession period, when river flows are dominated by the groundwater input (Q_G), it is undesirable for the river flow to fall below the long-term average of flow that is equalled or exceeded for 95% of the time.

For sustainable groundwater development, and in order to meet conflicting environmental (Q_G) and socioeconomic (Q_A) demands, it is highly desirable that these demands do not exceed the rate of groundwater recharge (equated to G_R). Three conditions can be defined in which: (i) $Q_A < G_R - Q_G$ and further groundwater resources are available for exploitation; (ii) $Q_A = G_R - Q_G$, in which case the safe yield has been achieved; and (iii) $Q_A > G_R - Q_G$ when the groundwater resources are over-exploited or mined. In the case of (iii), groundwater in support of abstractions

Table 8.1 Groundwater budget information for the High Plains Aquifer based on the United States Geological Survey's Regional Aquifer System Analysis groundwater model.* After Luckey et al. (1986).

Budget parameter	Northern High Plains	Central High Plains	Southern High Plains
Primary inflows (in cubic metres per year)			
Recharge from precipitation on rangeland and streams	5.98×10^8†	4.66×10^8†	1.97×10^8
Recharge from precipitation on agricultural land‡	2.89×10^9	–	1.43×10^9
Groundwater irrigation return (pumpage minus crop demand)	2.31×10^9	2.07×10^9	3.61×10^9
Recharge from other human activities (e.g. seepage from reservoirs and canals)	2.31×10^9	–	–
Recharge from other aquifers across subunit boundary	–	1.88×10^7§	–
Totals	8.11×10^9	2.55×10^9	5.24×10^9
Primary outflows (in cubic metres per year)			
Total pumpage	6.48×10^9	6.89×10^9‖	8.59×10^9
Discharge to streams and shallow water-table areas	2.87×10^9	4.15×10^8	–
Discharge along eastern boundary	–	6.97×10^7	1.05×10^8
Totals	9.35×10^9	7.37×10^9	8.70×10^9
Net residual	-1.24×10^9	-4.82×10^9	-3.46×10^9

* Assumptions: Inflow/Outflow values determined using 1960–1980 estimates; base of aquifer modelled as no-flow boundary; vertical flow in aquifer considered negligible on regional scale.
† Recharge distributed unevenly based on soil type.
‡ Additional recharge from precipitation on agricultural land because of changes in soil character due to tillage.
§ Flow only from northern and southern subunits to central subunit.
‖ Municipal and industrial pumpage is 3.2% of this amount.

is taken from aquifer storage with potential long-term impacts on groundwater levels.

8.2.2 Large-scale groundwater development schemes

The understanding of aquifer conditions and the compilation of a water balance are central to water resources management. An example of a groundwater budget for a large-scale aquifer system, in which the primary inflows and outflows to the aquifer are tabulated, is shown in Table 8.1 for the United States High Plains Aquifer. The High Plains Aquifer consists mainly of near-surface deposits of late Tertiary or Quaternary age forming one unconfined aquifer and underlies 450,660 km² in parts of eight States within the Great Plains physiographic province. The Ogallala Formation of Miocene age, which underlies 347,060 km² is the principal hydrogeological unit and consists of a heterogeneous sequence of clay, silt, sand and gravel. Use of the High Plains Aquifer as

a source of irrigation water has transformed the mid-section of the United States into one of the major agricultural regions of the world. Principal crops are cotton, alfalfa and grains, especially wheat, sorghum and maize. Grains provide feed for the 15 million cattle and the 4.25 million swine (1997) that are raised over the aquifer. In addition, the aquifer provides drinking water to 82% of the people who live within the aquifer boundaries (Dennehy et al. 2002).

According to Dennehy et al. (2002), groundwater flow in the High Plains Aquifer is generally from west to east, discharging naturally to springs and streams and is subject to evapotranspiration in areas where the water table is close to the land surface. Pumping from numerous irrigation wells is, however, the principal mechanism of groundwater discharge. Abstractions greatly exceed recharge in many areas, causing large declines in water levels; for example, declines of 30–43 m since the 1940s to 1980 in parts of Kansas, New Mexico, Oklahoma and Texas.

From the groundwater budget shown in Table 8.1, it is apparent that each region of the High Plains

Aquifer is in a deficit situation with outflows greater than inflows. In the southern and central regions, outflows are about twice the inflows. Prior to irrigation development, precipitation recharged the aquifer at an average rate of 15 mm a^{-1} and small quantities of water discharged to the springs and rivers. Irrigation represents a largely consumptive use of water and since development, groundwater abstraction has removed about 7% of the original total water volume from the aquifer. The water budget indicates that the decrease in storage would have been worse if 30–40% of the water pumped for irrigation had not infiltrated to the aquifer as irrigation return flows each year. Declining groundwater levels are a direct threat to the current way of life of the area, and Dennehy et al. (2002) detail attempts that are being made to introduce more water-efficient irrigation

and best-management farming practices including, ultimately, a shift from irrigated agriculture to dry-land farming with its attendant far-reaching implications for the local economy.

Although the High Plains Aquifer presents a case of non-sustainable development of groundwater, in practice there is no fundamental reason why the temporary over-exploitation of aquifer storage for a given benefit should not be allowed as part of a logical water resources management strategy as long as the groundwater system is sufficiently well understood to evaluate impacts. An example of the deliberate mining of groundwater resources for the benefit of a national economy is the Great Man-made River project (GMRP) in Libya (Box 8.1). The GMRP involves the abstraction of fossil groundwater recharged during a pluvial period of the last ice age

The Great Man-made River Project, Libya

BOX 8.1

Libya covers an area of some 1.8×10^6 km^2 bounded in the north by the Mediterranean coastline of approximately 1600 km in length (Fig. 1). The climate varies from Mediterranean along the coast with winter rainfall totals of about 270 mm in Tripoli and Benghazi to a desert climate in the south where rain seldom falls. Rainfall is higher at about 400 mm a^{-1} in the mountainous areas to the south of Tripoli and east of Benghazi. A total of 80% of Libya's population of about 4.7 million is concentrated in the coastal area with approximately 2.5 million residents in the major towns and cities of Tripoli and Benghazi. The majority (80%) of Libya's agricultural production is centred on the coastal plains and adjacent uplands but arable crops are only viable with irrigation. Importation of cereals, sheep meat and other food stuffs is costly in terms of foreign exchange and has given impetus to improving the productivity of the land through irrigation and creating employment opportunities in agriculture. However, the relatively shallow aquifers in the coastal areas and the pressure of competing municipal and industrial demands for water pose a limitation on further agricultural development.

During the 1960s, exploratory drilling for oil in the Libyan desert established the presence of extensive groundwater reserves in three to four major basins but an unwillingness of people to move to the Libyan desert to utilize this resource for irrigation gave rise to the State decision to initiate the Great Man-made River Project (GMRP), one of the world's major groundwater developments. The total capacity of the major basins, which is estimated to be 35,000 km^3, is immense and the only technical constraints on development are considered to be local ones of water quality and aquifer hydraulics.

The principal features of the GMRP are described by McKenzie and Elsaleh (1994) and include well-fields and conveyance systems comprising long lengths of very large diameter (4 m) pipelines required for the transfer of water to coastal districts. Phase I well-

fields at Sarir and Tazerbo (Fig. 1) are situated 381 km and 667 km south of Ajdabiya in the Sirt and Kufra Basins, respectively. The Sirt Basin contains a post-Eocene thickness of 1600 m of continental sands with increasing amounts of limestone found northwards. The important aquifer sediments of the Kufra Basin are Lower Devonian well-sorted, uncemented sand and sandstone. Regional groundwater contours indicate a small natural groundwater flow from south to north which is believed to be fed mainly from degradation of a mound of fossil groundwater located below the northern flank of the Tibisti mountains. This palaeogroundwater was recharged during a major pluvial period at the time of the last ice age between 14,000 and 38,000 years ago. Present recharge from rainfall is effectively zero in the central part of the Sirt Basin and Kufra Basin. Therefore, the eastern region well-field developments are dependent on groundwater mining with only a minor contribution from interception of throughflow (Pim & Binsariti 1994).

The Sarir well-field contains 136 production wells drilled at 450 m depth, each able to deliver 92 L s^{-1}. Pumping tests indicated that transmissivity ranges from 400 m^2 day^{-1} to 6000 m^2 day^{-1}. The pumped water quality ranges from 530 mg L^{-1} to 1367 mg L^{-1} of total dissolved solids with an average of 815 mg L^{-1}. At Tazerbo, 118 production wells are drilled at depths varying between 380 and 600 m depth, with each borehole able to deliver 102 L s^{-1}. Modelled drawdowns for the Tazerbo well-field are 95 m after 50 years with the greatest contribution (86%) to abstraction from vertical leakage (Pim & Binsariti 1994). Assuming an average annual demand of 10,000 m^3 ha^{-1} for irrigation water, the limit for irrigated agriculture for the Phase I supply is about 70,000 ha.

Phases II and III well-field developments in the western region are planned on a similar scale in the Hammadah Al Hamra and Murzuq Basins which lie north and south of the Gargaf uplift, respectively

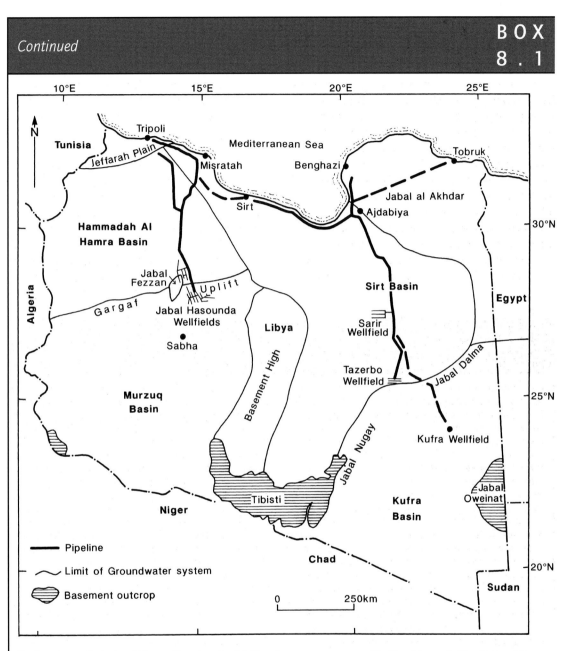

Fig. 1 Major aquifer basins of Libya and location of well-fields and conveyance systems of the Great Man-made River Project. After McKenzie and Elsaleh (1994) and Pim and Binsariti (1994).

(Fig. 1). Natural groundwater flows in the thick and complex sequence of layered and interconnected Palaeozoic and Mesozoic limestones and sandstones are generally from south to north, indicating that the southern mountains were a major recharge area during the last pluvial period. As in the eastern region, the source of present throughflow is considered to be the slowly depleting groundwater mound below the area of ancient recharge (Pim & Binsariti 1994).

The fully developed GMRP is expected to achieve a total groundwater abstraction of $6.18 \times 10^6 \, \mathrm{m^3 \, day^{-1}}$ and support approximately 200,000 ha of irrigated agricultural development with a minimum design life of 50 years. The total abstraction in 50 years is calculated to be $113 \, \mathrm{km^3}$ or 0.3% of the total groundwater resource. The estimated total cost of the GMRP is large at US$20 billion with an implementation period from the mid-1980s to 2005.

from beneath the Libyan desert and its transfer to coastal cities and towns. Clearly the groundwater supply is not a permanent, renewable resource but for a period of at least decades this fossil water can supply most of the population of Libya with an essential resource. As argued by Price (2002), the GMRP appears to meet the ethical requirements of groundwater mining, namely that: (i) evidence is available that pumping can be maintained for a long period; (ii) the negative impacts of development are smaller than the benefits; and (iii) the users and decision-makers are aware that the resource will be eventually depleted. However, Price (2002) is concerned that using non-renewable groundwater for inefficient agriculture in a region where there is effectively no recharge is not a sensible long-term plan.

8.2.3 Regional-scale groundwater development schemes

The above two examples of large-scale groundwater developments in support of irrigated agriculture tip the sustainability equation towards economic gain. At the regional scale, effective river basin management can develop the large storage volume in aquifers in conjunction with surface resources. This concept is not new and several schemes, principally the Thames, Great Ouse, Severn, Itchen and Waveney Schemes, were developed in England and Wales following the Water Resources Act of 1963 (see Section 1.7). Pilot studies were undertaken to assess the feasibility of regulating rivers by pumping groundwater into them for abstraction for water supply in the lower reaches, while maintaining an acceptable flow in the river to meet all other environmental requirements (Downing 1993).

As an example of a river regulation scheme, the Great Ouse Groundwater Scheme in eastern England, included two well groups in a pilot area of 71.5 km^2 in the River Thet catchment (Fig. 8.2): riverside wells and a second group more remote from the river on higher ground. A control area was established in the adjacent River Wissey catchment. Eighteen boreholes were drilled into the Chalk aquifer and yielded 70×10^3 m^3 day^{-1}. When all the wells were pumped for 250 days the pumping rate was three times the average infiltration across the

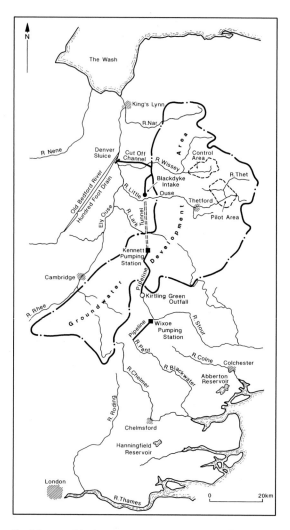

Fig. 8.2 Map of the location of the original development area of the Great Ouse Groundwater Scheme showing the various linked components of the current Ely Ouse Essex Water Transfer Scheme.

pilot area. In 6 months all the baseflow to the River Thet was intercepted and groundwater levels fell below the river bed over 80% of its length. Groundwater levels fell by an average of 2.5 m with stable conditions achieved after 200 days due to induced groundwater flow across the eastern boundary of the pilot area. In that the riparian zone is underlain by peats, silts and clays with a relatively low permeability, leakage through the river bed was 10% of the

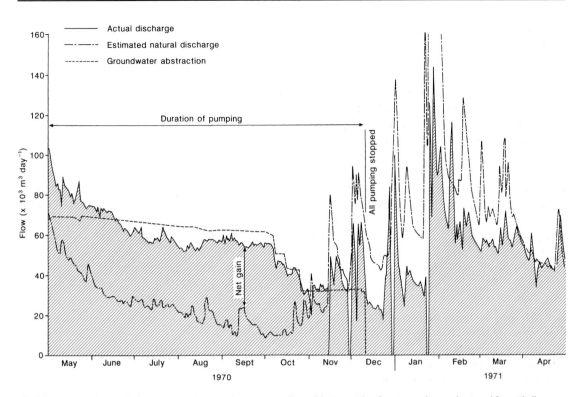

Fig. 8.3 Hydrograph records for 1970–1971 showing the net gain in flow of the River Thet from groundwater abstracted from Chalk boreholes in the pilot area of the Great Ouse Groundwater Scheme (see Fig. 8.2 for location). As shown, the net gain is the difference between the actual discharge with groundwater abstraction and the estimated natural discharge without abstraction. As a result of measurements taken for identical 24-hour periods without allowance for travel times (or temporary storage effects), occasional zero flows (or even apparent losses) can arise. After Backshall et al. (1972).

groundwater pumped. Concerns were expressed that pumping the Chalk aquifer could affect the Breckland Meres situated within the pilot area. Theses meres are groundwater-fed lakes of notable scientific and ecological importance. However, drawdown of the groundwater level by more than 1 m due to extensive pumping in 1970 was relatively limited within the pilot area (Backshall et al. 1972).

The success of a river regulation scheme can be expressed by the net gain to the river. The actual net gain at any given time depends on the aquifer properties and the distance of boreholes from natural spring discharges. During 1970 and 1971, and as shown in Fig. 8.3, the net gain in the Great Ouse Groundwater Scheme varied, but during summer periods values were about 70% of the quantity pumped. The scheme was therefore successful in maintaining river flows at a high proportion of the mean flow. After pumping stopped, river flows and groundwater

levels steadily recovered towards natural levels during the following winter. Natural river flows were re-established in time for the beginning on the following baseflow recession period (Fig. 8.3). The delay in recovery is a feature of river regulation schemes and is due to winter groundwater recharge first replenishing the groundwater storage depleted by pumping during the summer regulation period. By definition, the net gain to the river over a long period of years must be zero if environmental impacts are not to be permanent.

Full development of the Great Ouse Groundwater Scheme was envisaged to include almost 350 new boreholes distributed over an area of 2500 km². The possibility of developing the Chalk groundwater resource in conjunction with surface runoff was estimated to yield 360×10^3 m³ day^{-1} of water for export to demand areas outside of the basin. The drawdown of the Chalk aquifer for the fully developed scheme

Fig. 8.4 Transfer of water from the Ely Ouse River to the Cut-Off Channel at Denver. Water transferred from here travels to Abberton and Hanningfield reservoirs in support of water supply and river flows in the drier south-east of England (Fig. 8.2).

was predicted to be 4 m in a drought period with a 1 in 50-year return period. Such a large engineering programme of works was not carried forward partly because of conflicts with other users for Chalk groundwater, for example irrigation water for valuable vegetable crops, and also the change in philosophy towards protection of the environment, especially the Breckland Meres, from the effects of groundwater drawdown. However, the boreholes drilled for the pilot study are today incorporated in a much larger scheme, the Ely Ouse Essex Water Transfer Scheme, to provide groundwater to the large public water supply demand in the south-east of England.

The Ely Ouse Essex Water Transfer Scheme enables the transfer of surplus water from the Ely Ouse (the River Great Ouse) to the heads of Essex rivers in the south-east of the Anglian region (Fig. 8.2), thereby making extra water available to the Essex rivers. The county of Essex experiences conditions of low effective precipitation of less than 125 mm a^{-1}, yet has a large and expanding population to the north of London. One of the great merits of the scheme is that it augments existing reservoir capacity, thus avoiding the loss of agricultural land to create new reservoirs. Under the scheme, surplus water from the eastern part of the catchment, including Chalk groundwater resources available for regulation of the westward flowing Chalk rivers (the Rivers Lark, Little Ouse and Thet), is transferred to the flood protection scheme Cut-Off Channel at Denver (Fig. 8.4). At this point, all

upstream river needs have been met and so any surplus water that would otherwise be lost to tidal waters and eventually to the Wash is potentially available for transfer. The Impounding Sluice gate at Denver is designed to enable the water level in the river channel to be raised approximately 0.6 m, thereby producing a reversal of flow. The water is sent in a reverse direction approximately 25 km south-east to the Blackdyke Intake. Here it is drawn off into a 20-km-long tunnel which terminates at Kennett whereupon the water is pumped through a 14.3-km-long pipeline to the River Stour at Kirtling Green. Part of this discharge is drawn off at Wixoe, 13.7 km downstream and pumped 10.3 km to the River Pant. The water transferred from Denver travels 141 km to Abberton reservoir and 148 km to Hanningfield reservoir. For two-thirds of these distances, use is made of exiting watercourses.

8.2.4 Artificial storage and recovery schemes

As surface water and groundwater schemes reach full development, the final stage is artificial recharge where water, often treated wastewater, is recharged through basins and returned to the aquifer. Although practised in other countries, often on an uncontrolled basis and potentially threatening longer-term groundwater quality (for example in China and Mexico; Foster et al. 1999), artificial recharge is not typically practised in the United Kingdom where treated water is returned to rivers from sewage treatment works. A relatively recent development of artificial recharge is aquifer storage and recovery (ASR) to meet peak demands for water. The ASR technique, shown schematically in Fig. 8.5, works on the principle of using boreholes to recharge drinking water-quality water into aquifers and to subsequently recover the stored water from the same boreholes during times of peak demand or drought periods. Such schemes operate by displacing the native groundwater, effectively creating an underground reservoir of near drinking water-quality water. The volume of water recovered from the aquifer for supply purposes is generally close to but not more than the volume injected (Eastwood & Stanfield 2001). ASR schemes are therefore considered to offer a sustainable means of groundwater development.

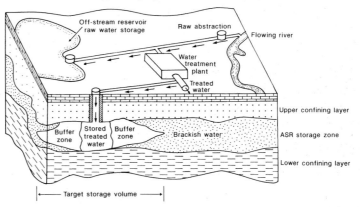

Wet season - aquifer storage replenishment

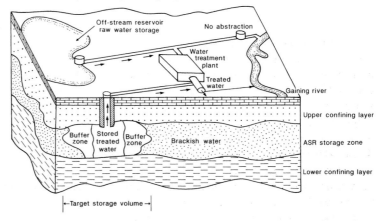

Dry season - aquifer storage depletion

Fig. 8.5 Schematic diagram showing the operation of an aquifer storage and recovery (ASR) scheme. ASR involves storage of available, principally surface or drinking water-quality water during the wet season through boreholes completed in a brackish-water aquifer, with subsequent retrieval from these same boreholes during dry periods. The freshwater forms a 'bubble' within the aquifer around the ASR well and can be retrieved when needed. The hydrogeological characteristics of a successful ASR storage zone include: moderate aquifer hydraulic conductivity; confinement above and below by low permeability strata; and water quality as fresh as possible to limit mixing with the native brackish water. After the United States Geological Survey (http://sofia.usgs.gov/sfrsf/hydrology/ASR/storagebig.jpg).

The criteria for deciding how much water can be recovered from an ASR scheme are normally based on water quality parameters. The recovery efficiency is calculated as the volume of water recovered when the water quality parameter is reached compared to the volume of water injected. It is not uncommon for recovery efficiencies to approach 100%. Full details of ASR programme development, system design and technical and non-technical issues are presented by Pyne (1995).

By the mid-1990s more than 20 ASR schemes were operational in the United States with many more planned. ASR is popular in that improvements in water quality in the native aquifer result from the injection of high-quality water, which then allows these aquifers to be used for supply purposes at a lower cost than other resource options. One ambitious

scheme is the application of ASR technology as part of the Everglades Restoration Project in South Florida (United States Geological Survey 2002). Between 300 and 330 ASR boreholes are planned with a combined capacity exceeding 6×10^6 m^3 day^{-1}, storing freshwater in the deep brackish Upper Floridan Aquifer (the ASR zone) using surface water that is currently discharged to the tide during the wet season. The stored water will be recovered during drought periods to sustain delivery of adequate high quality water to the Everglades (Box 8.6). ASR has not previously been implemented on this scale and key uncertainties include: the compatibility of the injected water with the aquifer water; effects of large volumes of injected water on the confining unit; efficiency in terms of how much water will be recovered; and the effects of the recovered water on the environment. To answer these concerns, a phased approach to ASR implementation is being adopted to evaluate its feasibility and effectiveness as a regional water storage option.

Attempts at ASR in the United Kingdom have met with mixed success. A full-scale trial of ASR on a confined Chalk aquifer in Dorset in the south of England proved that there were no detrimental environmental effects; for any injection and recovery scenario, the impacts are absorbed by aquifer storage. However, a fluoride concentration of 2 mg L^{-1}, equal to half of the background groundwater fluoride concentration, led to disappointingly low recovery efficiencies (<15%) (Eastwood & Stanfield 2001). In contrast to the Dorset scheme, ASR has been successfully operated in the confined Chalk and Tertiary Basal Sands aquifer in North London (Box 8.2).

8.2.5 Riverbank filtration schemes

In many countries of the world, alluvial aquifers hydraulically connected to a water course are preferred sites for drinking water production given the relative ease of shallow groundwater exploitation, the generally high production capacity and the proximity to demand areas (Doussan et al. 1997). Although proximity to a river can ensure significantly higher recharge and pumping rates, water quality problems may be encountered during exploitation of riverbank well-fields (Bertin & Bourg 1994). Even with these problems, groundwater derived from infiltrating river water provides 50% of potable supplies in the Slovak Republic, 45% in Hungary, 16% in Germany and 5% in The Netherlands. In Germany, riverbank filtration supplies 75% of the water supply to the City of Berlin and is the principal source of drinking water in Düsseldorf, situated on the Rhine (Box 8.3).

In the United States, the water supply industry has adopted the broadly defined regulatory concept of 'groundwater under the direct influence' (GWUDI) of surface water (variably defined and implemented in response to local conditions by each State, Tribe or other regulatory agent). Groundwater sources in this category are considered at risk of being contaminated with surface water-borne pathogens (specifically disinfection-resistant pathogenic protozoa such as

---

The North London Artificial Recharge Scheme

BOX 8.2

The North London Artificial Recharge Scheme and its relationship to existing surface water resources in the Lea Valley are shown in Fig. 1. In average rainfall years, flows in the Rivers Lea and Thames, with the associated pumped-storage reservoirs, are sufficient to meet current demands. Normally there is surplus water which can be used to increase aquifer storage. During a drought, when river flows and associated storage levels in the reservoirs become critical, the stored groundwater can be abstracted for supply. The abstracted water from the Enfield–Haringey boreholes is discharged to the New River, an aqueduct built in 1613, where it is transferred to the Coppermills water treatment works (Fig. 1). All groundwater, including water abstracted from the Lea Valley wells and boreholes and discharged directly into the surface reservoirs, is blended with raw surface water and treated at Coppermills, thus minimizing capital and operational costs. This is an important consideration in that the scheme has been designed for infrequent use with long periods of relatively small-scale recharge, followed by shorter periods of large-scale abstraction (O'Shea & Sage 1999).

The North London scheme utilizes fully treated drinking water as the source of the gravity-fed artificial recharge water, via the normal distribution system. The recharge water quality is similar to the background groundwater in the aquifer. The Enfield–Haringey boreholes vary in depth from 80 to 130 m and can provide a total yield of 90×10^3 m^3 day^{-1}. The Lea Valley wells and boreholes can supply 60×10^3 m^3 day^{-1}. Hence, the design yield is 150 m^3 day^{-1} for a drought period of 200 days and is expected to give only small declines in regional groundwater levels.

During the dry years of the 1990s, the North London Artificial Recharge Scheme was used on several occasions in 1995, 1996 and 1997 to support low river and reservoir levels in the Lea Valley. Cycles of abstraction and recharge from June to November 1997 recorded individual daily rates of abstraction averaging about 100 $\times 10^3$ m^3 day^{-1}. This abstraction rate allowed a decrease in support for the Lea Valley system from the River Thames which, in turn, decreased the rate of decline in the Thames stored-water system, while conserving aquifer storage. In total during this period, 10.7 $\times$ 10^6 m^3 were withdrawn from groundwater storage in North London, equivalent to 25% of the useable capacity of the Lea valley reservoirs (O'Shea & Sage 1999).

This innovative artificial storage and recovery scheme is therefore considered successful in providing good quality water with the environmental benefit of balancing groundwater abstraction with natural and artificial recharge with no net effect on long-term groundwater levels. In addition, the confined nature of the Chalk and Basal Sands aquifer ensures that abstraction has no impact upon the overlying river system.

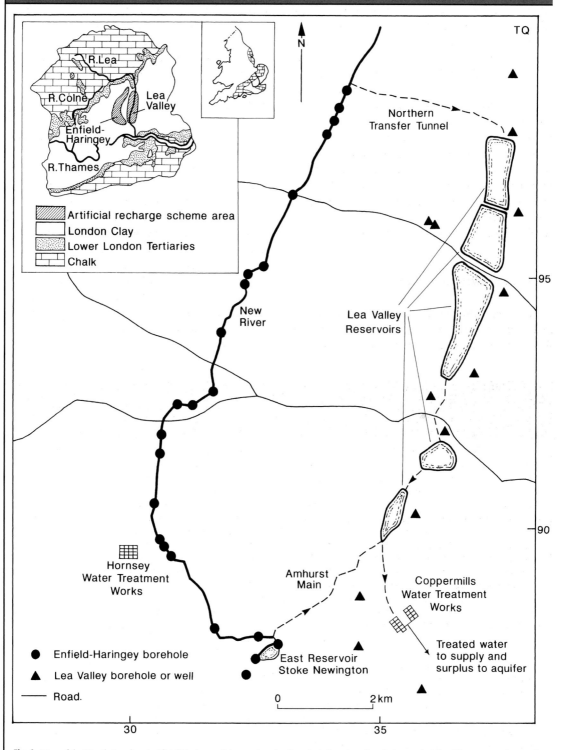

Fig. 1 Map of the North London Artificial Recharge Scheme showing borehole locations in relation to the New River, strategic transfer water mains and Lea Valley reservoirs. After O'Shea and Sage (1999).

Cryptosporidium). Current efforts to address the regulation of riverbank filtration focus on the removal of microbial pathogens and are contained in the Proposed Long Term 2 Enhanced Surface Water Treatment Rule (LT2ESWTR) of the United States Environmental Protection Agency (2001).

Riverbank filtration can occur under natural conditions or be induced by lowering the groundwater table below the surface water level by abstraction from adjacent boreholes. Typical flow conditions associated with different types of riverbank filtration schemes are shown in Fig. 8.6. For the quantitative and qualitative management of riverbank filtration systems, the catchment zones, infiltration zones, mixing proportions in the pumped raw water, flowpaths and flow velocities of the bank filtrate need to be

known. Flow conditions during riverbank filtration are commonly described using interpretations of water-level measurements and hydrogeological modelling. An important factor is the formation of the colmation layer at the interface between surface water and groundwater. This layer has a reduced hydraulic conductivity due to clogging from the input and precipitation of sediment particles, micro-organisms and colloids, precipitation of iron and manganese oxyhydroxides and calcium carbonates as well as gas bubbles. Schubert (2002) reported that the permeability of clogged areas varies with the dynamic hydrology and cannot be regarded as constant, particularly following periods of flooding. The hydraulic conductivity of the river bed is therefore a principal factor determining the volume of bank filtrate.

Riverbank filtration at the Düsseldorf waterworks, River Rhine, Germany

BOX 8.3

The Düsseldorf waterworks has been using riverbank filtration since 1870 and is the most important source for public water supply in this densely populated and industrialized region (Fig. 1). There have been several threats to this supply in the last few decades that have

included poor river water quality, heavy clogging of the river bed and accidental pollution, all of which have been overcome (Schubert 2002). Until about 1950, the Düsseldorf riverbank filtration scheme, like others in the Lower Rhine region, experienced

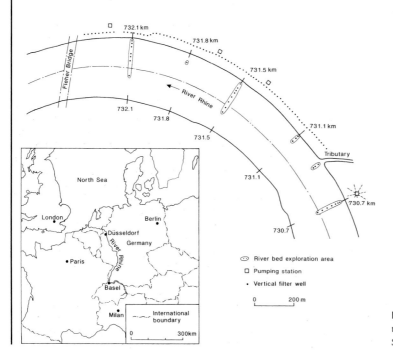

○＿ River bed exploration area
□ Pumping station
·   Vertical filter well

0   200 m

Fig. 1 Location map of the River Rhine at the Flehe waterworks, Düsseldorf. After Schubert (2002).

BOX
8.3

Continued

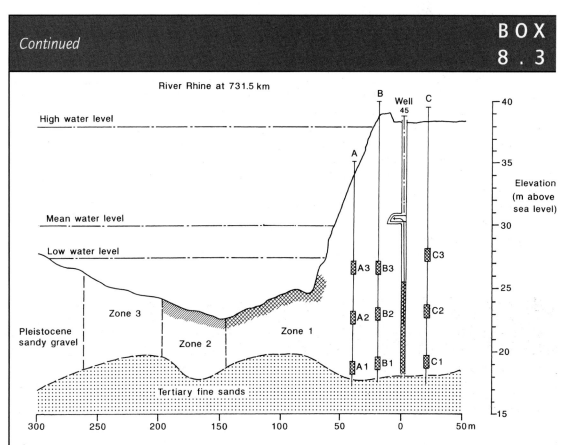

River Rhine at 731.5 km

Fig. 2 Vertical section through the Pleistocene sandy gravel aquifer below the River Rhine in front of the Flehe waterworks, Düsseldorf, showing the location of monitoring well installations A, B and C at position 731.5 km (see Fig. 1) and the three zones of classification of the river bed. After Schubert (2002).

good quality that permitted drinking water production without further treatment, other than disinfection. After 1950, the quality of the river water began to deteriorate gradually as the impacts of increasing quantities and insufficient treatment of industrial and municipal effluents became apparent. The consequences of these discharges were a noticeable drop in the oxygen concentration, an increase in the load of organic pollutants and the development of anoxic conditions in the adjacent aquifer. Moreover, clogging of the river bed with particulate organic matter threatened the well yield of the riverbank filtration scheme. At this time, the pumped well water had to be treated to remove iron, manganese and ammonium and, furthermore, micropollutants. These changes prompted field studies to understand the riverbank filtration process and to better manage the water supply through the development of calibrated numerical models for the simulation of flow and transport.

An investigation of the river bed conditions and flow and transport processes was carried out in 1987 at the Flehe waterworks, Düsseldorf. As shown in Fig. 1, the production wells are situated on an outer bend of the River Rhine between 730.7 and 732.1 km

down-river of the source. There is one horizontal collector well at 730.7 km situated approximately 80 m from the waterline (at mean discharge) and 70 vertical filter wells connected by siphon pipes that form a well gallery parallel to the riverbank at about 50 m from the waterline. The total length of the well gallery is 1400 m with the pumping rate during the field studies roughly constant between 3.0 and $3.4 \times 10^4 \, m^3 \, day^{-1}$. The aquifer consists of Pleistocene sandy gravel sediments with a hydraulic conductivity of between 4×10^{-3} and $2 \times 10^{-2} \, m \, s^{-1}$ and a thickness of approximately 20 m. The aquifer is overlain by a 0.5–2 m thick meadow loam and is underlain by nearly impermeable Tertiary fine sands. The riverbank slope is coated by a 0.5 m thick clay layer above the mean water level of the river and is protected by basalt blocks below the mean water level.

With the aid of a diving cabin, the 1987 field study identified three different zones on the river bed (Fig. 2). Zone 1 nearest the production wells is 80 m wide and is composed of fixed ground that is fully clogged by suspended solids that form a silt layer. This silt layer is formed by mechanical clogging and is almost impermeable

BOX
8 . 3

Continued

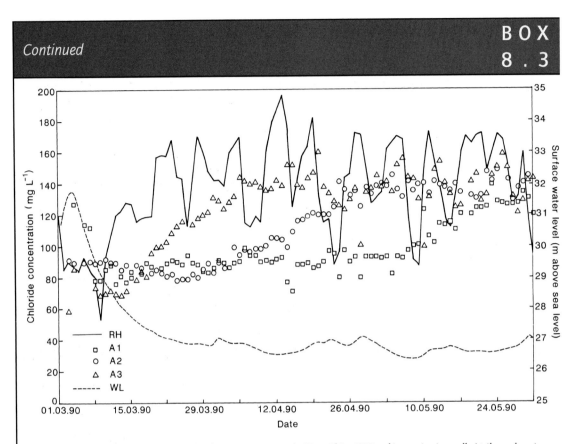

Fig. 3 Graph showing the fluctuation in chloride concentrations in the River Rhine (RH) and in monitoring wells A1 (lower layer), A2 (middle layer) and A3 (upper layer) at Flehe waterworks, Düsseldorf at position 731.5 km (see Fig. 1). WL, water level. After Schubert (2002).

with a hydraulic conductivity of 10^{-8} m s^{-1}. No chemical clogging due to mineral precipitation was observed under the aerobic conditions present in the aquifer. Zone 2 is also fixed ground but is only partially clogged allowing good permeability for infiltrating river water. This zone is 50–80 m wide and has a hydraulic conductivity of 3×10^{-3} m s^{-1}. Zone 3, the region between the middle of the river and the opposite bank, is formed by moveable ground which is shaped by normal river flow but mainly by flood events. The hydraulic conductivity in Zone 3 is higher than the other zones with a value in the range 4×10^{-3} to 2×10^{-2} m s^{-1}.

Insight into the flow processes is provided by data for chloride that acts as a tracer for the bank filtrate. Figure 3 shows the results from a later investigation between March and May 1990 when nearly steady-state conditions characterized the river level (line WL in Fig. 3). Observed weekly variations in chloride concentrations result due to inputs from salt mining and its effluents. Chloride concentration data for the River Rhine (RH) and sampling points A1, A2 and A3 in monitoring well A (see Fig. 2 for position) show a clear succession; flow first appears in A3 (the upper layer),

secondly in A2 (the middle layer) and thirdly in A1 (the lower layer). Under the prevailing steady-state conditions, the travel times can be calculated; infiltrated river water takes approximately 10 days to flow to A3, 30 days to reach A2 and 60 days to pass A1 (Schubert 2002). The important finding from these field studies is the marked age stratification of the bank filtrate between the river and wells.

The effect of age stratification is such that water pumped from a bank filtrate well (for example, Well 45 in Fig. 2) is of mixed composition and this explains the almost total equalization of the fluctuating solute concentrations between the river and production wells. This well-known balancing effect provides safe conditions for further water treatment and is also significant in mitigating the peaks of accidental river water pollution. It is for this reason that the Düsseldorf waterworks has been able to overcome extreme conditions with respect to poor river water quality between 1950 and 1975 and also to withstand the Sandoz accident involving a pesticides manufacturing plant in Basel in November 1986 (Schubert 2002).

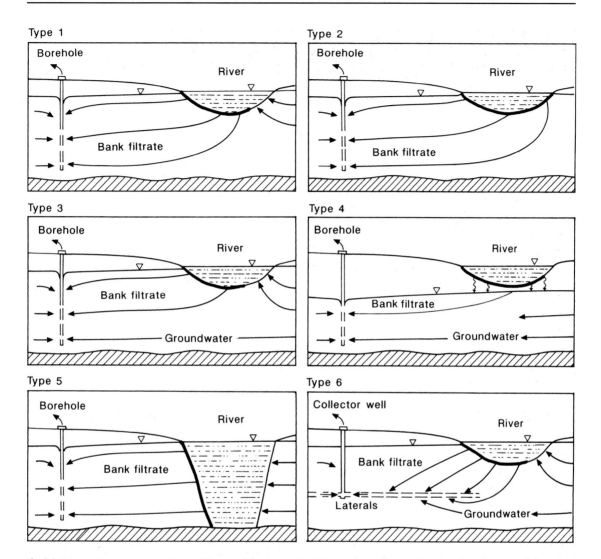

Fig. 8.6 Schematic representation of types of flow conditions at riverbank filtration sites. The majority of riverbank filtration schemes are of Type 1. Groundwater flow beneath the river (Types 3, 4 and 6) is typically neglected at most sites. The formation of unsaturated conditions beneath the river occurs if groundwater abstraction rates are not adapted to the hydraulic conductivity of the river bed or if the hydraulic conductivity of the river bed material becomes clogged due to surface water pollution inputs (Type 4). At some sites, the river bed cuts into the confining layer (Type 5). Collector wells are used with laterals at different depths, of different lengths and directions. Type 6 gives only one example with a lateral towards the river. After Hiscock and Grischek (2002).

Compared with conventional surface water abstraction, the natural attenuation processes of riverbank filtration can provide the following advantages: elimination of suspended solids, particles, biodegradable compounds, bacteria, viruses and parasites; part elimination of adsorbable compounds; and the equilibration of temperature changes and concentrations of dissolved constituents in the bank filtrate.

Undesirable effects of riverbank filtration on water quality can include increases in hardness, ammonium and dissolved iron and manganese concentrations and the formation of hydrogen sulphide and other malodorous sulphur compounds as a result of changing redox conditions (Fig. 8.7).

The beneficial attenuation processes result mainly from mixing, biodegradation and sorption processes

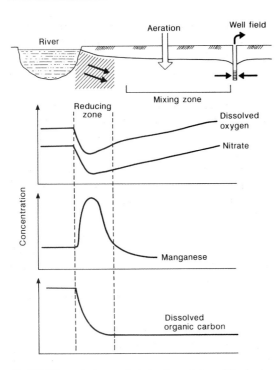

Fig. 8.7 Schematic diagram depicting the evolution of dissolved oxygen, nitrate, dissolved manganese and dissolved organic carbon along a flowpath during riverbank filtration. Dissolved oxygen becomes significantly depleted in the river bed sediments after a few metres of infiltration. Under these anoxic conditions, the microbial activity of denitrifying bacteria further decreases the groundwater redox potential leading to mobilization of the surface coatings of manganese and iron oxy-hydroxides causing a significant reduction in water quality. With further distance from the river bed, microbial activity decreases as a result of a decline in available electron donors. If the groundwater abstraction is non-continuous, or there are strong fluctuations in the river water level, a zone could potentially develop near the well where riverbank filtrate or groundwater flow temporarily results in fluctuations in microbial activity and redox conditions that can affect manganese and iron reduction and precipitation. After Tufenkji et al. (2002).

within two main zones: the biologically active colmation layer, where intensive degradation and adsorption processes occur within a short residence time; and along the main flowpath between the river and abstraction borehole where degradation rates and sorption capacities are lower and mixing processes greater. In general, the distance between production wells and the river or lake bank is more than 50 m with typical travel times of between 20 and 300 days (Grischek et al. 2002). Travel times based on measurements of specific conductance ranged from approximately 20 hours to 3 months at the site of a riverbank filtration scheme in south-west Ohio (Sheets et al. 2002), with shorter and more consistent travel times obtained under conditions of continuous pumping.

Diffuse sources of contamination in catchment runoff, especially from agricultural activities, can adversely affect river water quality and therefore bank filtrate. The study by Grischek et al. (1998) demonstrated the potential for denitrification at a sand and gravel aquifer site on the River Elbe in eastern Germany where both dissolved and solid organic carbon within the aquifer act as electron donors. Verstraeten et al. (2002) reported changes in concentrations of triazine and acetamide herbicides at a well-field on the River Platte in Nebraska and showed that parent compounds were reduced by 76% of the river water value (with a third of this due to riverbank filtration) but that increases in concentrations of specific metabolite compounds were identified after riverbank filtration and ozonation treatment.

Polar organic molecules are an increasingly problematic class of contaminants for riverbank filtration schemes and include several pharmaceutically active compounds (PhACs) that are discharged almost unchanged from municipal sewage treatment plants. Heberer (2002) reported monitoring studies carried out in Berlin where PhACs such as clofibric acid (a blood lipid regulator used in human medical care), diclofenac and ibuprofen were detected at individual concentrations of up to several $\mu g \, L^{-1}$ in groundwater samples from aquifers near to contaminated water courses. It is therefore apparent that several drug residues are not eliminated during recharge through the subsoil (Heberer 2002).

For the future optimization and protection of riverbank filtration schemes, further research is required to quantify chemical reaction rates and microbial degradation, especially in the river bed, and to include the effects of pH and redox controls, the behaviour and importance of biofilms, the fate of micropollutants and persistent compounds such as PhACs, and the mobility, adsorption and inactivation of viruses, pathogens and protozoa. The development of risk quantification methods, for example identification of relevant compounds and metabolites, and appropriate alarm systems are also required to ensure the long-term sustainability of riverbank filtration schemes (Hiscock & Grischek 2002).

8.2.6 Horizontal well schemes

It has long been recognized that well yields can be increased by driving horizontal tunnels (adits) below the water table which radiate away from a well or borehole shaft. Systems of adits, typically 1.2 m wide and 1.8 m high, are associated with many large groundwater sources in the Chalk of south-east England, Belgium and the Netherlands. The water supply to Brighton on the south coast of England includes 13.6 km of adits and in east London there are 18 km. Generally, the adits were driven to intersect the principal fissure or fracture directions in the Chalk aquifer (Downing et al. 1993). Groundwater flow in an adit may be pipe or open channel flow. Adits in the Chalk of south-east England are normally full of water contained under pressure. In this situation, Darcy's law (see eq. 2.5) is not applicable and alternative methods are required for modelling flow in aquifer–adit systems (Zhang & Lerner 2000).

More recently, and with advances in drilling technology, horizontal and slanted wells have been investigated for various hydrogeological situations (Chen et al. 2003; Park & Zhan 2003) and also for environmental applications such as vapour extraction in contaminated aquifers (Plummer et al. 1997; Zhan & Park 2002). Horizontal wells have screened sections that can be positioned parallel to the horizontal flow direction. These wells have several advantages, including: interception of vertical components of groundwater flow; greater control over the dynamics of the water table; better contact between well screens and horizontal aquifer units; easier drilling operations close to ground surfaces that are obstructed by infrastructure (airport runways, roads, buildings, etc.); and the possibility of installing long screen sections in aquifers of limited thickness.

8.3 Groundwater abstraction and river flows

The link between groundwater and river flows is fundamental to conserving the riparian environment yet is one of the more difficult hydrogeological situations to predict. This difficulty is due to the complex nature of river–groundwater interactions and uncertainties in the nature of the hydraulic connection at a particular location. Hence, making decisions about the impacts of groundwater abstraction on rivers is technically challenging.

An important determining factor of the flux of water between a river and an aquifer is the degree of connection between the river and aquifer as controlled by the material properties of the river bed and river bank sediments, and the extent to which the channel of the river intersects the saturated part of the aquifer. In general, and as shown in Fig. 8.8, there are three types of hydrogeological situation that lead to flow between an aquifer and a river.

The rate of change in river flow and the attenuation of short-term fluctuations in river flow, for example a flood event, can be strongly influenced by storage of water in the floodplain deposits surrounding a river. As the river stage and groundwater level increase, the extra water saturates the alluvial sediments and fills the available bank storage. When water levels recede this bank storage is released and can have a short-term beneficial effect in alleviating the immediate impact of adjacent abstractions on the river flow.

In the long term, and usually within one or two years for boreholes a few hundred metres from a river, groundwater abstraction will deplete the river flow at a rate equal to the pumping rate. As shown in Fig. 8.9a, the river flow depletion consists of two components:

1 interception of flow that would otherwise reach the river;
2 induced recharge from the river.

In general, the depletion of river flow caused by pumping increases with time and will increase more rapidly the closer the abstraction point is to the river. The degree of depletion is also dependent on the aquifer properties of transmissivity and storage coefficient.

A dramatic illustration of the impacts of groundwater abstraction in depleting river flows is the case of the River Colne valley, north of London (Fig. 8.10). In the first half of the twentieth century, the substantial growth of residential areas here and elsewhere on the outskirts of London was supplied by direct groundwater abstractions from the underlying Chalk aquifer. This rapid development had marked effects on certain rivers, particularly the Rivers Ver and Misbourne, tributaries of the River Colne, in the

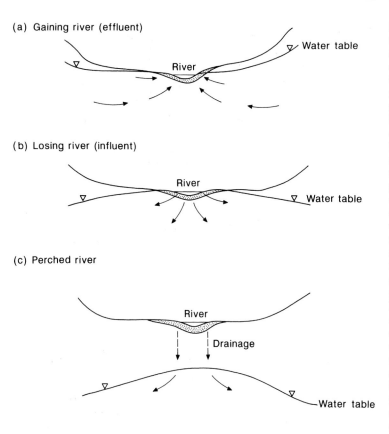

(a) Gaining river (effluent)

River

Water table

(b) Losing river (influent)

River

Water table

(c) Perched river

River

Drainage

Water table

Fig. 8.8 Three hydrogeological situations that represent flow between an aquifer and a river. (a) The water table in the aquifer is above the river stage and there is potential for flow from the aquifer to the gaining river (effluent condition) with the flux generally proportional to the difference between the elevations of the water table and river stage. (b) The water table is below the river stage and the losing river (influent condition) potentially loses water to the aquifer with the flux also generally proportional to the difference between the elevations of the river stage and water table. (c) A common situation is shown in which a partially penetrating river (where the saturated aquifer extends beneath the river) experiences a declining water table below the base of the river. In this situation, water will drain from the perched river under gravity with a unit head gradient. The unit head gradient creates a limiting infiltration rate such that river losses will not increase as the water table falls further. In each case shown, the nature of the river–aquifer interaction will also depend on the properties of the river bed sediments. Sediments with very low permeability can result in a significant resistance to flow. After Kirk and Herbert (2002).

Chiltern Hills. In the River Ver catchment, about 75% of the average annual recharge to the Chalk is allocated to licensed groundwater abstraction which is now almost taken up (Owen 1991). Approximately half of the abstracted water is exported to supply areas outside of the catchment and is therefore effectively lost. Of the remainder used within the catchment, effluent returns via sewage treatment works are typically in the lower reaches of the River Colne valley and therefore are unavailable for supporting river flows higher up in the catchment.

The effects of the large demand for water, especially from those boreholes situated towards the head of the River Colne valley, has been to dry up those springs at the source of the perennial rivers (Fig. 8.9b shows the hydraulic mechanism). In the case of the River Ver, the upper 10 km section of originally perennial or intermittent river is now normally dry (Fig. 8.10b). Further downstream, the remaining perennial section experiences much reduced flows

and this once typical Chalk stream has suffered substantial environmental degradation with major changes in riparian habitat, the loss of naturally sustained fisheries, the loss of watercress farming and reductions in the general amenity value of the river (Owen 1991).

The above example illustrates the need for careful management of catchment water resources and the need to be able to predict potential environmental impacts. However, without very significant effort towards field investigation and numerical modelling (see Section 5.9), it is often difficult to evaluate the impacts of abstractions on rivers with any degree of confidence. A not unusual limitation is the availability of accurate data, particularly for the physical properties of the river bed and river bank sediments. As discussed in the next section, a further approach is to employ an analytical solution in the hydrogeological assessment of river flow depletion caused by groundwater abstraction.

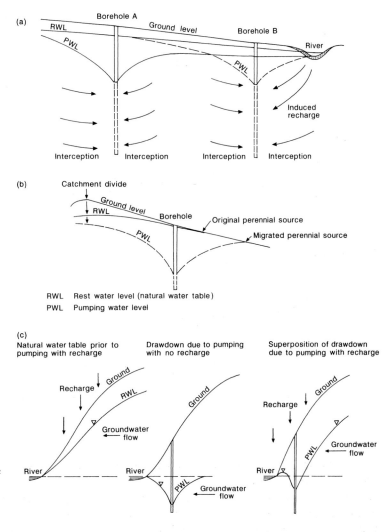

Fig. 8.9 Cross-sections to illustrate the impacts of groundwater abstraction on a groundwater-fed river. In (a) river flow is depleted by the effects of interception and induced recharge and in (b) the perennial source of the river migrates downstream in response to a drawdown in the water table (after Owen 1991). In (c) the principle of superposition is illustrated as applied to abstraction close to a river (after Kirk & Herbert 2002).

8.3.1 Analytical solutions of river flow depletion

Analytical solutions simplify the aquifer properties, often assigning single representative values to the transmissivity and storage parameters. It is also implicit that the aquifer can be characterized by a single value of groundwater level away from the influence of the abstraction point. The effect of abstraction is to create a cone of depression around the borehole or well. Using analytical solutions, this drawdown is simplified by adopting assumptions of idealized radial flow to calculate the shape of the drawdown zone that can be superimposed on the distribution

of head due to the natural behaviour of the system (Fig. 8.9c).

A discussion of the application of analytical solutions for the evaluation of stream flow depletion caused by pumping is presented in the classic papers by Jenkins (1968, 1970). A more recent review of analytical solutions is contained in Kirk and Herbert (2002). The method presented by Jenkins (1968) enables the estimation of the total depletion of stream flow as a function of time due to nearby abstraction with the following assumptions:

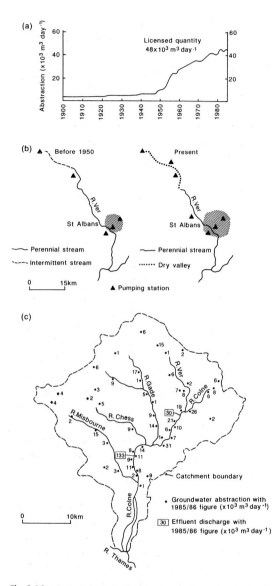

Fig. 8.10 The history of Chalk groundwater abstractions in the River Ver catchment, north London, and the impacts on river flow including: (a) a graph of annual groundwater abstractions; (b) a sketch of the reduction in length of the perennial section of the River Ver; (c) a location map of the River Ver tributary in the Colne catchment. After Owen (1991).

1 the aquifer is isotropic, homogeneous, semi-infinite in areal extent and bounded by an infinite, straight, fully penetrating stream;
2 water is released instantaneously from aquifer storage;

3 the borehole or well fully penetrates the aquifer;
4 the pumping rate is steady;
5 the residual effects of previous pumping are negligible.

The model further assumes that the aquifer is confined or that, for an unconfined aquifer, the water-table drawdown is negligible compared to the saturated aquifer thickness (in other words, the transmissivity remains constant). The temperature of the stream is assumed to be constant and equal to the temperature of the groundwater. A cross-section through the idealized conceptual model is shown in Fig. 8.11.

The mathematical solution to the problem shown in Fig. 8.11 gives the rate of stream depletion as a proportion of the groundwater abstraction rate as follows:

$$\frac{q}{Q} = \operatorname{erfc}\left(\frac{1}{2\tau}\right) \qquad \text{eq. 8.3}$$

where τ is a dimensionless length scale for the system given by:

$$\tau = \sqrt{\frac{tT}{L^2 S}} \qquad \text{eq. 8.4}$$

where T is the aquifer transmissivity, S is the aquifer storage coefficient (specific yield for unconfined aquifer approximations), L is the perpendicular distance of the borehole or well to the line of the river, Q is the abstraction rate of the borehole or well, q is the rate of stream flow depletion, t is time and erfc is the complementary error function (see Appendix 8).

Similarly, the volume of stream depletion, v, as a proportion of the groundwater abstraction rate is given by:

$$\frac{v}{Qt} = 4i^2\operatorname{erfc}\left(\sqrt{\frac{1}{2\tau}}\right) \qquad \text{eq. 8.5}$$

where v is the volume of stream depletion during time, t, and i^2erfc is the second repeated integral of the error function.

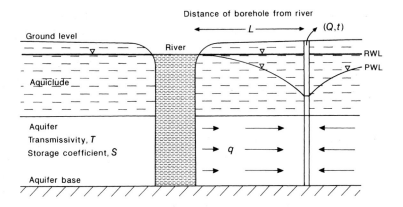

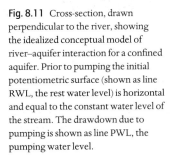

Fig. 8.11 Cross-section, drawn perpendicular to the river, showing the idealized conceptual model of river–aquifer interaction for a confined aquifer. Prior to pumping the initial potentiometric surface (shown as line RWL, the rest water level) is horizontal and equal to the constant water level of the stream. The drawdown due to pumping is shown as line PWL, the pumping water level.

Although the assumptions behind the solution of equations 8.3 and 8.5 are an over-simplification of reality, analytical results can provide rough estimates of the local impacts of abstraction on river flow and the timescales over which flow depletion occurs. By neglecting river bed and river bank sediments and assuming full aquifer penetration, the impact of pumping on the stream flow is over-estimated and the time delay between abstraction starting and the impact of pumping on stream flow is underestimated. A further assumption that prior to pumping the initial potentiometric surface is horizontal and equal to the constant water level of the stream (Fig. 8.11) is, in fact, not a major limitation (Kirk & Herbert 2002), but Wilson (1993, 1994) presented steady-state analytical solutions for two-dimensional, vertically integrated models of induced infiltration from surface water bodies for various combinations of aquifer geometry in the presence of ambient aquifer flow.

Wallace et al. (1990) extended the approach of Jenkins (1968) to show that pumping impacts may develop over several annual cycles. For cases where the stream depletion impacts develop over long timescales due to either the distance between the borehole and river, the type of aquifer properties or the possible role of river deposits, the maximum impact in later years may exceed the maximum depletion in the first year. Such delayed impacts are potentially an important catchment management consideration in order to avoid future low river flows.

A new analytical solution for estimation of drawdown and stream depletion under conditions that are more representative of those in natural systems (that is, finite-width stream of shallow penetration adjoining an aquifer of limited lateral extent) is presented by Butler et al. (2001). The solution shows that the conventional assumption of a fully penetrating stream can lead to significant over-estimation of stream depletion (>100%) in many practical situations, depending on the value of the stream leakance parameter and the distance from the pumping well to the stream. An important assumption underlying this new solution is that the penetration of the stream channel is negligible relative to aquifer thickness, although an approximate extension to the method provides reasonable results for the range of relative penetrations found in most natural systems (up to 85%); Butler et al. (2001).

To assist in the practical application of analytical solutions, stream depletion caused by groundwater abstraction can be readily calculated using dimensionless type curves and tables. Jenkins (1968) presented a number of worked examples, including computations of the rate of stream depletion for the pumping and following non-pumping periods, the volume of water induced by pumping and the effects (both rate and volume of stream depletion) of any selected pattern of intermittent pumping. An example calculation is given in Box 8.4.

8.3.2 Catchment resource modelling of river flow depletion

Local-scale impacts of groundwater abstraction on river flows can be investigated with the above

---

<table>
<tr><td>

Computation of the rate and volume of stream depletion by boreholes and wells

</td><td>

B O X
8 . 4

</td></tr>
</table>

The following example illustrates the application of an analytical solution to the problem of stream flow depletion caused by groundwater abstraction. The solution uses one of the type curves and tables presented by Jenkins (1968) to assist the calculation of the rate and volume of stream flow depletion.

The problem to be solved is as follows. A new borehole is to be drilled for supporting municipal water supply from an unconfined alluvial aquifer close to a stream. The alluvial aquifer has a transmissivity of 3×10^{-3} m^2 s^{-1} and a specific yield equal to 0.20. To protect the riparian habitat, the borehole should be located at a sufficient distance from the stream so that downstream of the new source stream flow depletion should not exceed a volume of 5000 m^3 during the dry season. The dry season is typically about 200 days in duration. The borehole is to be pumped continuously at a rate of 0.03 m^3 s^{-1} during the dry season only. During the wet season, recharge is sufficient to replace groundwater storage depleted by pumping in the previous dry season; hence the residual effects on steam flow during the following non-pumping period can be neglected.

The problem requires us to find the minimum allowable distance between the borehole and stream using the following given information:

v, volume of stream flow depletion during time $t_p = 5000$ m^3

t_p, total time of pumping = 200 days

Q, net pumping rate = 0.03 m^3 s^{-1}

T, transmissivity = 3×10^{-3} m^2 s^{-1}

S, specific yield = 0.20

Qt, net volume of water pumped = $(0.03$ m^3 s$^{-1}) \times (200$ days$) \times (86,400$ seconds per day$) = 5.184 \times 10^5$ m^3

As a first step, compute v/Qt, the dimensionless ratio of the volume of stream depletion to volume of water pumped for the pumping period of interest which, in our example is equal to $5000/5.184 \times 10^5 = 0.01$. Now, with this value of v/Qt, and using Curve B shown in Fig. 1, or the table of values given in Appendix 7, find the value of t/F where F is the stream depletion factor (arbitrarily defined by Jenkins (1968) as the time co-ordinate of the point where the volume of stream depletion, v, is equal to 28% of the volume of water pumped on a curve of v against t). If the system meets the assumptions of the idealized conceptual model shown in Fig. 8.11, then $F = L^2S/T$, where L is the perpendicular distance from the abstraction borehole to the stream.

In our example, for $v/Qt = 0.01$, $t/F = 0.12$ (from Curve B) and thus:

$$t/F = 0.12 = tT/L^2S$$

Rearranging and substituting values for T, S and t gives:

$$L^2 = (200 \times 86,400 \times 3 \times 10^{-3})/(0.12 \times 0.20) = 2.16 \times 10^6 \text{ m}^2$$

Hence, L, the required distance between the borehole and stream to avoid environmental impacts = 1470 m.

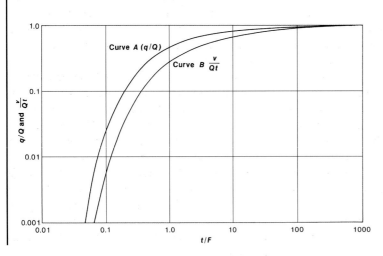

Fig. 1 Type curves to determine the rate and volume of stream depletion by boreholes and wells. After Jenkins (1968).

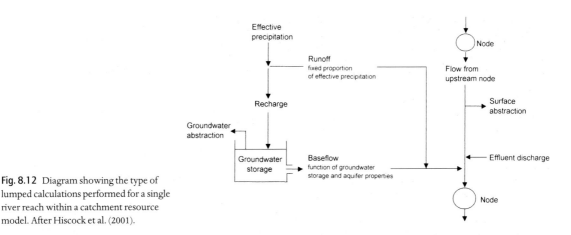

Fig. 8.12 Diagram showing the type of lumped calculations performed for a single river reach within a catchment resource model. After Hiscock et al. (2001).

analytical solutions or by numerical models where data availability permits. A further quantitative method is to naturalize the river flow record to remove the effects of catchment abstractions and discharges. The derived (naturalized) and actual (historic) river flow records can then be conveniently presented as flow duration curves and compared to assess the environmental impacts of catchment water resources management.

The derivation of a naturalized flow duration curve and simulation of river–aquifer interaction can be computed using a simple numerical catchment resource model. For example, Hiscock et al. (2001) developed a resource model to assess the impacts of surface water and groundwater abstractions on lowland river flows in eastern England. The resource model simulated river flows from the basic components of baseflow (aquifer discharge) and surface flows and included the net effect of surface water and groundwater abstractions, thus enabling the naturalization of measured river flows. Calibration of the resource model was achieved against historic flows prior to naturalization of the river flow record. The main output from the model was the construction of flow duration curves at selected points in a river using predicted mean weekly flows.

The catchment resource model is an example of a lumped model that depends on summing all inputs to and outputs from defined river reaches within the catchment or subcatchment (Fig. 8.12). The inflow to a single reach during a chosen time step is calculated as the sum of surface runoff, baseflow, abstractions

and effluent returns distributed uniformly over the length of the reach. In each catchment, the underlying aquifer is represented as a single storage cell into which recharge is added and from which abstractions and baseflow are subtracted. The concept of catchment averaging greatly simplifies the modelling of aquifer behaviour which, in calculating net aquifer storage, apportions the effects of groundwater abstractions to predictions of baseflow output. By approximating the aquifer area to an equivalent rectangular area, baseflow is related to storage within the cell by the empirical relationship, $T/L^2 S$, controlled by the aquifer transmissivity, T, storage coefficient, S, and the distance, L, between the river and catchment boundary. As with river abstractions, groundwater abstractions are assumed to be distributed uniformly over the entire aquifer area.

Using their resource model, Hiscock et al. (2001) obtained reasonable results when comparing simulated and observed 95 percentile 7-day flows (the flows equalled or exceeded for 95% of the time) as the calibration target for lowland rivers in eastern England. As to be expected, predictive errors were caused by the simplification of adopting a single river reach (with no in-channel storage) and a single aquifer storage cell for the underlying Chalk aquifer, and also the adoption of a simple representation of runoff independent of antecedent soil moisture conditions.

With a catchment resource model, the naturalization of river flows is achieved by setting all net abstractions to zero. Then, the arithmetic difference

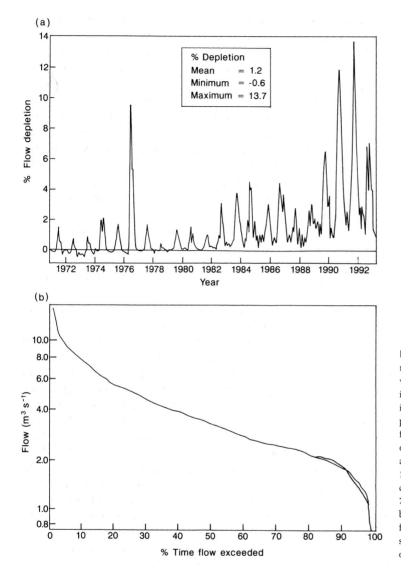

(a)

% Depletion
Mean   = 1.2
Minimum = -0.6
Maximum = 13.7

(b)

Fig. 8.13 Catchment resource model results illustrating the predicted effects of water resources management on flows in the River Wensum at Costessey Mill in Norfolk, eastern England. (a) The percentage depletion of mean weekly flows are shown due to the net effect of surface water and groundwater abstractions and discharges for the period 1971–1992. A negative percentage depletion indicates a gain in flow. (b) The 7-day flow duration curves are shown for both the naturalized and observed flows for the period 1971–1992. The upper curve shows natural flows, the lower curve observed flows. After Hiscock et al. (2001).

between the historic net abstractions and nil net abstractions produces a series of net effects of abstractions and discharges. An example is shown in Fig. 8.13a for the River Wensum in Norfolk, presented as percentage depletions of mean weekly flows. The predicted maximum depletion in the mean weekly flows of 14% occurred in the drought of 1989–1992. In Fig. 8.13b the modelled historic and naturalized surface flow characteristics are shown as flow duration curves. Again, in the case of the River Wensum, the impacts of net abstractions are small, with the

higher percentage flow depletion occurring when river flows are very low.

The catchment resource model was also used to predict the impact of an increase in spray irrigation to the total licensed amount from both surface water and groundwater sources (an important water use in the predominantly agricultural area of eastern England). For this scenario, the maximum potential depletion of the long-term (1971–1992) 95 percentile 7-day flow in the River Wensum is predicted to be 17%. Compared with flows during the drought of 1989–1992 when mean weekly flows in the river fell to $<0.8 \text{ m}^3 \text{ s}^{-1}$, the predicted flow depletion is approximately 30% of this mean. By highlighting spray irrigation, the catchment resource model illustrates the future challenge of managing water resources in eastern England with the additional factor of climate change (see Section 8.5). Unless farmers adapt, the expected warmer and drier summers in this area are likely to lead to a greater demand for irrigation water in direct conflict with other water users and the need for water for environmental protection.

8.4 Wetland hydrogeology

The global extent of wetlands is estimated to be from 7 to $8 \times 10^6 \text{ km}^2$ and, compared to other ecosystems, are an extremely productive part of the landscape with an estimated average annual production of 1.125 kg $C \text{ m}^{-2} \text{ a}^{-1}$ (Mitsch et al. 1994). The relatively high productivity and biological diversity of wetlands support an important landscape role in nutrient recycling, species conservation and plant and animal harvest. Although very much smaller in extent compared to marine habitats, inland water habitats exhibit greater variety in their physical and chemical characteristics. Wetlands, with their often abundant and highly conspicuous bird species, are protected by national and international agreements and legislation. Notable wetland protected areas include the Moremi Game Reserve in the Okavango Delta, Botswana, the Camargue National Reserve in France, the Keoladeo (Bharatpur) National Park in India, Doñana National Park in Spain and the Everglades National Park in the United States (Groombridge & Jenkins 2000). Inland water ecosystems are unusual in that an international convention, the 1975 Convention of Wetlands of International Importance especially as Water-fowl Habitat (the Ramsar Convention; Navid 1989), is dedicated specifically to them. Inland water habitats can be divided into running or **lotic** systems (rivers) and standing or **lentic** systems (lakes and ponds). Wetlands are typically heterogeneous habitats of permanent or seasonal shallow water dominated by large aquatic plants and broken into diverse microhabitats occupying transitional areas between terrestrial and aquatic

habitats (Groombridge & Jenkins 2000). The four major wetland habitat types are bogs, fens, marshes and swamps. Bogs are peat-producing wetlands in moist climates where organic matter has accumulated over long periods. Water and nutrient input is entirely through precipitation. Bogs are typically acid and deficient in nutrients and are often dominated by *Sphagnum* moss. Fens are peat-producing wetlands that are influenced by soil nutrients flowing through the system and are typically supplied by mineral-rich groundwater. Grasses and sedges, with mosses, are the dominant vegetation. Marshes are inundated areas with emergent herbaceous vegetation, commonly dominated by grasses, sedges and reeds, which are either permanent or seasonal and are fed by groundwater or river water, or both. Swamps are forested freshwater wetlands on waterlogged or inundated soils where little or no peat accumulation occurs. Like marshes, swamps may be either permanent or seasonal.

Various attempts have been made to classify wetlands and a variety of subdivisions have been recognized based on broad features such as substratum type, base status, nutrient status and water source, water level and successional stage. The development of the main wetland habitat categories and terms, in relation to the main ecological gradients, has been reviewed by Wheeler and Proctor (2000). Other approaches include hydrological and hydrogeological classifications based on the main inflows and outflows of water, flowpaths and water-level fluctuations (Lloyd et al. 1993; Gilvear & McInnes 1994) and a hydromorphological (or hydrotopographical) classification based on the shape of the wetland and its situation with respect to apparent sources of water (Goode 1977). A simplification of the hydrogeological classification is shown in Fig. 8.14 to illustrate the influence of topography, geology and water source in maintaining wetlands.

A change in the factors controlling the source of water to a wetland can have potentially devastating consequences for the fen community, particularly a change in flow direction and volume. An example of the impact of groundwater abstraction on the freshwater habitat of a valley fen and the measures taken to restore the fen is given in Box 8.5. When land drainage and competing demands for water for wetlands, agriculture and public supply conflict,

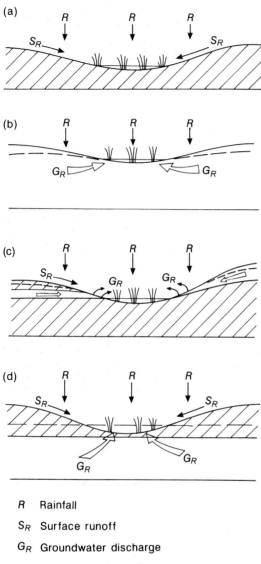

R   Rainfall

S_R Surface runoff

G_R Groundwater discharge

— —   Potentiometric surface

☐   Permeable strata (aquifer)

▨   Low permeability strata (aquitard)

Fig. 8.14 Simple hydrogeological classification of wetland types. (a) Surface runoff is fed by rainfall and collects in a topographic hollow (for example, valley bottom, pingo or kettle hole) underlain by a low permeability layer. (b) Rainfall recharge to an unconfined aquifer supports a wetland in a region of low topography and groundwater discharge. (c) Superficial deposits, both unconfined and semiconfined, and underlain by a low permeability layer, contribute groundwater seepage in addition to surface water runoff. (d) Surface water runoff is in addition to artesian groundwater discharge from a semiconfined aquifer.

widescale destruction of wetland habitat can occur, as illustrated graphically by the Florida Everglades (Box 8.6).

8.5 Climate change and groundwater resources

The global climate is undoubtedly changing. Instrumental records dating back to 1860 show that the globally averaged surface air temperature has risen by about 0.6°C since the beginning of the twentieth century, with about 0.4°C of this warming occurring since the 1970s (Fig. 8.15). The year 1998 was the warmest year recorded and 2003 the third warmest. Globally, the 1990s were the warmest decade in the last 100 years and it is likely that the last 100 years was the warmest century in the last millennium (Hulme et al. 2002). Other evidence for changes in global climate include more intense rainfall events over many Northern Hemisphere mid- to-high latitude land areas and a near world-wide decrease in mountain glacier extent and ice mass.

In central England, the thermal growing season for plants has lengthened by 1 month since 1900 and winters over the last 200 years have become wetter relative to summers throughout the United Kingdom. Also, a larger proportion of winter precipitation in all regions now falls on heavy rainfall days than was the case 50 years ago. Around the United Kingdom, and adjusting for natural land movements, average sea level is now about 10 cm higher than the level in 1900 (Hulme et al. 2002).

Climate change is influenced by both natural and human causes. The Earth's climate varies naturally as a result of interactions between the ocean and atmosphere, changes in the Earth's orbit, fluctuations in incoming solar radiation and volcanic activity. The main human cause is probably the increasing emissions of 'greenhouse' gases such as carbon dioxide, methane, nitrous oxide and chlorofluorocarbons. Currently, about 6.5×10^9 t a^{-1} of carbon are emitted globally into the atmosphere, mostly through the burning of fossil fuels. Changes in land use, including the clearance of tropical rainforest, contribute a further net emission of $1–2 \times 10^9$ t a^{-1}. Increasing concentrations of 'greenhouse' gases in the atmosphere in the last 200 years (Table 8.2) have trapped outgoing long-wave radiation in the lower atmosphere,

Redgrave and Lopham Fen is an internationally important British calcareous valley fen situated on the Norfolk and Suffolk border in the peat-filled headwaters of the River Waveney (see Figs 1 & 2 for location and general aspect). The fen, covering 123 ha, is the largest fen of its type in lowland Britain and was declared a Ramsar site in 1991. The largest part of the fen is covered by shallow peat supporting a complex mosaic of reed and sedge beds, mixed species fen and spring flushes. The fen is noted for its rare and precarious community of fen raft spiders. For nearly 40 years, the fen experienced substantial ecological change, principally due to a change in the groundwater flow regime relating to an adjacent water company borehole.

The general geology of the fen consists of Cretaceous Chalk covered by glacial till, sands and gravels. The Chalk surface is incised by a deep buried channel which is thought to be about 1 km wide in the vicinity of the fen. With reference to Fig. 8.14, the fen is a combination of wetland types (c) and (d). Before the late 1950s, calcareous and nutrient-poor water rose under artesian pressure from the semiconfined Chalk aquifer and seeped into the fen both around the fen margins and within the peats (Fig. 3a). The extreme heterogeneity of the superficial Quaternary deposits resulted in great spatial variation in the quantity of rising Chalk water. The interaction of base-poor water from marginal sands with the calcareous and acid peats produced local variation in soil chemistry that supported a diverse mosaic of fen plant communities of high botanical interest.

In 1957, two Chalk abstraction boreholes were installed adjacent to the fen (see Fig. 1 for location) for public water supplies

and licensed to abstract 3600 m^3 day^{-1} in 1965. Warby's Drain and the River Waveney were deep-dredged at this time, substantially increasing channel capacity, with a sluice at the downstream end of Redgrave Fen installed to control outflows. As shown schematically in Fig. 3b, the operation of the water company source led to the elimination of vertical groundwater seepage and the frequent drying out of Warby's Drain. The normal condition of perennial, high water levels with Chalk groundwater discharging through the fen,

Fig. 2 General aspect of Redgrave and Lopham Fen looking north-east across Great Fen from the position of the Sluice (see Fig. 1 for location).

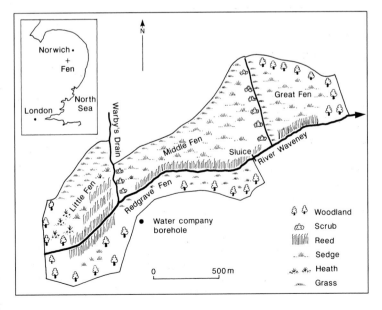

Fig. 1 Location and site map of Redgrave and Lopham Fen in East Anglia showing the position of the former operating water company borehole.

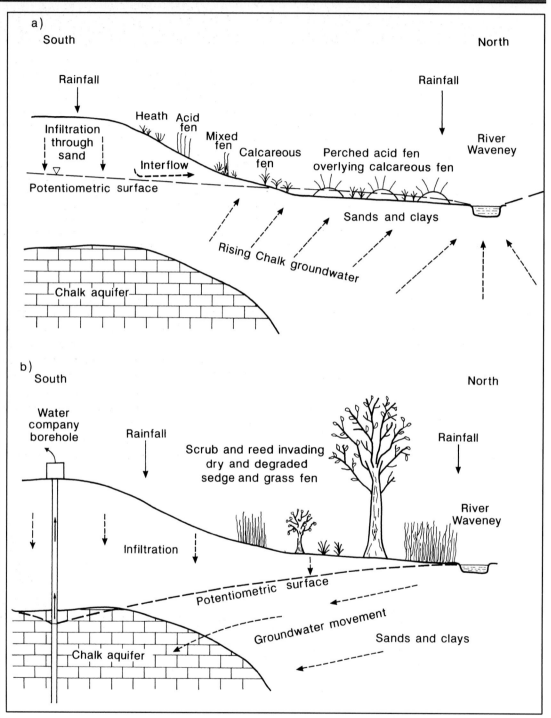

Fig. 3 Schematic cross-section through Redgrave and Lopham Fen illustrating groundwater and ecological conditions (a) before groundwater pumping and (b) after several years of groundwater pumping from the water company borehole (see Fig. 1 for location). After Burgess (2002).

BOX
8.5

Continued

thus maintaining a **soligenous** hydrology (where wetness of the site is maintained by water flow through soil), was replaced by a seasonal downward movement of surface water. The hydrology of the fen had now become controlled by rainfall patterns and river levels thus producing a **topogenous** hydrology (where wetness is maintained by the valley topography). During the summer, the fen dried out more frequently with groundwater heads reduced to a metre below the fen surface. Test pumping and radial flow modelling suggested that about a quarter of the pumped groundwater was at the expense of spring flow into the fen (Burgess 2002).

These hydrogeological changes caused by groundwater abstraction were matched to a deterioration of the flora and fauna at the site (Harding 1993). From a comparative study of botanical records, Harding (1993) showed that great changes had occurred to the ecological character of the fen as a result of the drying out, namely

the invasion of scrub. The reduction in the water table altered the balance of competition towards dry fen species and the expansion of *Phragmites* and *Molinia*, which are tolerant of low water levels, while previously dominant species such as *Cladium* and *Schoenus* contracted. The loss of calcareous and base-poor seepage water and the increased fertility from the sudden release of large amounts of stored nitrates through peat wastage under a lower water level also benefited *Phragmites*.

To reverse the environmental damage, the groundwater pumping was relocated to a borehole 3.5 km east and downstream of the fen that became operational in 1999. The total cost of the replacement supply was of the order of £3.3 million (US$4.8 million), which included the cost of the investigation, source works, pipeline and restoration work on the fen, principally the removal of scrub and the regeneration of peat areas.

leading to an increase in global temperature (Fig. 8.16). Opposite effects that act to cool the climate include other atmospheric pollutants such as sulphate aerosols that absorb and scatter incoming solar radiation back to space. To counter global warming, and under the Kyoto Protocol, countries of the European Union are committed to an 8% reduction in emissions from 1990 levels of a 'basket' of six gases, including carbon dioxide, methane and nitrous oxide by 2008–2012.

The hydrological cycle is an integral part of the climate system and is therefore involved in many of

the interactions and feedback loops that give rise to the complexities of the system (Askew 1987). Climate change during the next 100 years is expected to lead to an **intensification** of the global hydrological cycle and have major impacts on regional water resources (IPCC 1998). A summary of the likely impacts of climate change on natural hydrological systems is shown in Fig. 8.17. The potential water resources impacts of climate change are generally negative, such as a shorter precipitation season and an increase in hydrological extremes such as floods and droughts

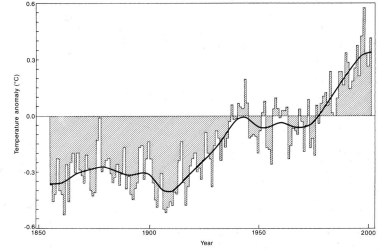

Fig. 8.15 The observed increase in globally averaged surface air temperature. Anomalies are relative to the 1961–1990 average. The smoothed curve is created using a 21-point binomial filter giving near-decadal averages. Two main warming periods are noticeable between about 1910 and 1945 and between 1976 and 2000. The rate of global warming from 1976 to 2000 was about twice as fast (but interannually more variable) than for the period 1910 to 1945. After IPCC (2001).

The Florida Everglades comprises part of the south Florida ecosystem that formed during the last several thousand years during the Holocene epoch. The ecosystem consists primarily of wetlands and shallow-water habitats set in a subtropical environment (McPherson & Halley 1997). The south Florida region is underlain by a thick sequence of shallow marine carbonate sediments deposited from the Cretaceous through to the early Tertiary as a carbonate platform. Younger Tertiary deposits consist of shallow marine sandy limestone, marls and sands. As shown in Fig. 1, the marine carbonate sediments contain three major aquifer systems: the Floridan; the intermediate; and the surficial. The surficial aquifer system includes the highly permeable Biscayne aquifer. The Biscayne aquifer is more than 60 m thick under parts of the Atlantic Coastal Ridge and wedges out about 65 km to the west in the Everglades. The shallow aquifer of southwest Florida is about 40 m thick along the Gulf Coast and wedges out in the eastern Big Cyprus Swamp. The surficial aquifer system is recharged by abundant rainfall that under natural conditions favoured the expansion of coastal and freshwater wetlands during the Holocene and the deposition of thick layers of peat.

Wetlands are the predominant landscape feature of south Florida. Before development of the area, the natural functioning of the wetlands depended on several weeks of flooded land following the wet season. For example, the Kissimmee–Okeechobee–Everglades catchment, an area of about 23,000 km^2, once extended as a single hydrological unit from present-day Orlando to Florida Bay, about 400 km to the south (Fig. 2). In the northern half of the catchment, the Kissimmee River and other tributaries drained slowly through large areas of wetlands into Lake Okeechobee, a shallow lake of about 1900 km^2. The lake periodically spilled water south into the Everglades, a vast wetland of about 12,000 km^2. Under high water level conditions, water in the Everglades moved slowly to the south by sheet flow, thus forming the area known as the River of Grass before discharging into Florida Bay and the Gulf of Mexico.

The Everglades was formerly a complex mosaic of wetland plant communities and landscapes with a central core of peatland that extended from Lake Okeechobee to mangrove forest that borders Florida Bay. The peatland was covered by a swamp forest along the southern shore of Lake Okeechobee and by a vast plain of monotypic sawgrass to the south and east of the swamp forest. Further southeast, the sawgrass was broken by sloughs and small tree islands in Shark River Slough and Hillsborough Lake Slough (Fig. 3).

Prior to development, water levels in the Everglades fluctuated over a wider range but water management has tended to reduce peak and minimum water levels and to lessen flooding and drought (Fig. 4). Water management, principally drainage for agricultural development, has altered most of south Florida and caused severe environmental changes including large losses of soil through oxidation and subsidence, degradation of water quality, nutrient enrichment, contamination by pesticides and mercury, fragmentation of the landscape, large losses of wetlands and wetland functions, and

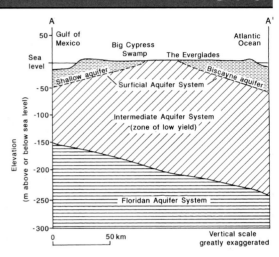

Fig. 1 Generalized hydrogeological cross-section of south Florida (the line of the section is shown in Fig. 2) showing the three major aquifer systems. After McPherson and Halley (1997).

widespread invasion by exotic species. Additionally, the large and growing human population and the agricultural development in the region are in intense competition with the natural system for freshwater resources (McPherson & Halley 1997).

Public water supplies for the 5.8 million people in south Florida are abstracted from shallow aquifers, the most productive and widespread of which are the Biscayne aquifer in the south-east and the shallow aquifer in the south-west (Fig. 1). Freshwater abstractions within south Florida were about 15.6×10^6 m^3 day^{-1} in 1990 with most of this water used for public supply (22%) and agriculture (67%). Groundwater supplied 94% (3.3×10^6 m^3 day^{-1}) of the water used for public supply in 1990. Water abstracted for agricultural purposes is divided between groundwater and surface water. In 1990, groundwater accounted for 4.7×10^6 m^3 day^{-1} and surface water accounted for 5.7×10^6 m^3 day^{-1} of the agricultural requirement.

To contribute to the restoration of the Everglades, a major effort is required to understand the hydrology, geology and ecology of the Everglades and to monitor modifications to the land drainage and flood control structures. Better land management to improve water quality and the development of more sustainable water supplies are also an integral part of the solution. As part of the Central and Southern Florida Project Comprehensive Review Study (Restudy), aquifer storage and recovery (ASR) is being developed on a large scale in the Upper Floridan aquifer (see Section 8.2.4). It is envisaged that ASR technology as a regional water resources option will provide great benefits to agricultural and urban users and to the environment.

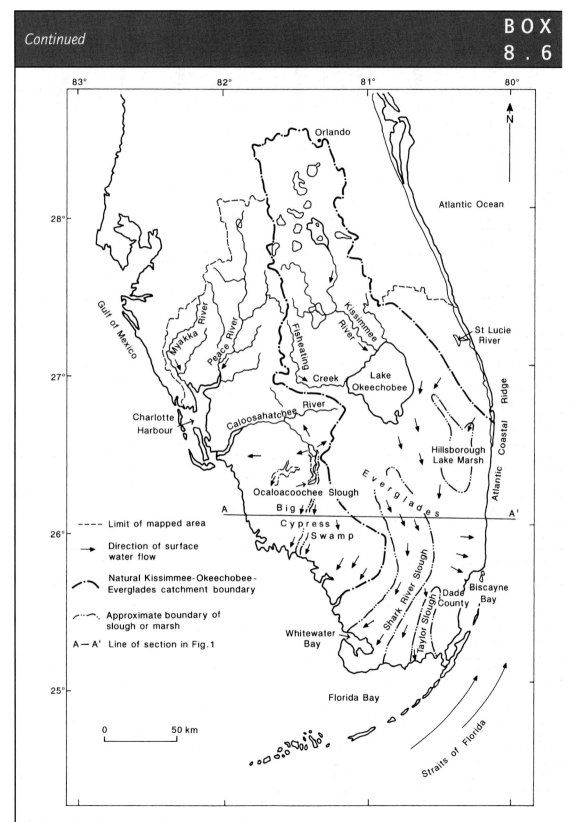

Fig. 2 Hydrological features and the natural direction of surface water and coastal water flows under natural conditions in south Florida. After McPherson and Halley (1997).

BOX
8 . 6

Continued

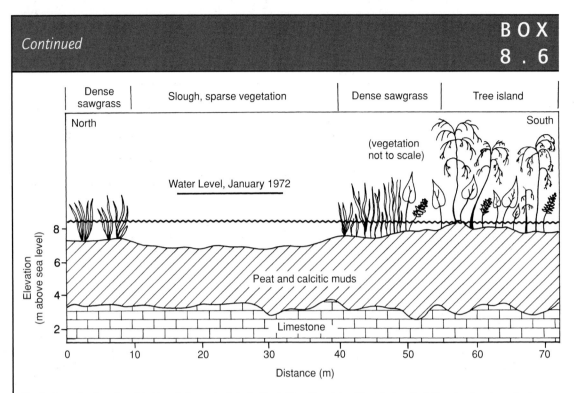

Fig. 3 Generalized section of the Everglades wetlands in the Shark River Slough (see Fig. 2 for location). Peat develops in wetlands that are flooded for extensive periods during the year and calcitic muds develop in wetlands where the periods of flooded landed are shorter and limestone is near the surface. The Everglades has been a dynamic environment with numerous shifts between marl- and peat-forming marshes and between sawgrass marshes and water-lily sloughs. After McPherson and Halley (1997).

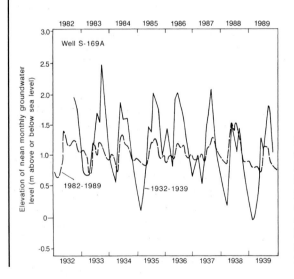

Fig. 4 Long-term hydrograph showing water level fluctuations at a well in southern Dade County (see Fig. 2 for approximate location), 1932–1939 and 1982–1989. Drainage of the Everglades began in the early 1880s and continued into the 1960s with the purpose of reducing the risk of flooding and drought and so opening land for agricultural development south of Lake Okeechobee. The effect of water management in the Everglades has been to reduce peak and minimum water levels in the Everglades, as illustrated in the well hydrograph. After McPherson and Halley (1997).

Table 8.2 Major sources, concentrations and residence times of important 'greenhouse' gases in the atmosphere. After IPCC (2001).

Greenhouse gas	Pre-industrial concentration (1750)	Present concentration (1998)	Residence time	Annual rate of increase*	Major sources
Water vapour	3000 ppm	3000 ppm	10–15 days	n.a.	Oceans
CO_2 (carbon dioxide)	~280 ppm	~365 ppm	5–200 years†	1.5 ppm a^{-1}‡	Combustion of fossil fuels, deforestation
CH_4 (methane)	~700 ppb	1745 ppb	12 years§	7.0 ppb a^{-1}‡	Rice production, cattle rearing, industry
N_2O (nitrous oxide)	~270 ppb	314 ppb	114 years§	0.8 ppb a^{-1}	Agriculture, industry, biomass burning
CFC-11 (chlorofluorocarbon-11)	0	268 ppt	45 years	−1.4 ppt a^{-1}	Aerosols, refrigeration
HFC-23 (hydrofluorocarbon-23)	0	14 ppt	260 years	0.55 ppt a^{-1}	Industrial byproduct
CF_4 (perfluoromethane)	40 ppt	80 ppt	>50,000 years	1 ppt a^{-1}	Aluminium industry

n.a., not applicable.

* Rate is calculated over the period 1990 to 1999.

† No single lifetime can be defined for CO_2 because of the different rates of uptake by different removal processes.

‡ Rate has fluctuated between 0.9 ppm a^{-1} and 2.8 ppm a^{-1} for CO_2 and between 0 and 13 ppb a^{-1} for CH_4 over the period 1990–1999.

§ This lifetime has been defined as an 'adjustment' time that takes into account the indirect effect of the gas on its own residence time.

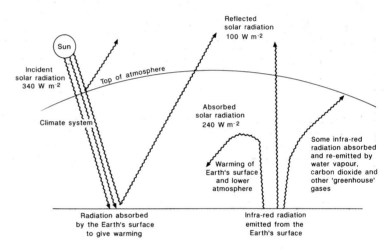

Fig. 8.16 Schematic representation of the global radiation budget. Averaged over the globe, there is 340 W m^{-2} of incident solar radiation at the top of the atmosphere. The climate system absorbs 240 W m^{-2} of solar radiation, so that under equilibrium conditions it must emit 240 W m^{-2} of infra-red radiation. The carbon dioxide radiative forcing for a doubling of carbon dioxide concentrations constitutes a reduction in the emitted infra-red radiation of 4 W m^{-2} producing a heating of the climate system known as global warming. This heating effect acts to increase the emitted radiation in order to re-establish the Earth's radiation balance. After IPCC (1990).

(Box 8.7 gives a classification and assessment of drought severity). A shorter precipitation season, possibly coupled with heavier precipitation events and a shift from snow precipitation to rainfall, would generate larger volumes of runoff over shorter time intervals. More runoff would occur in winter and less runoff would result from spring snowmelt. This complicates the storage and routing of floodwater, both for the purpose of protecting the human environment as well as for meeting water-supply targets. It may also complicate the conjunctive use of surface water and groundwater, as the opportunity for groundwater recharge is reduced under these conditions. In tropical latitudes, water resources are not likely to suffer changes under the predicted climate change impacts. Water quality is also affected by changes in temperature, rainfall and sea level rise that affect the volume of river flow and the degree of saline intrusion (Loáiciga et al. 1996).

Hydrological models of varying degrees of complexity in representing current and future climatic conditions provide an objective approach to estimating hydrological responses to climate change. The linking of physically based hydrological models to output from global circulation models (GCMs) enables the study of a variety of climate change effects (Conway 1998). A note of caution is required in dealing with the scale effects which complicate the

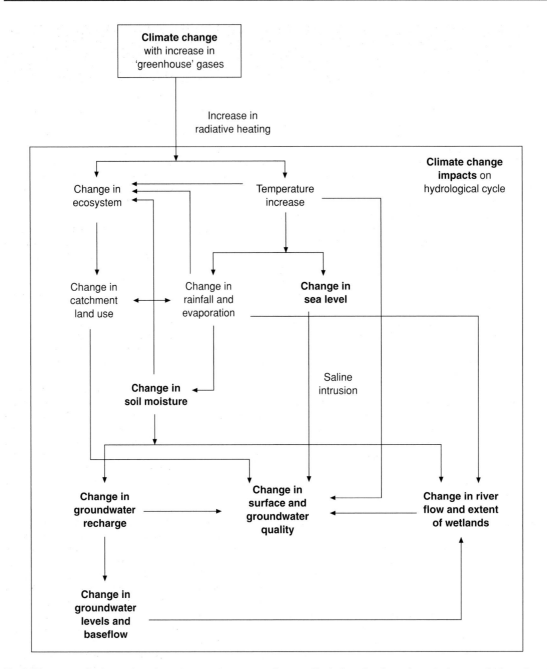

Fig. 8.17 Impacts of increasing 'greenhouse' gas concentrations on the natural hydrological cycle emphasizing changes in hydrogeological conditions. After Arnell (1996).

Droughts and floods are characterized by the extremes of the frequency, intensity and amounts of precipitation (Trenberth et al. 2003). Interestingly, there is no single definition of drought. Beran and Rodier (1985) considered, without specifying when a period of dry weather becomes a drought, that the chief characteristic of a drought is a decrease of water availability in a particular period over a particular area. Meteorological drought is defined in terms of a deficit of precipitation. Agricultural drought relates mostly to deficiency of soil moisture, while hydrological drought relates to deficiencies in, for example, lake levels, river flows and groundwater levels. The greatest severity and extent of drought in the United States occurred during the Dust Bowl years of the 1930s, particularly during 1934 and 1936. The decades of the 1950s and 1960s were also characterized by episodes of widespread, severe drought, while the 1970s and 1980s as a whole were unusually wet. Drought conditions that began in the west of the United States in 1998 persisted in many areas through to the summer of 2003 (Trenberth et al. 2004). In the United Kingdom, long time-series data of daily precipitation and monthly groundwater level provide an indicator of hydrological drought periods, as illustrated by comparing the Central England annual precipitation record and the mean annual Chalk groundwater level for Chilgrove House in southern England (Fig. 1).

Globally, drought areas increased more than 50% throughout the twentieth century, largely due to the drought conditions over the Sahel and Southern Africa during the latter part of the century, while changes in wet areas were relatively small (Trenberth et al. 2004). The most significant cause of drought worldwide is the El Niño Southern Oscillation (ENSO) which also emphasizes the concurrent nature of floods and droughts, with droughts favoured in some areas during an ENSO event, while wet areas are favoured in others. These areas tend to switch during La Niña in the tropics and subtropics. Palaeoclimatological studies show evidence of dramatic changes in drought and the hydrological cycle over many parts of the world, with droughts lasting several decades not uncommon (Box 5.1). The full range of drought variability is probably much larger than has been experienced in the last 100 years (Trenberth et al. 2004).

The shortage of water experienced by water users and the aquatic environment during a drought is a complex issue. Potentially, the full characterization of events that vary regionally and temporally can involve the assessment of a wide range of hydrological indices (for example precipitation amount, river flows, surface reservoir storage and groundwater levels). According to Mawdsley et al. (1994), there are essentially two types of users impacted by the shortage of water. Firstly, there are those who are affected directly by a deficit of precipitation, possibly compounded by high evaporation, and secondly, those consumers who are affected indirectly due to the way in which water storage facilities are managed during a drought. Following from this, and including environmental water requirements, it is recommended that droughts are classified either as environmental or water supply droughts. An environmental drought measures the significance of the drought for those water uses directly affected by a shortage of precipitation, for example aquatic ecology, fisheries, low river flows, reduced spring flows and groundwater levels, agriculture and horticulture. A water supply drought measures the significance for those indirectly affected, for example the risk of water demand reductions imposed on domestic and industrial water consumers. In many cases, a drought will be classified as having both environmental and water supply impacts (Mawdsley et al. 1994).

The communication of the severity of a drought can be simply presented according to its intensity and duration and classified as moderate, serious or severe (Table 1). The duration of an environmental drought is of particular relevance to groundwater resources. If a region is largely dependent on groundwater, and if the available aquifer storage is relatively insensitive to droughts only lasting one recharge period, then this situation is likely to result in a short environmental drought without significant implications for water supplies and the aquatic environment. However, if the drought is of long duration during which there is a reduction in seasonal recharge intensity, then a moderate drought may result with significant implications for water supplies, water levels and river flows dependent on groundwater.

A quantitative measure of environmental droughts is to calculate a drought severity index. In the United States, the Palmer Drought Severity Index (PDSI), as devised by Palmer (1965), is used to represent the severity of dry and wet periods based on weekly or monthly temperature and precipitation data as well as the soil water holding capacity at a location (Dai et al. 1998). Areas experiencing a severe drought score a PDSI of -4.0 while areas with severe moisture surplus score $+4.0$. Between these two extremes, 11 categories of wet and dry conditions are defined.

A more generally applicable drought severity index (DSI), which may be considered for different hydrological data, is to calculate an accumulated monthly deficit relative to the mean for a standard period (Bryant et al. 1992). In this approach, based on available long-term hydrological records, a drought is considered to end when, for example, the 3-monthly precipitation total exceeds the 3-monthly mean for these months. It must be noted that the choice of termination criterion must be considered carefully since different rules can produce different impressions of drought severity that are potentially inappropriate for some long-duration events. A disadvantage of this approach is that a DSI based only on precipitation data does not directly indicate the impact on the environment since this will depend on the antecedent soil moisture conditions in a catchment. Instead, drought severity may be better determined using effective precipitation, or selected local river flow and groundwater level data, even though such records may only be available for a short time relative to precipitation records. Potentially, effective precipitation (the balance between precipitation and evapotranspiration, eq. 8.2) could be a useful indicator of environmental drought, particularly in regions with significant groundwater resources.

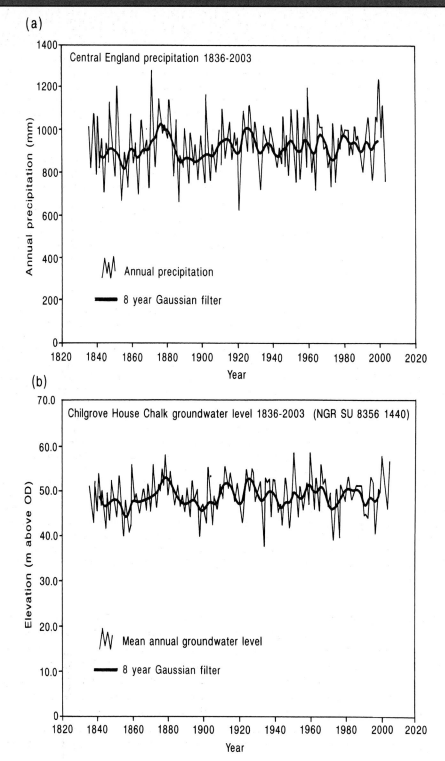

(a)

Central England precipitation 1836-2003

Annual precipitation (mm)

Annual precipitation

8 year Gaussian filter

Year

(b)

Chilgrove House Chalk groundwater level 1836-2003 (NGR SU 8356 1440)

Elevation (m above OD)

Mean annual groundwater level

8 year Gaussian filter

Year

BOX
8.7

Continued

Table 1 Classification scheme for environmental droughts and consequences for groundwater resources. After Mawdsley et al. (1994).

Class of drought	Duration	Return period	Groundwater impacts
Moderate	Short	5–20 years	Reduced spring and river flows; drying out of floodplain areas
	Long*	5–20 years	Reduced spring and river flows; drying out of floodplain areas and wetlands; well yields may decrease
Serious	Short	20–50 years	Reduced spring and river flows; wetlands and ponds dry up
	Long*	20–50 years	Reduced spring and river flows; rivers become influent; wetlands and ponds dry up; saline intrusion in coastal aquifers
Severe	Short	>50 years	Springs and rivers dry up; wetlands and ponds dry up; well yields decrease as groundwater levels fall
	Long*	>50 years	Springs and rivers dry up; wetlands and ponds dry up; well yields fail as groundwater levels fall substantially; saline intrusion in coastal aquifers

* Longer than one groundwater recharge season.

For the prediction of climate change impacts on future minimum groundwater levels across the southern half of England, Bloomfield et al. (2003) applied a statistical method, based on a multiple linear regression model of monthly rainfall totals for a given period against values of minimum annual groundwater levels for the same period, to synthetic rainfall from climate change scenarios to model changes in future annual minimum groundwater levels. In general, the results showed that there is a small reduction in annual minimum groundwater levels for a specific return period and that changes in the seasonality and frequency of extreme events could lead to an increase in the frequency and intensity of groundwater droughts in some areas of the United Kingdom. Bloomfield et al. (2003) concluded that the Chalk aquifer in southern and eastern England might be most susceptible to these effects.

Fig. 1 (*opposite*) Comparison of (a) the Central England annual precipitation record (http://www.met-office.gov.uk/research/hadleycentre/CR_data/Monthly/HadEWP_act.txt) with (b) the mean annual Chalk groundwater level record at Chilgrove House (http://www.ceh.ac.uk/data/NWA.htm) for the period 1836–2003. The data are presented with a smoothed line calculated using an 8-year Gaussian filter (Fritts 1976) applied as a moving average to the annual values. Statistical analysis of the record for Chilgrove House gives mean and median groundwater levels of 48.81 and 48.96 m above Ordnance Datum (OD), respectively, with a standard deviation of 4.03 m. The record indicates that the first and second highest groundwater levels (58.49 and 58.31 m above OD) occurred in 1960 and 1951, respectively. The two lowest groundwater levels (38.48 and 39.51 m above OD) occurred in 1934 and 1973, respectively, giving a maximum range of groundwater level fluctuation over the length of the record of 20.01 m. Applying the formula of Gringorten (1963) for recurrence interval or return period, T, where $T = (n + 0.12)/(m - 0.44)$ with n equal to the number of events and m equal to the event ranking, events ranked first and second in the mean annual groundwater level series have return periods of 300 and 108 years, respectively. Comparison of the records shown in (a) and (b) highlights the hydrological droughts that occurred in the 1840s, 1850s, 1900s, 1940s, 1970s and 1990s.

prediction of regional hydrological changes. Runoff and precipitation at regional scales are highly variable, with 10- to 20-year averages commonly fluctuating in the range ±25% of their long-term means. At shorter timescales the problem is likely to be worse, as inherent hydrological variability increases over a shorter time average (Loáiciga et al. 1996).

An early example of hydrological simulation under climate change is presented by Vaccaro (1992), who studied the sensitivity of groundwater recharge estimates for a semi-arid basin located on the Columbia Plateau, Washington, USA, to historic and projected climatic regimes. Recharge was estimated for pre-development and current (1980s) land use conditions using a daily energy-soil-water balance model. A synthetic daily weather generator was used to simulate lengthy sequences with parameters estimated from subsets of the historical record that were unusually

Table 8.3 Scenarios of climate and groundwater use change effects considered for the Edwards Aquifer, south-central Texas. After Loáiciga (2003).

	Scenario			
	I Base	**II** Climate change effect	**III** Groundwater use change effect	**IV** Total effects
Climate (recharge)	$R_{1978-89}$*	$R_{1978-89}(2 \times CO_2 / 1 \times CO_2)$†	$R_{1978-89}$*	$R_{1978-89}(2 \times CO_2 / 1 \times CO_2)$
Groundwater use	1978–1989 use‡	1978–1989 use	2050 use§	2050 use

* Historical recharge (R) during 1978–1989 (mean = 0.949×10^9 m^3 a^{-1}).
† Historical recharge scaled to $2 \times CO_2$ climate conditions.
‡ Average groundwater use between 1978 and 1989 = 0.567×10^9 m^3 a^{-1}.
§ Groundwater use forecast for 2050 = 0.784×10^9 m^3 a^{-1}.

wet and unusually dry. Estimated recharge in the basin was found to be sensitive to climatic variability in the historical record, especially the precipitation variability. For GCM-projected climate changes to carbon dioxide doubling, the variability in the estimated annual recharge was less than that estimated from the historic data. In addition, the median annual recharge, for the case of a climate scenario averaged from three different GCM simulation runs, was less than 75% of the median annual recharge for the historical simulation. This large change reflects the potential importance of climate on groundwater recharge.

In a more recent study, Loáiciga et al. (2000) and Loáiciga (2003) generated climate change scenarios from scaling factors derived from several GCMs to assess the likely impacts of aquifer pumping on the water resources of the Edwards Balcones Fault Zone aquifer system in south-central Texas. This karst aquifer, formed in the Edwards limestone formation, is one of the most productive regional aquifers in the United States and is a primary source of agricultural and municipal water supplies. The Edwards Aquifer has been identified as one of the regional catchments most vulnerable to climate change impacts in the United States. Groundwater recharge to the aquifer takes place almost exclusively as stream seepage, with groundwater discharge directed to major springs that support an important and threatened groundwater ecosystem. To explore the vulnerability of these springs to combined pumping and climate change impacts, several pumping scenarios were combined with $2 \times CO_2$ climate scenarios. The climate change scenarios were linked to the surface

hydrology to provide future recharge values which were then used as input to a numerical groundwater flow model. The scenarios and model results are shown in Tables 8.3 and 8.4 and summarize the effects on spring flows in the Edwards Aquifer. In the Comal Springs and, to a lesser extent, the San Marcos Springs, there is a simulated increase in spring flow under a doubling of CO_2 for an unchanged groundwater use condition (Scenario II). For the case of Scenario III (base climate and increased groundwater use) there are negative impacts on spring flow. Combining the effects of both climate change and increased groundwater use (Scenario IV) simulates a serious depletion of 73% in spring flow for the Comal springs but only a marginal decrease (1%) for the San Marcos springs. This study highlights that the primary threat to groundwater use in the Edwards Aquifer is from the potential rise in groundwater use caused by population growth and not from climate change, although protracted droughts under climate change will accentuate the competition between human and ecological water uses. Hence, and as concluded by Loáiciga (2003), aquifer strategies in the Edwards Aquifer must be adapted to climate variability and climate change.

Uncertainty is inherent in the development of climate change scenarios due to unknown future emissions of 'greenhouse' gases and aerosols, uncertain global climate sensitivity and the difficulty of simulating the regional characteristics of climate change. It is because of these uncertainties that the regional changes in climate derived from GCM experiments are termed **scenarios** or projections and cannot be considered predictions. Comprehensive reviews of

Table 8.4 Simulated minimum spring flows in the Edwards Aquifer, south-central Texas, for climate and groundwater use change effects listed in Table 8.3. After Loáiciga (2003).

Climate change and groundwater use scenario	Edwards Aquifer springs	
	Comal	San Marcos
I	4.84	4.84
II	12.7 (+162%)	5.67 (+17%)
III	0 (−100%)	3.79 (−22%)
IV	1.31 (−73%)	4.79 (−1%)

Notes:

Spring flows are given in 10^6 m^3 $month^{-1}$.

The numbers in parentheses represent the percentage increase (+) or decrease (−) caused by a scenario relative to the base condition (I).

climate change scenarios, including standards and construction, can be found in Carter and La Rovere (2001). The approach of the United Kingdom Climate Change Impacts Programme (UKCIP) (Hulme et al. 2002) is to present four alternative scenarios of climate change for the United Kingdom that span a reasonable range of possible future climates. The scenarios are labelled Low (L), Medium-low (ML), Medium-high (MH) and High (H) and refer to different greenhouse gas emissions scenarios and changes in global temperature for the 2020s, 2050s and 2080s.

In a study of the impacts of climate change on groundwater resources in the Chalk aquifer of eastern England, Yusoff et al. (2002) developed scaling factors for changes in precipitation (P) and potential evapotranspiration (PE). The factors, as percentage changes in P and PE, were defined by comparing the monthly average P and PE values for a control run of the United Kingdom Hadley Centre HadCM2 model with the monthly average values for future MH and ML climate change scenarios defined for the 2020s and 2050s. In this approach, it is assumed that the monthly factors can be applied equally to each year of the observed historical record to obtain calculated future groundwater recharge values. The derived recharge values were then used as input to a numerical groundwater model calibrated against an historical record of groundwater levels and river baseflow. A limitation of this linked GCM-hydrological modelling approach is that the climate variability represented within the historic record is preserved in the future scenarios. However, the approach gives a general indication of the possible range of changes in hydrological regimes.

The most noticeable and consistent result of the climate change impact simulations carried out by Yusoff et al. (2002) was a decrease in groundwater recharge expected in autumn for all scenarios as a consequence of the smaller amount of summer precipitation and increased autumn potential evapotranspiration. For the 2050MH scenario, these conditions lead to a 42% increase in autumn soil moisture deficit and a 26% reduction in recharge. Hence, eastern England can expect longer and drier summers and a delay in the start of groundwater recharge in the autumn and winter period. The drier conditions will have relatively little effect on summer groundwater levels (generally a 1–2% decrease), but a modelled decrease of up to 14% in autumn baseflow volume for the 2050MH scenario indicates that Chalk groundwater-fed rivers may show environmental impacts with potential conflicts with other water demands.

Other scenarios for climate change impacts on groundwater relate to water quality and the balance of ocean salinity. Younger et al. (2002) examined the possibility that carbonate aquifers may act as a possible sink (or source) for atmospheric carbon dioxide and therefore have important consequences for the calcium carbonate content, or hardness, of groundwater. Younger et al. (2002) modelled increases in calcium concentrations of ≤10 mg L^{-1} for two European carbonate aquifers over a 50-year simulation period to 2045. These increases are negligible in water resources terms but draw attention to the possibility that the world's carbonate aquifers may represent a sink for atmospheric carbon dioxide and a slowing of global warming over long time scales.

Direct groundwater discharges to the world's oceans and seas from inland catchments are estimated to be 2220 km^3 a^{-1} (see Section 1.5). Zektser and Loaiciga (1993) argued that a hypothetical 10% increase in global precipitation of 2 mm a^{-1} for greenhouse warming would provide an additional direct groundwater discharge of approximately 222 km^3 a^{-1}. Although this additional flux is small, the salt load, with a dissolved solids content of 585 mg L^{-1} (see Section 1.5), would add a further 1.3×10^8 t a^{-1} of salts to the world's oceans and seas. Whether this additional salt load would increase the salinity of the oceans is uncertain and depends on the balance of water inputs to and from the oceans under an intensified hydrological cycle and any volume change in the oceans as a result of thermal expansion and ice melting. It should also be considered that these effects of changes in direct groundwater flow might only stabilize over timescales of hundreds to thousands of years.

8.6 FURTHER READING

Arnell, N. (1996) *Global Warming, River Flows and Water Resources.* John Wiley, Chichester.

Arnell, N. (2002) *Hydrology and Global Environmental Change.* Prentice Hall, Harlow.

Burke, J.J. & Moench, M.H. (2000) *Groundwater and Society: resources, tensions and opportunities.* United Nations Department of Economic and Social Affairs and Institute for Social and Environmental Transition, United Nations, New York.

Cook, H.F. (1998) *The Protection and Conservation of Water Resources: a British perspective.* John Wiley, Chichester.

Hiscock, K.M., Rivett, M.O. & Davison, R.M. (eds) (2002) *Sustainable Groundwater Development.* Geological Society, London, Special Publications, **193**.

IPCC (2001) *Climate Change 2001. The scientific basis.* A Report of Working Group I of the Intergovernmental Panel on Climate Change. Cambridge University Press, Cambridge.

Keddy, P.A. (2000) *Wetland Ecology: principles and conservation.* Cambridge University Press, Cambridge.

Mitsch, W.J. & Gosselink, J.G. (2000) *Wetlands*, 3rd edn. John Wiley, New York.

Conversion factors

Appendix 1

Quantity	Unit	SI equivalent
Length	1 yard (yd)	0.914 m
	1 foot (ft)	0.305 m
	1 inch (in)	25.4 mm
	1 mile	1.609 km
Area	1 yd^2	0.836 m^2
	1 ft^2	9.290×10^{-2} m^2
	1 in^2	6.452×10^2 mm^2
	1 mile^2	2.590 km^2
	1 acre	4.407×10^3 m^2
	1 hectare (ha)	1×10^4 m^2
Volume	1 ft^3	2.832×10^{-2} m^3
	1 UK gallon	4.546×10^{-3} m^3
	1 US gallon	3.784×10^{-3} m^3
Mass	1 pound (lb)	4.536×10^{-1} kg
	1 UK ton	1.0165×10^3 kg
	1 US ton	9.072×10^2 kg
	1 tonne (t)	1000 kg
Density	1 lb ft^{-3}	16.0185 kg m^{-3}
Force and weight	1 poundal (pdl)	0.138 N
	1 poundal-force (lb$_f$)	4.448 N
Pressure	1 lb$_f$ in^{-2} (psi)	6.895 kPa
	1 lb$_f$ ft^{-2}	4.788×10^{-2} kPa
	1 mm Hg	133.322 Pa
	1 ft H$_2$O	2.989 kPa
	1 m H$_2$O	9.807 kPa
	1 kg cm^{-2}	98.067 kPa
	1 bar	100 kPa
	1 atmosphere	101.325 kPa

Quantity	Unit	SI equivalent
Velocity	1 ft s^{-1}	3.048×10^{-1} m s^{-1}
	1 km h^{-1}	2.778×10^{-1} m s^{-1}
	1 mile h^{-1}	4.470×10^{-1} m s^{-1}
Viscosity	1 lb.ft^{-1}s^{-1}	1.488 N s m^{-2}
	1 Poise	0.1 N s m^{-2}
Viscosity (kinematic)	1 stoke	10^{-4} m^{2}s^{-1}
Energy	1 kWh	3.6 MJ
	1 Btu	1.055 kJ
	1 Therm	105.506 MJ
	1 kcal	4.187 kJ
	1 foot poundal	4.214×10^{-2} J
Power	1 horse power	7.457×10^{-1} kW
	1 Btu ft^{-2}	11.357 kJ m^{-2}
	1 cal cm^{-2}	41.868 kJ m^{-2}
Discharge	ft^3 s^{-1}	2.832×10^{-2} m^3 s^{-1}
	US gallon min^{-1}	6.309×10^{-5} m^3 s^{-1}
Intrinsic permeability	ft^2	9.290×10^{-2} m^2
	darcy	9.870×10^{-13} m^2
Hydraulic conductivity	ft s^{-1}	3.048×10^{-1} m s^{-1}
	US gallon day^{-1} ft^{-2}	4.720×10^{-7} m s^{-1}
Transmissivity	ft^2 s^{-1}	9.290×10^{-2} m^2 s^{-1}
	US gallon day^{-1} ft^{-1}	1.438×10^{-7} m^2 s^{-1}

Properties of water in the range 0–100°C

Values of the density, viscosity, vapour pressure and surface tension for liquid water in the range 0–100°C. All values (except vapour pressure) refer to a pressure of 100 kPa (1 bar). Source: Lide, D.R. (ed.) (1991) *Handbook of Chemistry and Physics*, 72nd edn. CRC Press, Boca Baton.

Temperature (°C)	Density (kg m^{-3})	Viscosity ($\times 10^{-6}$ N s m^{-2})	Vapour pressure (kPa)	Surface tension (mN m^{-1})
0	999.84	1793	0.6113	75.64
10	999.70	1307	1.2281	74.23
20	998.21	1002	2.3388	72.75
30	995.65	797.7	4.2455	71.20
40	992.22	653.2	7.3814	69.60
50	988.03	547.0	12.344	67.94
60	983.20	466.5	19.932	66.24
70	977.78	404.0	31.176	64.47
80	971.82	354.4	47.373	62.67
90	965.35	314.5	70.117	60.82
100	958.40	281.8	101.325	58.91

The geological timescale

Appendix 3

Phanerozoic Eon

(544 Ma to Present)

The period of time between the end of the Precambrian Eon and today. The Phanerozoic Eon begins with the start of the Cambrian Period, 544 Ma. The period encompasses the period of abundant, complex life on the Earth.

Era	Period or System		Epoch or Series
Cenozoic (65 Ma to Present)	Quaternary (1.8 Ma to Present)		Holocene (8000 years to Present)
			Pleistocene (1.8 million to 8000 years)
	Tertiary (65–1.8 Ma)		Pliocene (5.3–1.8 Ma)
			Miocene (23.8–5.3 Ma)
			Oligocene (33.7–23.8 Ma)
			Eocene (55.5–33.7 Ma)
			Palaeocene (65–55.5 Ma)
Mesozoic (248–65 Ma)	Cretaceous (145–65 Ma)		
	Jurassic (213–145 Ma)		
	Triassic (248–213 Ma)		
Palaeozoic (544–248 Ma)	Permian (286–248 Ma)		
	Carboniferous (360–286 Ma)	Pennsylvanian* (325–286 Ma)	
		Mississippian* (360–325 Ma)	
	Devonian (410–360 Ma)		
	Silurian (440–410 Ma)		
	Ordovician (505–440 Ma)		
	Cambrian (544–505 Ma)		

Precambrian Eon

(Beginning of the Earth to 544 Ma)

All geological time before the beginning of the Palaeozoic Era. This includes about 90% of all geological time and spans the time from the beginning of the Earth, about 4500 Ma, to 544 Ma. The Precambrian Eon is usually divided into the Archaen (pre-2500 Ma) and the Proterozoic (the remainder), both being commonly called eons.

Source: Topinka (2001) http://vulcan.wr.usgs.gov/Glossary/geo_time_scale.html

Ma, million years before present.

* North American usage.

Symbols, atomic numbers and atomic weights of the elements

Appendix 4

Atomic number	Element	Symbol	Atomic weight	Atomic number	Element	Symbol	Atomic weight
1	Hydrogen	H	1.008	36	Krypton	Kr	83.80
2	Helium	He	4.003	37	Rubidium	Rb	85.47
3	Lithium	Li	6.939	38	Strontium	Sr	87.62
4	Beryllium	Be	9.012	39	Yttrium	Y	88.91
5	Boron	B	10.81	40	Zirconium	Zr	91.22
6	Carbon	C	12.01	41	Niobium	Nb	92.91
7	Nitrogen	N	14.01	42	Molybdenum	Mo	95.94
8	Oxygen	O	16.00	43	Technetium*	Tc	97
9	Fluorine	F	19.00	44	Ruthenium	Ru	101.07
10	Neon	Ne	20.18	45	Rhodium	Rh	102.91
11	Sodium	Na	22.99	46	Palladium	Pd	106.4
12	Magnesium	Mg	24.31	47	Silver	Ag	107.87
13	Aluminium	Al	26.98	48	Cadmium	Cd	112.41
14	Silicon	Si	28.09	49	Indium	In	114.82
15	Phosphorus	P	30.97	50	Tin	Sn	118.69
16	Sulphur	S	32.06	51	Antimony	Sb	121.75
17	Chlorine	Cl	35.45	52	Tellurium	Te	127.60
18	Argon	Ar	39.95	53	Iodine	I	126.90
19	Potassium	K	39.10	54	Xenon	Xe	131.30
20	Calcium	Ca	40.08	55	Caesium	Cs	132.91
21	Scandium	Sc	44.96	56	Barium	Ba	137.33
22	Titanium	Ti	47.90	57	Lanthanium	La	138.91
23	Vanadium	V	50.94	58	Cerium	Ce	140.12
24	Chromium	Cr	52.00	59	Praseodymium	Pr	140.91
25	Manganese	Mn	54.94	60	Neodymium	Nd	144.24
26	Iron	Fe	55.85	61	Promethium*	Pm	145
27	Cobalt	Co	58.93	62	Samarium	Sm	150.35
28	Nickel	Ni	58.71	63	Europium	Eu	151.96
29	Copper	Cu	63.54	64	Gadolinium	Gd	157.25
30	Zinc	Zn	65.37	65	Terbium	Tb	158.93
31	Gallium	Ga	69.72	66	Dysprosium	Dy	162.50
32	Germanium	Ge	72.59	67	Holmium	Ho	164.93
33	Arsenic	As	74.92	68	Erbium	Er	167.26
34	Selenium	Se	78.96	69	Thulium	Tm	168.93
35	Bromine	Br	79.91	70	Ytterbium	Yb	173.04

Atomic number	Element	Symbol	Atomic weight	Atomic number	Element	Symbol	Atomic weight
71	Lutetium	Lu	174.97	91	Protactinium*	Pa	231.04
72	Hafnium	Hf	178.49	92	Uranium*	U	238.03
73	Tantalum	Ta	180.95	93	Neptunium*	Np	237
74	Tungsten	W	183.85	94	Plutonium*	Pu	244
75	Rhenium	Re	186.21	95	Americium*	Am	243
76	Osmium	Os	190.2	96	Curium*	Cm	247
77	Iridium	Ir	192.22	97	Berkelium*	Bk	247
78	Platinum	Pt	195.09	98	Californium*	Cf	251
79	Gold	Au	196.97	99	Einsteinium*	Es	254
80	Mercury	Hg	200.59	100	Fermium*	Fm	253
81	Thallium	Tl	204.37	101	Mendelevium*	Md	256
82	Lead	Pb	207.19	102	Nobelium*	No	253
83	Bismuth	Bi	208.98	103	Lawrentium*	Lw	257
84	Polonium*	Po	209				
85	Astatine*	At	210				
86	Radon*	Rn	222				
87	Francium*	Fr	223				
88	Radium*	Ra	226.03				
89	Actinium*	Ac	227.03				
90	Thorium*	Th	232.04				

* Radioactive element.

Composition of seawater and rainwater

Appendix 5

A5.1 Seawater composition

Ocean water consists of a complex solution with a total salt content, or salinity, of 3.5% (usually stated as 35‰ or 35 parts per thousand). The composition of seawater is mainly controlled by a balance between the addition of dissolved material from river water and groundwater, and various processes of removal. Important removal processes include losses to accumulating sediments by precipitation, sorption and organic activity, and reactions with basalt at mid-ocean ridges. Of the dissolved materials listed in Table A5.1, the most abundant and most constant in concentration in all parts of the ocean are the

Table A5.1 Table of average abundance of elements in seawater. Sources: Goldberg (1963); Li (1991).

Element	Abundance (mg L^{-1})	Principal species
O	880,000	H_2O; $O_2(g)$; SO_4^{2-} and other anions
H	110,000	H_2O
Cl	18,800	Cl^-
Na	10,800	Na^+
Mg	1290	Mg^{2+}; $MgSO_4$
S	900	SO_4^{2-}
N	670	NO_3^-; NO_2^-; NH_4^+; $N_2(g)$; organic compounds
Ca	450	Ca^{2+}; $CaSO_4$
K	390	K^+
Br	67	Br^-
C	28	HCO_3^-; H_2CO_3; CO_3^{2-}; organic compounds
Sr	7.8	Sr^{2+}; $SrSO_4$
B	4.5	$B(OH)_3$; $B(OH)_2O^-$
Si	2.5	$Si(OH)_4$; $Si(OH)_3O^-$
F	1.3	F^-
Ar	0.6	$Ar(g)$
Li	0.18	Li^+
Rb	0.12	Rb^+
P	0.09	HPO_4^{2-}; $H_2PO_4^-$; PO_4^{3-}; H_3PO_4
I	0.058	IO_3^-; I^-
Ba	0.015	Ba^{2+}; $BaSO_4$
Mo	0.01	MoO_4^{2-}
U	0.0032	$UO_2(CO_3)_3^{4-}$
Al	0.003	
Fe	0.003	$Fe(OH)_3(s)$
V	0.0022	$VO_2(OH)_3^{2-}$
As	0.0017	$HAsO_4^{2-}$; $H_2AsO_4^-$; H_3AsO_4; H_3AsO_3

Table A5.1 *(cont'd)*

Element	Abundance (mg L^{-1})	Principal species
Ni	0.0005	Ni^{2+}; NiSO$_4$
Cr	0.0003	
Cs	0.0003	Cs$^+$
Kr	0.0003	Kr(g)
Zn	0.0003	Zn^{2+}; ZnSO$_4$
Cu	0.0002	Cu^{2+}; Cu SO$_4$
Mn	0.0002	Mn^{2+}; MnSO$_4$
Sb	0.0002	
Se	0.0002	SeO$_4^{2-}$
Ne	0.0001	Ne(g)
Ti	0.0001	
W	0.0001	WO$_4^{2+}$
Xe	0.0001	Xe(g)
Cd	8×10^{-5}	Cd^{2+}; CdSO$_4$
Zr	2×10^{-5}	
Nb	1×10^{-5}	
Tl	1×10^{-5}	Tl$^+$
Y	1×10^{-5}	
La	6×10^{-6}	
He	5×10^{-6}	He(g)
Au	4×10^{-6}	AuCl$_4^-$
Ge	4×10^{-6}	Ge(OH)$_4$; Ge(OH)$_3$O$^-$
Nd	4×10^{-6}	
Ag	3×10^{-6}	AgCl$_2^-$; AgCl$_3^{2-}$
Hf	3×10^{-6}	
Pb	3×10^{-6}	Pb^{2+}; PbSO$_4$
Ce	2×10^{-6}	
Dy	2×10^{-6}	
Ga	2×10^{-6}	
Ta	2×10^{-6}	
Yb	2×10^{-6}	
Co	1×10^{-6}	Co^{2+}; CoSO$_4$
Er	1×10^{-6}	
Gd	1×10^{-6}	
Pr	9×10^{-7}	
Sc	9×10^{-7}	
Sm	8×10^{-7}	
Sn	6×10^{-7}	
Ho	5×10^{-7}	
Hg	4×10^{-7}	HgCl$_3^-$; HgCl$_4^{2-}$
Lu	3×10^{-7}	
Pt	3×10^{-7}	
Tm	3×10^{-7}	
Be	2×10^{-7}	
Eu	2×10^{-7}	
Tb	2×10^{-7}	
In	1×10^{-7}	
Pd	7×10^{-8}	
Th	5×10^{-8}	
Bi	4×10^{-9}	
Ra	1×10^{-10}	Ra^{2+}; RaSO$_4$
Rn	6×10^{-17}	Rn(g)

conservative species Na^+, K^+, Ca^{2+}, Mg^{2+}, Cl^- and SO_4^{2-}. Some of the less abundant, non-conservative species, notably HCO_3^-, SiO_2 and the ions of N and P, participate in biological processes and therefore show widely varying concentrations depending on the local abundance of marine organisms and supply of organic carbon. Seawater pH is remarkably constant, normally in the range 7.8–8.4, and is buffered principally by inorganic reactions involving carbonate species (Krauskopf & Bird 1995).

A5.2 Rainwater composition

Rainwater can be described as a weakly acidic, dilute solution, with a pH in the range 4–6 and a total salt content of just a few milligrammes per litre. Evaporation into the atmosphere results in separation of water molecules from dissolved salts in surface waters. The resulting water vapour ultimately condenses to form rain, and the overall process can be viewed as purification by natural distillation. However, solid particles and gases in the atmosphere are dissolved in rainwater resulting in a wide range in chemical composition, as well as variation in pH. Broadly, and as shown in Tables A5.2 and A5.3, rainwater species derived from terrestrial sources are mainly dominated by Ca^{2+}, K^+, NH_4^+ and NO_3^-, and from marine sources the main species are Cl^-, Na^+, Mg^{2+} and SO_4^{2-}. Elements in rain that result from rainout (determined by the composition of nucleating aerosols) show little change or a slight rise in concentration with time. By contrast, elements contributed by washout (determined by the composition of soluble trace gases) exhibit a sharp decrease in concentration with time as the air is essentially cleaned during the rainfall event (Berner & Berner 1987).

Table A5.2 Chemical composition of rainwater samples from land, marine, island and coastal sites. All concentrations and pH are given as mean values. Source: Cornell (1996).

Location (no. of samples)	Concentrations (μmol L^{-1})									
	pH	NH$_4^+$	Na$^+$	K$^+$	Ca^{2+}	Mg^{2+}	NO$_3^-$	Cl$^-$	SO$_4^{2-}$	NssSO$_4^{2-}$
Land sites:										
Norwich, UK ($n = 25$) 52°38′N 1°17′E	4.7	31.6	55.0	5.4	11.3	15.3	37.0	48.0	46.5	43.2
Norfolk, UK ($n = 12$) 52°50′N 1°0′E	5.0	35.8	76.4	3.0	5.9	n.d.	39.7	27.9	71.4	79.2
Fichtelberg, Czech Republic ($n = 5$) 50°10′N 12°0′E	4.0	9.4	3.0	2.0	4.9	2.1	10.1	8.2	39.7	39.4
Cullowhee-NC, USA ($n = 8$) 34°54′N 82°24′W	4.9	2.1	8.4	9.2	8.9	2.0	6.3	11.4	18.1	19.4
Maraba-PA, Brazil ($n = 5$) 5°20′S 49°5′W	4.7	54.7	59.5	2.9	2.1	6.2	21.5	17.0	6.6	10.6
Marine, island and coastal sites:										
North Atlantic ($n = 8$) 50°60′N ~30°W and 38°52′N ~30°W	4.9	0.6	856.1	19.8	12.7	74.1	3.1	754.5	144.7	11.0
BBSR, Bermuda ($n = 18$) 32°35′N 8°25′W	5.0	8.0	58.2	2.6	6.9	3.1	2.9	88.2	35.8	25.8
Recife, Brazil ($n = 9$) 8°0′S 35°0′W	5.3	4.9	41.5	2.7	1.9	5.3	3.6	35.6	15.5	13.0
Tahiti ($n = 16$) 17°37′S 149°27′W	5.2	2.1	138.9	1.8	2.5	4.9	0.6	48.6	3.8	3.2

n.d., not determined; NssSO$_4^{2-}$, concentration of SO$_4^{2-}$ from sources other than sea salt.

Table A5.3 Volume-weighted mean concentrations of trace elements in rain water from various sites in and around the central southern North Sea. Source: Kane et al. (1994).

Element	Concentration ($\mu g\ L^{-1}$)		
	Pellworm Island, Germany	Mannington, Norfolk, UK	North Sea
Fe	88	17	31
Mn	3.8	2.7	3.6
Cu	2.3	4.0	1.0
Zn	13	2.9	7.6
Pb	4.0	4.1	3.5
Cd	0.7	0.25	0.08
Na	–	1200	82,100
Ca	388	126	3396
NO_3^-	–	1240	3236
SO_4^{2-}	–	2790*	3273†
NH_4^+	–	4070	668

* Predominantly non-sea-salt SO_4^{2-}.
† Non-sea-salt SO_4^{2-}.

A5.3 REFERENCES

Berner, E.K. & Berner, R.A. (1987) *The Global Water Cycle: geochemistry and environment.* Prentice Hall, Englewood Cliffs, New Jersey, 62–70.

Cornell, S.E. (1996) Dissolved Organic Nitrogen in Rainwater. PhD Thesis, University of East Anglia, Norwich, 134–138.

Goldberg, E.D. (1963) The oceans as a chemical system. In: *The Sea. Ideas and Observations on progress in the study of the sea.* Vol. 2. *The Composition of Sea-water; comparative and descriptive oceanography* (ed. M.N. Hill). Wiley-Interscience, New York, 3–25.

Kane, M.M., Rendell, A.R. & Jickells, T.D. (1994) Atmospheric scavenging processes over the North Sea. *Atmospheric Environment* **28**, 2523–2530.

Krauskopf, K.B. & Bird, D.K. (1995) *Introduction to Geochemistry*, 3rd edn. McGraw-Hill, New York, 309–317.

Li, Y.-H. (1991) Distribution patterns of the elements in the ocean: a synthesis. *Geochimica et Cosmochimica Acta* **55**, 3223–3240.

Values of $W(u)$ for various values of u

Appendix 6

u	u values								
	1.0	2.0	3.0	4.0	5.0	6.0	7.0	8.0	9.0
$\times 1$	0.219	0.049	0.013	0.0038	0.0011	0.00036	0.00012	0.000038	0.000012
$\times 10^{-1}$	1.82	1.22	0.91	0.70	0.56	0.45	0.37	0.31	0.26
$\times 10^{-2}$	4.04	3.35	2.96	2.68	2.47	2.30	2.15	2.03	1.92
$\times 10^{-3}$	6.33	5.64	5.23	4.95	4.73	4.54	4.39	4.26	4.14
$\times 10^{-4}$	8.63	7.94	7.53	7.25	7.02	6.84	6.69	6.55	6.44
$\times 10^{-5}$	10.94	10.24	9.84	9.55	9.33	9.14	8.99	8.86	8.74
$\times 10^{-6}$	13.24	12.55	12.14	11.85	11.63	11.45	11.29	11.16	11.04
$\times 10^{-7}$	15.54	14.85	14.44	14.15	13.93	13.75	13.60	13.46	13.34
$\times 10^{-8}$	17.84	17.15	16.74	16.46	16.23	16.05	15.90	15.76	15.65
$\times 10^{-9}$	20.15	19.45	19.05	18.76	18.54	18.35	18.20	18.07	17.95
$\times 10^{-10}$	22.45	21.76	21.35	21.06	20.84	20.66	20.50	20.37	20.25
$\times 10^{-11}$	24.75	24.06	23.65	23.36	23.14	22.96	22.81	22.67	22.55
$\times 10^{-12}$	27.05	26.36	25.96	25.67	25.44	25.26	25.11	24.97	24.86
$\times 10^{-13}$	29.36	28.66	28.26	27.97	27.75	27.56	27.41	27.28	27.16
$\times 10^{-14}$	31.66	30.97	30.56	30.27	30.05	29.87	29.71	29.58	29.46
$\times 10^{-15}$	33.96	33.27	32.86	32.58	32.35	32.17	32.02	31.88	31.76

$$W(u), \text{ the well function,} = \int_u^\infty \frac{e^{-u}du}{u} = [-0.5772 - \log_e u + u - u^2/2.2! + u^3/3.3! - u^4/4.4! + \ldots] \qquad \text{eq. A6.1}$$

where

$$u = \frac{r^2 S}{4Tt} \qquad \text{eq. A6.2}$$

Values of q/Q and v/Qt corresponding to selected values of t/F for use in computing the rate and volume of stream depletion by wells and boreholes

Appendix

t/F	q/Q	v/Qt	t/F	q/Q	v/Qt
0	0	0	1.9	0.608	0.409
0.07	0.008	0.001	2.0	0.617	0.419
0.10	0.025	0.006	2.2	0.634	0.438
0.15	0.068	0.019	2.4	0.648	0.455
0.20	0.114	0.037	2.6	0.661	0.470
0.25	0.157	0.057	2.8	0.673	0.484
0.30	0.197	0.077	3.0	0.683	0.497
0.35	0.232	0.097	3.5	0.705	0.525
0.40	0.264	0.115	4.0	0.724	0.549
0.45	0.292	0.134	4.5	0.739	0.569
0.50	0.317	0.151	5.0	0.752	0.587
0.55	0.340	0.167	5.5	0.763	0.603
0.60	0.361	0.182	6.0	0.773	0.616
0.65	0.380	0.197	7	0.789	0.640
0.70	0.398	0.211	8	0.803	0.659
0.75	0.414	0.224	9	0.814	0.676
0.80	0.429	0.236	10	0.823	0.690
0.85	0.443	0.248	15	0.855	0.740
0.90	0.456	0.259	20	0.874	0.772
0.95	0.468	0.270	30	0.897	0.810
1.0	0.480	0.280	50	0.920	0.850
1.1	0.500	0.299	100	0.944	0.892
1.2	0.519	0.316	600	0.977	0.955
1.3	0.535	0.333			
1.4	0.550	0.348			
1.5	0.564	0.362			
1.6	0.576	0.375			
1.7	0.588	0.387			
1.8	0.598	0.398			

Source: Jenkins, C.T. (1968) Computation of rate and volume of stream depletion by wells. In: *Techniques of Water-Resources Investigations of the United States Geological Survey*. Book 4, *Hydrologic Analysis and Interpretation*. United States Department of the Interior, Washington, Chap. D1.

Values of error function, erf (β), and complementary error function, erfc (β), for positive values of β

Appendix 8

β	erf (β)	erfc (β)	β	erf (β)	erfc (β)
0	0	1.0	1.4	0.952285	0.047715
0.05	0.056372	0.943628	1.5	0.966105	0.033895
0.1	0.112463	0.887537	1.6	0.976348	0.023652
0.15	0.167996	0.832004	1.7	0.983790	0.016210
0.2	0.222703	0.777297	1.8	0.989091	0.010909
0.25	0.276326	0.723674	1.9	0.992790	0.007210
0.3	0.328627	0.671373	2.0	0.995322	0.004678
0.35	0.379382	0.620618	2.1	0.997021	0.002979
0.4	0.428392	0.571608	2.2	0.998137	0.001863
0.45	0.475482	0.524518	2.3	0.998857	0.001143
0.5	0.520500	0.479500	2.4	0.999311	0.000689
0.55	0.563323	0.436677	2.5	0.999593	0.000407
0.6	0.603856	0.396144	2.6	0.999764	0.000236
0.65	0.642029	0.357971	2.7	0.999866	0.000134
0.7	0.677801	0.322199	2.8	0.999925	0.000075
0.75	0.711156	0.288844	2.9	0.999959	0.000041
0.8	0.742101	0.257899	3.0	0.999978	0.000022
0.85	0.770668	0.229332			
0.9	0.796908	0.203092			
0.95	0.820891	0.179109			
1.0	0.842701	0.157299			
1.1	0.880205	0.119795			
1.2	0.910314	0.089686			
1.3	0.934008	0.065992			

$$\mathrm{erf}\,(\beta) = \frac{2}{\pi} \int_{0}^{\beta} e^{-\varepsilon^2}\, d\varepsilon.$$

$\mathrm{erf}\,(-\beta) = -\mathrm{erf}\,(\beta).$

$\mathrm{erfc}\,(\beta) = 1 - \mathrm{erf}\,(\beta).$

$\mathrm{erfc}\,(-\beta) = 1 + \mathrm{erf}\,(\beta).$

Drinking water quality standards and Lists I and II Substances

A9.1 Drinking water quality guidelines and standards

Parameter	World Health Organization guideline value	European Union parametric value	United States Environmental Protection Agency primary and secondary standards	
			NPDWR	
Microbiological parameters			**MCLG**	**MCL or TT**
E. coli or thermotolerant coliform bacteria	0 in 100 mL sample	0 in 100 mL sample	–	–
Total coliforms* (including faecal coliform and *E. coli*)	–	0 in 100 mL sample	0	No more than 5% of samples total coliform-positive in a month
Enterococci	–	0 in 100 mL sample	–	
Cryptosporidium	–	–	0	TT
Giardia lamblia	–	–	0	TT
Heterotrophic plate count	–	–	n/a	TT
Legionella	–	–	0	TT
Viruses (enteric)	–	–	0	TT
Turbidity†	5 NTU	n/a	n/a	TT
Disinfection byproducts				
Bromate	0.025 mg L^{-1}	0.01 mg L^{-1}	0 mg L^{-1}	0.010 mg L^{-1}
Chlorite	0.2	–	0.8	1.0
Haloacetic acids	<0.1	–	n/a	0.060
Total trihalomethanes	‡	0.1	n/a	0.080
Disinfectants			**MRDLG**	**MRDL**
Chloramines (as Cl$_2$)	3 mg L^{-1}	–	4 mg L^{-1}	4.0 mg L^{-1}
Chlorine (as Cl$_2$)	5	–	4	4.0
Chlorine dioxide (as ClO$_2$)	n/a	–	0.8	0.8
Radionuclides			**MCLG**	**MCL or TT**
Alpha particles	0.1 Bq L^{-1}	–	0	15 pCi L^{-1}
Beta particles	1 Bq L^{-1}	–	0	4 millirems a^{-1}
Radium-226, -228 (combined)	–	–	0	5 pCi L^{-1}
Uranium	2 (P) µg L^{-1}	–	0	30 µg L^{-1}
Tritium	–	100 Bq L^{-1}	–	–
Total indicative dose	–	0.10 mSv a^{-1}	–	–
Pesticides			**MCLG**	**MCL or TT**
Alachlor	20 µg L^{-1}	0.1 µg L^{-1} for an individual pesticide (except where indicated); 0.5 µg L^{-1} for the sum of all individual pesticides	0 µg L^{-1}	2 µg L^{-1}
Aldicarb	10		–	–
Aldrin/dieldrin	0.03		–	–
Atrazine	2		–	–
			3	3

Parameter	World Health Organization guideline value	European Union parametric value	United States Environmental Protection Agency primary and secondary standards	
			NPDWR	
Bentazone	300	–	–	–
Carbofluran	7	–	40	40
Chlordane	0.2	–	0	2
Chlorotoluron	30	–	–	–
Cyanazine	0.6	–	–	–
DDT	2	–	–	–
1,2-Dibromo-3-chloropropane	1	–	0	0.2
1,2-Dibromoethane	0.4–15(P)	–	–	–
2,4-Dichlorophenoxyacetic acid (2,4-D)	30	–	70	70
1,2-Dichloropropane (1,2-DCP)	40(P)	–	0	5
1,3-Dichloropropane	NAD	–	–	–
1,3-Dichlorpropene	20	–	–	–
Diquat	10(P)	–	20	20
Heptachlor and heptachlorepoxide	0.03	0.03	0	0.4 and 0.2
Hexachlorobenzene	1	–	0	1
Isoproturon	9	–	–	–
Lindane	2	–	0.2	0.2
MCPA	2	–	–	–
Methoxychlor	20	–	40	40
Metolachlor	10	–	–	–
Molinate	6	–	–	–
Pendimethalin	20	–	–	–
Pentachlorophenol	9(P)	–	0	1
Permethrin	20	–	–	–
Propanil	20	–	–	–
Pryidate	100	–	–	–
Simazine	2	–	4	4
Terbuthylazine (TBA)	7	–	–	–
Trifluralin	20	–	–	–
2,4-DB	90	–	–	–
Dichlorprop	100	–	–	–
Fenoprop	9	–	–	–
MCPB	NAD	–	–	–
Mecoprop	10	–	–	–
2,4,5-T	9	–	–	–
Organic constituents			MCLG	MCL or TT
Acrylamide	$0.5\,\mu g\,L^{-1}$	$0.10\,\mu g\,L^{-1}$	$0\,\mu g\,L^{-1}$	TT
Benzene	10	1.0	0	$5\,\mu g\,L^{-1}$
Benzo[a]pyrene (PAHs)	0.7	0.10 (sum of specified PAHs)	0	0.2
Carbon tetrachloride	2	–	0	5
Chlorobenzene	–	–	100	100
Dalapon	–	–	200	200
o-Dichlorobenzene	1000	–	600	600
p-Dichlorobenzene	300	–	75	75
1,2-Dichloroethane	30	3.0	0	5
1,1-Dichloroethylene	30	–	7	7
cis-1,2-Dichloroethylene	50	–	70	70
trans-1,2-Dichloroethylene	–	–	100	100
Dichloromethane	20	–	0	5
Di(2-ethylhexyl) adipate	80	–	400	400
Di(2-ethylhexyl) phthalate	8	–	0	6

Parameter	World Health Organization guideline value	European Union parametric value	United States Environmental Protection Agency primary and secondary standards	
			NPDWR	
Dinoseb	–	–	7	7
Dioxin (2,3,7,8-TCDD)	–	–	0	0.00003
Endothall	–	–	100	100
Endrin	–	–	2	2
Epichlorohydrin	–	0.10	0	TT
Ethylbenzene	300	–	700	700
Ethylene dibromide	–	–	0	0.05
Fluoranthene	U	–	–	–
Glyphosate	U	–	700	700
Hexachlorocyclopentadiene	–	–	50	50
Microcystin-LR	1(P)	–	–	–
Oxamyl (Vydate)	–	–	200	200
Polychlorinated biphenyls (PCBs)	–	–	0	0.5
Picloram	–	–	500	500
Styrene	20	–	100	100
Tetrachloroethene	40	10	0	5
Toluene	700	–	1000	1000
Toxaphene	–	–	0	3
2,4,5-TP (Silvex)	–	–	50	50
1,2,4-Trichlorobenzene	20 (total)	–	70	70
1,1,1-Trichloroethane	2000 (P)	–	200	200
1,1,2-Trichloroethane	–	–	3	5
Trichloroethene	70 (P)	10	0	5
Vinyl chloride	5	0.5	0	2
Xylenes (total)	500	–	10000	10000
Inorganic constituents			MCLG	MCL or TT
Antimony	0.005 (P) mg L^{-1}	0.005 mg L^{-1}	0.006 mg L^{-1}	0.006 mg L^{-1}
Arsenic	0.01 (P)	0.01	0	0.01
Asbestos (fibre > 10 μm)	U	–	7 MFL	7 MFL
Barium	0.7	–	2	2
Beryllium	NAD	–	0.004	0.004
Boron	0.05 (P)	1.0	–	–
Cadmium	0.003	0.005	0.005	0.005
Chromium (total)	0.05 (P)	0.05	0.1	0.1
Copper	2 (P)	2	1.3	TT (action level = 1.3)
Cyanide (as free cyanide)	0.07	0.05	0.2	0.2
Fluoride	1.5	1.5	4.0	4.0
Lead	0.01	0.01	0	TT (action level = 0.015)
Molybdenum	0.07	–	–	–
Mercury	0.001	0.001	0.002	0.002
Nickel	0.02 (P)	0.02	–	–
Nitrate (as N)	11.3 (acute)	11.3	10	10
Nitrite (as N)	0.91 (acute), 0.06 (P) (chronic)	0.15	1	1
Selenium	0.01	0.01	0.05	0.05
Thallium	–	–	0.0005	0.002
Indicator parameters			NSDWR	
Aluminium	0.2 mg L^{-1}	0.2 mg L^{-1}	0.05–0.2 mg L^{-1}	–
Ammonium	1.5	0.5	–	–

Parameter	World Health Organization guideline value	European Union parametric value	United States Environmental Protection Agency primary and secondary standards	
			NSDWR	–
Chloride	250	250	250	–
Clostridium perfringens	–	0 in 100 mL	–	–
Colour	15 TCU	n/a/c	15 colour units	–
Conductivity	–	2500 μS cm^{-1} at 20°C	–	–
			–	–
Dissolved oxygen	n/a	–	–	–
Hardness	n/a	–	–	–
Hydrogen sulphide	0.05 mg L^{-1}	–	–	–
Hydrogen ion concentration (pH)	n/a	≥6.5 and ≤9.5	6.5–8.5	–
Iron	0.3 mg L^{-1}	0.2 mg L^{-1}	0.3 mg L^1	–
Manganese	0.5 (P)	0.05	0.05	–
Odour	n/a	n/a/c	3 threshold odour number	–
				–
Oxidizability	–	5.0 mg L^{-1} O$_2$	–	–
Silver	U	–	0.1 mg L^1	–
Sodium	200 mg L^{-1}	200 mg L^{-1}	–	–
Sulphate	250	250	250	–
Taste	n/a	n/a/c	–	–
Temperature	n/a	–	–	–
Tin	U	–	–	–
Colony count 22°	–	n/a/c	–	–
Total organic carbon	–	n/a/c	–	–
Total dissolved solids	1000 mg L^{-1}	–	500 mg L^1	–
Zinc	3	–	5	–

Notes and abbreviations:

MCL, maximum contaminant level: the highest level of a contaminant that is allowed in drinking water. MCLs are set close to MCLGs as feasible using the best available treatment technology and taking cost into consideration. MCLs are enforceable standards.

MCLG, maximum contaminant level goal: the level of a contaminant in drinking water below which there is no known or expected risk to health. MCLGs allow for a margin of safety and are non-enforceable public health goals.

MFL, × 10^6 fibres L^{-1}.

MRDL, maximum residual disinfectant level: the highest level of a disinfectant allowed in drinking water. There is evidence that addition of a disinfectant is necessary for control of microbial contaminants.

MRDLG, maximum residual disinfectant level goal: the level of a drinking water disinfectant below which there is no known or expected risk to health. MRDLGs do not reflect the benefits of the use of disinfectants to control microbial contaminants.

n/a, not applicable (should be acceptable).

n/a/c, no abnormal change.

NAD, no adequate data to permit recommendation of a health-based guideline value.

NPDWR, National Primary Drinking Water Regulations (or primary standards): legally enforceable standards applied to public water systems. Primary standards protect public health by limiting the levels of contaminants in drinking water.

NSDWR, National Secondary Drinking Water Regulations: non-enforceable guidelines regulating contaminants that may cause cosmetic effects (such as skin or tooth discolouration) or aesthetic effects (such as taste, odour or colour) in drinking water.

NTU, nephelometric turbidity unit.

(P), provisional guideline value.

TCU, true colour unit.

TT, Treatment technique: a required process intended to reduce the level of a contaminant in drinking water.

U, Not hazardous to human health at concentrations normally found in drinking water.

* Coliforms are naturally present in the environment and are not a health threat in itself; the parameter is used to indicate whether other potentially harmful bacteria are present.

† Turbidity is a measure of cloudiness of water that is used to indicate water quality and filtration effectiveness.

‡ The sum of the ratio of the concentration of each to its respective guideline value should not exceed 1.

A9.1.1 Sources

Council of European Communities (1998) *Directive on the Quality of Water Intended for Human Consumption (98/83/ EC).* Official Journal of the European Communities, L330, Brussels.

United States Environmental Protection Agency (2002) *National Primary and Secondary Drinking Water Regulations.* Report EPA-816-F-02-013. http:www.epa.gov/safewater/ mcl.html.

World Health Organization (1996) *Guidelines for Drinking- water Quality,* 2nd edn, Vol. 2, *Health criteria and other supporting information; Addendum to volume 2* (1998). World Health Organization, Geneva. http:www.who.int/ water_sanitation_health/GDWQ/Summary_tables/.

A9.2 List I and II substances as defined by EC Groundwater Directive (80/68/EEC)

A9.2.1 List I of families and groups of substances

These substances should be prevented from being discharged into groundwater.

List I contains the individual substances which belong to the families and groups of substances specified below, with the exception of those that are considered inappropriate to List I on the basis of a low risk toxicity, persistence and bioaccumulation.

Such substances which, with regard to toxicity, persistence and bioaccumulation are appropriate to List II, are to be classed in List II.

1 Organohalogen compounds and substances which may form such compounds in the aquatic environment.

2 Organophosphorus compounds.

3 Organotin compounds.

4 Substances which possess carcinogenic, mutagenic or teratogenic properties in or via the aquatic environment.[1]

5 Mercury and its compounds.

6 Cadmium and its compounds.

7 Mineral oils and hydrocarbons.

8 Cyanides.

[1] Where certain substances in List II are carcinogenic, mutagenic or teratogenic they are included in Category 4 of List I.

A9.2.2 List II of families and groups of substances

Discharges of these substances into groundwater should be minimized.

List II contains the individual substances and the categories of substances belonging to the families and groups of substances listed below which could have a harmful effect on groundwater.

1 The following metalloids and metals and their compounds:

Zinc	Tin
Copper	Barium
Nickel	Beryllium
Chrome	Boron
Lead	Uranium
Selenium	Vanadium
Arsenic	Cobalt
Antimony	Thallium
Molybdenum	Tellurium
Titanium	Silver

2 Biocides and their derivatives not appearing in List I.

3 Substances which have a deleterious effect on the taste and/or odour of groundwater and compounds liable to cause the formation of such substances in such water and to render it unfit for human consumption.

4 Toxic or persistent organic compounds of silicon and substances which may cause the formation of such compounds in water, excluding those which are biologically harmless or are rapidly converted in water into harmless substances.

5 Inorganic compounds of phosphorus and elemental phosphorus.

6 Fluorides.

7 Ammonia and nitrates.

A9.2.3 Sources

Council of European Communities (1980) *Directive on the Protection of Groundwater Against Pollution Caused by Certain Dangerous Substances (80/68/EEC).* Official Journal of the European Communities, L20, Brussels.

Environment Agency (1998) *Policy and Practice for the Protection of Groundwater.* Environment Agency, Bristol.

Review questions and exercises

Appendix 10

A10.1 Hydrological cycle

1 Explain the relative size of the store of groundwater in the global hydrological cycle and its significance in terms of water and material fluxes to the world's oceans.

2 Given that the average annual fluxes of precipitation and evaporation over the global land surface are 0.110×10^6 km^3 and 0.073×10^6 km^3, respectively, and that the storage of water in rivers and shallow groundwater (above a depth of ~750 m) is 0.0017×10^6 km^3 and 4.2×10^6 km^3, respectively, answer the following:

(a) What is the balance of average annual precipitation and evaporation (the effective precipitation or total runoff) from the global land surface?

(b) If the total runoff calculated in (a) is partitioned between surface runoff (94%) and groundwater discharge (6%), estimate the residence (flushing) times of water held in storage in rivers and shallow groundwater.

(c) Why is the figure obtained in (b) for groundwater residence time only likely to give a very rough estimate?

(d) An estimate of the direct flux of groundwater to the oceans is 2×10^3 km^3 a^{-1}. Assuming an average total dissolved solids content for groundwater of 585 mg L^{-1}, estimate and comment on the annual material flux to the oceans from groundwater.

A10.2 Physical hydrogeology, groundwater potential and Darcy's law

1 Write concise definitions of the following, briefly explaining the significance of each in hydrogeology:

(a) Porosity.
(b) Effective porosity.
(c) Voids ratio.
(d) Specific yield.
(e) Storativity.
(f) Hydraulic conductivity.
(g) Transmissivity.
(h) Reynolds number.

2 Discuss how topographical and geological setting, both in terms of the variety of rock types and geological structure, influence aquifer conditions and groundwater flow regimes.

3 For an unconfined aquifer condition in arid, frigid and humid climatic settings, discuss how groundwater flow and surface drainage networks are connected.

4 From first principles, derive the following expression for the groundwater potential at any point in a porous material:

$$\Phi = gh$$

where Φ is groundwater potential, g is acceleration due to gravity and h is hydraulic head measured above an arbitrary datum. Also show that the hydraulic head at a point is equal to the sum of the pressure head, ψ, and the elevation head, z.

5 The following field notes were taken at a nest of piezometers installed in close proximity to each other at a site:

	Piezometer		
	a	b	c
Elevation at surface (m above sea level)	350	350	350
Depth of piezometer (m)	115	80	40
Depth to water (m)	37	28	21

If A, B, C refer to the points of measurement at the tips of piezometers a, b, c, calculate the following:

  (a) The hydraulic head at A, B and C (m).
  (b) The pressure head at A, B and C (m).
  (c) The elevation head at A, B and C (m).
  (d) The hydraulic gradients between A and B, and B and C.

Is the piezometer nest in a recharge area or a discharge area?

6 Define Darcy's law and discuss its applicability to groundwater flow in porous materials.

7 Give a brief classification of voids in rocks and sediments and describe the relationships between the voids, the hydraulics of groundwater flow and the bulk flow patterns in the following:

  (a) Unconsolidated alluvial sediments.
  (b) Jointed porous sandstone.
  (c) Fractured granitic rock.
  (d) Karstic limestone.

8 A sand filter is enclosed within a cylinder 2 m long and with a 50 cm internal diameter. The cylinder is inclined such that the axis of the upper end lies 75 cm above the axis of the lower end. Water enters the apparatus at one end and flows out of the other end. Sensors monitor the pressure as the water enters and leaves the apparatus. The sensors are also separated by a 75-cm elevation difference.

If the lower pressure reading is 120.5 kPa and the upper reading is 117.6 kPa, is the water flowing up or down through the cylinder? Assume a density of water of 1 g cm^{-3}, so that 1 m of water exerts a pressure of 9.81 kPa.

If the apparatus operates at a temperature of 20°C and the specific discharge was measured as 4.0×10^{-5} m s^{-1}, answer the following:

  (a) What is the hydraulic conductivity of the sand?
  (b) What is the intrinsic permeability of the sand?
  (c) What volume of water could the apparatus transmit in a day?
  (d) If the operating temperature were to be increased to 40°C, what pressure differential would be required to keep the flow rate equal to that at 20°C?

Values of the density and viscosity of water at different temperatures are given in Appendix 2.

9 As part of a site investigation of the area shown in Fig. A10.1, 13 boreholes have been drilled. The area is underlain by a bed of sand above which is a low-permeability clay. Figure A10.1 shows structure contours of the contact between the sand and this clay. Three of the boreholes were deep enough to prove the existence of a second low-permeability clay beneath the sand. A single-borehole tracer test conducted at borehole E gave a value of hydraulic conductivity for the sand of 1 m day^{-1}.

Prepare a short explanation of the hydrogeology of the area using the above information. Illustrate your account by first completing a copy of the map showing contours of the groundwater level and then the cross-section showing geological structure and the potentiometric surface. Indicate the parts of the area in which the groundwater is recharged to and discharged from the ground surface, and the unconfined and confined parts of the sand aquifer. Estimate the transmissivity of the sand and the discharge of groundwater from the mapped area to the length of stream C–D.

It has been proposed to drain the boggy ground adjacent to the stream by means of ditches. Comment on the likely success or failure of this approach, giving your reasons.

10 Figure A10.2 shows longitudinal and transverse sections of an alluvial valley formed in a low-permeability bedrock. The average temperature of groundwater in the valley aquifer is 20°C. Continuous recharge to the upstream section of the valley creates uniform, steady groundwater flow in a down-gradient direction. The alluvium has an intrinsic permeability of 30 darcies and an effective porosity of 0.16. If the static water level elevation in Well 1 is 83.4 m relative to datum, and in Well 2, 82.9 m relative to datum, calculate the following:

  (a) The groundwater discharge (in m^3 day^{-1}) across the width of the aquifer.
  (b) Relative to your answer in part (a), what is the percentage change in groundwater discharge if the average groundwater temperature was decreased by 50%?
  (c) For an average groundwater temperature of 20°C, and stating your assumptions, how long would it take for a contaminant introduced in Well 1 to reach Well 2?

Note: the gravitational acceleration, $g = 9.81$ m s^{-2}. Values of the density and viscosity of water at different temperatures are given in Appendix 2.

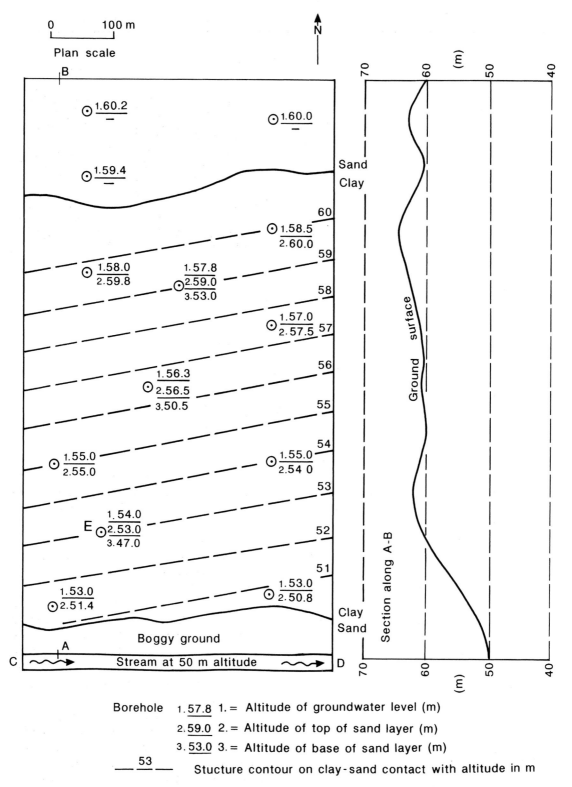

Fig. A10.1 Site investigation area of a sand aquifer.

(a)   Longitudinal section

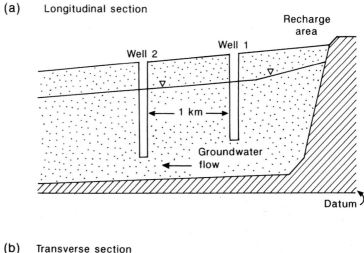

(b)   Transverse section

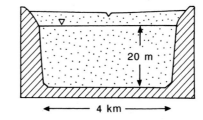

Fig. A10.2 Longitudinal and transverse sections of an alluvial valley.

11 Figure A10.3 shows a generalized section along the direction of flow in a confined sandstone aquifer with shale confining beds. The topographic scarp between A and B lies parallel to the strike of the aquifer. Springs and seepages occur frequently towards the base of the scarp in a zone which is parallel with the strike of the aquifer. The marshy pools lie on the surface trace of a fault which is also parallel to the strike.

Consider a strip of aquifer in the plane of the diagram. If the hydraulic conductivity of the sandstone is 200 m day^{-1}, find the following:

(a) The discharge per unit width of the spring zone, S.

(b) The discharge per unit width of the marshy pools zone.

(c) The transmissivity of the fault plane beneath the spring zone.

(d) The transmissivity of the fault plane beneath the marshy pools.

On a copy of Fig. A10.3, sketch the flow pattern of the aquifer. You can assume that the springs and marshy pools are the only outlet for groundwater. Write short notes with your answers stating the assumptions you are making and justifying your calculations.

12

(a) A vertical fissure in a limestone aquifer is 2 m high and 0.1 m wide. A tracer moves along the fissure at a velocity of 100 m h^{-1}. Is the flow laminar or turbulent?

(b) A well 0.5 m in diameter in a confined sand aquifer 5 m thick is pumped at 50 L s^{-1}. If the sand has a median grain size of 1.25 mm, estimate the Reynolds number at the following distances from the centre of the well: 0.25 m (i.e. the outer boundary of the well screen), 0.5 m, 1.0 m, 2.0 m, 4.0 m and 10.0 m.

Using your answers obtained in part (b), plot a graph of Reynolds number against radius. Within what radius of the well would you expect non-Darcian flow to occur and briefly explain why?

13 Write an essay on how and why changes in storage occur in confined and unconfined aquifers in response to both natural and artificial causes.

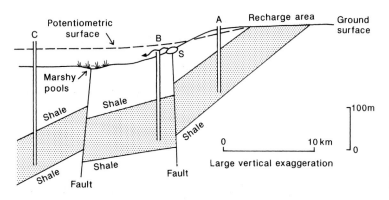

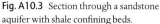

Fig. A10.3 Section through a sandstone aquifer with shale confining beds.

14 The Salt River Valley in central Arizona is an alluvial valley forming an unconfined aquifer covering an area of about 100,000 ha. The alluvial aquifer is heavily pumped but receives little replenishment from groundwater recharge. Groundwater abstractions total about 500×10^6 m^3 a^{-1} with an associated water-table decline of about 3 m a^{-1}. Assuming no groundwater recharge, estimate the specific yield of the alluvial aquifer?

15 Explain the basic principles of steady groundwater flow in porous materials starting from a consideration of fluid potential. Demonstrate how variations in hydraulic conductivity between different geological formations lead to groundwater flow lines being refracted at their boundaries. Discuss how such refraction patterns can be used to classify layered systems of geological formations into aquifers and aquitards, and describe the basis for classifying aquifers into confined, semiconfined (leaky) and unconfined.

16 Figure A10.4 is a flow net of seepage under a dam through an isotropic sandstone with a porosity of 20%. A conservative tracer injected at P takes 5 days to reach Q.

(a) What are the assumptions used to construct the flow net?

(b) On a copy of Fig. A10.4, sketch and label piezometric levels in the nested piezometers at A, B and C.

(c) What is the hydraulic conductivity of the sandstone?

(d) Leaky drums of hazardous aqueous waste were dumped in the reservoir at sites D and E. How long will it take for pollution to seep out on the downgradient side of the dam from each of the two sites?

(e) Why may the actual breakthrough times of the pollutant differ from those predicted?

A10.3 Chemical hydrogeology

1 Describe three graphical methods of hydrochemical data presentation and define the term **hydrochemical facies**, explaining the value of this concept in determining groundwater flow patterns.

2 Describe the classic sequence of Chebotarev (1955) to explain the chemical evolution of natural waters along a groundwater flowpath. Illustrate your answer with diagrams relating to at least one specific example.

3 Compare and contrast the carbonate chemistry of groundwaters evolving under open-system and closed-system conditions with respect to dissolved CO_2.

4 Figure A10.5 shows the geographical location and topography of Kuwait. Fresh and brackish groundwater fields are shown together with a new area known as the Al-Wafra well-field. The stratigraphic sequence of Kuwait is divided into clastic sediments known as the Kuwait Group and the underlying carbonate sediments of the Hasa Group. The Dammam aquifer occurs in the uppermost part of the Hasa Group and consists of 200 m of porous limestones. The top of the aquifer is marked by the presence of a hard, siliceous layer and shale horizon. Table A10.1 is a summary of the mean and standard deviation of hydrochemical data for the Dammam limestone aquifer.

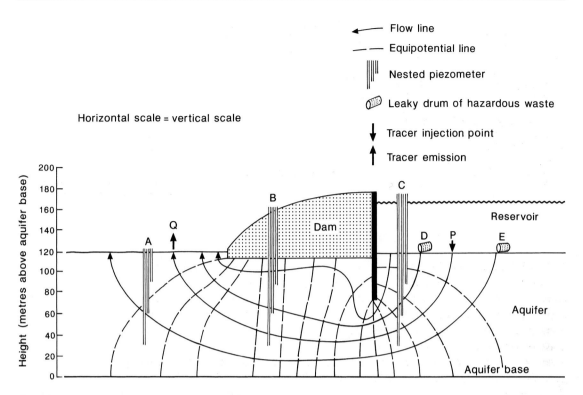

Fig. A10.4 Flow net of seepage under a dam.

(a) From inspection of the concentration data given in Table A10.1, assign a hydrochemical water type to the groundwater in the Dammam limestone aquifer.

(b) What reasons might explain the high Ca^{2+}, Na^+, HCO_3^- and SO_4^{2-} concentrations in the aquifer?

(c) Calculate logarithmic values of saturation indices for calcite ($CaCO_3$) and halite ($NaCl$) using the mean concentration data, and predict whether these two minerals have reached chemical equilibrium in the Dammam limestone groundwater.

Note: Thermodynamic equilibrium constants in pure water at 25°C and 1 bar total pressure: $K_{HCO3} = 10^{-10.33}$; $K_{calcite} = 10^{-8.48}$; $K_{halite} = 10^{0.773}$.

5 Consider the following two redox half-reactions involving the reduction of nitrate (NO_3^-) to gaseous nitrogen (N_2) and sulphate (SO_4^{2-}) to bisulphide (HS^-):

$$^1/_5NO_3^- + {}^6/_5H^+ + e^- = {}^1/_{10}N_{2(g)} + {}^3/_5H_2O$$

$$^1/_8SO_4^{2-} + {}^9/_8H^+ + e^- = {}^1/_8HS^- + {}^1/_2H_2O$$

A groundwater with nitrate and sulphate concentrations of 2×10^{-3} mol L^{-1} and 1×10^{-3} mol L^{-1}, respectively, and a pH of 8 experiences denitrification followed by sulphate reduction. The reactions produce a partial pressure of dissolved nitrogen of 1×10^{-3} bar and a bisulphide concentration of 1×10^{-3} mol L^{-1}. For an ideal system, which can be described by reversible thermodynamics, calculate the pe and Eh values produced by each of these redox half-reactions. Assume a temperature of 25°C and that the activities of the dissolved species are equal to their concentrations.

If, in the process of denitrification, ferrous (Fe^{2+}) iron is oxidized to ferric (Fe^{3+}) iron (eq. 3.29 for pe^o = +13.0), calculate the amount of ferric iron produced by this redox reaction in an aquifer containing 1×10^{-5} mol L^{-1} of ferrous iron.

Note:

$$pe = pe^o - \frac{1}{n}\log_{10}\frac{[\text{reductants}]}{[\text{oxidants}]}$$

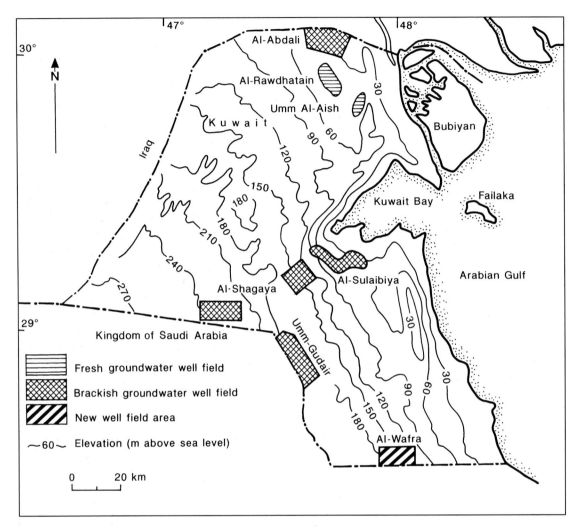

Fig. A10.5 Location and topographic map of Kuwait. After Saleh et al. (1999).

where pe is equal to $-\log_{10}[e^-]$ and describes the electron activity; pe° is equal to $\log_{10}K$, where K is an equilibrium constant; and n is the number of electrons involved in the redox half-reaction. Values of pe° are given in Table 3.10. $Eh = \dfrac{0.059}{n} pe$ at 25°C.

6 Fresh groundwater in coastal areas is typically dominated by Ca^{2+} and HCO_3^- ions, as a result of calcite dissolution. Consequently, cation exchangers in the aquifer mostly have Ca^{2+} adsorbed on their surfaces. In seawater Na^+ and Cl^- are the dominant ions, and sediment in contact with seawater will adsorb Na^+. Given this information, write two equations that illustrate the expected concentrations of Na^+ and Ca^{2+} resulting from simple mixing and cation exchange between freshwater and seawater when: (a) seawater intrudes a coastal freshwater aquifer; (b) freshwater flushes a saline aquifer.

The hydrochemical data in Table A10.2 are for samples of freshwater, seawater and a mixed $CaCl_2$ groundwater obtained for a coastal aquifer experiencing seawater intrusion. Using these data, calculate the quantities of Na^+ and Ca^{2+} involved in cation exchange during seawater intrusion.

Table A10.1 Kuwait Dammam limestone aquifer hydrochemical data. Source: Saleh, A., Al-Ruwaih, F. & Shehata, M. (1999) Hydrogeochemical processes operating within the main aquifers of Kuwait. *Journal of Arid Environments* **42**, 195–209.

Parameter	Concentration	
	Mean	Standard deviation
TDS	5980	1123
pH	7.4	0.2
K^+	56	9.7
Na^+	1180	305.2
Ca^{2+}	540	80.7
Mg^{2+}	164	22.4
HCO_3^-	173	7.1
Cl^-	2365	614.2
SO_4^{2-}	1032	65.6

Note: concentration units in mg L^{-1} except pH.

Briefly discuss your results, stating any assumptions you make in your calculations.

7 Figure A10.6 shows a location map of the Fife and Kinross area of Scotland. Northern Fife is underlain by Devonian Old Red Sandstone and southern Fife by Carboniferous rocks. To the north of the Old Red Sandstone outcrop are the Ochil Hills which consist

Table A10.2 Hydrochemical data for mixing of freshwater and seawater.

Ion	Seawater	Mixed $CaCl_2$ groundwater	Freshwater
Na^+	485.0	374.8	0
Ca^{2+}	10.7	10.0	3.0
Cl^-	566.0	440.0	0

All concentrations in mmol L^{-1}.

of low-permeability volcanic rocks. To the south, a roughly east–west-trending belt through Central Fife is formed of sandstone with conglomerate, including the Knox Pulpit Formation, the most productive aquifer in the Old Red Sandstone (capable of borehole yields in excess of 30 L s^{-1}). The Knox Pulpit Formation is a fine- to medium-grained and weakly cemented sandstone. There is also a narrow coastal strip of sandstone faulted against the volcanic rocks to the north of the Ochil Hills and west of easting 30.

The Carboniferous rocks of southern Fife are a complex sequence of lithologies interspersed with intrusive and extrusive igneous rocks. In east Fife there are sandstones and subordinate shales, thin limestones, seatearths and coal. In central Fife there

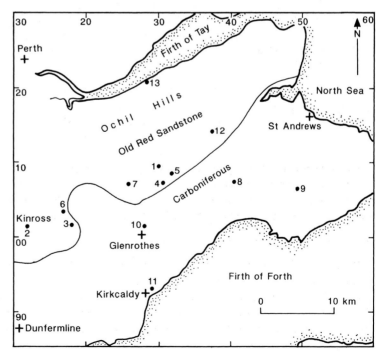

Fig. A10.6 Location map of the Fife and Kinross area of Scotland showing sampling sites. After Robins (1990b).

Table A10.3 Hydrochemical data for Fife and Kinross. Source: Robins, N.S. (1990b) Groundwater chemistry of the main aquifers in Scotland. *Report of the British Geological Survey, Vol. 18, No. 2.*

Sample number	Geological formation	Borehole depth (m)/spring	E.C. (μS cm⁻¹)	Temp (°C)	pH	Eh (mV)	Concentration (mg L⁻¹)								Diss. O₂	Total Fe	Tritium (TU)
							Ca^{2+}	Mg^{2+}	Na^+	K^+	HCO_3^-	SO_4^{2-}	Cl^-	NO_3-N			
1	ORS	13	510	8.5	7.7	+390	29	30	16	0.4	113	39	35	18	8.0	<0.3	nd
2	ORS	100	508	9.0	7.4	+350	46	24	16	2.3	205	23	35	4	6.0	<0.3	nd
3	ORS	Spring	329	9.0	7.6	+345	37	16	8	1.1	157	16	14	3	10.0	<0.3	nd
4	ORS	80	828	9.5	7.7	+400	53	39	24	2.0	187	53	41	12	3.0	<0.3	27.6
5	ORS	70	495	9.0	7.7	+400	49	25	18	1.7	173	46	42	9	6.1	<0.3	nd
6	ORS	120	343	8.0	7.2	+370	48	12	7	2.4	124	22	16	8	11.8	<0.3	nd
7	ORS	75	283	8.5	7.6	+390	42	11	7	1.6	137	19	12	2	sat	<0.3	19.6
8	Carb	40	865	8.5	6.8	−90	100	60	13	4.1	323	203	26	3	0	1.12	nd
9	Carb	69	635	8.0	6.6	+270	76	30	17	1.5	234	113	20	1	0	<0.3	nd
10	Carb	50	325	10.0	7.7	+110	45	19	12	2.4	194	59	14	<0.1	1.0	<0.3	nd
11	Carb	130	1560	13.0	6.6	+210	105	107	47	12.5	591	374	42	<0.1	0	4.90	nd
12	ORS	60	965	9.0	7.3	+320	61	43	48	3.7	274	71	83	0.2	6.2	<0.1	6.0
13	ORS	86	1040	10.0	7.4	+130	115	15	72	6.9	210	308	76	0.1	3.9	0.10	nd

ORS, Old Red Sandstone; Carb, Carboniferous; nd, not determined; sat, saturated.

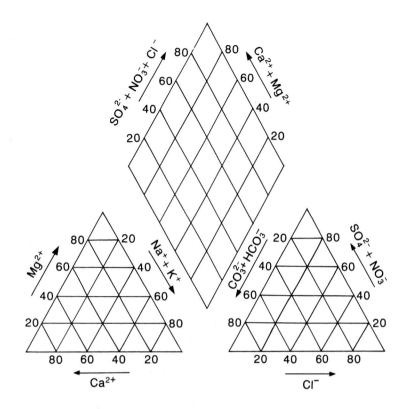

Fig. A10.7 Trilinear (Piper) diagram.

are sandstones, mudstones, subordinate limestones and coals. In south and west Fife, sandstones, fireclays and some coals are present with a single basal limestone. All these strata contain some groundwater and there is often hydraulic continuity across poorly permeable horizons via cracks, joints and old workings. Borehole yields up to 5 L s^{-1} are not uncommon.

Table A10.3 details the analytical hydrochemical data for 13 samples collected from the sites shown in Fig. A10.6. Using these data, complete a copy of the trilinear (Piper) diagram given in Fig. A10.7 for the 13 samples given in Table A10.3.

Based on your Piper diagram, and using all the available data, provide both a description and explanation of the hydrochemical characteristics of groundwaters found in Fife and Kinross.

A10.4 Environmental isotope hydrogeology

1 Explain the meaning of the age of groundwater and indicate the range of ages encountered in porous geological materials.

2 Discuss the application of ^{14}C and ^{3}H radiometric dating methods in hydrogeological investigations. Pay particular attention to basic principles, the appropriate methods required to correct for 'dead' carbon (^{14}C method) and the limitations of each technique.

3 Discuss how the stable isotope ratios ^{18}O/^{16}O and ^{2}H/^{1}H and noble gases (Ne, Ar, Kr and Xe) can be used to characterize the recharge history and subsequent evolution of groundwaters. Illustrate your answer with appropriate examples.

4 Llandrindod Wells in mid-Wales was an important spa town in Victorian times. The mineralized spring waters, of varying composition, all lie in close proximity to each other in the River Irthon valley in an area known as the Rock Park and centred on the Rock Park Fault. The geology of both the immediately local and recharge areas is dominated by the Builth Volcanics, a low porosity sequence that is extensively fractured and faulted. The relief of the area is moderate with the springs cropping out at 190 m above sea level, and the surrounding hills rising to 300 m in the immediate vicinity and 400 m at distances of 10 km in all directions. A schematic cross-section is given in Fig. A10.8.

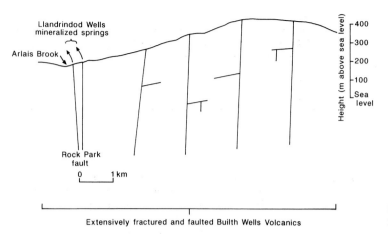

Fig. A10.8 Schematic cross-section of the Builth Volcanics, Llandrindod Wells.

Table A10.4 Stable isotope and hydrochemical data for Llandrindod Wells. Source: Merrin, P.D. (1996) *A Geochemical Investigation into the Nature and Occurrence of Groundwaters from the Llandrindod Wells Ordovician Inlier, Wales*. MSc Thesis, University of East Anglia, Norwich.

Spring	$\delta^{18}O$ (‰$_{VSMOW}$)	δ^2H (‰$_{VSMOW}$)	Na^+ (mg L^{-1})	Cl^- (mg L^{-1})
Chalybeate spring	−8.15	−54	1565	3890
Saline spring	−7.85	−51.5	1105	2950
Sulphur spring	−7.20	−47	490	1270
Magnesium spring	−8.10	−53.5	183	310
Eye Well 1	−6.50	−47.2	1325	3670
Eye Well 2	−5.45	−43	1530	4220
Eye Well 3	−6.65	−44	62	130
Arlais Brook (surface stream)	−6.80	−43	80	140
Local precipitation	−6.70	−43	2	4
Seawater	0	0	10,500	19,000

The hydrogeochemistry ($\delta^{18}O$, δ^2H, Na^+ and Cl^-) of selected springs, local precipitation and surface waters are given in Table A10.4. Using these data, present a model that accounts for the origin and subsequent modification of the spring water compositions and the relationships between them. Use a copy of Fig. A10.8 to provide a conceptual model for the evolution of the mineral spring waters.

5 As part of a hydrogeological investigation of a porous, fissured limestone aquifer both stable isotope and radioisotope data have been collected at three sites (Table A10.5). Site 1 is located on the aquifer outcrop and Sites 2 and 3 are located downgradient in the confined section of the aquifer. Assuming a $\delta^{13}C_{lithic}$ value of +2‰ and a $\delta^{13}C_{biogenic}$ value of −26‰, and using the information in Table A10.5, calculate:

(a) The corrected groundwater ages at the three sites.

(b) An indicative groundwater velocity in the aquifer.

Comment on the groundwater age dates and the palaeohydrogeology of the limestone aquifer as indicated by the stable isotope ($\delta^{18}O$ and δ^2H) data.

A10.5 Stream gauging, infiltration measurements and groundwater recharge estimation

1 Discuss the range of applications of, and operational constraints on, various methods that are available for measuring stream discharge.

2 Provide a description of the physical storage and movement of water in the soil zone and explain why

Table A10.5 Stable isotope and radioisotope data for a limestone aquifer.

Site	Distance from outcrop (km)	Tritium (TU)	^{14}C (pmc)	δ^{13}C (‰$_{PDB}$)	δ^{18}O (‰$_{VSMOW}$)	δ^{2}H (‰$_{VSMOW}$)
1	0	57	60.8	−13.2	−7.1	−46.0
2	15.1	4	4.1	−4.5	−7.4	−48.0
3	16.3	0	2.2	−0.9	−7.9	−51.0

an understanding of infiltration is important in hydrology.

3 Why is knowledge of groundwater recharge important in the estimation of catchment water resources? Describe three methods for calculating average annual groundwater recharge, highlighting the advantages and disadvantages of each.

4 Write concise definitions for the following hydrological parameters and describe methods for the determination of the parameters listed as (a), (b) and (c):

  (a) Volumetric moisture content.
  (b) Tension head.
  (c) Infiltration capacity.
  (d) Field capacity.

5 As part of a groundwater investigation, the discharge of a karst spring was measured in a channel a short distance downstream of its point of emergence. The following data were collected at a well-maintained, horizontal, sharp-crested, thin-plate weir:

Width of crest, $L = 1.67$ m
Number of side constrictions, $n = 2$
Ventilation of falling water = free
Head above crest of weir, measured on a stage board calibrated in centimetres and positioned in the approach pool, $H = 0.12$ m
Velocity of approach = low

  Determine the spring discharge, Q, in the channel from these data, using the following stage-discharge relationship:

$$Q = 1.84 \, (L - 0.1 \, nH)H^{3/2}$$

  Downstream of the weir, the data given in Table A10.6 were collected using a well-calibrated current meter in a straight, uniform part of the channel with a

Table A10.6 Current metering data.

Object measured	Distance of object from left bank (m)	Current meter depth below surface (m)	Water velocity (m s^{-1})
Current meter	0.12	0.30	0.015
Current meter	0.38	0.31	0.035
Current meter	0.62	0.32	0.08
Current meter	0.88	0.31	0.12
Current meter	1.12	0.30	0.23
Current meter	1.38	0.30	0.25
Current meter	1.62	0.32	0.15
Current meter	1.88	0.29	0.10
Current meter	2.12	0.28	0.07
Current meter	2.38	0.26	0.04
Right bank	2.55	–	–

roughly rectangular cross-section. Calculate the discharge in the channel from these measurements, showing your working.

  Briefly discuss the magnitudes of error involved in these two sets of measurements, and determine whether the discrepancy between them is likely to be due to random or systematic error.

6 The accuracy of a stage-discharge calibration at a current metered gauging station is being checked by dilution gauging.

  A slug of 15 kg of rhodamine WT tracer was injected into the river and samples were taken 1 km downstream. No tributaries entered the river over the reach. Given the results in Table A10.7, calculate the discharge for comparison with the value obtained from the rating curve (110 m^3 s^{-1}).

  What could account for the discrepancy between the two discharge values?

7 Two small catchments of similar area drain an upland region. One catchment is covered by grassland developed on a calcareous till deposit, and the

Table A10.7 Dilution gauging data.

	Time (h)								
	0	1	2	3	4	5	6	7	8
Concentration (ppb)	0	1	5	10	7	5	3	1	0

Table A10.8 Two-component mixing model data.

Water type	Cl⁻ (mg L⁻¹)	$\delta^{18}O$ (‰$_{VSMOW}$)
Rain event	2.0	−6.00
Grassland stream		
Total discharge	6.0	−7.00
Pre-event water	10.0	−8.00
Forest stream		
Total discharge	6.0	−7.00
Pre-event water	7.0	−7.25

Table A10.9 Clay soil volumetric moisture content and soil water potential measurements during July 2003.

Depth below ground (cm)	Volumetric moisture content, θ		Soil water potential, Φ* ($\times 10^3$ cm of water)	
	1.7.03	8.7.03	1.7.03	8.7.03
−5	0.356	0.349	−4.00	−4.20
−15	0.375	0.363	−3.50	−3.75
−25	0.394	0.378	−3.00	−3.40
−35	0.418	0.390	−2.30	−2.75
−45	0.428	0.390	−1.50	−2.00
−55	0.413	0.390	–	–
−70	0.399	0.387	–	–
−85	0.385	0.378	–	–

* Soil water potential data not recorded below 45 cm depth.

other is covered by forest developed on acidic soil. A field experiment, conducted during the wet season, has been completed to determine the component of total (peak) stream discharge (Q_T) contributed by storm runoff (event water, Q_E) and baseflow (pre-event water, Q_P) in each catchment. With reference to the chemical and stable isotope data given in Table A10.8, and applying a two-component mixing model, rearrange the following equation to express the ratio Q_P/Q_T in terms of the total discharge, event and pre-event stream water tracer concentrations (C_T, C_E and C_P, respectively):

$$C_T Q_T = C_P Q_P + C_E Q_E$$

and answer the following:

(a) If the total stream discharge measured at the outlets from the grassland and forest catchments during a rainfall event are 2.5 m³ s⁻¹ and 1.8 m³ s⁻¹, respectively, calculate values for the baseflow discharge components using, in turn, the chloride and $\delta^{18}O$ measurements given in Table A10.8. What proportion of each total flow is contributed by storm runoff?

(b) From the results obtained in part (a) above, comment on any observed differences in the hydrograph separations for the two catchments.

8 Table A10.9 contains information on volumetric moisture content and soil water potential recorded beneath a plot of ground of area 1 m² by means of a neutron probe and tensiometers. The soil is a homogeneous, structureless clay soil with a cover of short-rooting vegetation, and is 90 cm in depth.

(a) Plot the data provided on arithmetic graph paper and, having identified the position of the zero flux plane (ZFP), calculate the evaporation and drainage losses from the profile during the 7-day period separating the readings, assuming no rain fell during the period. Express your results in millimetres of water.

(b) For the depth range 30–40 cm, estimate the unsaturated hydraulic conductivity value for the clay soil on 8 July 2003.

(c) In the event of rainfall, what would be the effect of prior intense drying of the soil upon the accuracy of your method for calculating the drainage loss?

9 Table A10.10 gives the available rainfall and potential evapotranspiration data for a catchment with an area of 200 km². The catchment is in a remote area with currently no abstraction from the underlying unconfined sand and gravel aquifer. Land use in the catchment is predominantly grassland. A single observation borehole in the catchment recorded an increase in the level of the water table of 2.75 m during the winter recharge period 2002–2003.

Table A10.10 Rainfall (P) and potential evapotranspiration (PE) data.

	Month											
	J	F	M	A	M	J	J	A	S	O	N	D
2002												
P (mm)	65	69	34	14	29	66	38	66	128	73	142	39
PE (mm)	17	21	37	90	110	105	112	88	54	31	16	20
2003												
P (mm)	90	32	78	43	73	23	32	18	128	28	71	34
PE (mm)	16	17	33	60	84	124	122	102	54	24	8	7

(a) Assuming a root constant for short-rooted vegetation of 75 mm, use the Penman–Grindley method to calculate the groundwater recharge (hydrological excess water, HXS) expected for the period 2002–2003. Start your calculation on 1 April 2002 when the soils in the catchment were at field capacity (i.e. soil moisture deficit, SMD = 0).

(b) If the recharge calculated in (a) above entirely replenishes the sand and gravel aquifer, calculate a storage coefficient value for the aquifer and estimate the change in groundwater storage volume that occurred over the catchment.

(c) A river draining the catchment has an average annual flow of 1.843 $m^3 s^{-1}$. As a measure of the baseflow index of this river, calculate the fraction of the 2002–2003 groundwater recharge relative to the average annual flow.

(d) In a fuller assessment of the catchment water balance, what other considerations would you include in estimating the safe yield of future groundwater abstractions?

10 Agricultural production in the coastal plain of North Africa is supported by drip-feed irrigation. As part of a new development, an area of 26,000 ha is to be irrigated using groundwater pumped from an inland desert area from an unconfined aquifer with a recharge area of 200 km^2. From the information below, and using the chloride budget method, evaluate the sustainable supply of water available for agricultural use in units of litres per hectare of agricultural land per day (L ha^{-1} day^{-1}).
Notes:

Mean annual precipitation: 120.1 mm a^{-1}.

Mean concentration of chloride in rainwater: 4.33 mg L^{-1}.

Mean concentration of chloride in interstitial pore water in the unsaturated zone: 8.00 mg L^{-1}.

If the unsaturated zone in the recharge area is 78 m thick, estimate the residence time of water in the unsaturated zone, if the average moisture content is 0.65%.

A10.6 Groundwater resources, pumping tests and stream depletion analysis

1 How can an analysis of river and well hydrographs assist in the assessment of catchment water resources?

2 Describe the repercussions of aquifer over-exploitation and discuss how different methods of aquifer management can address the impacts.

3 You are instructed to prepare a report on the groundwater resources of an area comprising (i) folded limestones overlain in part by (ii) horizontal thick basaltic lava flows. The area is cut by a mature drainage system with (iii) sandy point-bar and silty mud overbank deposits. What type of aquifers might you expect to occur in the area? What type of information would you expect to find in a hydrogeological report on the area which would aid you in this task?

4 Explain the principles by which the transmissivity and storativity of an aquifer can be determined by the Theis method of pumping test analysis. In your answer, describe how changes in storage occur in confined and unconfined aquifers.

Table A10.11 Constant discharge pumping test data.

Time since start of pumping (min)	Drawdown (m)
7	0.15
7.5	0.16
8	0.17
8.5	0.19
9	0.21
9.5	0.23
10	0.25
11	0.30
12	0.36
13	0.42
14	0.48
15	0.55
16	0.62
17	0.68
18	0.75
19	0.82
20	0.89
22	1.03
24	1.16
26	1.29
28	1.42
30	1.54
35	1.84
40	2.10
44.5	2.32

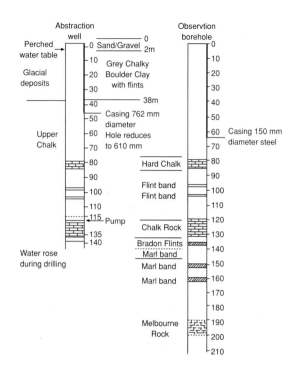

Fig. A10.9 Geological logs of two Chalk boreholes, Norfolk, England.

5 A borehole penetrates a 25 m thick, confined aquifer and is pumped at a steady rate of 0.2 m³ s⁻¹ until a steady-state cone of depression is formed. At this time, two observation boreholes, 350 m and 1000 m away, have water levels of 29.50 m and 30.84 m above the top of the aquifer, respectively. Assuming that the initial piezometric surface was horizontal, estimate the transmissivity and hydraulic conductivity of the aquifer.

6 Describe the essential procedures for carrying out a two-well, constant rate, time-variant pumping test in the field. Show how the measurements that are necessary relate to the requirements and assumptions of using the Theis formula for interpreting the results.

The data in Table A10.11 were obtained during a short pumping test of a confined Chalk aquifer at a site in Norfolk, England. Figure A10.9 shows geological logs of the two boreholes used. The distance from the abstraction borehole to the observation borehole, r, is 218 m and the abstraction borehole discharge, Q,

was 295 m³ h⁻¹. Use the Theis method to determine the transmissivity and storativity of the Chalk aquifer at this site. To plot the field curve you will need a sheet of log 3 cycles by 4 cycles graph paper. A type curve for copying is provided as Fig. 5.33.

7 A river support borehole abstracts groundwater from a lower, confined limestone aquifer which is successively overlain by a low permeability clay aquitard and an upper, unconfined limestone aquifer. A site plan (a) and generalized cross-section (b) are shown in Fig. A10.10. The support borehole was test pumped at a constant discharge of 2000 m³ day⁻¹, with the groundwater fed into the adjacent river. Two observation boreholes, situated at a distance of 165 m from the support borehole, and positioned in the lower limestone aquifer (borehole A) and upper limestone aquifer (borehole B), produced the drawdown data given in Table A10.12.

(a) Using the Cooper–Jacob method of pumping test analysis, and the drawdown versus time data for observation borehole A (Table A10.12),

(a) Site plan

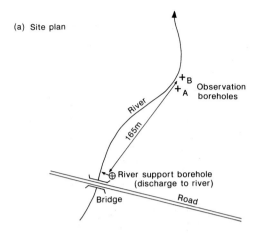

(b) Generalized cross-section

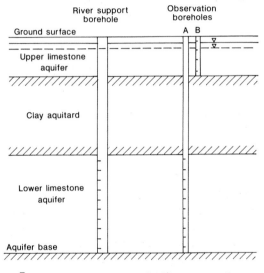

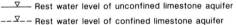

——▽—— Rest water level of unconfined limestone aquifer

——▽—— Rest water level of confined limestone aquifer

Fig. A10.10 Site plan and generalized cross-section for a layered limestone aquifer system.

calculate the transmissivity and storativity values for the confined limestone aquifer.

(b) The record of drawdown in observation borehole B shows that the unconfined limestone aquifer experienced a rise in water level during the period of the pumping test. Provide an explanation of the possible cause of this response at borehole B.

8 The water supply to a large town is normally supplied by a surface storage reservoir. However, during dry summer months, the reservoir is unreliable and

Table A10.12 River support borehole pumping test data.

Time since start of test (s)	Drawdown* (m)	
	Borehole A	Borehole B
0	0	0
120	0.315	0
240	0.555	0
360	0.774	−0.005
480	0.953	−0.007
600	1.105	−0.009
720	1.245	−0.010
840	1.370	−0.011
960	1.485	−0.015
1080	1.595	−0.015
1200	1.693	−0.017
1500	1.915	−0.020
1800	2.115	−0.020
2100	2.295	−0.025
2400	2.495	−0.029
2700	2.615	−0.030
3000	2.755	−0.030
3300	2.895	−0.031
3600	3.015	−0.032
3900	3.145	−0.033
4200	3.255	−0.035
4500	3.365	−0.035
4800	3.485	−0.036
5100	3.585	−0.040
5400	3.685	−0.040

* A negative drawdown indicates a rise in water level.

it is planned to switch the water supply to a new well-field capable of pumping 5×10^4 m^3 day^{-1} from a highly productive sandstone aquifer. The sandstone is unconfined and has a transmissivity of 1500 m^2 day^{-1} and a specific yield of 0.10. The well-field is to be situated in the floodplain of a major river that is in direct hydraulic contact with the aquifer. To avoid damage to the river ecosystem, it is essential that the river flow is not reduced by more than 2×10^4 m^3 day^{-1} as a result of leakage caused by groundwater abstraction. The well-field is to be operated for 90 days each summer.

Using the above information and the table of values provided in Appendix 7, calculate a safe distance for the position of the centre of the well-field from the river in order to avoid the effects of stream depletion, explaining the assumptions and limitations of your answer.

A10.7 Contaminant hydrogeology

1 In the context of contaminant hydrogeology, provide brief definitions of the following terms:
   (a) Advection.
   (b) Hydrodynamic dispersion.
   (c) Retardation.
   (d) Sorption isotherm.

2 Provide a definition of the hydrodynamic dispersion coefficient, D_l, where l denotes the longitudinal direction, and explain the meaning of the terms given in the following equation:

$$D_l = \alpha_l \bar{v} + D^\star$$

3 Describe how the lithological characteristics of different rock types can affect the transport and attenuation of dissolved, reactive contaminant species.

4 Describe the main features and patterns of groundwater contamination around point sources of pollutants, such as landfill sites, wastewater lagoons or spillage sites. What general recommendations for landfill site selection and management would you advise to limit such pollution?

5 Choosing either septic tanks or municipal landfill sites, discuss the importance of redox processes in the development of pollution plumes that may occur down-gradient of the point of groundwater contamination.

6 Provide, with illustrations, a general description of the physicochemical processes that influence the migration of hydrocarbons and organic solvents in porous materials.

7 Explain the hydrophobic sorption model in predicting the attenuation of organic compounds in saturated, porous materials, outlining the applications and limitations of the model.

8 Outline the principal industrial and agricultural activities that present a threat to groundwater quality. In your answer, describe the nature of each contaminant source (whether point or diffuse pollution) and the potential for natural attenuation once contamination has occurred.

9 Devise a strategy for the protection of groundwater resources from diffuse sources of agricultural pollution explaining the benefits and disadvantages of your approach.

10 Define what is meant by the **risk** of groundwater pollution. What action would you take to protect groundwater resources from the risk of groundwater pollution in (a) an industrialized country and (b) an agrarian, developing country?

11 A former industrial site is to be developed as a retail and entertainment complex known as Waterside. The site is at a distance of 3 km from a river and there is concern that 'hotspots' of contaminated groundwater in the underlying sand aquifer will migrate towards the river if left undisturbed. However, commercial pressures are such that groundwater remediation at the site will only be considered if a contaminant concentration of greater than or equal to 1% of the original source concentration is likely to reach the river within the 30-year lifespan of the Waterside development. Stating your assumptions, use the following site information to determine whether you would recommend remediation.

Hydraulic gradient between Waterside and river $= 1 \times 10^{-4}$

Hydraulic conductivity of sand aquifer = 100 m day^{-1}

Effective porosity of sand aquifer = 0.115

Longitudinal dispersivity, α_l, of sand aquifer = 100 m

Note: Values of the complementary error function, erfc(β), for positive values of β are given in Appendix 8.

12 Figure A10.11 shows a cross-section through a former sand and gravel quarry that is now used as a landfill for the co-disposal of municipal and industrial wastes. The site is covered daily with low-permeability clay and leachate develops within the waste. Although the site has an impermeable base, horizontal movement of leachate through the sand and gravel aquifer has been induced by groundwater abstraction from a borehole located at a distance of 100 m from the site. The borehole completely penetrates the aquifer. The aquifer comprises three layers, each with homogeneous and isotropic hydraulic properties. Table A10.13 details the aquifer characteristics of the three layers.

Analysis of the leachate reveals that the chlorinated solvent trichloroethene (TCE) is present as a dissolved phase. Assuming steady-state groundwater flow conditions and a partition coefficient between organic carbon and water, K_{oc}, for TCE of 150 cm^3 g^{-1}, find the following:

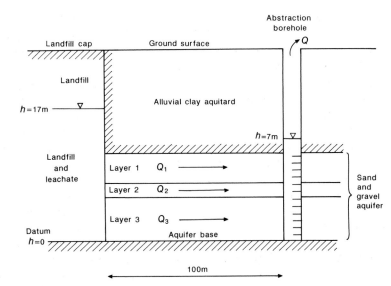

Fig. A10.11 Cross-section of a former sand and gravel quarry now operated as a landfill site.

Table A10.13 Layered sand and gravel aquifer characteristics.

	Hydraulic conductivity, K (m day^{-1})	Layer thickness, d (m)	Porosity, n	Bulk density, ρ_b (g cm^{-3})	Fraction of organic carbon, f_{OC}
Layer 1 (sand)	10	2	0.20	1.60	0.015
Layer 2 (gravel)	50	1	0.40	1.50	0.015
Layer 3 (fine sand)	2	3	0.15	1.65	0.015

(a) A distribution coefficient, K_d, and retardation factor, R_t, for each of the three aquifer layers.

(b) The breakthrough time for TCE in each layer at the abstraction borehole.

(c) The resulting concentration breakthrough curve for TCE arriving at the borehole.

(d) The breakthrough time and number of pore volume flushes of the layered aquifer that have occurred once the concentration of TCE in the borehole is half of the original concentration in the leachate (i.e. $C/C_o = 0.5$).

13 A site has been located in an aquitard that is suitable for the deep disposal of radioactive waste. Figure A10.12 shows the position of the engineered disposal site within a low permeability siltstone layer confined between two very low permeability layers composed of crystalline rock. The siltstone has been subjected to minor fracturing that imparts a small secondary porosity and hydraulic conductivity of 0.01 and 1×10^{-5} m s^{-1}, respectively. A large groundwater head is developed in the high topography above the position of the disposal site such that there is the potential for groundwater flow in the siltstone layer to migrate to the ground surface where the aquitard reaches outcrop.

Residents in the adjacent valley fear that radioactive plutonium, ^{239}Pu, will contaminate local groundwater supplies for future generations. Using the information given in Fig. A10.12, assess the likelihood that a measurable concentration of ^{239}Pu will reach the position of P2 close to the shallow water table in the valley. Comment on any assumptions you make in your calculation and whether you regard your answer as a conservative estimate or not.

14 A railroad tanker with a cargo of trichloroethene (TCE) has derailed, spilling 35,000 L of the solvent

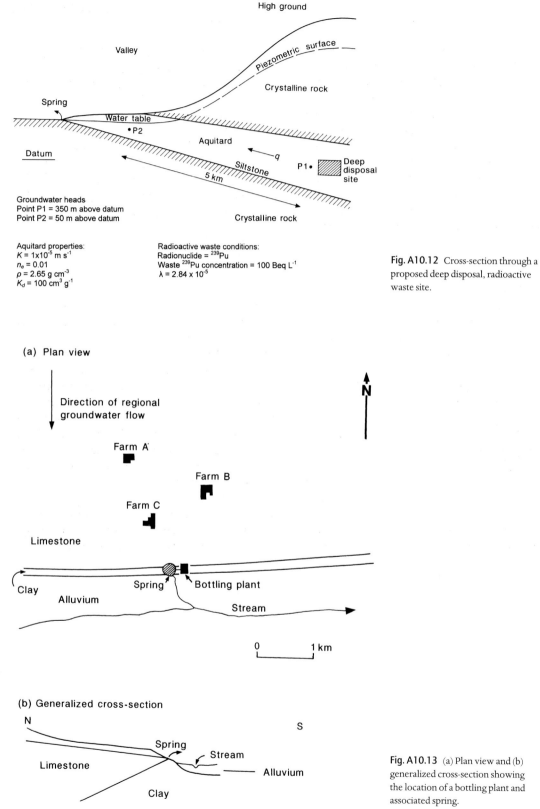

High ground

Valley

Piezometric surface

Crystalline rock

Spring

Water table

• P2

Aquitard

←—q

Datum

Siltstone

P1 • ▨ Deep disposal site

5 km

Crystalline rock

Groundwater heads
Point P1 = 350 m above datum
Point P2 = 50 m above datum

Aquitard properties:
$K = 1 \times 10^{-5}$ m s^{-1}
$n_e = 0.01$
$\rho = 2.65$ g cm^{-3}
$K_d = 100$ cm^3 g^{-1}

Radioactive waste conditions:
Radionuclide = ^{239}Pu
Waste ^{239}Pu concentration = 100 Beq L^{-1}
$\lambda = 2.84 \times 10^{-5}$

Fig. A10.12 Cross-section through a proposed deep disposal, radioactive waste site.

(a) Plan view

Direction of regional groundwater flow

N

Farm A

Farm B

Farm C

Limestone

Clay

Alluvium

Spring Bottling plant

Stream

0   1 km

(b) Generalized cross-section

N   S

Spring

Stream

Limestone

Alluvium

Clay

Fig. A10.13 (a) Plan view and (b) generalized cross-section showing the location of a bottling plant and associated spring.

directly on an unconfined alluvial sand aquifer. The solvent has entered the water table and a contaminant plume of dissolved TCE has developed in the direction of groundwater flow. The groundwater flow is uniform and horizontal under a hydraulic gradient of 0.005 and discharges to a river 400 m from the site of the derailment. The aquifer has a hydraulic conductivity of 2×10^{-3} m s^{-1}, an effective porosity of 0.25, an organic carbon content of 0.5% and a bulk density of 2.10 g cm^{-3}. Ignoring dispersion of the dissolved contaminant in the alluvial sand aquifer, answer the following, explaining any assumptions you make:

(a) What is the maximum concentration of dissolved TCE expected in the aquifer?

(b) What is the average linear groundwater velocity in the sand aquifer?

(c) Estimate the average velocity of the contaminated groundwater and the time taken for dissolved TCE to arrive at the river.

Notes:

Solubility of TCE at 20°C = 1×10^3 mg L^{-1}.

$\log_{10} K_{OW}$ for TCE at 25°C = 2.42.

$K_d = K_{OC} \cdot f_{OC}$.

$\log_{10} K_{OC} = 0.49 + 0.72 \log_{10} K_{OW}$ (Schwarzenbach & Westall (1981) equation).

15 Figure A10.13 shows the plan view and a generalized cross-section centred on a Bottling Plant taking water from a spring that issues from a limestone aquifer at the junction with an underlying clay aquitard. The limestone has a transmissivity of 5000 m^2 day^{-1} and an average thickness of 20 m. On average, the

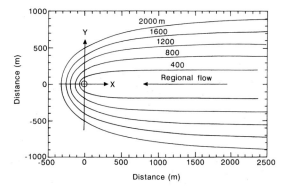

Fig. A10.14 Type curves showing the capture zones of a single pumping well located at point (0, 0) for various values of Q/bq_a.

Bottling Plant abstracts 2.5×10^3 m^3 of water each day from the spring for process use. In addition, and under the abstraction licence agreement, the Bottling Plant must leave 0.5×10^3 m^3 day^{-1} of spring water to compensate the downstream water course. Recently, the spring water has shown increasing concentrations of ammonium and biological oxygen demand, causing fear that the source is vulnerable to contamination from one or more of the three dairy farms located upgradient from the spring. The regional groundwater hydraulic gradient is 6×10^{-4}, oriented in a north to south direction.

Using the capture zone type curves given in Fig. A10.14, and stating your assumptions, determine which, if any, of the three dairy farms may be posing a threat to the security of the Bottling Plant.

References

Adak, G.K., Long, S.M. & O'Brien, S.J. (2002) Trends in indigenous foodborne disease and deaths, England and Wales: 1992–2000 *Gut* **51**, 832–841.

ADAS (2003) *Assessment of the Effectiveness of the Nitrate Sensitive Areas Scheme in Reducing Nitrate Concentrations.* Technical report for R&D project P2-267/U/2. Environment Agency, Bristol.

Albu, M., Banks, D. & Nash, H. (1997) *Mineral and Thermal Groundwater Resources.* Chapman & Hall, London.

Allen-King, R.M., Grathwohl, P. & Ball, W.P. (2002) New modeling paradigms for the sorption of hydrophobic organic chemicals to heterogeneous carbonaceous matter in soils, sediments, and rocks. *Advances in Water Resources* **25**, 985–1016.

Aller, L., Bennett, T., Lehr, J.H., Petty, R.J. & Hackett, G. (1987) *DRASTIC: a standardized system for evaluating ground water pollution potential using hydrogeologic settings.* NWWA/EPA Series. EPA/600/2-87/035. Washington, DC.

Ambroggi, R.P. (1977) Underground reservoirs to control the water cycle. *Scientific American* **236**(5), 21–27.

Anderson, M., Low, R. & Foot, S. (2002) Sustainable groundwater development in arid, high Andean basins. In: *Sustainable Groundwater Development* (eds K.M. Hiscock, M.O. Rivett & R.M. Davison). Geological Society, London, Special Publications **193**, 133–144.

Anderson, M.P. & Woessner, W.W. (1992) *Applied Groundwater Modeling: simulation of flow and advective transport.* Academic Press, San Diego.

Andrews, J.E., Brimblecombe, P., Jickells, T.D., Liss, P.S. & Reid, B.J. (2004) *An Introduction to Environmental Chemistry,* 2nd edn. Blackwell Science, Oxford.

Andrews, J.N. & Lee, D.J. (1979) Inert gases in groundwater from the Bunter sandstone of England as indicators of age and palaeoclimatic trends. *Journal of Hydrology* **41**, 233–252.

Andrews, J.N., Burgess, W.G., Edmunds, W.M., Kay, R.L.F. & Lee, D.J. (1982) The thermal springs of Bath. *Nature* **298**, 339–343.

Appelo, C.A.J. & Postma, D. (1994) *Geochemistry, Groundwater and Pollution.* A.A. Balkema, Rotterdam.

Aravena, R. & Robertson, W.D. (1998) Use of multiple isotope tracers to evaluate denitrification in ground water: study of nitrate from a large-flux septic system plume. *Ground Water* **36**, 975–982.

Arnell, N. (1996) *Global Warming, River Flows and Water Resources.* John Wiley, Chichester.

Askew, A.J. (1987) Climate change and water resources. *IAHS Publications* **168**, 421–430.

ASTM (2001) *Risk-based Corrective Action (RBCA) (E2081-00) Standard Guide for Risk-based Corrective Action.* American Society for Testing and Materials, Philadelphia, Pennsylvania.

Atkinson, T.C. (1977) Diffuse flow and conduit flow in limestone terrain in the Mendip Hills, Somerset (Great Britain). *Journal of Hydrology* **35**, 93–110.

Atkinson, T.C. & Davison, R.M. (2002) Is the water still hot? Sustainability and the thermal springs at Bath, England. In: *Sustainable Groundwater Development* (eds K.M. Hiscock, M.O. Rivett & R.M. Davison). Geological Society, London, Special Publications **193**, 15–40.

Atkinson, T.C. & Smart, P.L. (1981) Artificial tracers in hydrogeology. In: *A Survey of British Hydrogeology 1980.* Royal Society, London, 173–190.

Atkinson, T.C., Smith, D.I., Lavis, J.J. & Whitaker, R.J. (1973) Experiments in tracing underground waters in limestones. *Journal of Hydrology* **19**, 323–349.

Back, W., Rosenshein, J.S. & Seaber, P.R. (eds) (1988) *Hydrogeology. The geology of North America,* vol. O-2. The Geological Society of North America, Boulder, Colorado.

Backshall, W.F., Downing, R.A. & Law, F.M. (1972) Great Ouse Groundwater Study. *Water and Water Engineering* **June**, 3–11.

Baker, J.R., Mihelcic, J.R., Luehrs, D.C. & Hickey, J.P. (1997) Evaluation of estimation methods for organic carbon normalized sorption coefficients. *Water Environment Research* **69**, 136–145.

Banwart, S.A., Evans, K.A. & Croxford, S. (2002). Predicting mineral weathering rates at field scale for mine water risk assessment. In: *Mine Water Hydrogeology and Geochemistry* (eds P.L. Younger & N.S. Robins). Geological Society, London, Special Publications **198**, 137–157.

Barker, R.D. (1986) Surface geophysical techniques. In: *Groundwater: occurrence, development and protection* (ed. T.W. Brandon). Institution of Water Engineers and Scientists, London, pp. 271–314.

Barker, R.D. & Harker, D. (1984) The location of the Stour buried tunnel-valley using geophysical techniques. *Quarterly Journal of Engineering Geology* **17**, 103–115.

Barrett, M.H., Hiscock, K.M., Pedley, S., Lerner, D.N., Tellam, J.H. & French, M.J. (1999) Marker species for identifying urban groundwater recharge sources: a review and case study in Nottingham UK. *Water Research* **33**, 3083–3097.

Bateman, A.S., Hiscock, K.M. & Atkinson, T.C. (2001) Qualitative risk assessment using tracer tests and groundwater modelling in an unconfined sand and gravel aquifer. In: *New Approaches Characterizing Groundwater Flow* (eds K.-P. Seiler & S. Wohnlich). Swets & Zeitlinger, Lisse, 251–255.

Beaumont, P. (1968) Qanats on the Varamin Plain, Iran. *Transactions of the Institution of British Geographers* **45**, 169–179.

Beesley, K. (1986) Downhole geophysics. In: *Groundwater: occurrence, development and protection* (ed. T.W. Brandon). Institution of Water Engineers and Scientists, London, pp. 315–352.

Bekins, B., Rittmann, B. & MacDonald, J. (2001a) Natural attenuation strategy for groundwater cleanup focuses on demonstrating cause and effect. *Eos* **82**, 53, 57–58.

Bekins, B.A., Cozzarelli, I.M., Godsy, E.M., Warren, E., Essaid, H.I. & Tuccillo, M.E. (2001b) Progression of natural attenuation processes at a crude oil spill site: II. Controls on spatial distribution of microbial populations. *Journal of Contaminant Hydrology* **53**, 387–406.

Bennett, C.E. (1969) *Frontinus: the stratagems and the aqueducts of Rome.* Harvard University Press, Cambridge, Massachusetts.

Bentley, H.W., Phillips, F.M., Davis, S.N., Habermehl, M.A., Airey, P.L., Calf, G.E., Elmore, D., Gove, H.E. & Torgersen, T. (1986) Chlorine 36 dating of very old groundwater: The Great Artesian Basin, Australia. *Water Resources Research* **22**, 1991–2001.

Beran, M. & Rodier, J.A. (1985) Hydrological aspect of drought. *Studies and Report in Hydrology* **39**. UNESCO-WMO, Paris.

Berg, M., Tran, H.C., Nguyen, T.C., Pham, H.V., Schertenleib, R. & Giger, W. (2001) Arsenic contamination of groundwater and drinking water in Vietnam: a human health threat. *Environmental Science & Technology* **35**, 2621–2626.

Bergmire-Sweat, D., Wilson, K., Marengo, L., Lee, Y.M., MacKenzie, W.R., Morgan, J., Von Alt, K., Bennett, T., Tsang, V.C.W. & Furness, B. (1999) Cryptosporidiosis in Brush Creek: describing the epidemiology and causes of a large outbreak in Texas, 1998. In: *Proceedings of the International Conference on Emerging Infectious Diseases, Milwaukee, Wisconsin.* American Water Works Association, Denver, Colorado.

Berner, E.K. & Berner, R.A. (1987) *The Global Water Cycle: geochemistry and environment.* Prentice-Hall, Englewood Cliffs, New Jersey.

Bertin, C. & Bourg, A.C.M. (1994) Radon-222 and chloride as natural tracers of the infiltration of river water into an alluvial aquifer in which there is significant river/groundwater mixing. *Environmental Science & Technology* **28**, 794–798.

Besien, T.J., Williams, R.J. & Johnson, A.C. (2000) The transport and behaviour of isoproturon in unsaturated chalk cores. *Journal of Contaminant Hydrology* **43**, 91–110.

Beven, K. & Germann, P. (1982) Macropores and water flow in soil. *Water Resources Research* **18**, 1311–1325.

Beyerle, U., Aeschbach-Hertig, W., Hofer, M., Imboden, D.M., Baur, H. & Kipfer, R. (1999) Infiltration of river water to a shallow aquifer investigated with $^3H/^3He$, noble gases and CFCs. *Journal of Hydrology* **220**, 169–185.

Binley, A., Daily, W. & Ramirez, A. (1997) Detecting leaks from environmental barriers using electrical current imaging. *Journal of Environmental and Engineering Geophysics* **2**, 11–19.

Binley, A., Cassiani, G., Middleton, R. & Winship, P. (2002a) Vadose zone flow model parameterisation using cross-borehole radar and resistivity imaging. *Journal of Hydrology* **267**, 147–159.

Binley, A., Winship, P., West, J., Pokar, M. & Middleton, R. (2002b) Seasonal variation of moisture content in unsaturated sandstone inferred from borehole radar and resistivity profiles. *Journal of Hydrology* **267**, 160–172.

Bishop, P.K., Burston, M.W., Lerner, D.N. and Eastwood, P.R. (1990) Soil gas surveying of chlorinated solvents in relation to groundwater pollution studies. *Quarterly Journal of Engineering Geology* **23**, 255–265.

Biswas, A.K. (1972) *History of Hydrology.* North-Holland, Amsterdam.

Blaney, H.F. & Criddle, W.D. (1962) Determining consumptive use and irrigation water requirements. *United States Department of Agriculture Technical Bulletin* **1275**.

Bloomfield, J.P., Gaus, I. & Wade, S.D. (2003) A method for investigating the potential impacts of climate-change scenarios on annual minimum groundwater levels. *Journal of the Chartered Institution of Water and Environmental Management* **17**, 86–91.

Blowes, D.W., Ptacek, C.J., Benner, S.G., McRae, C.W.T., Bennett, T.A. & Puls, R.W. (2000) Treatment of inorganic contaminants using permeable reactive barriers. *Journal of Contaminant Hydrology* **45**, 123–137.

Boni, C., Bono, P. & Capelli, G. (1986) *Hydrogeological Scheme of Central Italy. Sheet 1 (A. Hydrogeological Map).* Memoir of the Geological Society of Italy **XXXV**, Rome.

Bono, P. & Boni, C. (2001) Water supply of Rome in antiquity and today. In: *Springs and Bottled Waters of the World: ancient history, source, occurrence, quality and use* (eds P.E. LaMoreaux & J.T. Tanner). Springer-Verlag, Berlin, 200–210.

Bottrell, S.H., Moncaster, S.J., Tellam, J.H., Lloyd, J.W., Fisher, Q.J. & Newton, R.J. (2000) Controls on bacterial sulphate reduction in a dual porosity aquifer system: the Lincolnshire Limestone aquifer, England. *Chemical Geology* **169**, 461–470.

Boulton, N.S. (1963) Analysis of data from non-equilibrium pumping tests allowing for delayed yield from storage. *Proceedings of the Institution of Civil Engineers* **26**, 469–482.

Bouwer, H. (1989) The Bouwer and Rice slug test – an update. *Ground Water* **27**, 304–309.

Bouwer, H. (1991) Simple derivation of the retardation equation and application to preferential flow and macrodispersion. *Ground Water* **29**, 41–46.

Bouwer, H. & Rice, R.C. (1976) A slug test for determining hydraulic conductivity of unconfined aquifers with completely or partially penetrating wells. *Water Resources Research* **12**, 423–428.

Brassington, R. (1998). *Field Hydrogeology*, 2nd edn. John Wiley, Chichester.

British Geological Survey (1978) *1 : 100,000 Hydrogeological Map of the South West Chilterns and the Berkshire and Marlborough Downs.* Natural Environment Research Council. British Geological Survey, Keyworth, Nottingham.

British Geological Survey (1988) *1 : 625,000 Hydrogeological Map of Scotland.* Natural Environment Research Council. British Geological Survey, Keyworth, Nottingham.

British Geological Survey (1990) *1 : 50,000 Map of the Carnmenellis Granite. Hydrogeological, hydrogeochemical and geothermal characteristics.* Natural Environment Research Council. British Geological Survey, Keyworth, Nottingham.

British Geological Survey (1994) *1 : 250,000 Hydrogeological Map of Northern Ireland.* Natural Environment Research Council. British Geological Survey, Keyworth, Nottingham.

Brown, M.C. & Ford, D.C. (1971) Quantitative tracer methods for investigation of karst hydrologic systems.

Transactions of the Cave Research Group of Great Britain **13**, 37–51.

Bryan, K. (1919) Classification of springs. *Journal of Geology* **27**, 552–561.

Bryant, S.J., Arnell, N.W. & Law, F.M. (1992) The long-term context for the current hydrological drought. In: *Proceedings of the Institution of Water and Environmental Management Conference on the Management of Scarce Water Resources.*

BSI (1988) *British Standard Guide for Geophysical Logging of Boreholes for Hydrogeological Purposes.* British Standards Institution, London.

BSI (1992) *Code of Practice for Test Pumping Water Wells. BS 6316: 1992.* British Standards Institution, London.

BSI (1998) *Measurement of Liquid Flow in Open Channels – general guidelines for the selection of method. BS ISO/TR 8363: 1997.* British Standards Institution, London.

Burgess, D.B. (2002) Groundwater resource management in eastern England: a quest for environmentally sustainable development. In: *Sustainable Groundwater Development* (eds K.M. Hiscock, M.O. Rivett & R.M. Davison). Geological Society, London, Special Publications **193**, 53–62.

Burgess, W.G., Edmunds, W.M., Andrews, J.N., Kay, R.L.F. & Lee, D.J. (1980) *Investigation of the Geothermal Potential of the UK. The hydrogeology and hydrochemistry of the thermal water in the Bath–Bristol Basin.* Institute of Geological Sciences, London.

Burgess, W.G., Burren, M., Perrin, J., Mather, S.E. & Ahmed, K.M. (2002) Constraints on sustainable development of arsenic-bearing aquifers in southern Bangladesh. Part 1: a conceptual model of arsenic in the aquifer. In: *Sustainable Groundwater Development* (eds K.M. Hiscock, M.O. Rivett & R.M. Davison). Geological Society, London, Special Publications **193**, 145–163.

Butler, J.J., Zlotnik, V.A. & Tsou, M.-S. (2001) Drawdown and stream depletion produced by pumping in the vicinity of a partially penetrating stream. *Ground Water* **39**, 651–659.

Carey, M.A., Fretwell, B.A., Mosley, N.G. & Smith, J.W.N. (2002) *Guidance on the Use of Permeable Reactive Barriers for Remediating Contaminated Groundwater.* National Groundwater and Contaminated Land Centre report NC/01/51. Environment Agency, Bristol.

Carlyle, H.F., Tellam, J.H. & Parker, K.E. (2004) The use of laboratory-determined ion exchange parameters in the predictive modelling of field-scale major cation migration in groundwater over a 40-year period. *Journal of Contaminant Hydrology* **68**, 55–81.

Carter, T.R. & La Rovere, E.L. (2001) Developing and applying scenarios. In: *Climate Change 2001: impacts, adaptation and vulnerability* (eds J. McCarthy, O.F. Canziani,

N. Leary, D.J. Dokken & K.S. White). Cambridge University Press, Cambridge, 145–190.

Cedergren, H.R. (1967) *Seepage, Drainage and Flow Nets*. J. Wiley, New York.

Champ, D.R., Gulens, J. & Jackson, R.E. (1979) Oxidation-reduction sequences in ground water flow systems. *Canadian Journal of Earth Sciences* **16**, 12–23.

Chapellier, D. (1992) *Well Logging in Hydrogeology*. A.A. Balkema, Rotterdam.

Chebotarev, I.I. (1955) Metamorphism of natural water in the crust of weathering. *Geochimica et Cosmochimica Acta* **8**, 22–48, 137–170, 198–212.

Chen, C.X., Wan, J.W. & Zhan, H.B. (2003) Theoretical and experimental studies of coupled seepage-pipe flow to a horizontal well. *Journal of Hydrology* **281**, 159–171.

Chiang, W.-H. & Kinzelbach, W. (2001) *3D-Groundwater modeling with PMWIN: a simulation system for modeling groundwater flow and pollution*. Springer-Verlag, Berlin.

CL:AIRE (2002) Introduction to an integrated approach to the investigation of fractured rock aquifers contaminated with non-aqueous phase liquids. *Technical Bulletin* **TB1**, Contaminated Land: Applications in Real Environments, London.

Christiansen, E.A., Acton, D.F., Long, R.J., Meneley, W.A. & Sauer, E.K. (1981) *Fort Qu'Appelle Geolog. The Valleys – past and present*. Interpretive Report No. 2. The Saskatchewan Research Council, Canada.

Christensen, T.H., Kjeldsen, P., Bjerg, P.L., Jensen, D.L., Christensen, J.B., Baun, A., Albrechtsen, H.-J. & Heron, G. (2001) Biogeochemistry of landfill leachate plumes. *Applied Geochemistry* **16**, 659–718.

Clark, L. (1977) The analysis and planning of step drawdown tests. *Quarterly Journal of Engineering Geology* **10**, 125–143.

Clark, L. (1984) Groundwater development of the Chepstow Block: a study of the impact of domestic waste disposal on a karstic limestone aquifer in Gwent, South Wales. In: *Proceedings of the IAH International Groundwater Symposium on Groundwater Resources Utilization and Contaminant Hydrogeology* (ed. R. Pearson). Atomic Energy of Canada Ltd., Manitoba, Canada **II**, 300–309.

Clark, L. (1988) *The Field Guide to Water Wells and Boreholes*. Geological Society of London Professional Handbook Series. Open University Press, Milton Keynes.

Clark, L. & Gomme, J. 1992. Pesticides in a Chalk catchment in Eastern England. *Hydrogéologie* **4**, 169–174.

Clark, L. & Sims, P.A. (1998) Investigation and clean-up of jet-fuel contaminated groundwater at Heathrow International Airport, UK. In: *Groundwater Contaminants and their Migration* (eds J. Mather, D. Banks, S. Dumpleton and M. Fermor). Geological Society, London, Special Publications **128**, 147–157.

Commission of the European Communities (2003) *Proposal for a Directive of the European Parliament and of the Council on the Protection of Groundwater against Pollution*. (2003/0210 (COD).) Commission of the European Communities, Brussels.

Conway, D. (1998) Recent climate variability and future climate change scenarios for Great Britain. *Progress in Physical Geography* **22**, 350–374.

Cooper, H.H. & Jacob, C.E. (1946) A generalized graphical method for evaluating formation constants and summarizing well field history. *Transactions of the American Geophysical Union* **27**, 526–534.

Council of European Communities (1980) *Directive on the Protection of Groundwater against Pollution Caused by Certain Dangerous Substances*. (80/68/EEC.) Official Journal of the European Communities, L20. Brussels.

Council of European Communities (1991) *Directive Concerning the Protection of Waters against Pollution Caused by Nitrates from Agricultural Sources*. (91/676/EEC.) Official Journal of the European Communities, L375. Brussels.

Council of European Communities (1998) *Directive Relating to the Quality of Water Intended for Human Consumption*. (98/83/EC.) Official Journal of the European Communities, L330. Brussels.

Council of European Communities (2000) *Directive Establishing a Framework for Community Action in the Field of Water Policy*. (2000/60/EC.) Official Journal of the European Communities, L327. Brussels.

Cozzarelli, I.M., Bekins, B.A., Baedecker, M.J., Aiken, G.R., Eganhouse, R.P. & Tuccillo, M.E. (2001) Progression of natural attenuation processes at a crude oil spill site: I Geochemical evolution of the plume. *Journal of Contaminant Hydrology* **53**, 369–385.

Craig, H. (1961) Isotopic variations in meteoric water. *Science* **133**: 1702–1703.

Craun, G.F. (1985) A summary of waterborne illness transmitted through contaminated groundwater. *Journal of Environmental Health* **48**, 122–127.

Craun, G.F., Greathouse, D.G. & Gunderson, D.H. (1981) Methemoglobin levels in young children consuming high nitrate well water in the United States. *International Journal of Epidemiology* **10**, 309–317.

Cronin, A.A., Taylor, R.G., Powell, K.L., Barrett, M.H., Trowsdale, S.A. & Lerner, D.N. (2003) Temporal trends in the depth-specific hydrochemistry and sewage-related microbiology of an urban sandstone aquifer, Nottingham, United Kingdom. *Hydrogeology Journal* **11**, 205–216.

Custodio, E. (ed.) (1987) *Groundwater Problems in Coastal Areas: a contribution to the International Hydrological Programme*. Studies and Reports in Hydrology 45. UNESCO, Paris.

Dai, A., Trenberth, K.E. & Karl, T. (1998) Global variations in droughts and wet spells: 1900–1995. *Geophysical Research Letters* **25**, 3367–3370.

Dalton, J. (1799) *Experiments and Observations to Determine Whether the Quantity of Rain and Dew is Equal to the Quantity of Water Carried off by Rivers and Raised by Evaporation, with an Enquiry into the Origin of Springs.* Literary & Philosophical Society, Manchester.

Darcy, H. (1856) *Les fontaines publiques de la ville de Dijon.* Victor Dalmont, Paris.

Darling, W.G. & Talbot, J.C. (2003) The O & H stable isotopic composition of fresh waters in the British Isles. 1. Rainfall. *Hydrology and Earth System Sciences* **7**, 163–181.

Darling, W.G., Edmunds, W.M. & Smedley, P.L. (1997) Isotopic evidence for palaeowaters in the British Isles. *Applied Geochemistry* **12**, 813–829.

Darling, W.G., Bath, A.H. & Talbot, J.C. (2003) The O & H stable isotopic composition of fresh waters in the British Isles. 2. Surface waters and groundwater. *Hydrology and Earth System Sciences* **7**, 183–195.

Davis, S.N. & De Wiest, R.J.M. (1966) *Hydrogeology.* John Wiley & Sons, New York.

Davison, R.M., Prabnarong, P., Whittaker, J.J. & Lerner, D.N. (2002) A probabilistic management system to optimize the use of urban groundwater. In: *Sustainable Groundwater Development* (eds K.M. Hiscock, M.O. Rivett & R.M. Davison). Geological Society, London, Special Publications **193**, 265–276.

De Breuck, W. (ed.) (1991) *Hydrogeology of Salt Water Intrusion. International Contributions to Hydrogeology,* vol. 11. Verlag Heinz Heise, Hannover.

DEFRA (2002) *Guidelines for Farmers in NVZs – England.* http://www.defra.gov.uk/environment/water/quality/nitrate/nvz.htm. Department for Environment, Food & Rural Affairs, London.

Dennehy, K.F., Litke, D.W. & McMahon, P.B. (2002) The High Plains Aquifer, USA: groundwater development and sustainability. In: *Sustainable Groundwater Development* (eds K.M. Hiscock, M.O. Rivett & R.M. Davison). Geological Society, London, Special Publications **193**, 99–119.

Dennis, F., Andrews, J.N., Parker, A., Poole, J. & Wolf, M. (1997) Isotopic and noble gas study of Chalk groundwater in the London Basin, England. *Applied Geochemistry* **12**, 763–773.

Dent, B.B. (2002) *The Hydrogeological Context of Cemetery Operations and Planning in Australia.* PhD Thesis, University of Technology, Sydney.

Department of the Environment (1995) *Landfill Design, Construction and Operational Practice.* Waste Management Paper 26B. HMSO, London.

Devitt, D.A., Evans, R.B., Jury, W.A., Starks, T.R., Eklund, B., Gnolson, A. & Van Ee, J.J. (1987) *Soil Gas Sensing for Detection and Mapping of Volatile Organics.* National Water Well Association, Ohio.

Dhar, R.K., Biswas, B.K., Samanta, G., Mandal, B.K., Chakraborti, D., Roy, S., Jafar, A., Islam, A., Ara, G., Kabir, S., Khan, A.W., Ahmed, S.A. & Hadi, S.A. (1997) Groundwater arsenic calamity in Bangladesh. *Current Science* **73**, 48–59.

Dissanayake, C.B. (1991) The fluoride problem in the groundwater of Sri Lanka – environmental management and health. *International Journal of Environmental Studies* **38**, 137–156.

Dissanayake, C.B., Senaratne, A. & Weerasooriya, V.R. (1982) Geochemistry of well water and cardiovascular diseases in Sri Lanka. *International Journal of Environmental Studies* **19**, 195–203.

Domenico, P.A. & Schwartz, F.W. (1998) *Physical and Chemical Hydrogeology,* 2nd edn. John Wiley & Sons, New York.

Doussan, C., Poitevin, G., Ledoux, E. & Detay, M. (1997) River bank filtration: modelling of the changes in water chemistry with emphasis on nitrogen species. *Journal of Contaminant Hydrology* **25**, 129–156.

Downing, R.A. (1993) Groundwater resources, their development and management in the UK: an historical perspective. *Quarterly Journal of Engineering Geology* **26**, 335–358.

Downing, R.A., Price, M. & Jones, G.P. (1993) The making of an aquifer. In: *The Hydrogeology of the Chalk of North-West Europe* (eds R.A. Downing, M. Price & G.P. Jones). Clarendon Press, Oxford, 1–13.

Downing, R.A., Smith, D.B., Pearson, F.J., Monkhouse, R.A. & Otlet, R.L. (1977) The age of groundwater in the Lincolnshire Limestone, England and its relevance to the flow mechanism. *Journal of Hydrology* **33**, 201–216.

DPHE (1999) *Groundwater Studies for Arsenic Contamination in Bangladesh. Final report, rapid investigation phase.* Department of Public Health Engineering, Government of Bangladesh. British Geological Survey & MottMacDonald.

DPHE (2000) *Groundwater Studies for Arsenic Contamination in Bangladesh. Supplemental data to final report, rapid investigation phase.* http://www.bgs.ac.uk/arsenic/Bangladesh/home.htm. Department of Public Health Engineering, Government of Bangladesh. British Geological Survey.

Drever, J.I. (1988) *The Geochemistry of Natural Waters,* 2nd edn. Prentice-Hall, Englewood Cliffs, New Jersey.

Driscoll, F.G. (1986) *Groundwater and Wells,* 2nd edn. Johnson Filtration Systems, St Paul, Minnesota.

Durand, P., Neal, M. & Neal, C. (1993) Variations in stable oxygen isotope and solute concentrations in small sub-mediterranean montane streams. *Journal of Hydrology* **144**, 283–290.

Durov, S.A. (1948) Natural waters and graphic representation of their composition. *Dokl. Akad. Nauk SSSR* **59**, 87–90.

Eastwood, J.C. & Stanfield, P.J. (2001) Key success factors in an ASR scheme. *Quarterly Journal of Engineering Geology and Hydrogeology* **34**, 399–409.

Edmunds, M. (1991) Groundwater recharge in the West African Sahel. *NERC News* **April**, 8–10.

Edmunds, W.M. (2001) Palaeowaters in European coastal aquifers – the goals and main conclusions of the PALAEAUX project. In: *Palaeowaters in Coastal Europe: evolution of groundwater since the late Pleistocene* (eds W.M. Edmunds & C.J. Milne). Geological Society, London, Special Publications **189**, 1–16.

Edmunds, W.M. & Gaye, C.B. (1994) Estimating the spatial variability of groundwater recharge in the Sahel using chloride. *Journal of Hydrology* **156**, 47–59.

Edmunds, W.M. & Walton, N.G.R. (1983) The Lincolnshire Limestone – hydrogeochemical evolution over a ten-year period. *Journal of Hydrology* **61**, 201–211.

Edmunds, W.M., Andrews, J.N., Burgess, W.G., Kay, R.L.F. & Lee, D.J. (1984) The evolution of saline and thermal groundwaters in the Carnmenellis granite. *Mineralogical Magazine* **48**, 407–424.

Edmunds, W.M., Kay, R.L.F. & McCartney, R.A. (1985) Origin of saline groundwaters in the Carnmenellis Granite (Cornwall, England): natural processes and reaction during hot dry rock reservoir circulation. *Chemical Geology* **49**, 287–301.

Elliot, T., Andrews, J.N. & Edmunds, W.M. (1999) Hydrochemical trends, palaeorecharge and groundwater ages in the fissured Chalk aquifer of the London and Berkshire Basins, UK. *Applied Geochemistry* **14**, 333–363.

Environment Agency (1998) *Policy and Practice for the Protection of Groundwater*. Environment Agency, Bristol.

European Environment Agency (2003) *Europe's Water: an indicator-based assessment. Summary*. European Environment Agency, Copenhagen.

Ferris, J.G., Knowles, D.B., Browne, R.H. & Stallman, R.W. (1962) Theory of aquifer tests. *United States Geological Survey Water Supply Paper* 1536-E.

Fetter, C.W. (1999) *Contaminant hydrogeology*, 2nd edn. Prentice-Hall, New Jersey.

Ford, D.C. & Williams, P.W. (1989) *Karst Geomorphology and Hydrology*. Unwin Hyman, London.

Ford, M. & Tellam, J.H. (1994) Source, type and extent of inorganic contamination within the Birmingham urban aquifer system, UK. *Journal of Hydrology* **156**, 101–135.

Foster S.S.D. (1985) Groundwater protection in developing countries. In: *Theoretical Background, Hydrogeology and Practice of Groundwater Protection Zones* (eds G. Matthess, S.S.D. Foster & A.C. Skinner). Verlag Heinz Heise, Hanover, 167–200.

Foster, S., Morris, B., Lawrence, A. & Chilton, J. (1999) Groundwater impacts and issues in developing cities – an introduction. In: *Groundwater in the Urban Environment: selected city profiles* (ed. J. Chilton). A.A. Balkema, Rotterdam, 3–16.

Foster, S., Garduno, H., Evans, R., Olson, D., Tian, Y., Zhang, W. & Han, Z. (2004) Quaternary Aquifer of the North China Plain – assessing and achieving groundwater resource sustainability. *Hydrogeology Journal* **12**, 81–93.

Freeze, R.A. & Cherry, J.A. (1979) *Groundwater*. Prentice-Hall, Englewood Cliffs, New Jersey.

Freeze, R.A. & Witherspoon, P.A. (1967) Theoretical analysis of regional groundwater flow. 2. Effect of water-table configuration and subsurface permeability variation. *Water Resources Research* **3**, 623–634.

Fritts, H.C. (1976) *Tree Rings and Climate*. Academic Press, London.

Fuge, R., Laidlaw, I.M.S., Perkins, W.T. & Rogers, K.P. (1991) The influence of acid mine and spoil drainage on water quality in the mid-Wales area. *Environmental Geochemistry and Health* **13**, 70–75.

Fukada, T., Hiscock, K.M. & Dennis, P.F. (2004) A dual-isotope approach to the nitrogen hydrochemistry of an urban aquifer. *Applied Geochemistry* **19**, 709–719.

Fulford, J.M., Thibodeaux, K.G. & Kaehrle, W.R. (1993) *Comparison of Current Meters Used for Stream Gaging*. http://water.usgs.gov/osw/pubs/CompCM.pdf.

Gale, J.E. (1982) *Assessing Permeability Characteristics of Fractured Rock*. Geological Society of America, Boulder, Colorado, Special Paper **189**, 163–181.

Gandy, C.J. & Younger, P.L. (2003) Effect of a clay cap on oxidation of pyrite within mine spoil. *Quarterly Journal of Engineering Geology & Hydrogeology* **36**, 207–215.

Gardner, C.M.K., Bell, J.P., Cooper, J.D., Darling, W.G. & Reeve, C.E. (1991) Groundwater recharge and water movement in the unsaturated zone. In: *Applied Groundwater Hydrology: a British perspective* (eds R.A. Downing & W.B. Wilkinson). Clarendon Press, Oxford, 54–76.

Garrels, R.M. & Christ, C.L. (1965) *Solutions, Minerals and Equilibria*. Harper & Row, New York.

Garrels, R.M., Mackenzie, F.T. & Hunt, C. (1975) *Chemical Cycles and the Global Environment: assessing human influences*. Kaufman, Los Altos, California.

Gauntlett, R.B. & Craft, D.G. (1979) *Biological Removal of Nitrate from River Water*. Report TR 98. Water Research Centre, Medmenham, Buckinghamshire.

George, M.A. (1998) *High Precision Stable Isotope Imaging of Groundwater Flow Dynamics in the Chalk Aquifer Systems of Cambridgeshire and Norfolk.* PhD Thesis, University of East Anglia, Norwich, 164–165.

Gerstl, Z. (1990) Estimation of organic chemical sorption by soils. *Journal of Contaminant Hydrology* **6**, 357–375.

Gilvear, D.J. & McInnes, R.J. (1994) Wetland hydrological vulnerability and the use of classification procedures: a Scottish case study. *Journal of Environmental Management* **42**, 403–414.

Girvin, D.C. & Scott, A.J. (1997) Polychlorinated biphenyl sorption by soils: measurement of soil-water partition coefficients at equilibrium. *Chemosphere* **35**, 2007–2025.

Goldich, S.S. (1938) A study in rock-weathering. *Journal of Geology* **46**, 17–58.

Gomme, J., Shurvell, S., Hennings, S.M. & Clark, L. (1992) Hydrology of pesticides in a Chalk catchment: groundwaters. *Journal of the Institution of Water and Environmental Management* **6**, 172–178.

Goode, D. (1977) Peatlands. In: *A Nature Conservation Review*, vol. 1 (ed. D.A. Ratcliffe). Cambridge University Press, Cambridge, pp. 249–287.

Goss, M.J., Barry, D.A.J., Rudolph, D.L. (1998) Contamination in Ontario farmstead wells and its association with agriculture. I. Results from drinking water wells. *Journal of Contaminant Hydrology* **32**, 63–90.

Gosselin, D.C., Headrick, J., Tremblay, R., Chen X. & Summerside, S. (1997) Domestic well water quality in rural Nebraska: focus on nitrate-nitrogen, pesticides, and coliform bacteria. *Ground Water Monitoring and Remediation* **17**, 77–87.

Green, A.R., Feast, N.A., Hiscock, K.M. & Dennis, P.F. (1998) Identification of the source and fate of nitrate contamination of the Jersey bedrock aquifer using stable nitrogen isotopes. In: *Groundwater Pollution, Aquifer Recharge and Vulnerability* (ed. N.S. Robins). Geological Society, London, Special Publications **130**, 23–35.

Griffin, A.R. & Mather, J.D. (1998) Landfill disposal of urban wastes in developing countries: balancing environmental protection and cost. In: *Geohazards in Engineering Geology* (eds J.G. Maund & M. Eddleston). Geological Society, London, Engineering Geology Special Publications **15**, 339–348.

Griffiths, D.H. & King, R.F. (1981) *Applied Geophysics for Geologists and Engineers: the elements of geophysical prospecting*, 2nd edn. Pergamon Press, Oxford.

Grindley, J. (1967) The estimation of soil moisture deficits. *Meteorological Magazine* **96**(1137), 97–108.

Grindley, J. (1969) The calculation of actual evaporation and soil moisture deficits over specified catchment areas. *Hydrological Memoir 38*. Meteorological Office, Bracknell.

Gringorten, I.I. (1963) A plotting rule for extreme probability paper. *Journal of Geophysical Research* **68**, 813–814.

Grischek, T., Hiscock, K.M., Metschies, T., Dennis, P.F. & Nestler, W. (1998) Factors affecting denitrification during infiltration of river water into a sand aquifer in Saxony, eastern Germany. *Water Research* **32**, 450–460.

Grischek, T., Schoenheinz, D., Worch, E. & Hiscock, K. (2002) Bank filtration in Europe – an overview of aquifer conditions and hydraulic controls. In: *Management of Aquifer Recharge for Sustainability* (ed. P.J. Dillon). A.A. Balkema Publishers, Lisse.

Groombridge, B. & Jenkins, M.D. (2000) *Global Biodiversity: Earth's living resources in the 21st century.* UNEP – World Conservation Monitoring Centre, World Conservation Press, Cambridge.

Guenther, W.B. (1975) *Chemical Equilibrium. A practical introduction for the physical and life sciences.* Plenum Press, New York.

Habermehl, M.A. (1980) The Great Artesian Basin, Australia. *BMR Journal of Australian Geology and Geophysics* **5**, 9–38.

Habermehl, M.A. & Lau, J.E. (1997) *1 : 2 500 000 Map of the Hydrogeology of the Great Artesian Basin, Australia.* Australian Geological Survey Organization, Canberra, ACT.

Haines, T.S. & Lloyd, J.W. (1985) Controls on silica in groundwater environments in the United Kingdom. *Journal of Hydrology* **81**, 277–295.

Halley, E. (1691) On the circulation of the vapors of the sea and the origin of springs. *Philosophical Transactions of the Royal Society* **17**(192), 468–473.

Hanshaw, B.B. & Back, W. (1979) Major geochemical processes in the evolution of carbonate-aquifer systems. *Journal of Hydrology* **43**, 287–312.

Hantush, M.S. (1956) Analysis of data from pumping tests in leaky aquifers. *Transactions of the American Geophysical Union* **37**, 702–714.

Harding, M. (1993) Redgrave and Lopham Fens, East Anglia, England: a case study of change in flora and fauna due to groundwater abstraction. *Biological Conservation* **66**, 35–45.

Haslam, S., Sinker, C. & Wolseley, P. (1975) British water plants. *Field Studies* **4**, 243–351.

Hassett, J.J., Banwart, W.L. & Griffen, R.A. (1983) Correlation of compound properties with sorption characteristics of nonpolar compounds by soil and sediments: concepts and limitations. In: *Environmental and Solid Wastes Characterization, Treatment and Disposal* (eds C.W. Francis & S.I. Auerbach). Butterworth Publishers, Newton, Massachusetts.

Headworth, H.G. & Fox, G.B. (1986) The South Downs Chalk aquifer: its development and management. *Journal of the Institution of Water Engineers* **40**, 345–361.

Heaton, T.H.E. (1986) Isotopic studies of nitrogen pollution in the hydrosphere and atmosphere: a review. *Chemical Geology* **59**, 87–102.

Heberer, T. (2002) Tracking persistent pharmaceutical residues from municipal sewage to drinking water. *Journal of Hydrology* **266**, 175–189.

Hecky, R.E. & Kilham, P. (1988) Nutrient limitation of phytoplankton in freshwater and marine environments: a review of recent evidence on the effects of enrichment. *Limnology and Oceanography* **33**, 796–822.

Hem, J.D. (1959) Study and interpretation of the chemical characteristics of natural water. *United States Geological Survey Water Supply Paper* 1473.

Hem, J.D. (1985) Study and interpretation of the chemical characteristics of natural water, 3rd edn. *United States Geological Survey Water Supply Paper* 2254.

Herbert, R. & Kitching, R. (1981) Determination of aquifer parameters from large-diameter dug well pumping tests. *Ground Water* **19**, 593–599.

Hess, K.M., Davis, J.A., Kent, D.B. & Coston, J.A. (2002) Multispecies reactive tracer test in an aquifer with spatially variable chemical conditions, Cape Cod, Massachusetts: dispersive transport of bromide and nickel. *Water Resources Research* **38**, 10.1029/2001WR000945.

Hillel, D. & Baker, R.S. (1988) A descriptive theory of fingering during infiltration into layered soils. *Soil Science* **146**, 51–56.

Hiscock, K.M. (1982) *Hydraulic Properties of the Chalk at Ludgershall Sewage Treatment Works.* BSc Thesis, University of East Anglia, Norwich.

Hiscock, K.M. (1993) The influence of pre-Devensian glacial deposits on the hydrogeochemistry of the Chalk aquifer system of north Norfolk UK. *Journal of Hydrology* **144**, 335–369.

Hiscock, K.M. & Grischek, T. (2002) Attenuation of groundwater pollution by bank filtration. *Journal of Hydrology* **266**, 139–144.

Hiscock, K.M. & Paci, A. (2000) Groundwater resources in the Quaternary deposits and Lower Palaeozoic bedrock of the Rheidol catchment, west Wales. In: *Groundwater in the Celtic Regions: studies in hard rock and Quaternary hydrogeology* (eds N.S. Robins & B.D.R. Misstear). Geological Society, London, Special Publications **182**, 141–155.

Hiscock, K.M., Lloyd, J.W. & Lerner, D.N. (1991) Review of natural and artificial denitrification of groundwater. *Water Research* **25**, 1099–1111.

Hiscock, K.M., Dennis, P.F., Saynor, P.R. & Thomas, M.O. (1996) Hydrochemical and stable isotope evidence for the extent and nature of the effective Chalk aquifer of north Norfolk, UK. *Journal of Hydrology* **180**, 79–107.

Hiscock, K.M., Lister, D.H., Boar, R.R. & Green, F.M.L. (2001) An integrated assessment of long-term changes in the hydrology of three lowland rivers in eastern England. *Journal of Environmental Management* **61**, 195–214.

Hiscock, K.M., Rivett, M.O. & Davison, R.M. (2002) Sustainable groundwater development. In: *Sustainable Groundwater Development* (eds K.M. Hiscock, M.O. Rivett & R.M. Davison). Geological Society, London, Special Publications **193**, 1–14.

Hobson, G. (1993) Practical use of tracers in hydrogeology. *Geoscientist* **4**, 26–27.

Hocking, G., Wells, S.L. & Ospina, R.I. (2000) Deep reactive barriers for remediation of VOCs and heavy metals. In: *Chemical Oxidation and Reactive Barriers: remediation of chlorinated and recalcitrant compounds* (eds G.B. Wickramanayake, A.R. Gavaskar & A.S.C. Chen). Battelle Press, Columbus, Ohio, 307–314.

Höhener, P., Werner, D., Balsiger, C. & Pasteris, G. (2003) Worldwide occurrence and fate of chlorofluorocarbons in groundwater. *Critical Reviews in Environmental Science and Technology* **33**, 1–29.

Holman, I.P. & Hiscock, K.M. (1998) Land drainage and saline intrusion in the coastal marshes of northeast Norfolk. *Quarterly Journal of Engineering Geology* **31**, 47–62.

Holman, I.P., Hiscock, K.M. & Chroston, P.N. (1999) Crag aquifer characteristics and water balance for the Thurne catchment, northeast Norfolk. *Quarterly Journal of Engineering Geology* **32**, 365–380.

Holme, R. (2003) Drinking water contamination in Walkerton, Ontario: positive resolutions from a tragic event. *Water Science and Technology* **47**, 1–6.

Howard, K.W.F. & Beck, P.J. (1993) Hydrogeochemical implications of groundwater contamination by road deicing chemicals. *Journal of Contaminant Hydrology* **12**, 245–268.

Howard, K.W.F. & Lloyd, J.W. (1983) Major ion characterization of coastal saline ground waters. *Ground Water* **21**, 429–437.

Hubbert, M.K. (1940) The theory of groundwater motion. *Journal of Geology* **48**, 785–944.

Hulme, M., Turnpenny, J. & Jenkins, G. (2002) *Climate Change Scenarios for the United Kingdom. The UKCIP02 Briefing Report.* Tyndall Centre for Climate Change, University of East Anglia, Norwich; Hadley Centre for Climate Prediction and Research, UK Meteorological Office.

Hutson, S.S., Barber, N.L., Kenny, J.F., Linsey, K.S., Lumia, D.S. & Maupin, M.A. (2004) Estimated use of water in the United States in 2000. *United States Geological Survey Circular* 1268.

Hvorslev, M.J. (1951) Time lag and soil permeability in ground water observations. *United States Army Corps of Engineers Waterways Experimentation Station Bulletin* 36.

IAEA/WMO (1998) *Global Network of Isotopes in Precipitation*. The GNIP Database. http://isohis.iaea.org/.

Ineson, J. & Downing, R.A. (1963) Changes in the chemistry of groundwaters of the Chalk passing beneath argillaceous strata. *Bulletin of the Geological Survey Great Britain* 20, 176–192.

Institute of Geological Sciences (1977) *1 : 625,000 Hydrogeological Map of England and Wales*. Natural Environment Research Council.

IPCC (1990) *Climate Change. The Intergovernmental Panel on Climate Change Scientific Assessment* (eds J.T. Houghton, G.J. Jenkins & J.J. Ephraums). Cambridge University Press, Cambridge.

IPCC (1998) The regional impacts of climate change: an assessment of vulnerability. In: *A Special Report of the IPCC WGII* (eds R.T. Watson, M.C. Zinyowera & R.H. Moss). Intergovernmental Panel on Climate Change. Cambridge University Press, Cambridge.

IPCC (2001) *Climate Change 2001. The Scientific Basis. A Report of Working Group I of the Intergovernmental Panel on Climate Change*. Cambridge University Press, Cambridge.

Iribar, V., Carrera, J., Custodio, E. & Medina, A. (1997) Inverse modelling of seawater intrusion in the Llobregat delta deep aquifer. *Journal of Hydrology* 198, 226–244.

Jackson, R.E. (ed.) (1980) *Aquifer Contamination and Protection*. UNESCO, Paris.

Jacob, C.E. (1940) On the flow of water in an elastic artesian aquifer. *Transactions of the American Geophysical Union* 22, 574–586.

Jacob, C.E. (1950) Flow of groundwater. In: *Engineering Hydraulics* (ed. H. Rouse). John Wiley, New York.

Javandel, I. & Tsang, C.-F. (1986) Capture-zone type curves: a tool for aquifer cleanup. *Ground Water* 24, 616–625.

Jenkins, C.T. (1968) Computation of rate and volume of stream depletion by wells. In: *Techniques of Water-Resources Investigations of the United States Geological Survey. Hydrologic analysis and interpretation*, book 4. United States Department of the Interior, Washington, chap. D1, 17 pp.

Jenkins, C.T. (1970) Techniques for computing rate and volume of stream depletion by wells. *Ground Water* 6, 37–46.

Johnson, T.L., Scherer, M.M. & Tratnyek (1996) Kinetics of halogenated organic compound degradation by iron metal. *Environmental Science & Technology* 30, 2634–2640.

Jones, J.A.A. (1997) *Global Hydrology: processes, resources and environmental management*. Addison Wesley Longman, Harlow, Essex.

Jones, K. & Moon, G. (1987) *Health, Disease and Society: a critical medical geography*. Routledge and Kegan Paul, London, pp. 134–140.

Karami, G.H. & Younger, P.L. (2002) Analysing step-drawdown tests in heterogeneous aquifers. *Quarterly Journal of Engineering Geology and Hydrogeology* 35, 295–303.

Karickhoff, S.W., Brown, D.S. & Scott, T.A. (1979) Sorption of hydrophobic pollutants on natural sediments. *Water Research* 13, 241–248.

Karst Working Group (2000) *The Karst of Ireland*. Karst Working Group, Geological Survey of Ireland, Dublin.

Kearey, P. & Brooks, M. (1991) *An Introduction to Geophysical Exploration*, 2nd edn. Blackwell Science, Oxford.

Kemper, K.E. (2004) Groundwater – from development to management. *Hydrogeology Journal* 12, 3–5.

Kenaga, E.E. & Goring, C.A.I. (1980) Relationship between water solubility, soil sorption, octanol-water partitioning, and concentration of chemicals in biota. In: *Aquatic Toxicology ASTM Special Technical Publication 707* (eds J.G. Eaton, P.R. Parrish & A.C. Hendricks). American Society for Testing and Materials, Philadelphia, PA, 78–115.

Kendall, C. (1998) Tracing nitrogen sources and cycling in catchments. In: *Isotope-Tracers in Catchment Hydrology* (eds C. Kendall and J.J. McDonnel). Elsevier, Amsterdam, 534–569.

Khan, S. & Rushton, K.R. (1996) Reappraisal of flow to tile drains I. Steady state response. *Journal of Hydrology* 183, 351–366.

Kirk, S. & Herbert, A.W. (2002) Assessing the impact of groundwater abstractions on river flows. In: *Sustainable Groundwater Development* (eds K.M. Hiscock, M.O. Rivett & R.M. Davison). Geological Society, London, Special Publications 193, 211–233.

Kirkaldy, J.F. (1954) *General Principles of Geology*. Hutchinson's Scientific and Technical Publications, London, 284–286.

Kitching, R. & Shearer, T.R. (1982) Construction and operation of a large undisturbed lysimeter to measure recharge to the chalk aquifer, England. *Journal of Hydrology* 58, 267–277.

Kitching, R., Edmunds, W.M., Shearer, T.R., Walton, N.G.R. & Jacovides, J. (1980) Assessment of recharge to aquifers. *Hydrological Sciences Bulletin* 25, 217–235.

Klein, H. & Hull, J.E. (1978) Biscayne Aquifer, Southeast Florida. *United States Geological Survey Water-Resources Investigation 78-107*.

Knight, M.J. & Dent, B.B. (1998) Sustainability of waste and groundwater management systems. In: *Proceedings of the Congress of the International Association of Hydrogeologists on Groundwater: Sustainable Solutions, Melbourne.

Kobayashi, J. (1957) On geographical relations between the chemical nature of river water and death rate from apoplexy. *Berichte des Ohara Instituts für Landwirtschaftliche Biologie, Okayama University* **11**, 12–21.

Kolpin, D.W., Kalkhoff, S.J., Goolsby, D.A., Sneck-Fahrer, D.A. & Thurman, E.M. (1997) Occurrence of selected herbicides and herbicide degradation products in Iowa's ground water, 1995. *Ground Water* **35**, 679–688.

Kolpin, D.W., Barbash, J.E. & Gilliom, R.J. (2000) Pesticides in ground water of the United States, 1992–1996. *Ground Water* **38**, 858–863.

Konikow, L.F. & Bredehoeft, J.D. (1978) Computer model of two-dimensional solute transport and dispersion in groundwater. *Techniques of Water-Resources Investigations of the United States Geological Survey*, book 7. Scientific Software Group, Washington, DC, chap. C2.

Korom, S.F. (1992) Natural denitrification in the saturated zone: a review. *Water Resources Research* **28**, 1657–1668.

Krauskopf, K.B. & Bird, D.K. (1995) *Introduction to Geochemistry*, 3rd edn. McGraw-Hill, New York.

Krothe, N.C. & Bergeron, M.P. (1981) Hydrochemical facies in a Tertiary basin in the Milligan Canyon area, Southwest Montana. *Ground Water* **19**, 392–399.

Krumbein, W.C. & Monk, G.D. (1943) Permeability as a function of the size parameters of unconsolidated sand. *Transactions of the American Institution of Mining and Metallurgy Engineers* **151**, 153–163.

Kruseman, G.P. & de Ridder, N.A. (1990) *Analysis and Evaluation of Pumping Test Data*, 2nd edn. Pudoc Scientific Publishers, Wageningen, The Netherlands.

Ku, H.F.H. (1980) Ground-water contamination by metal-plating wastes, Long Island, New York, USA. In: *Aquifer Contamination and Protection* (ed. R.E. Jackson). UNESCO, Paris, 310–317.

Kung, K-J.S. (1990a) Preferential flow in a sandy vadose zone: 1. Field observation. *Geoderma* **46**, 51–58.

Kung, K-J.S. (1990b) Preferential flow in a sandy vadose zone: 2. Mechanism and implications. *Geoderma* **46**, 59–71.

Lake, I.R., Lovett, A.A., Hiscock, K.M., Betson, M., Foley, A., Sünnenberg, G., Evers, S. & Fletcher, S. (2003) Evaluating factors influencing groundwater vulnerability to nitrate pollution: developing the potential of GIS. *Journal of Environmental Management* **68**, 315–328.

Langmuir, D. (1971) The geochemistry of some carbonate groundwaters in Central Pennsylvania. *Geochimica et Cosmochimica Acta* **35**, 1023–1045.

Lawrence, A.R. & Foster, S.S.D. (1986) Denitrification in a limestone aquifer in relation to the security of low-nitrate groundwater supplies. *Journal of the Institution of Water Engineers and Scientists* **40**, 159–172.

Leeson, J., Edwards, A., Smith, J.W.N. & Potter, H.A.B. (2003) *Hydrogeological Risk Assessment for Landfills and the Derivation of Groundwater Control and Trigger Levels*. Environment Agency, Bristol.

Lerner, D.N. & Teutsch, G. (1995) Recommendations for level-determined sampling in wells. *Journal of Hydrology* **171**, 355–377.

Lerner, D.N., Issar, A.S. & Simmers, I. (1990) Groundwater recharge: a guide to understanding and estimating natural recharge. In: *International Contributions to Hydrogeology*, vol. 8. Verlag Heinz Heise, Hannover.

Lewis, D.C., Kriz, G.J. & Burgy, R.H. (1966) Tracer dilution sampling technique to determine hydraulic conductivity of fractured rock. *Water Resources Research* **2**, 533–542.

Livingstone, S., Franz, T. & Guiguer, N. (1995) Managing ground-water resources using wellhead protection programs. *Geoscience Canada* **22**, 121–128.

Lloyd, J.W. & Heathcote, J.A. (1985) *Natural Inorganic Hydrochemistry in Relation to Groundwater: an introduction*. Clarendon Press, Oxford.

Lloyd, J.W., Williams, G., Foster, S.S.D., Ashley, R.P. & Lawrence, A.R. (1991) Urban and industrial groundwater pollution. In: *Applied Groundwater Hydrology: A British Perspective* (eds R.A. Downing & W.B. Wilkinson). Clarendon Press, Oxford, 134–148.

Lloyd, J.W., Tellam, J.H., Rukin, N. & Lerner, D.N. (1993) Wetland vulnerability in East Anglia; a possible conceptual framework and generalised approach. *Journal of Environmental Management* **37**, 87–102.

Loáiciga, H.A. (2003) Climate change and ground water. *Annals of the Association of American Geographers* **93**, 30–41.

Loáiciga, H.A., Valdes, J.B., Vogel, R., Garvey, J. & Schwarz, H. (1996) Global warming and the hydrologic cycle. *Journal of Hydrology* **174**, 83–127.

Loáiciga, H.A., Maidment, D.R. & Valdes, J.B. (2000) Climate-change impacts in a regional karst aquifer, Texas, USA. *Journal of Hydrology* **227**, 173–194.

Love, A.J., Herczeg, A.L., Sampson, L., Cresswell, R.G. & Fifield, L.K. (2000) Sources of chloride and implications for ^{36}Cl dating of old groundwater, southwestern Great Artesian Basin, Australia. *Water Resources Research* **36**, 1561–1574.

Lucas, H.C. & Robinson, V.K. (1995) Modelling of rising groundwater levels in the Chalk aquifer of the London Basin. *Quarterly Journal of Engineering Geology*, **28**, S51–S62.

Lucas, J. (1877) *Hydrogeological Survey. Sheet 1 (South London)*. Stanford, London.

Luckey, R.R., Gutentag, E.D., Heimes, F.J. & Weeks, J.B. (1986) Digital simulation of ground-water flow in the

High Plains aquifer in parts of Colorado, Kansas, Nebraska, New Mexico, Oklahoma, South Dakota, Texas, and Wyoming. *United States Geological Survey Professional Paper* 1400-D.

McArthur, J.M., Ravenscroft, P., Safiulla, S. & Thirwall, M.F. (2001) Arsenic in groundwater: testing pollution mechanisms for sedimentary aquifers in Bangladesh. *Water Resources Research*, **37**, 109–117.

McDonald, M.G. & Harbaugh, A.W. (1988) A modular three-dimensional finite-difference ground-water flow model. In: *Techniques of Water-Resources Investigations of the United States Geological Survey*, book 6. Scientific Software Group, Washington, DC, chap. A1.

Mackay, D., Shiu, W.-Y. & Ma, K.-C. (1997) *Illustrated Handbook of Physical-Chemical Properties and Environmental Fate for Organic Chemicals*. Vol. I: *Monoaromatic hydrocarbons, chlorobenzenes, and PCBs*. Vol. II: *Polynuclear aromatic hydrocarbons, polychlorinated dioxins, dibenzofurans*. Vol. III: *Volatile organic chemicals*. Vol. IV: *Oxygen, nitrogen, and sulfur containing compounds*. Vol. V: *Pesticide chemicals*. Lewis Publishers, Boca Raton, FL.

Mackay, D.M. & Cherry, J.A. (1989) Groundwater contamination: pump-and-treat remediation. *Environmental Science & Technology* **23**, 630–636.

Mackay, D.M., Freyberg, D.L. & Roberts, P.V. (1986) A natural gradient experiment on solute transport in a sand aquifer 1. Approach and overview of plume movement. *Water Resources Research* **22**, 2017–2029.

McKenzie, H.S. & Elsaleh, B.O. (1994) The Libyan Great Man-Made River Project Paper 1. Project overview. *Proceedings of the Institution of Civil Engineers – Water Maritime and Environment* **106**, 103–122.

McPherson, B.F. & Halley, R. (1997) The South Florida environment – a region under stress. *United States Geological Survey Circular* 1134.

Mallin, M.A. & Cahoon, L.B. (2003) Industrialized animal production – a major source of nutrient and microbial pollution to aquatic ecosystems. *Population and Environment* **24**, 369–385.

Marsh, T.J. & Davies, P.A. (1983) The decline and partial recovery of groundwater levels below London. *Proceedings of the Institution of Civil Engineers, Part 1*, **74**, 263–276.

Marsily, G. (1986) *Quantitative Hydrogeology: groundwater hydrology for engineers*. Academic Press, San Diego, CA.

Mather, J. (1998) From William Smith to William Whitaker: the development of British hydrogeology in the nineteenth century. In: *Lyell: the Past is the Key to the Present* (eds D.J. Blundell & A.C. Scott). Geological Society, London, Special Publications **143**, 183–196.

Maurits la Riviére, J.W. (1989) Threats to the world's water. *Scientific American* **September**, 48–55.

Mawdsley, J., Petts, G. & Walker, S. (1994) *Assessment of Drought Severity*. British Hydrological Society Occasional Paper No. 3. Institute of Hydrology, Wallingford.

Meinzer, O.E. & Hard, H.H. (1925) The artesian water supply of the Dakota sandstone in North Dakota, with special reference to the Edgeley Quadrangle. *United States Geological Survey Water Supply Paper* **520-E**, 73–95.

Meinzer, O.E. (1923) The occurrence of groundwater in the United States with a discussion of principles. *United States Geological Survey Water Supply Paper* 489.

Meinzer, O.E. (1928) Compressibility and elasticity of artesian aquifers. *Economic Geology* **23**, 263–291.

Meju, M.A., Denton, P. & Fenning, P. (2002) Surface NMR sounding and inversion to detect groundwater in key aquifers in England: comparisons with VES-TEM methods. *Journal of Applied Geophysics* **50**, 95–111.

Meju, M.A., Fontes, S.L., Oliveira, M.F.B., Lima, J.P.R., Ulugergerli, E.U. & Carrasquilla, A.A. (1999) Regional aquifer mapping using combined VES-TEM-AMT/EMAP methods in the semiarid eastern margin of Parnaiba Basin, Brazil. *Geophysics* **64**, 337–356.

Miles, R. (1993) Maintaining groundwater supplies during drought conditions in the Brighton area. *Journal of the Institution of Water and Environmental Management* **7**, 382–386.

Mitsch, W.J., Mitsch, R.H. & Turner, R.E. (1994) Wetlands of the Old and New Worlds: ecology and management. In: *Global Wetlands: Old World and New* (ed. W.J. Mitsch). Elsevier Science BV, Amsterdam, 3–56.

Monteith, J.L. (1965) Evaporation and the environment. *Proceedings of the Symposium of the Society for Experimental Biology* **19**, 205–234.

Monteith, J.L. (1985) Evaporation from land surfaces: progress in analysis and prediction since 1948. In: *Advances in Evaporation*. American Society of Agricultural Engineers, St Joseph, Michigan, 4–12.

Mook, W.G. (1980) Carbon-14 in hydrogeological studies. In: *Handbook of Environmental Isotope Geochemistry*, vol. 1, *The terrestrial environment, A* (eds P. Fritz & J. Ch. Fontes). Elsevier Scientific Publishing Company, Amsterdam, pp. 49–74.

Morgan-Jones, M. & Eggboro, M.D. (1981) The hydrogeochemistry of the Jurassic limestones in Gloucestershire, England. *Quarterly Journal of Engineering Geology* **14**, 25–39.

Morris, B.L. & Foster, S.S.D. (2000) *Cryptosporidium* contamination hazard assessment and risk management for British groundwater sources. *Water Science and Technology* **41**, 67–77.

Mühlherr, I.H., Hiscock, K.M., Dennis, P.F. & Feast, N.A. (1998) Changes in groundwater chemistry due to rising groundwater levels in the London Basin between 1963 and 1994. In: *Groundwater Pollution, Aquifer Recharge and Vulnerability* (ed. N.S. Robins). Geological Society, London, Special Publications **130**, 47–62.

Municipality of Aalborg (2001) *Sustainable Land Use in Ground Water Catchment Areas*. Technical Final Report and Layman's Report. EU LIFE Project Number LIFE97 ENV/DK/000347. Municipality of Aalborg, Denmark.

MWR (1992) *Water Resources Assessment for China*. Ministry of Water Resources, Water & Power Press, Beijing, China.

Nace, R.L. (ed.) (1971) Scientific framework of world water balance. *UNESCO Technical Papers in Hydrology* **7**.

National Academy of Sciences (1981) *The Health Effects of Nitrate, Nitrite and N-nitroso Compounds*. Part 1 of a two-part study by the Committee on Nitrite and Alternative Curing Agents in Food. National Academy Press, Washington, DC.

National Academy of Sciences (1994) *Alternatives for Ground Water Cleanup*. Report of the National Academy of Sciences Committee on Ground Water Cleanup Alternatives. National Academy Press, Washington, DC.

National Academy of Sciences (2000) *Natural Attenuation for Groundwater Remediation*. National Academy Press, Washington, DC.

National Rivers Authority (1992) *Policy and Practice for the Protection of Groundwater*. National Rivers Authority, Bristol.

Navid, D. (1989) The international law of migratory species: the Ramsar Convention. *Natural Resources Journal* **29**, 1001–1016.

NEGTAP (2001) *Transboundary Air Pollution: acidification, eutrophication and ground-level ozone in the UK*. Report prepared by the National Expert Group on Transboundary Air Pollution. Centre for Ecology and Hydrology, Edinburgh.

Nesbitt, H.W. & Young, G.M. (1984) Prediction of some weathering trends of plutonic and volcanic rocks based on thermodynamic and kinetic considerations. *Geochimica et Cosmochimica Acta* **48**, 1523–1534.

Neuman, S.P. (1975) Analysis of pumping test data from anisotropic confined aquifers considering delayed gravity response. *Water Resources Research* **11**, 329–342.

Nicholson, R.V., Cherry, J.A. & Reardon, E.J. (1983) Migration of contaminants in groundwater at a landfill: a case study. 6. Hydrogeochemistry. *Journal of Hydrology* **63**, 131–176.

Nickson, R.T., McArthur, J.M., Ravenscroft, P., Burgess, W.G. & Ahmed, K.M. (2000) Mechanism of arsenic release to groundwater, Bangladesh and West Bengal. *Applied Geochemistry*, **15**, 403–413.

Nomura, A. (1996) Stomach cancer. In: *Cancer Epidemiology and Prevention*, 2nd edn (eds D. Scottenfeld & J.F. Fraumeni). Oxford University Press, New York, 707–724.

Nordstrom, D.K. & Southam, G. (1997) Geomicrobiology of sulphide mineral oxidation. Geomicrobiology: interactions between microbes and minerals. *Reviews in Mineralogy* **35**, 361–390.

O'Hannesin, S.F. & Gillham, R.W. (1998) Long-term performance of an in situ 'iron wall' for remediation of VOCs. *Ground Water* **36**, 164–170.

O'Shea, M.J. & Sage, R. (1999) Aquifer recharge: an operational drought-management strategy in North London. *Journal of the Institution of Water and Environmental Management* **13**, 400–405.

Ogata, A. & Banks, R.B. (1961) A solution of the differential equation of longitudinal dispersion in porous media. *United States Geological Survey Professional Paper* 411-A.

Ortega-Guerrero, A., Cherry, J.A. & Rudolph, D.L. (1993) Large-scale aquitard consolidation near Mexico City. *Ground Water* **31**, 708–718.

Owen, M. (1991) Relationship between groundwater abstraction and river flow. *Journal of the Institution of Water and Environmental Management* **5**, 697–702.

Palmer, W.C. (1965) *Meteorological Drought*. Research Paper No. 45. Department of Commerce, Washington, DC.

Park, E. & Zhan, H.B. (2003) Hydraulics of horizontal wells in fractured shallow aquifer systems. *Journal of Hydrology* **281**, 147–158.

Parker, J.M., Young., C.P & Chilton, P.J. (1991) Rural and agricultural pollution of groundwater. In: *Applied Groundwater Hydrology: a British perspective* (eds R.A. Downing & W.B. Wilkinson). Clarendon Press, Oxford, 149–163.

Pedley, S. & Howard, G. (1997) The public health implications of microbiological contamination of groundwater. *Quarterly Journal of Engineering Geology* **30**, 179–188.

Penman, H.L. (1948) Natural evaporation from open water, bare soil and grass. *Proceedings of the Royal Society of London, Series A* **193**, 120–145.

Penman, H.L. (1949) The dependence of transpiration on weather and soil conditions. *Journal of Soil Science* **1**, 74–89.

Petts, J., Cairney, T. & Smith, M. (1997) *Risk-based Contaminated Land Investigation and Assessment*. John Wiley, Chichester.

Philip, J.R. (1969) Theory of infiltration. In: *Advances in Hydroscience*, vol. 5 (ed V.T. Chow). Academic Press, New York, 215–296.

Pim, R.H. & Binsariti, A. (1994) The Libyan Great Man-Made River Project Paper 2. The water resource. *Proceedings of the Institution of Civil Engineers – Water Maritime and Environment* **106**, 123–145.

Piper, A.M. (1944) A graphic procedure in the geochemical interpretation of water analyses. *Transactions of the American Geophysical Union* **25**, 914–923.

Plummer, C.R., Nelson, J.D. & Zumwalt, G.S. (1997) Horizontal and vertical well comparison for in situ air sparging. *Ground Water Monitoring and Remediation* **17**, 91–96.

Plummer, L.N. & Busenberg, E. (1982) The solubilities of calcite, aragonite and vaterite in CO_2-H_2O solutions between 0 and 90°C, and an evaluation of the aqueous model for the system $CaCO_3$-CO_2-H_2O. *Geochimica et Cosmochimica Acta* **46**, 1011–1040.

Plummer, L.N. & Busenberg, E. (2000) Chlorofluorocarbons. In: *Environmental Tracers in Subsurface Hydrology* (eds P. Cook & L. Herczeg). Kluwer Academic Publishers, Norwell, Massachusetts, 441–478.

Plummer, L.N., Parkhurst, D & Kosier, D.R. (1975) MIX2, a computer program for modeling chemical reactions in natural water. *United States Geological Survey Report* 61–75.

Plummer, L.N., Busenberg, E., Böhlke, J.K., Nelms, D.L., Michel, R.L. & Schlosser, P. (2001) Groundwater residence times in Shenandoah National Park, Blue Ridge Mountains, Virginia, USA: a multi-tracer approach. *Chemical Geology* **179**, 93–111.

Pocock, S.J., Shaper, A.G., Cook, D.G., Packham, R.F., Lacey, R.F., Powell, P. & Russell, P.F. (1980) British Regional Heart Study: geographic variations in cardiovascular mortality, and role of water quality. *British Medical Journal* **280**, 1243–1249.

Powell, K.L., Taylor, R.G., Cronin, A.A., Barrett, M.H., Pedley, S., Sellwood, J., Trowsdale, S. & Lerner, D.N. (2003) Microbial contamination of urban sandstone aquifers in the UK. *Water Research* **37**, 339–352.

Price, M. (2002) Who needs sustainability? In: *Sustainable Groundwater Development* (eds K.M. Hiscock, M.O. Rivett & R.M. Davison). Geological Society, London, Special Publications **193**, 75–81.

Price, M. & Williams, A. (1993) The influence of unlined boreholes on groundwater chemistry: a comparative study using pore-water extraction and packer testing. *Journal of the Institution of Water and Environmental Management* **7**, 651–659.

Price, M., Atkinson, T.C., Wheeler, D., Barker, J.A. & Monkhouse, R.A. (1989) Highway drainage to the Chalk aquifer: the movement of groundwater in the Chalk near Bricket Wood, Hertfordshire, and its possible pollution by drainage from the M25. *British Geological Survey Technical Report* WD/89/3.

Prickett, T.A., Naymik, T.G. & Lonnquist, C.G. (1981) A 'Random-Walk' solute transport model for selected groundwater quality evaluations. *Illinois State Water Survey Bulletin* 65.

Pugh, J.C. (1975) *Surveying for Field Scientists*. Methuen, London.

Pyne, R.D.G. (1995) *Groundwater Recharge and Wells: a guide to aquifer storage and recovery*. CRC Press, Boca Raton, Florida.

Rahman, A.A. & Ravenscroft, P. (eds) (2003) *Groundwater Resources and Development in Bangladesh: background to the arsenic crisis, agricultural potential and the environment*. University Press, Dhaka.

Rajasooriyar, L., Mathavan, V., Dharmagunawardhane, H.A. & Nandakumar, V. (2002) Groundwater quality in the Valigamam region of the Jaffna Peninsula, Sri Lanka. In: *Sustainable Groundwater Development* (eds K.M. Hiscock, M.O. Rivett & R.M. Davison). Geological Society, London, Special Publications **193**, 181–197.

Rajasooriyar, L.D. (2003) *A Study of the Hydrochemistry of the Uda Walawe Basin, Sri Lanka, and the Factors that Influence Groundwater Quality*. PhD Thesis, University of East Anglia, Norwich.

Rantz, S.E. et al. (1982) Measurement and computation of streamflow (2 vols). *United States Geological Survey Water Supply Paper* 2175.

Ravenscroft, P., McArthur, J.M. & Hoque, B.A. (2001) Geochemical and palaeohydrological controls on pollution of groundwater by arsenic. In: *Arsenic Exposure and Health Effects* (eds W.R. Chappell, C.O. Abernathy & R.L. Calderon). Elsevier Science, Amsterdam, 53–77.

Ravenscroft, P., Burgess, W.G., Ahmed, K.M., Burren, M. & Perrin, J. (2004) Arsenic in groundwater of the Bengal Basin, Bangladesh: distribution, field relations, and hydrogeological setting. *Hydrogeology Journal*, 10.1007/s10040-003-0314-0.

Richardson, J.P. & Nicklow, J.W. (2002) *In situ* permeable reactive barriers for groundwater contamination. *Soil and Sediment Contamination* **11**, 241–268.

Rivett, M.O. & Allen-King, R.M. (2003) A controlled field experiment on groundwater contamination by a multicomponent DNAPL: dissolved-plume retardation. *Journal of Contaminant Hydrology* **66**, 117–146.

Rivett, M.O., Lerner, D.N., Lloyd, J.W. & Clark, L. (1990) Organic contamination of the Birmingham aquifer, U.K. *Journal of Hydrology* **113**, 307–323.

Rivett, M.O., Feenstra, S. & Cherry, J.A. (2001) A controlled field experiment on groundwater contamination by multicomponent DNAPL: creation of the emplaced-source and overview of dissolved plume development. *Journal of Contaminant Hydrology* **49**, 111–149.

RIVM & RIZA (1991) *Sustainable Use of Groundwater: problems and threats in the European communities*. Report no. 600025001. National Institute of Public Health & Environmental Protection & Institute for Inland Water Management & Waste Water Treatment, Bilthoven, The Netherlands.

Roberts, P.V., Goltz, M.N. & Mackay, D.M. (1986) A natural gradient experiment on solute transport in a sand aquifer 3. Retardation estimates and mass balances for organic solutes. *Water Resources Research* **22**, 2047–2058.

Robertson, W.D., Cherry, J.A. & Sudicky, E.A. (1991) Ground-water contamination from two small septic systems on sand aquifers. *Ground Water* **29**, 82–92.

Robertson, W.D., Russell, B.M. & Cherry, J.A. (1996) Attenuation of nitrate in aquitard sediments of southern Ontario. *Journal of Hydrology* **180**, 267–281.

Robins, N.S. (1990a) *Hydrogeology of Scotland*. HMSO for the British Geological Survey, London.

Robins, N.S. (1990b) Groundwater chemistry of the main aquifers in Scotland. *British Geological Survey Report* 18/2.

Robins, N.S., Adams, B., Foster, S.S.D. & Palmer, R.C. (1994) Groundwater vulnerability mapping: the British perspective. *Hydrogèologie* **3**, 35–42.

Robinson, H.D. (1989) Development of methanogenic conditions within landfills. In: *Proceedings of the 2nd International Landfill Symposium, Sardinia*.

Rodvang, S.J. & Simpkins, W.W. (2001) Agricultural contaminants in Quaternary aquitards: a review of occurrence and fate in North America. *Hydrogeology Journal* **9**, 44–59.

Royal Society (1992) *Risk: analysis, perception and management*. Royal Society, London.

Rozanski, K. (1985) Deuterium and oxygen-18 in European groundwaters – links to atmospheric circulation in the past. *Chemical Geology (Isotope Geoscience Section)* **52**, 349–363.

Rozanski, K., Araguás-Araguás, L. & Gonfiantini, R. (1993) Isotopic patterns in modern global precipitation. *Geophysical Monograph of the American Geophysical Union* **78**, 1–36.

Rubin, J. (1966) Theory of rainfall uptake by soils initially drier than their field capacity and its applications. *Water Resources Research* **2**, 739–749.

Rushton, K.R. (1986) Groundwater models. In: *Groundwater: occurrence, development and protection* (ed. T.W. Brandon). Institution of Water Engineers and Scientists, London, pp. 189–228.

Rushton, K.R. (2003) *Groundwater Hydrology: conceptual and computational models*. John Wiley, Chichester.

Rushton, K.R. & Tomlinson, L.M. (1999) Total catchment conditions in relation to the Lincolnshire Limestone in South Lincolnshire. *Quarterly Journal of Engineering Geology* **32**, 233–246.

Rushton, K.R. & Ward, C. (1979) The estimation of groundwater recharge. *Journal of Hydrology* **41**, 345–361.

Saleh, A., Al-Ruwaih, F. & Shehata, M. (1999) Hydrogeochemical processes operating within the main aquifers of Kuwait. *Journal of Arid Environments* **42**, 195–209.

Sauty, J.-P. (1980) An analysis of hydrodynamic transfer in aquifers. *Water Resources Research* **16**, 145–158.

Schoelle, P.A. & Arthur, M.A. (1980) Carbon isotope fluctuations in Cretaceous pelagic limestones: potential stratigraphic and petroleum exploration tool. *American Association of Petroleum Geologists Bulletin* **64**, 67–87.

Schoeller, H. (1962) *Les Eaux Souterraines*. Maison et Cie, Paris.

Schoenheinz, D., Grischek, T., Worch, E., Bereznoy, V., Gutkin, I., Shebesta, A., Hiscock, K., Macheleidt, W. & Nestler, W. (2002) Groundwater pollution at a pulp and paper mill at Sjasstroj near Lake Ladoga, Russia. In: *Sustainable Groundwater Development* (eds K.M. Hiscock, M.O. Rivett & R.M. Davison). Geological Society, London, Special Publications **193**, 277–291.

Schubert, J. (2002) Hydraulic aspects of riverbank filtration – field studies. *Journal of Hydrology* **266**, 145–161.

Schwartz, F.W. & Muehlenbachs, K. (1979) Isotope and ion geochemistry of groundwaters in the Milk River Aquifer, Alberta. *Water Resources Research* **15**, 259–268.

Schwartz, F.W. & Zhang, H. (2003) *Fundamentals of Ground Water*. John Wiley, New York.

Schwarzenbach, R.P. & Westall, J. (1981) Transport of nonpolar organic compounds from surface water to groundwater. Laboratory sorption studies. *Environmental Science & Technology* **15**, 1360–1367.

Schwille, F. (1988) *Dense Chlorinated Solvents in Porous and Fractured Media: model experiments*. Lewis Publishers, Chelsea, Michigan.

Shapiro, S.D., LeBlanc, D., Schlosser, P. & Ludin, A. (1999) Characterizing a sewage plume using the ^{3}H-^{3}He dating technique. *Ground Water* **37**, 861–878.

Sharma, P.V. (1997) *Environmental and Engineering Geophysics*. Cambridge University Press, Cambridge.

Sharp, J.M. (1988) Alluvial aquifers along major rivers. In: *Hydrogeology. The geology of North America*, vol. O-2 (eds W. Back, J.S. Rosenshein & P.R. Seaber). The Geological Society of North America, Boulder, Colorado, 273–282.

Sheets, R.A., Darner, R.A. & Whitteberry, B.L. (2002) Lag times of bank filtration at a well field, Cincinnati, Ohio, USA. *Journal of Hydrology* **266**, 162–174.

Sherlock, R.L. (1962) *British Regional Geology: London and Thames Valley*, 3rd edn (reprint with minor additions). HMSO, London.

Simpson, B., Blower, T., Craig, R.N. & Wilkinson, W.B. (1989) *The Engineering Implications of Rising Groundwater*

Levels in the Deep Aquifer Beneath London. Construction Industry Research & Information Association, London, Special Publication 69.

Smart, P.L. & Laidlaw, I.M.S. (1977) An evaluation of some fluorescent dyes for water tracing. *Water Resources Research* **13**, 15–33.

Smedley, P.L. and Kinniburgh, D.G. (2002) A review of the source, behaviour and distribution of arsenic in natural waters. *Applied Geochemistry* **17**, 517–568.

Smith, D.B., Downing, R.A., Monkhouse, R.A., Otlet, R.L. & Pearson, F.J. (1976a) The age of groundwater in the Chalk aquifer of the London Basin. *Water Resources Research* **12**, 392–404.

Smith, D.I., Atkinson, T.C. & Drew, D.P. (1976b) The hydrology of limestone terrains. In: *The Science of Speleology* (eds T.D. Ford & C.H.D. Cullingford). Academic Press, London.

Smith, R.E. (2002) Infiltration theory for hydrologic applications. *Water Resources Monograph*, 15. American Geophysical Union, Washington, DC.

Snow, D.T. (1969) Anisotropic permeability of fractured media. *Water Resources Research* **5**, 1273–1289.

Solley, W.B., Pierce, R.R. & Perlman, H.A. (1998) Estimated use of water in the United States in 1995. *United States Geological Survey Circular* 1200.

Solomon, D.K., Schiff, S.L., Poreda, R.J. & Clarke, W.B. (1993) A validation of the $^3H/^3He$ method for determining groundwater recharge. *Water Resources Research* **29**, 2951–2962.

Spector, W.S. (1956) *Handbook of Biological Data.* Saunders, Philadelphia, Pennsylvania.

Spitz, K. & Moreno, J. (1996) *A Practical Guide to Groundwater and Solute Transport Modeling.* John Wiley, New York.

Sprinkle, C.L. (1989) Geochemistry of the Florida Aquifer System in Florida and in parts of Georgia, South Carolina, and Alabama. *United States Geological Survey Professional Paper* 1403-I.

Starr, R.C. & Cherry, J.A. (1994) In situ remediation of contaminated ground water: the funnel-and-gate system. *Ground Water* **32**, 465–476.

Stiff, H.A. (1951) The interpretation of chemical water analysis by means of patterns. *Journal of Petroleum Technology* **3**, 15–17.

Stumm, W. & Morgan, J.J. (1981) *Aquatic Chemistry: an introduction emphasizing chemical equilibria in natural waters*, 2nd edn. John Wiley, New York.

Stumm, W. & Wollast, R. (1990) Coordination chemistry of weathering: kinetics of the surface-controlled dissolution of oxide minerals. *Reviews of Geophysics* **28**, 53–69.

Stute, M., Forster, M., Frischkorn, H., Serejo, A., Clark, J.F., Schlosser, P., Broecker, W.S. & Bonani, G. (1995) Cooling of tropical Brazil (5°C) during the last glacial maximum. *Science* **269**, 379–383.

Sudicky, E.A. (1986) A natural gradient experiment on solute transport in a sand aquifer: spatial variability of hydraulic conductivity and its role in the dispersion process. *Water Resources Research* **22**, 2069–2082.

Talibudeen, O. (1981) Cation exchange in soils. In: *The Chemistry of Soil Processes* (eds D.J. Greenland & M.H.B. Hayes). John Wiley, Chichester, 115–177.

Tallaksen, L.M. (1995) A review of baseflow recession analysis. *Journal of Hydrology* **165**, 349–370.

Tardy, Y. (1971) Characterization of the principal weathering types by the geochemistry of waters from some European and African crystalline massifs. *Chemical Geology* **7**, 253–271.

Tellam, J.H. (1994) The groundwater chemistry of the Lower Mersey Basin Permo-Triassic Sandstone Aquifer system, UK: 1980 and pre-industrialisation-urbanisation. *Journal of Hydrology* **161**, 287–325.

Tellam, J.H. (1995) Hydrochemistry of the saline groundwaters of the lower Mersey Basin Permo-Triassic sandstone aquifer, UK. *Journal of Hydrology* **165**, 45–84.

Tellam, J.H., Lloyd, J.W. & Walters, M. (1986) The morphology of a saline groundwater body: its investigation, description and possible explanation. *Journal of Hydrology* **83**, 1–21.

Theis, C.V. (1935) The relation between the lowering of the piezometric surface and rate and duration of discharge of a well using groundwater storage. *Transactions of the American Geophysical Union* **2**, 519–524.

Thomas, M.R., Garthwaite, D.G. & Banham, A.R. (1997) *Pesticide Usage Survey Report 141: Arable Farm Crops in Great Britain 1996.* Ministry of Agriculture, Fisheries & Food and Scottish Office Agriculture, Environment & Fisheries Department.

Thompson, N., Barrie, I.A. & Ayles, M. (1981) *The Meteorological Office Rainfall and Evaporation Calculation System: MORECS.* Hydrological Memorandum No. 45. Meteorological Office, Bracknell.

Thornthwaite, C.W. (1948) An approach towards a rational classification of climate. *Geographical Reviews* **38**, 55–94.

Thurman, E.M. (1985) *Organic Geochemistry of Natural Waters.* Nijhoff-Junk, Dordrecht.

Todd, D.K. (1980) *Groundwater Hydrology*, 2nd edn. John Wiley, New York.

Tòth, J. (1963) A theoretical analysis of groundwater flow in small drainage basins. *Journal of Geophysical Research* **68**, 4795–4812.

Toynton, R. (1983) The relation between fracture patterns and hydraulic anisotropy in the Norfolk Chalk, England. *Quarterly Journal of Engineering Geology* **16**, 169–185.

Trainer, F.W. (1988) Plutonic and metamorphic rocks. In: *Hydrogeology. The geology of North America*, vol. O-2 (eds W. Back, J.S. Rosenshein & P.R. Seaber). The Geological Society of North America, Boulder, Colorado, 367–380.

Trenberth, K., Overpeck, J. & Soloman, S. (2004) Exploring drought and its implications for the future. *Eos* **85**, 27.

Trenberth, K.E., Dai, A., Rasmussen, R.M. & Parsons, D.B. (2003) The changing character of precipitation. *Bulletin of the American Meteorological Society* **84**, 1205–1217.

Truesdell, A.H. & Jones, B.R. (1973) WATEQ, *a computer program for calculating chemical equilibria of natural waters*. United States Geological Survey, National Technical Information Service, PB-220 464.

Tufenkji, N., Ryan, J.N. & Elimelech, M. (2002) The promise of bank filtration. *Environmental Science & Technology* **November 1**, 423A–428A.

UNESCO (1976) *1 : 1500,000 International Hydrogeological Map of Europe. Map B4 London*. Bundesanstalt für Geowissenschaften und Rohstoffe and UNESCO, Hannover.

UNESCO (1980) *1 : 1500,000 International Hydrogeological Map of Europe. Map B3 Edinburgh*. Bundesanstalt für Geowissenschaften und Rohstoffe and UNESCO, Hannover.

UNESCO/FAO (1973) *Irrigation, Drainage and Salinity: an international source book*. Hutchinson & Co. (Publishers), London.

United States Environmental Protection Agency (1991a) *Wellhead Protection Strategies for Confined-aquifer Settings*. EPA Report /570-9-91/008. Office of Ground Water and Drinking Water. Washington, DC.

United States Environmental Protection Agency (1991b) *Delineation of Wellhead Protection Areas in Fractured Rocks*. EPA Report /570-9-91/009. Office of Ground Water and Drinking Water. Washington, DC.

United States Environmental Protection Agency (1993) *Guidelines for Delineation of Wellhead Protection Areas*. EPA Report /440-5-93/001, Office of Ground Water, Office of Ground Water Protection. Washington, DC.

United States Environmental Protection Agency (1997) *Use of Monitored Natural Attenuation at Superfund, RCRA Corrective Action, and Underground Storage Tank Sites*. Directive 9200.4-17P. Office of Solid Waste and Emergency Response, Washington, DC.

United States Environmental Protection Agency (1999) *Safe Drinking Water Act, Section 1429, Ground Water Report to Congress*. Report EPA-816-R-99-016. Office of Water (4606), Washington DC.

United States Environmental Protection Agency (2001) *National Primary Drinking Water Regulations: long term enhanced surface water treatment rule*. Environmental Protection Agency, Washington, DC, 40 CFR Parts 9, 141 and 142: 217–231.

United States Geological Survey (2002) SOFIA-SFRSF-Hydrology – What is aquifer storage and recovery (ASR)? http://sofia.usgs.gov/sfrsf/rooms/hydrology/ASR/index.html.

United States National Research Council (1983) *Risk Assessment in the Federal Government: Managing the Process*. National Academy Press, Washington, DC.

Vaccaro, J.J. (1992) Sensitivity of groundwater recharge estimates to climate variability and change, Columbia Plateau, Washington. *Journal of Geophysical Research* **97**, 2821–2833.

Vaikmäe, R., Vallner, L., Loosli, H.H., Blaser, P.C. & Juillard-Tardent, M. (2001) Palaeogroundwater of glacial origin in the Cambrian-Vendian aquifer of northern Estonia. In: *Palaeowaters in Coastal Europe: evolution of groundwater since the late Pleistocene* (eds W.M. Edmunds & C.J. Milne). Geological Society, London, Special Publications **189**, 17–27.

van Beek, C.G.E.M. (2000) Redox processes active in denitrification. In: *Redox: Fundamentals, Processes and Applications* (eds J. Schüring, H.D. Schulz, W.R. Fischer, J. Böttcher & W.H.M. Duijnisveld). Springer-Verlag, Berlin, 152–160.

van Waegeningh, H.G. (1985) Protection of groundwater quality in porous permeable rocks. In: *Theoretical Background, Hydrogeology and Practice of Groundwater Protection Zones* (eds G. Matthess, S.S.D. Foster & A.C. Skinner). Verlag Heinz Heise, Hanover, 111–121.

Verhagen, B.T., Geyh, M.A., Fröhlich, K. & Wirth, K. (1991) *Isotope Hydrological Methods for the Quantitative Evaluation of Ground Water Resources in Arid and Semiarid Areas*. Research Report of the Federal Ministry of Economic Cooperation of the Federal Republic of Germany, Bonn.

Verschueren, K. (1983) *Handbook of Environmental Data on Organic Chemicals*, 2nd edn. Van Nostrand Reinhold, New York.

Verstraeten, I.M., Thurman, E.M., Lindsey, M.E., Lee, E.C. & Smith, R.D. (2002) Changes in concentrations of triazine and acetamide herbicides by bank filtration, ozonation, and chlorination in a public water supply. *Journal of Hydrology* **266**, 190–208.

Voss, C.I. (1984) A finite-element simulation model for saturated-unsaturated, fluid-density-dependent groundwater flow with energy transport or chemically-reactive single-species solute transport. In: *Water-Resources Investigations of the United States Geological Survey*. Report 84–4369. Washington, DC.

Wallace, R.B., Darama, Y. & Annable, M.D. (1990) Stream depletion by circle pumping of wells. *Water Resources Research* **26**, 1263–1270.

Walton, G. (1951) Survey of literature relating to infant methemoglobinemia due to nitrate-contaminated water. *American Journal of Public Health* **41**, 986–996.

Walton, N.R.G. (1981) *A Detailed Hydrogeochemical Study of Groundwaters from the Triassic Sandstone Aquifer of South-west England*. Institute of Geological Sciences, Report No. 81/5. HMSO, London.

Walton, W.C. (1960) Leaky artesian aquifer conditions in Illinois. *Illinois State Water Survey Report Investigation 39*.

Walton, W.C. (1987) *Groundwater Pumping Tests: design and analysis*. Lewis Publishers, Chelsea, MI & National Water Well Association, Dublin, OH.

Wang, H.F. & Anderson, M.P. (1982) *Introduction to Groundwater Modelling: finite difference and finite element methods*. Academic Press, San Diego, CA.

Ward, R.C. & Robinson, M. (2000) *Principles of Hydrology*, 4th edn. McGraw-Hill Publishing, Maidenhead, Berkshire.

Ward, R.S., Williams, A.T., Barker, J.A., Brewerton, L.J. & Gale, I.N. (1998) Groundwater tracer tests: a review and guidelines for their use in British aquifers. *British Geological Survey Report* WD/98/19.

Wellings, S.R. & Bell, J.P. (1982) Physical controls of water movement in the unsaturated zone. *Quarterly Journal of Engineering Geology* **15**, 235–241.

Wheeler, B.D. & Proctor, M.C.F. (2000) Ecological gradients, subdivisions and terminology of north-west European mires. *Journal of Ecology* **88**, 187–203.

Whitaker, W. & Reid, C. (1899) *The Water Supply of Sussex from Underground Sources*. Memoir of the Geological Survey. HMSO, London.

Wilhelm, H., Rabbel, W., Lüschen, E., Li, Y.-D. & Bried, M. (1994a) Hydrological aspects of geophysical borehole measurements in crystalline rocks of the Black Forest. *Journal of Hydrology* **157**, 325–347.

Wilhelm, S.R., Schiff, S.L. & Cherry, J.A. (1994b) Biogeochemical evolution of domestic waste water in septic systems: 1. Conceptual model. *Ground Water* **32**, 905–916.

Williams, G.M., Young, C.P. & Robinson, H.D. (1991) Landfill disposal of wastes. In: *Applied Groundwater Hydrology* (eds R.A. Downing & W.B. Wilkinson). Clarendon Press, Oxford, 114–133.

Williamson, A.K., Prudic, D.E. & Swain, L.A. (1989) Ground-water flow in the Central Valley, California. *United States Geological Survey Professional Paper* 1401-D.

Willocks, L., Crampin, A., Milne, L., Seng, C., Susman, M., Gair, R., Shafi, S., Wall, R., Wiggins, R. & Lightfoot, N.

(1999) A large outbreak of cryptosporidiosis associated with a public water supply from a deep chalk borehole. *Communicable Diseases and Public Health* **1**, 239–243.

Wilson, E.M. (1990) *Engineering Hydrology*, 4th edn. Macmillan, London.

Wilson, G, & Grace, H. (1942) The settlement of London due to the under drainage of the London Clay. *Journal of the Institution of Civil Engineers* **12**, 100–127.

Wilson, J.L. (1993) Induced infiltration in aquifers with ambient flow. *Water Resources Research* **29**, 3503–3512.

Wilson, J.L. (1994) Correction to 'Induced infiltration in aquifers with ambient flow'. *Water Resources Research* **30**, 1207.

Witherspoon, P.A., Wang, J.S.Y., Iwal, K. & Gale, J.E. (1980) Validity of cubic law for fluid flow in a deformable rock fracture. *Water Resources Research* **16**, 1016–1024.

WMO (1994) *Guide to Hydrological Practices*, 5th edn. WMO Publication number 168. World Meteorological Office, Geneva.

Wood, S.C., Younger, P.L. & Robins, N.S. (1999) Long-term changes in the quality of polluted minewater discharges from abandoned underground coal workings in Scotland. *Quarterly Journal of Engineering Geology* **32**, 69–79.

Worch, E., Grischek, T., Börnick, H. & Eppinger, P. (2002) Laboratory tests for simulating attenuation processes of aromatic amines in riverbank filtration. *Journal of Hydrology* **266**, 259–268.

World Health Organization (1994) *Guidelines for Drinking Water Quality. Vol. 1. Recommendations*, 2nd edn. World Health Organization, Geneva.

World Health Organization (2002) *World Health Organization Guidelines for Drinking Water Quality. Hardness*. http://www.who.int/water_sanitation_health/GDWQ/Chemicals/hardnfull.htm. July 2002. World Health Organization, Geneva.

Yong, R.N. & Phadungchewit, Y. (1993) pH influence on selectivity and retention of heavy-metals in some clay soils. *Canadian Geotechnical Journal* **30**, 821–833.

Young, M.E. 2002. Institutional development for sustainable groundwater management – an Arabian perspective. In: *Sustainable Groundwater Development* (eds K.M. Hiscock, M.O. Rivett & R.M. Davison). Geological Society, London, Special Publications **193**, 63–74.

Younger, P.L. (1995) Hydrogeochemistry of minewaters flowing from abandoned coal workings in County Durham. *Quarterly Journal of Engineering Geology* **28**, S101–S113.

Younger, P.L., Teutsch, G., Custodio, E., Elliot, T., Manzano, M. & Sauter, M. (2002) Assessments of the sensitivity to climate change of flow and natural water quality in four major carbonate aquifers of Europe. In: *Sustainable Groundwater Development* (eds K.M. Hiscock, M.O. Rivett & R.M. Davison). Geological Society, London, Special Publications **193**, 303–323.

Youngs, E.G. (1991) Infiltration measurements – a review. *Hydrological Processes* **5**, 309–320.

Yusoff, I., Hiscock, K.M. & Conway, D. (2002) Simulation of the impacts of climate change on groundwater resources in eastern England. In: *Sustainable Groundwater Development* (eds K.M. Hiscock, M.O. Rivett & R.M. Davison). Geological Society, London, Special Publications **193**, 325–344.

Zektser, I.S. & Loáiciga, H.A. (1993) Groundwater fluxes in the global hydrologic cycle: past, present and future. *Journal of Hydrology* **144**, 405–427.

Zhan, H.B. & Park, E. (2002) Vapor flow to horizontal wells in unsaturated zones. *Soil Science Society of America Journal* **66**, 710–721.

Zhang, B.Y. & Lerner, D.N. (2000) Modeling of ground water flow to adits. *Ground Water* **38**, 99–105.

Zhang, X.C. & Norton, L.D. (2002) Effect of exchangeable Mg on saturated hydraulic conductivity, disaggregation and clay dispersion of disturbed soils. *Journal of Hydrology* **260**, 194–205.

Zheng, C. (1990) *MT3D a Modular Three-dimensional Transport Model for Simulation of Advection, Dispersion and Chemical Reactions of Contaminants in Groundwater Systems*. S.S. Papadopulos & Associates, Rockville, Maryland.

Zobell, C.E. & Grant, C.W. (1942) Bacterial activity in dilute nutrient solutions. *Science* **96**, 189.

Sources

The author and Blackwell Publishing gratefully acknowledge the following sources in the publication of this book. Figure numbers for *Hydrogeology: Principles and Practice* are shown in bold thus, **Fig. 4.7**.

Albu, M., Banks, D. & Nash, H. (1997) Mineral and thermal groundwater resources. Chapman & Hall, London. Fig. 3.13 for **Fig. 4.7** With kind permission of Springer Science and Business Media.

Anderson, M., Low, R. & Foot, S. (2002) Sustainable groundwater development in arid, high Andean basins. In: *Sustainable Groundwater Development* (eds K.M. Hiscock, M.O. Rivett & R.M. Davison). Geological Society, London, Special Publications **193**, 133–144. Figs 4, 5, 6 for **Box 5.6 Fig. 1**

Aravena, R. & Robertson, W.D. (1998) Use of multiple isotope tracers to evaluate denitrification in ground water: study of nitrate from a large-flux septic system plume. *Ground Water* **36**, 975–982. Fig. 1 for **Fig. 6.24** Reprinted from *Ground Water* with permission of the National Ground Water Association. Copyright 1998.

Arnell, N. (1996) *Global warming, river flows and water resources.* John Wiley & Sons Ltd., Chichester. Fig. 2.6 for **Fig. 8.17** © John Wiley & Sons Limited. Reproduced with permission.

Banwart, S.A., Evans, K.A. & Croxford, S. (2002). Predicting mineral weathering rates at field scale for mine water risk assessment. In: *Mine water hydrogeology and geochemistry* (eds P.L. Younger & N.S. Robins). Geological Society, London, Special Publications **198**, 137–157. Fig. 1 for **Box 6.5 Fig. 1**

Bateman, A.S., Hiscock, K.M. & Atkinson, T.C., 2001. Qualitative risk assessment using tracer tests and groundwater modelling in an unconfined sand and gravel aquifer. In: *New Approaches Characterizing Groundwater Flow* (eds K.-P. Seiler & S. Wohnlich). Swets & Zeitlinger, Lisse, 251–255. Fig. 3 for **Box 5.5 Fig. 2**

Bekins, B., Rittmann, B. & MacDonald, J. (2001a) Natural attenuation strategy for groundwater cleanup focuses on demonstrating cause and effect. *Eos* **82**, 53, 57–58. Fig. 1 for **Fig. 7.2** Reproduced by permission of the American Geophysical Union.

Bekins, B.A., Cozzarelli, I.M., Godsy, E.M., Warren, E., Essaid, H.I. & Tuccillo, M.E. (2001b) Progression of natural attenuation processes at a crude oil spill site: II. Controls on spatial distribution of microbial populations. *Journal of Contaminant Hydrology* **53**, 387–406. Fig. 2 for **Box 7.4 Fig. 1**

Berner, E.K. & Berner, R.A. (1987) *The Global Water Cycle: Geochemistry and Environment.* Prentice-Hall, Inc., Englewood Cliffs, New Jersey. Fig. 2.1 for **Fig. 1.5** Reprinted by permission of Pearson Education, Inc. Upper Saddle River, NJ.

Beyerle, U., Aeschbach-Hertig, W., Hofer, M., Imboden, D.M., Baur, H. & Kipfer, R. (1999) Infiltration of river water to a shallow aquifer investigated with $^{3}H/^{3}He$, noble gases and CFCs. *Journal of Hydrology* **220**, 169–185. Fig. 2 for **Fig. 4.11** With permission from Elsevier.

Bishop, P.K., Burston, M.W., Lerner, D.N. and Eastwood, P.R. (1990) Soil gas surveying of chlorinated solvents in relation to groundwater pollution studies. *Quarterly Journal of Engineering Geology* **23**, 255–265. Fig. 5 for **Fig. 3.12**

Brassington, R. (1998). *Field hydrogeology* (2nd edn). John Wiley & Sons, Chichester. Fig. 2.1 for **Fig. 5.1**, Fig. 6.1 for **Fig. 5.2**, Fig. 7.9 for **Fig. 5.19** © John Wiley & Sons Limited. Reproduced with permission.

British Geological Survey (1978) *1:100 000 Hydrogeological Map of the South West Chilterns and the Berkshire and Marlborough Downs.* Natural Environment Research Council. **Fig. 2.21a** Reproduced by permission of the British Geological Survey. © NERC. All rights reserved. IPR/55–29C.

British Geological Survey (1990) *1:50 000 Map of The Carnmenellis Granite. Hydrogeological, hydrogeochemical and geothermal characteristics*. Natural Environment Research Council. **Box 3.10 Fig. 1** Reproduced by permission of the British Geological Survey. © NERC. All rights reserved. IPR/55–29C.

Brown, M.C. & Ford, D.C. (1971) Quantitative tracer methods for investigation of karst hydrologic systems. *Transactions of the Cave Research Group of Great Britain* **13**, 37–51. Fig. 1 for **Fig. 5.38** Reproduced with permission from the Cave Research Group of Great Britain.

Burgess, D.B. (2002) Groundwater resource management in eastern England: a quest for environmentally sustainable development. In: *Sustainable Groundwater Development* (eds K.M. Hiscock, M.O. Rivett & R.M. Davison). Geological Society, London, Special Publications **193**, 53–62. Fig. 3 for **Box 8.5 Fig. 3**

Champ, D.R., Gulens, J. & Jackson, R.E. (1979) Oxidation-reduction sequences in ground water flow systems. *Canadian Journal of Earth Sciences* **16**, 12–23. Fig. 4 for **Fig. 3.21** Reproduced with permission from NRC Research Press.

CL:AIRE (2002) Introduction to an integrated approach to the investigation of fractured rock aquifers contaminated with non-aqueous phase liquids. *Technical Bulletin* **TB1**, Contaminated Land: Applications in Real Environments, London. **Fig. 6.14** Reproduced with permission from CL:AIRE.

Clark, L. and Sims, P.A. (1998) Investigation and clean-up of jet-fuel contaminated groundwater at Heathrow International Airport, UK. In: *Groundwater contaminants and their migration* (eds J. Mather, D. Banks, S. Dumpleton and M. Fermor). Geological Society, London, Special Publications **128**, 147–57. Fig. 4 for **Box 7.2 Fig. 1**, Fig. 6 for **Box 7.2 Fig. 2**, Fig. 7 for **Box 7.2 Fig. 3**

Darling, W.G., Bath, A.H. & Talbot, J.C. (2003) The O & H stable isotopic composition of fresh waters in the British Isles. 2. Surface waters and groundwater. *Hydrology and Earth System Sciences* **7**, 183–195. Fig. 6 for **Fig. 4.2** Reproduced with permission of the European Geosciences Union.

DEFRA (2002) *Map of Nitrate Vulnerable Zones in England and Wales*. Her Majesty's Stationery Office. **Fig. 7.11** Reproduced with permission of HMSO.

DPHE (1999) *Groundwater Studies for Arsenic Contamination in Bangladesh. Final Report, Rapid Investigation Phase*. Department of Public Health Engineering, Government of Bangladesh. British Geological Survey & Mott-MacDonald. **Box 7.6 Fig. 1a, 1b** Reproduced by permission of the British Geological Survey. © NERC. All rights reserved. IPR/55–29C.

Domenico, P.A. Schwartz, F.W. (1998) *Physical and Chemical Hydrogeology*, 2nd edn. John Wiley & Sons, New York. Fig 12.7 for Fig 6.9. This material is used by permission of John Wiley & Sons, Inc.

Edmunds, W.M. & Walton, N.G.R. (1983) The Lincolnshire Limestone – hydrogeochemical evolution over a ten-year period. *Journal of Hydrology* **61**, 201–211. Fig. 2 for **Fig. 3.19** With permission from Elsevier.

Edmunds, W.M. 2001. Palaeowaters in European coastal aquifers – the goals and main conclusions of the PALAEAUX project. In: *Palaeowaters in Coastal Europe: evolution of groundwater since the late Pleistocene* (eds W.M. Edmunds & C.J. Milne). Geological Society, London, Special Publications **189**, 1–16. Fig. 6 for **Fig. 4.12a**

Edmunds, W.M., Kay, R.L.F. & McCartney, R.A. (1985) Origin of saline groundwaters in the Carnmenellis Granite (Cornwall, England): natural processes and reaction during hot dry rock reservoir circulation. *Chemical Geology* **49**, 287–301. Fig. 2 for **Box 3.10 Fig. 2** With permission from Elsevier.

Fetter, C.W. (1999) *Contaminant hydrogeology* (2nd edn). Prentice-Hall, Inc., New Jersey. Fig. 5.5 for **Fig. 6.10**, Fig. 5.6 for **Fig. 6.11**, Fig. 5.26 for **Fig. 6.13** Reprinted by permission of Pearson Education, Inc. Upper Saddle River, NJ.

Fetter, C.W. (2001) *Applied hydrogeology* (4th edn). Pearson Higher Education, New Jersey. Figs 3.27, 3.28 for **Fig. 2.6** Reprinted by permission of Pearson Education, Inc. Upper Saddle River, NJ.

Ford, M. & Tellam, J.H. (1994) Source, type and extent of inorganic contamination within the Birmingham urban aquifer system, UK. *Journal of Hydrology* **156**, 101–135. Fig. 4a for **Fig. 6.18** With permission from Elsevier.

Foster, S., Garduno, H., Evans, R., Olson, D., Tian, Y., Zhang, W. & Han, Z. (2004) Quaternary Aquifer of the North China Plain – assessing and achieving groundwater resource sustainability. *Hydrogeology Journal* **12**, 81–93. Fig. 1 for **Box 1.2 Fig. 1**, Fig. 2 for **Box 1.2 Fig. 2** Reproduced with permission of Springer-Verlag.

Freeze, R.A. & Cherry, J.A. (1979) *Groundwater*. Prentice-Hall, Inc., Englewood Cliffs, New Jersey. Fig. 2.8 for **Fig. 2.7**, Fig. 7.11 for **Fig. 3.16**, Fig. 9.4 for **Fig. 6.5**, Fig. 9.1 for **Fig. 6.7**, Figs 9.9 and 9.16 for **Fig. 6.16**, Fig. 6.17 for **Fig. 6.17** Reprinted by permission of Pearson Education, Inc. Upper Saddle River, NJ.

Green, A.R., Feast, N.A., Hiscock, K.M. & Dennis, P.F. (1998) Identification of the source and fate of nitrate contamination of the Jersey bedrock aquifer using stable nitrogen isotopes. In: *Groundwater Pollution, Aquifer Recharge and Vulnerability* (ed. N.S. Robins). Geological Society, London, Special Publications **130**, 23–35. Fig. 4 for **Box 6.6 Fig. 1**, Fig. 5 for **Box 6.6 Fig. 2**

Guenther, W.B. (1975) *Chemical Equilibrium. A Practical Introduction for the Physical and Life Sciences*. Plenum Press, New York. Fig. 11.1 for **Fig. 3.14**

Habermehl, M.A. (1980) The Great Artesian Basin, Australia. *BMR Journal of Australian Geology and Geophysics* **5**, 9–38. Fig. 1 for **Box 2.11 Fig. 1**, Fig. 18 (top) for **Box 2.11 Fig. 2**, Fig 18 (bottom) for **Box 2.11 Fig. 3**, Fig. 12 for **Box 2.11 Fig. 4**

Hanshaw, B.B. & Back, W. (1979) Major geochemical processes in the evolution of carbonate-aquifer systems. *Journal of Hydrology* **43**, 287–312. Fig. 2 for **Fig. 3.3** With permission from Elsevier.

Headworth, H.G. & Fox, G.B. (1986) The South Downs Chalk aquifer: its development and management. *Journal of the Institution of Water Engineers* **40**, 345–361. Fig. 2d for **Fig. 6.28** Reproduced with permission of the Chartered Institute of Water and Environmental Management.

Hess, K.M., Davis, J.A., Kent, D.B. & Coston, J.A. (2002) Multispecies reactive tracer test in an aquifer with spatially variable chemical conditions, Cape Cod, Massachusetts: Dispersive transport of bromide and nickel. *Water Resources Research* **38**, 10.1029/2001WR000945. Figs 2a, 8a for **Fig. 6.6** Reproduced by permission of the American Geophysical Union.

Hiscock, K.M. & Grischek, T. (2002) Attenuation of groundwater pollution by bank filtration. *Journal of Hydrology* **266**, 139–144. Fig. 1 for **Fig. 8.6** With permission from Elsevier.

Hiscock, K.M. (1993) The influence of pre-Devensian glacial deposits on the hydrogeochemistry of the Chalk aquifer system of north Norfolk, UK. *Journal of Hydrology* **144**, 335–369. Fig. 5 for **Fig. 3.17** With permission from Elsevier.

Hiscock, K.M., Dennis, P.F., Saynor, P.R. & Thomas, M.O. (1996) Hydrochemical and stable isotope evidence for the extent and nature of the effective Chalk aquifer of north Norfolk, UK. *Journal of Hydrology* **180**, 79–107. Fig. 13 for **Fig. 4.4** With permission from Elsevier.

Hiscock, K.M., Lister, D.H., Boar, R.R. & Green, F.M.L. (2001) An integrated assessment of long-term changes in the hydrology of three lowland rivers in eastern England. *Journal of Environmental Management* **61**, 195–214. Fig. 7 for **Fig. 8.12**, Figs 8c, 10c for **Fig. 8.13** With permission from Elsevier.

Hiscock, K.M., Rivett, M.O. & Davison, R.M. (2002) Sustainable groundwater development. In: *Sustainable Groundwater Development* (eds K.M. Hiscock, M.O. Rivett & R.M. Davison). Geological Society, London, Special Publications **193**, 1–14. Fig. 1 for **Fig. 1.1**

Iribar, V., Carrera, J., Custodio, E. & Medina, A. (1997) Inverse modelling of seawater intrusion in the Llobregat delta deep aquifer. *Journal of Hydrology* **198**, 226–244. Fig. 1a for **Box 6.7 Fig. 1**, Fig. 1b for **Box 6.7 Fig. 2**, Fig. 3b for **Box 6.7 Fig. 3**, Fig. 1c for **Box 6.7 Fig. 4** With permission from Elsevier.

Kirk, S. & Herbert, A.W. (2002) Assessing the impact of groundwater abstractions on river flows. In: *Sustainable Groundwater Development* (eds K.M. Hiscock, M.O. Rivett

& R.M. Davison). Geological Society, London, Special Publications **193**, 211–233. Fig. 1 for **Fig. 8.8**, Fig. 2c for **Fig. 8.9c**

Krothe, N.C. & Bergeron, M.P. (1981) Hydrochemical facies in a Tertiary basin in the Milligan Canyon area, Southwest Montana. *Ground Water* **19**, 392–399. Figs 4, 8 for **Fig. 3.10**

Ku, H.F.H. (1980) Ground-water contamination by metal-plating wastes, Long Island, New York, USA. In: *Aquifer contamination and protection* (ed. R.E. Jackson). UNESCO, Paris, pp. 310–317. Fig. II – 8.1, Fig. II – 8.4 (top section) for **Box 6.2 Fig. 1** Reproduced by permission of UNESCO.

Lake, I.R., Lovett, A.A., Hiscock, K.M., Betson, M., Foley, A., Sunnenberg, G., Evers, S. & Fletcher, S. (2003) Evaluating factors influencing groundwater vulnerability to nitrate pollution: developing the potential of GIS. *Journal of Environmental Management* **68**, 315–328. Fig. 2 for **Fig. 7.5**

Lawrence, A.R. & Foster, S.S.D. (1986) Denitrification in a limestone aquifer in relation to the security of low-nitrate groundwater supplies. *Journal of the Institution of Water Engineers and Scientists* **40**, 159–172. Fig. 3 for **Box 3.8 Fig. 1** Reproduced with permission of the Chartered Institute of Water and Environmental Management.

Livingstone, S., Franz, T. & Guiguer, N. (1996) Managing ground-water resources using wellhead protection programs. *Geoscience Canada* **22**, 121–128. Fig. 1 for **Fig. 7.7**, Fig. 4 for **Fig. 7.8** Reproduced with permission from NRC Research Press.

Lloyd, J.W. & Heathcote, J.A. (1985) *Natural Inorganic Hydrochemistry in Relation to Groundwater: an introduction.* Clarendon Press, Oxford. Fig. 8.14 for **Fig. 4.6** By permission of Oxford University Press.

Lucas, H.C. & Robinson, V.K. (1995) Modelling of rising groundwater levels in the Chalk aquifer of the London Basin. *Quarterly Journal of Engineering Geology*, **28**, S51–S62. Fig. 4 for **Box 2.4 Fig. 1**

Marsh, T.J. & Davies, P.A. (1983) The decline and partial recovery of groundwater levels below London. *Proceedings of the Institution of Civil Engineers, Part 1*, **74**, 263–276. Figs 4,7 for **Box 2.5 Fig. 2**, Fig. 10 for **Box 2.5 Fig. 3** Reproduced with permission of the Institution of Civil Engineers.

Maurits la Riviere, J.W. (1989) Threats to the world's water. *Scientific American* **September 1989**, 48–55. **Fig. 1.6** Reproduced with permission of George Retseck Illustration.

McKenzie, H.S. & Elsaleh, B.O. (1994) The Libyan Great Man-Made River Project Paper 1. Project overview. *Proceedings of the Institution of Civil Engineers – Water Maritime and Environment* **106**, 103–122. Fig. 4 for **Box 8.1**

Fig. 1 Reproduced with permission of the Institution of Civil Engineers.

Morgan-Jones, M. & Eggboro, M.D. (1981) The hydrogeochemistry of the Jurassic limestones in Gloucestershire, England. *Quarterly Journal of Engineering Geology* **14**, 25–39. Figs 1, 2 for **Box 3.5 Fig. 1**, Figs 5a, 5b for **Box 3.5 Fig. 2**, Fig. 6 for **Box 3.5 Fig. 3**

O'Hannesin, S.F. & Gillham, R.W. (1998) Long-term performance of an in situ 'iron wall' for remediation of VOCs. *Ground Water* **36**, 164–170. Figs 1a, 1b for **Box 7.3 Fig. 1**, Fig. 2 for **Box 7.3 Fig. 2**

O'Shea, M.J. & Sage, R. (1999) Aquifer recharge: an operational drought-management strategy in North London. *Journal of the Institution of Water and Environmental Management* **13**, 400–405. Fig. 2 for **Box 8.2 Fig. 1** Reproduced with permission of the Chartered Institute of Water and Environmental Management.

Owen, M. (1991) Relationship between groundwater abstraction and river flow. *Journal of the Institution of Water and Environmental Management* **5**, 697–702. Figs 1, 2 for **Figs 8.9a, 8.9b**, Figs 3, 4 for **Fig. 8.10** Reproduced with permission of the Chartered Institute of Water and Environmental Management.

Parker, J.M., Young, C.P. & Chilton, P.J. (1991) Rural and agricultural pollution of groundwater. In: *Applied Groundwater Hydrology: a British perspective* (eds R.A. Downing & W.B. Wilkinson). Clarendon Press, Oxford, 149–163. Figs 10.1a and 10.5 for **Fig. 4.10** By permission of Oxford University Press.

Pim, R.H. & Binsariti, A. (1994) The Libyan Great Man-Made River Project Paper 2. The water resource. *Proceedings of the Institution of Civil Engineers – Water Maritime and Environment* **106**, 123–145. Fig. 1 for **Box 8.1 Fig. 1** Reproduced with permission of the Institution of Civil Engineers.

Pocock, S.J., Shaper, A.G., Cook, D.G., Packham, R.F., Lacey, R.F., Powell, P. & Russell, P.F. (1980) British Regional Heart Study: geographic variations in cardiovascular mortality, and role of water quality. *British Medical Journal* **280**, 1243–1249. Fig. 4 for **Box 6.1 Fig. 1** Reproduced with permission from the BMJ Publishing Group.

Rivett, M.O., Feenstra, S. & Cherry, J.A. (2001) A controlled field experiment on groundwater contamination by multicomponent DNAPL: creation of the emplaced-source and overview of dissolved plume development. *Journal of Contaminant Hydrology* **49**, 111–149. Fig. 11 for **Box 6.4 Fig. 4** With permission from Elsevier.

Rivett, M.O., Lerner, D.N., Lloyd, J.W. & Clark, L. (1990) Organic contamination of the Birmingham aquifer, U.K. *Journal of Hydrology* **113**, 307–323. Figs 4, 5 for **Fig. 6.19** With permission from Elsevier.

Roberts, P.V., Goltz, M.N. & Mackay, D.M. (1986) A natural gradient experiment on solute transport in a sand aquifer 3. Retardation estimates and mass balances for organic solutes. *Water Resources Research* **22**, 2047–2058. Fig. 1 for **Box 6.4 Fig. 1**, Fig. 5 for **Box 6.4 Fig. 2**, Fig. 8 for **Box 6.4 Fig. 3** Reproduced by permission of the American Geophysical Union.

Robins, N.S. (1990) Groundwater chemistry of the main aquifers in Scotland. *Report of the British Geological Survey*, Vol. 18, No. 2. Fig. 13 for **Fig. A10.6** Reproduced by permission of the British Geological Survey. © NERC. All rights reserved. IPR/55–44C.

Rozanski, K. (1985) Deuterium and oxygen-18 in European groundwaters – links to atmospheric circulation in the past. *Chemical Geology (Isotope Geoscience Section)* **52**, 349–363. Fig. 3 for **Fig. 4.1** With permission from Elsevier.

Rushton, K.R. & Tomlinson, L.M. (1999) Total catchment conditions in relation to the Lincolnshire Limestone in South Lincolnshire. *Quarterly Journal of Engineering Geology* **32**, 233–246. Fig. 3 for **Box 2.10 Fig. 1**

Saleh, A., Al-Ruwaih, F. & Shehata, M. (1999) Hydrogeochemical processes operating within the main aquifers of Kuwait. *Journal of Arid Environments* **42**, 195–209. Fig. 1 for **Fig. A10.5** With permission from Elsevier.

Schubert, J. (2002) Hydraulic aspects of riverbank filtration – field studies. *Journal of Hydrology* **266**, 145–161. Fig. 3 for **Box 8.3 Fig. 1**, Fig. 6 for **Box 8.3 Fig. 2**, Fig. 15 for **Box 8.3 Fig. 3** With permission from Elsevier.

Schwartz, F.W. & Muehlenbachs, K. (1979) Isotope and ion geochemistry of groundwaters in the Milk River Aquifer, Alberta. *Water Resources Research* **15**, 259–268. Fig. 6 for **Fig. 3.5** Reproduced by permission of the American Geophysical Union.

Sherlock, R.L. (1962) *British Regional Geology: London and Thames Valley* (3rd ed Reprint with Minor Additions). HMSO, London. Fig. 24 for **Box 2.5 Fig. 1** Reproduced by permission of the British Geological Survey. © NERC. All rights reserved. IPR/56–10C.

Smith, D.B., Downing, R.A., Monkhouse, R.A., Otlet, R.L. & Pearson, F.J. (1976) The age of groundwater in the Chalk aquifer of the London Basin. *Water Resources Research* **12**, 392–404. Figs 1, 3, 5 for **Fig. 4.8** Reproduced by permission of the American Geophysical Union.

Smith, D.I., Atkinson, T.C. & Drew, D.P. (1976) The hydrology of limestone terrains. In: *The Science of Speleology* (eds T.D. Ford & C.H.D. Cullingford). Academic Press, London. **Box 2.2 Fig. 1**

Starr, R.C. & Cherry, J.A. (1994) In situ remediation of contaminated ground water: the funnel-and-gate system. *Ground Water* **32**, 465–476. Fig. 1 for **Fig. 7.1**

Stumm, W. & H. Morgan J.J. (1981). *Aquatic Chemistry: An introduction emphasizing chemical equilibria in natural waters*, 2nd edn. John Wiley, New York. Fig 7.8b for Box 3.9 Fig.1 This material is used by permission of John Wiley & Sons, Ltd.

Stute, M., Forster, M., Frischkorn, H., Serejo, A., Clark, J.F., Schlosser, P., Broecker, W.S. & Bonani, G. (1995)

Cooling of tropical Brazil (5°C) during the last glacial maximum. *Science* **269**, 379–383. Fig. 3 for **Fig. 4.12b** © AAAS. Reprinted with permission.

Tardy, Y. (1971) Characterization of the principal weathering types by the geochemistry of waters from some European and African crystalline massifs. *Chemical Geology* **7**, 253–271. Figs 1, 2, 3 for **Fig. 3.23** With permission from Elsevier.

Tellam, J.H. (1994) The groundwater chemistry of the Lower Mersey Basin Permo-Triassic Sandstone Aquifer system, UK: 1980 and pre-industrialisation-urbanisation. *Journal of Hydrology* **161**, 287–325. Fig. 5 (top) for **Box 3.7 Fig. 1**, Fig. 6 for **Box 3.7 Fig. 2** With permission from Elsevier.

Todd, D.K. (1980) *Groundwater Hydrology*, 2nd edn. John Wiley, New York. Fig. 14.6 for Fig 6.27 This material is used by permission of John Wiley & Sons, Inc.

Tufenkji, N., Ryan, J.N. & Elimelech, M. (2002) The promise of bank filtration. *Environmental Science & Technology*, 423A–428A. Fig. 2 for **Fig. 8.7** © American Chemical Society. Reprinted with permission.

Vaikmae, R., Vallner, L., Loosli, H.H., Blaser, P.C. & Juillard-Tardent, M. 2001. Palaeogroundwater of glacial origin in the Cambrian-Vendian aquifer of northern Estonia. In: *Palaeowaters in Coastal Europe: evolution of groundwater since the late Pleistocene* (eds W.M. Edmunds & C.J. Milne). Geological Society, London, Special Publications **189**, 17–27. Fig. 2 for **Fig. 4.5**

van Beek, C.G.E.M. (2000) Redox processes active in denitrification. In: *Redox: Fundamentals, Processes and Applications* (eds J. Schuring, H.D. Schulz, W.R. Fischer, J. Bottcher & W.H.M. Duijnisveld). Springer-Verlag, Berlin, 152–160. Fig. 12.1 for **Box 3.9 Fig. 2**, Fig. 12.3 for **Box 3.9 Fig. 3** Reproduced with permission of Springer-Verlag.

van Waegeningh, H.G. (1985) Protection of groundwater quality in porous permeable rocks. In: *Theoretical background, hydrogeology and practice of groundwater protection zones* (eds G. Matthess, S.S.D. Foster & A.C. Skinner). Verlag Heinz Heise, Hanover, 111–21. Fig. 3.3 for **Fig. 7.6** Reproduced with permission of the International Association of Hydrogeologists.

Wilhelm, S.R., Schiff, S.L. & Cherry, J.A. (1994) Biogeochemical evolution of domestic waste water in septic systems: 1. Conceptual model. *Ground Water* **32**, 905–916. Fig. 1 for **Fig. 6.23**

Williams, G.M., Young, C.P. & Robinson, H.D. (1991) Landfill disposal of wastes. In: *Applied Groundwater Hydrology* (eds R.A. Downing & W.B. Wilkinson). Clarendon Press, Oxford, 114–133. Figs 8.1, 8.2, 8.3, 8.8, 8.13 and 8.14 for **Figs 6.20, 6.21 and 6.22** By permission of Oxford University Press.

Wilson, E.M. (1990) *Engineering Hydrology*, 4th edn. Macmillan, London. Figs 7.7 and 7.8 for Fig 5.29 With permission from Palgrave Macmillian.

Wood, S.C., Younger, P.L. & Robins, N.S. (1999) Long-term changes in the quality of polluted minewater discharges from abandoned underground coal workings in Scotland. *Quarterly Journal of Engineering Geology* **32**, 69–79. Figs 2, 4 for **Box 6.5 Fig. 2**

Worch, E, Grischek, T., Bornick, H. & Eppinger, P. (2002) Laboratory tests for simulating attenuation processes of aromatic amines in riverbank filtration. *Journal of Hydrology* **266**, 259–268. Fig. 7 for **Fig. 6.8** With permission from Elsevier.

Younger, P.L. (1995) Hydrogeochemistry of minewaters flowing from abandoned coal workings in County Durham. *Quarterly Journal of Engineering Geology* **28**, S101–S113. Fig. 1 for **Box 6.5 Fig. 3**

Index